Mathematical Morphology in Geomorphology and GISci

Mathematical Morphology in Geomorphology and GISci

B. S. Daya Sagar

CRC Press
Taylor & Francis Group
Boca Raton London New York

CRC Press is an imprint of the
Taylor & Francis Group, an **informa** business
A CHAPMAN & HALL BOOK

CRC Press
Taylor & Francis Group
6000 Broken Sound Parkway NW, Suite 300
Boca Raton, FL 33487-2742

First issued in paperback 2018

ISBN-13: 978-1-4398-7200-0 (hbk)
ISBN-13: 978-1-138-37459-1 (pbk)

Library of Congress Cataloging-in-Publication Data

Daya Sagar, B. S.
    Mathematical morphology in geomorphology and GISci / Behara Seshadri Daya Sagar.
      pages cm
    Includes bibliographical references and index.
    ISBN 978-1-4398-7200-0 (hardcover : alk. paper) 1. Geomorphology--Mathematical models. I. Title.

GB400.42.M33D39 2013
551.4101'51--dc23   2013010451

Visit the Taylor & Francis Web site at
http://www.taylorandfrancis.com

and the CRC Press Web site at
http://www.crcpress.com

To my wife, Latha, and my sons, Saketh and Sriniketh

Contents

List of Symbols and Notations

IR^2	two-dimensional space
Z	set of all integers
Z^n	discrete n-dimensional space
E	Euclidean space
Z^n	Euclidean discrete n-dimensional space
$X, A, B, M, S...$	subsets of IR^2
M^c	complement of M in Z^2
$x, a, b, m, ...$	points of Z^2; elements of vector points of IR^2, i.e., a point in the 2-D space
$m \in M$	element m belongs to M
$m \notin M$	m does not belong to M
$\varnothing$	empty set
$\cup, \cap, \setminus$	logical union, logical intersection, and logical difference
$\subseteq$	improper subset
$\subset$	subset
$S \cup X$	union of S and X
$S \cap X$	intersection of S and X
$S \setminus X$	set difference of S and X
$M \cup X$	union of M and X
$M \cap X$	intersection of M and X
$M \setminus X$	set difference of M and X
$\cup, \cap, \setminus, \subseteq$	logical union, logical intersection, logical difference, and improper subset
$A(\cdot)$	finite set of cardinality
n	iteration/cycle number (or radius of structuring element, where $n = 0, 1, 2, ..., N$)
nB	nth-size structuring element symmetric w.r.t. origin at center
$1B$	primitive element with origin at center, and radius 1
NB	largest size of structuring element
$\oplus, \ominus, o$	symbols for dilation, erosion, and opening
$X \ominus B$	erosion of X by B where B is symmetric
$X \oplus B$	dilation of X by B where B is symmetric
D	fractal dimension
$L_{\|}$	longitudinal length
$L_{\perp}$	transverse length
H	exponent derived from $L_{\|}$ and $L_{\perp}$
h	exponent derived from $L_{\|}$ and A
h_X	exponent h for water bodies
h_M	exponent h for zones of influence

H_X	exponent H for water bodies
H_M	exponent H for zones of influence
N_X	number of water bodies
A_X	area of water bodies
N_M	number of zones of influence
A_M	area of zones of influence
f	function of basin represented as digital topographic image
j	index: representing threshold value—$j = 0, 1, 2, ..., J$
i	index: representing isolated threshold set—$i = 0, 1, 2, ..., J$
J	maximum nonnegative intensity (elevation) value
S^c	complement of S in $\|R^d$
S_s	S shifts by $\overrightarrow{os}$ (o = origin of $\|R^d$)
∂S	boundary of S
C_{flow}	channel flow
NC_{flow}	nonchannel flow
TB_{flow}	flow in the tidal basin
$\rho(X_{ij})$	minimum distance (Hausdorff distance) between sets X_i and X_j
N	Total number of sets in a cluster set
(X_i, X_j)	sets X_i, X_j of a cluster set X
$(X_i, X_j) = (X_{ij})$	
$P(X_{ij})$	Length of the boundary being shared between the sets X_i, X_j
$d(X_{ij})$	Dilation distance between the sets X_i, X_j
$C(X_{ij})$	Contextuality between the sets X_i, X_j
$\rho(X_{ij})$	Hausdorff distance between the sets X_i, X_j
P_{max}	Maximum boundary length that the state shares with other (adjacent) states
d_{max}	Maximum distance between any two sets of a cluster
$H/P(X_{ij})$	Spatial complexity with respect to $P(X_{ij})$
$H/d(X_{ij})$	Spatial complexity with respect to $d(X_{ij})$
$H/C(X_{ij})$	Spatial complexity with respect to $C(X_{ij})$
(SX_i^p)	Strategic set in terms of perimeter
(SX_i^d)	Strategic set in terms of distance
(SX_i^C)	Strategic set in terms of contextuality
(SH_i^p)	Strategic set in terms of spatial complexity with respect to perimeter (P)
(SH_i^d)	Strategic set in terms of spatial complexity with respect to distance (d)
(SH_i^C)	Strategic set in terms of spatial complexity with respect to contextuality (C)

$M(X)$	medial axis of sets X
$\varphi_{\pi_{\theta+}}$	half-plane closing at specific orientation determined by π_θ
$CH(X)$	convex hull of set X
$CH(X)$	$X \Rightarrow$ set X is convex
$\rho(X_1, X_2)$	minimum distance (Hausdorff distance) between sets X_i and X_j
$\beta(X_1, X_2)$	"between" space between sets X_1 and X_2
Z	set of all integers
E	Euclidean space
X_i, X_j, B	subsets of E
n	iteration number (or radius of structuring element, where n = 0, 1, 2, …, N)
(X_i, X_j)	sets of a cluster
$d(X_{ij})$	dilation distance between the sets X_i and X_j
d_{max}	maximum distance between any two sets of a cluster
N	number of limbs in a symmetric fold (three for the fold type I and two for the fold type II)
L	rigid length of the fold limb and rigid length of sand dine slipface
d	distance of the vertical projection of the upright symmetric fold and sand dune base length
D	fractal dimension $\left(\dfrac{LogN}{Log\left(\dfrac{d}{L}\right)} \right)$
D_T	topological dimension, 0, in 1-D space, 1 in 2-D space, and 2 in 3-D space
α_t	normalized fractal dimensions (NFDs) at discrete time interval $(0 < \alpha_t < 1)$, $\alpha_t = D - D_T$
θ	interlimb angle (IA) $(\theta > 60° < 180°$ for the symmetric fold type I; $\theta > 90° < 180°$ for the fold type II, and also for pyramidal sand dune)
θ^*	attractor interlimb angle (AIA) and attractor interslip face angle of pyramidal sand dune
λ	constant stress $(1 < \lambda < 4)$
λ_t	time-dependent stress parameter
μ	strength of stress modulation (SSM) parameter to compute time-dependent stress parameter $(1 < \mu < 40 < \# < 1)$

Foreword

Various branches of physics and earth sciences have been studied from the point of view of mathematical morphology and its applications, but for the first time a whole book is devoted to a morphological approach to structural geology. In this sense, it fits in with a long tradition since the two founders of mathematical morphology were both mining engineers.

The various problems addressed by Prof. B. S. Daya Sagar are a matter for geomorphology and dynamics at various scales, and concern water bodies, lakes, sand dunes, and relief structures. A common style governs the method he elaborates for studying these different domains. He has selected, indeed, some notions of mathematical morphology, well suited with his purpose, and that he uses with a remarkable virtuosity, with a typical personal touch. Classically, mathematical morphology builds geometrical descriptors along three main lines. The first two stem from the notions of dilation and of connection respectively, and are deterministic, unlike the third one, which is devoted to models of random sets.

Here, the theory of morphology derives practically exclusively from the dilation branch. But these tools—dilation, ultimate erosion, granulometries, etc., which usually serve as image filters—are viewed in the present case in an original manner: they turn out to provide types of "structural harmonics" and lead to decompose the structures under study. Several informative examples illustrate this point. In Chapter 7, the complex non-network spaces of basins are thus transformed into partitions of simpler convex polygonal classes. Similarly, in Chapter 4, the topological networks of fluvial or tidal systems are extracted and reduced to simplified versions for their modeling. Or again, in Chapter 9, a series of isotropic dilations model the evolution of water bodies under flooding due to peak stream flow discharge, whereas the dual erosions model the drought due to low-stream flow discharge.

The twofold background of structural geology and image analysis makes the book rich in interpretation models of the physics. The relation between these two fields is successively analyzed in the context of applications like pattern retrieval, pattern analysis, spatial reasoning, and simulations. The latter aspect, particularly developed in the book, allows the reader to visualize complex dynamic physical processes by means of synthetic data and by simulation of the behavior of water bodies, or dunes, where the dynamic changes are interpreted in terms of sequences of morphological operations.

The book is intended for an audience of geomorphologists, and great care is taken in introducing the morphological notions in a pedagogical way. We hope that the numerous examples will allow engineers and researchers in structural geology to exercise their creative faculties and to find new formulations of their own problems.

Jean Serra
Laboratoire d'Informatique Gaspard Monge
Université Paris-Est
Paris, France

Preface

Data related to terrestrial (geomorphologic) phenomena at spatial and temporal intervals are now available in numerous formats, facilitating visualization at spatiotemporal intervals. Availability of such data from a wide range of sources in a variety of formats poses challenges to the Earth informatics community. The utility and application of such data could be substantially enhanced through related technologies developed in the recent past.

Functions, sets, and skeletons are mathematical representations of surfaces, planes, and networks, respectively. From the point of geophysical context, the surfaces include topographic surfaces, cloud surfaces, and surfaces that possess uneven values at different spatial positions. Sets include water bodies, planar views of catchments, clouds, and threshold elevation regions. Networks include channel and ridge connectivity networks as well as dendritic structures and loop-like watershed lines. I address the description, representation, simulation, and quantitative characterization of geomorphologically relevant functions, sets, and skeletons. In this book, digital topographic surfaces are considered as functions. Such functions are taken as the basis for decomposing planar views of catchments, threshold elevation regions, and water bodies. All these geomorphologically relevant features are mathematically described with mathematically viable decomposition procedures. Characterization of these features represented with functions, sets, and skeletons present several challenges. The aim of this book is to address these challenges by mathematical means that have not hitherto been employed in a geophysical context.

This book explains how mathematical morphology could be employed to essentially deal with quantitative morphologic and scaling analyses of terrestrial phenomena and processes. It provides information on (i) the *retrieval, analysis,* and *modeling and simulation* of spatial phenomena of terrestrial importance and (ii) the applications of mathematical morphology (an advanced spatial statistical tool, popular in image processing and image analysis) to essentially deal with quantitative, morphologic, and scaling analyses of certain *geomorphologically relevant functions, sets, and skeletons* (in other words, terrestrial phenomena and processes). The motivation to propose this book is to show how and why mathematical morphology is a better choice to deal with the four aspects.

The *first aspect* includes the *retrieval* of complex topological connectivity networks of channels and ridges from digital elevation models (DEMs) by employing nonlinear morphological transformations that take advantage of curvatures over the terrain for this purpose. In contrast to other recent works, which have focused on extraction of channel networks via algorithms that fail to precisely extract networks from nonhilly regions (e.g., tidal regions), we have provided approaches for simultaneous extraction of both channel and ridge networks via algorithms that can be generalized to both hilly and nonhilly terrains.

The *second aspect* of this book is on *analysis* of terrestrial surfaces and associated features to quantitatively characterize the spatiotemporal terrestrial complexity via scale-invariant measures that explain the commonly shared physical mechanisms involved in terrestrial phenomena and processes. Such characterization highlighted the evidence of self-organization via scaling laws—in networks, hierarchically decomposed subwatersheds, and water bodies and their zones of influence, which evidently belong to different universality classes—which are in excellent agreement with geomorphologic laws such as Horton's laws, Hurst's exponents, Hack's exponent, and other power laws given in nongeoscientific contexts. This aspect is further extended based on intuitive arguments that these universal scaling laws possess limited utility in exploring possibilities to relate them with geomorphologic processes. These arguments form the basis to provide alternative methods that yield scale-invariant but shape-dependent power laws.

The *third aspect* is on *modeling* the geomorphologic processes, in discrete space, under perturbations caused due to cascading forces (flood–drought, expansion–contraction, uplift–erosion, protruding–flattening, and shortening–amplification) in nonlinear fashion mimicking realistic situations. This third aspect on modeling provided unique contributions on network simulations, laws of geomorphic structures under the perturbations created through interplay between numerical simulations and graphic analysis, and understanding spatial and/or temporal behaviors of certain evolving and dynamic geomorphic phenomena.

The *fourth aspect* is on possible application of mathematical morphology in quantitative spatial reasoning tasks and in spatial interpolations. Such applications of mathematical morphology provide insights into GISci.

B. S. Daya Sagar
Bangalore, India

Acknowledgments

I am fortunate to have found many kind people who have helped me in my professional career in numerous ways. These include my friends, teachers, mentors, external mentors, collaborators, critiques, reviewers, editors, colleagues, students, employers, and family members. Their support and help came in the form of sharing formal and informal experiences; teaching and guidance; willingness to participate in discussions; academic visits; organizing conferences, workshops, and schools; handling technical manuscripts; making corrections; creating a conducive environment; and affection. I am grateful to all of them, as follows:

External mentors: B. L. Deekshatulu, Jean Serra, Arthur Cracknell, Alan Wilson, Vladimir Gontar, Gabor Korvin, and Benoit Mandelbrot.

Teachers and supervisors: B. S. Prakasa Rao, S. V. L. N. Rao, V. R. R. M. Babu, R. V. Rama Rao, V. Venkateswara Rao, and E. Amminedu.

Employers: Bimal Roy (director of Indian Statistical Institute) and Sankar Pal (distinguished scientist and former director of Indian Statistical Institute); Gauth Jasmon (former president of Multimedia University, Malaysia, and current vice chancellor of the University of Malaya); Chuah Hean-Teik (former vice president of Multimedia University, Malaysia, and current president of Universiti Tunku Abdul Rahman, Malaysia); and Lim Hock (former director of the Centre for Remote Imaging Sensing and Processing and current director of Temasek Laboratories, The National University of Singapore) for creating a highly conducive academic environment.

My collaborators, PhD students, co-guest editors and coauthors of journal papers: Laurent Najman, C. Babu Rao, G. Rangarajan, Daniele Veneziano, Jean Serra, Gabor Korvin, Lim Hock, VC Koo, Rajesh, Ashok, Pratap, Rajashekhara, Saroj Meher, Baldev Raj, C. Babu Rao, M. Venu, G. Gandhi, K. S. R. Murthy, D. Srinivas, B. S. Prakasa Rao, Tay Lea Tien, Lim Sin Liang, Alan Tan Wee Chet, Radhakrishnan, Teo Lay Lian, B. Venkatesh, Uma Devi, L. Chockalingam, Koo Voon Chet, S. Dinesh, H.M. Rajashekara, N. Rajesh, S. Ashok Vardhan, Pratap Vardhan, Arun Kumar, N. Rama Rao, and several others.

Editors: Arthur Cracknell, Paolo Gamba, James Famigliatti, Andrea Rinaldo, Kelin Whipple, William Emery, Michale Sonis, Vladimir Gontar, Paul Curran, Giles Foody, Peter Atkinson, Daniel Merriam, Mike Ed Hohn, Petros Maragos, Mohammad El Naschie, and Hideki Takayasu.

Reviewers and critiques: It is with great pleasure that I acknowledge the support and encouragement given by Philippos Pomonis, Jean Cousty, Christian Lantuejoul, Daniele Veneziano, Jayanth Banavar, Prasad Patnaik, Jean-Claude Thill, Paolo Gamba, Petros Maragos, Vlad Nikora, B. K. Sahu, K. V. Subbarao, Murugesu Sivapalan, Vijay Gupta, Bellie Sivakumar,

Qiuming Cheng, Vera Pawlowsky-Glahn, Robert Marschallinger, Peter Atkinson, John (Jack) Schuenemeyer, Wolfgang-Martin Boerner, Frits Agterberg, Ricardo Olea, and several anonymous reviewers who provided useful comments and suggestions on the papers that I have published with my collaborators, students, and colleagues and that are parts of this book.

Colleagues: B. B. Chaudhuri, B. P. Sinha, Bhargab Bhattacharya, Malay Kumar Kundu, Babhatosh Chanda, Subhas Nandy, C. A. Murthy, I. K. Ravichandra Rao, N. S. S. Narayana, K. S. Raghavan, N. S. N. Sastry, Saroj Kumar Meher, Sasthi Ghosh, A. R. D. Prasad, M. Krishnamurthy, Devika, and Somnath Ray.

Friends: M. Venu, K. S. R. Murthy, D. Srinivas, G. Gandhi, J. Kiran, Charles Omoregie, and Bala Venkatesh.

Publishers who granted permissions: Taylor & Francis Group Publishers, Elsevier Science Publishers, Springer, IEEE, American Geophysical Union, Indian Academy of Science, World Scientific Publishers, and Hindawi Publishers.

Grants, fellowships, and/or memberships of societies: Financial support to carry out this work was provided by Indian Statistical Institute, the Department of Science and Technology, the Council of Scientific and Industrial Research, the Board of Research in Nuclear Science, MOSTI-Malaysia, EMCAB, ICT-Asia-French-Govt Fund, Indian Geophysical Union, the International Association for Mathematical Geosciences, and IEEE Geoscience and Remote Sensing Society.

I am grateful to Aastha Sharma, commissioning editor, Marsha Pronin, project coordinator, and Robert Sims, project editor at Taylor & Francis Group, and Remya Divakaran at SPi-Global, for their support, advice, and patience. I am thankful to Putta Raju, Anupama, and Kalyan Raman for their secretarial help.

My late parents supported me during hard times, especially during 1987–2001, but could not live to see my professional success. A lot of help has been rendered to me by my brothers Satish, Shekhar, Swarup, and Sanjay, and my only sister Sowjanya. I would like to express my gratitude to my wife Latha for her understanding, patience, and love. I cannot imagine life without her.

Author

B. S. **Daya Sagar** was educated in St Anthony School, Visakhapatnam, Government Arts College, and the Andhra University, India, where he studied Earth sciences. He received his BSc in 1987 from Shree Durga Prasad Saraf College of Arts and Applied Sciences and his MSc in 1991 from the College of Engineering. He then received his PhD in 1994 from Andhra University for his thesis, "Applications of Remote Sensing, Mathematical Morphology, and Fractals to Study Certain Surface Water Bodies."

From 1991 to 1992, he was a project assistant for the project "PC-based image processing system" funded by the Ministry of Human Resource Development in the Department of Geoengineering, Andhra University College of Engineering; from 1992 to 1994, he served as a senior research fellow at the Council of Scientific and Industrial Research (CSIR); from 1994 to 1995, he was a research associate at CSIR; in 1997, a served as a research scientist/principal investigator in a Scheme for Extramural Research for Young Scientists funded by the Ministry of Science and Technology; in 1998, he was a senior research associate at CSIR; from 1998 to 2001, he served as a Gr-A research scientist at the Centre for Remote Imaging Sensing and Processing in the National University of Singapore. He was appointed associate professor of the Faculty of Engineering and Technology in Multimedia University, Malaysia, in 2001 and as deputy chairman at the Centre for Applied Electromagnetics in 2003, where he served until 2007. Since 2007, he has been an associate professor at Indian Statistical Institute, Bangalore, and since 2009, he has been serving as founding head of Systems Science and Informatics Unit, a unit that has been established as one of the five constituent units of the Computer and Communication Sciences Division of Indian Statistical Institute.

Sagar authored the book *Qualitative Models of Certain Discrete Natural Features of Drainage Environment* (Allied Publishers Pvt. Limited, 2005). He has edited six theme issues—"Mathematical Geosciences," "Quantitative Image Morphology," "Fractals in Geophysics," "Surficial Mapping," "Spatial Information Retrieval, Analysis Reasoning and Modeling," and "Filtering and Segmentation with Mathematical Morphology"—for the *Journal of Mathematical Geosciences, International Journal of Pattern Recognition and Artificial Intelligence, Chaos Solitons & Fractals, IEEE Geoscience and Remote Sensing Letters, International Journal of Remote Sensing,* and *IEEE Journal on Selected Topics of Signal Processing.* He has also written more than 80 papers, out of which 55 papers appeared in international journals. He has served as editor of *Discrete Dynamics in Nature and Society: Multidisciplinary Research and Review Journal* since 2003. His research interests include mathematical morphology and fractal geometry and their applications in relation to several aspects of terrestrial surface and associated features, as well as geomorphologic information retrieval, quantitative geomorphometry, allometry, granulometry, scaling analysis, quantitative spatial reasoning, and spatiotemporal modeling. He is also interested in conducting research both in basic and applied fields of "mathematical morphology with an emphasis on complex terrestrial geomorphologic phenomena and processes" and in teaching different subjects, including remote sensing, digital image processing, and applications of mathematical morphology, at graduate and PhD levels. He has also delivered lectures both in India and abroad.

The key links that Prof. Sagar has shown between (i) pattern retrieval, (ii) pattern analysis, (iii) simulation and modeling, and (iv) spatial reasoning and their importance in understanding spatiotemporal behaviors of several terrestrial phenomena and processes was a significant success. For *retrieval* of topologically unique geomorphologic features from both fluvial and tidal regions, he has developed generalized algorithms. He has shown evidence of self-organization in several terrestrial phenomena and processes via scaling laws and has observed their limited utility to distinguish between the geomorphologic basins possessing topologically invariant networks. Based on such an observation, he has provided approaches to derive shape-dependent but scale-invariant indices for better terrestrial *analysis*. Sagar has developed a *fractal-skeletal channel network model* that can exhibit various empirical features that the random model cannot. Through discrete simulations based on interplay between numeric and graphic analyses, he has shown various behavioral phases that geomorphologic systems such as water bodies, folds, dunes, and landscapes traverse. Recently, he developed novel methods for *spatial interpolation and spatial reasoning* to visualize spatiotemporal behavior, generate contiguous maps, and to identify strategically significant set(s). His work has spurred interdisciplinary activity and has yielded insights into quantitative geomorphology and spatiotemporal GISci.

As a deputy chairman of the Centre for Applied Electromagnetics at Multimedia University, Malaysia, Sagar guided a group of young researchers who developed algorithms for surficial mapping and terrestrial characterization. Six students who worked under his supervision were awarded PhDs. As head of Systems Science and Informatics Unit (SSIU) that was set up in 2009 at Indian Statistical Institute, he was responsible for setting up the Spatial Informatics Lab. He also took the initiative of setting up the Spatial Informatics Research Group, which provides a forum for researchers, engineers, and practitioners in all applications that involve spatial information. He has organized three international conferences and several short courses related to spatial informatics. He is also a member of various examination committees, administrative committees, recruitment committees, and board of studies and has been a coordinator for various subjects that he taught to undergraduate students. He has been on adjudicating panels for about ten PhD students and numerous master's students. He also secured funding from the Government of India, the French Government, and the Malaysian Government during 1995–2009.

Sagar was elected as a member of New York Academy of Sciences in 1995, as a fellow of Royal Geographical Society in 2000, as a senior member of IEEE Geoscience and Remote Sensing Society in 2003, and as a fellow of the Indian Geophysical Union in 2011. He is also a member of American Geophysical Union since 2003, the International Association for Mathematical Geosciences since 2006, and the Association for Computing Machinery since 2008. The Andhra Pradesh Academy of Sciences awarded him the Dr. Balakrishna Memorial Award in 1995. He was also awarded the Krishnan Medal of the Indian Geophysical Union in 2002. In 2011, he was a recipient of the Georges Matheron Award with Lectureship of the International Association for Mathematical Geosciences.

1

Introduction

A basic understanding of many geophysical and engineering challenges across multiple spatial and/or temporal scales of terrestrial phenomena and processes is among the greatest of challenges facing contemporary sciences and engineering. This book is to show the importance of mathematical morphology (Matheron 1975, Serra 1982) in geomorphology and geographic information science (GISci). Important links among the key aspects like pattern retrieval, pattern analysis, spatial reasoning, and simulation and modeling for understanding spatiotemporal behaviors of several of terrestrial phenomena and processes are shown. The key links that were shown between those aspects are summarized in this book. To address these intertwined topics, various original algorithms and modeling techniques that are mainly based on mathematical morphology, fractal geometry, and chaos theory have been developed and their utilities have been demonstrated. In order to develop models, synthetic data sets and realistic data such as remotely sensed data are considered.

Surficial Features

Terrestrial surfaces of Earth and Earth-like planets exhibit variations across spatiotemporal scales. Recent advancements in remote sensing technologies that take the advantage of wavelength bands of wide ranging electromagnetic spectra paved a way to properly sense the terrestrial–oceanic–atmospheric fields. Data with respect to terrestrial phenomena are available in multiple spatial and temporal scales. Such data are acquired by various mechanisms such as physical surveys, remote sensing satellites, etc. Proper approaches to represent such data in a mode useful for further processing to prepare thematic maps are available. This book mainly addresses the feature retrieval from remotely sensed data, and analysis, reasoning, and modeling phenomena that are retrieved from multiple spatial and temporal data. The phenomena that were addressed in these investigations include small water bodies (SWBs), channel networks, watersheds, sand dunes, and sand stone porous media.

- Landscape (combination of watersheds), watershed (combination of subwatersheds).
- Hierarchical decomposition of landscape—multiscaling.

- Watershed(s)—catchment basins.
- Topographic depressions—water bodies.
- Landscapes as random functions—mathematical representations.
- Description of landscapes in discrete space and its importance.
- Landscape—a combination of watersheds, subwatersheds, topographic depressions, valleys, and ridges.
- Why do we need mathematical morphology to treat terrestrial surfaces?
- Functions, sets, and skeletons as terrestrial surfaces, threshold decomposed features, and geophysical networks.
- Mathematical morphology to deal with geophysical information retrieval, analysis, reasoning, and modeling.

Spatial Data

Availability of spatial data—for natural, anthropogenic, and socioeconomic phenomena studies—from such a wide range of sources and a variety of formats opens new horizons to the geomorphology and GISci communities. In relation to spatial information, schematically there are four aspects that are also four challenges that these scientific communities face. They have to retrieve this information, which supposes to segment the space in homogeneous zones according to some criteria. This often implies filtering steps. They must analyze the selected regions, i.e., associate with them certain significant numbers and numerical functions, such as size distributions. They have to implicate and apply the aforementioned geometrical descriptors in more general reasoning of some specific context, such as "what is the best place to locate a hospital or to trace a road?" And, sometimes, they have to conceive random or deterministic models for synthesizing the results of the analysis phase in order to make forecast evolutions. The studies that follow in this book heavily rely on the ideas stemmed from mathematical morphology (Matheron 1975, Serra 1982, Najman and Talbot 2010). As a matter of fact, many map algebraic operations on maps (Tomlin 1983) involved in GISci-related analyses can be performed via mathematical morphology (e.g., Soille 1999, Pullar 2001, Stell 2007).

For *retrieval* of topologically unique geomorphologic features from both fluvial and tidal regions, generalized algorithms were developed. In Chapters 5 and 6, evidence of self-organization in several terrestrial phenomena and processes via scaling laws was shown, and based on the observation of their limited utility to distinguish between the geomorphologic basins possessing topologically invariant networks, novel approaches are shown to derive shape-dependent but scale-invariant indexes for better terrestrial *analysis*. Fractal–skeletal channel network (F-SCN) *model* can exhibit various empirical features that the random model cannot. Through discrete simulations based on interplay between numeric and graphic analyses, various behavioral phases that geomorphologic systems—such as water bodies, folds, dunes, and landscapes—traverse were shown. Novel methods for *spatial interpolation and spatial reasoning* to visualize spatiotemporal behavior, generate contiguous maps, and identify strategically significant set(s) were proposed. This book spurs interdisciplinary activity that has implications and would yield insights for quantitative geomorphology and spatiotemporal GISci.

General Organization of the Book

This book provides the following:

A brief introduction along with the general organization of this book is given in this chapter. In Chapter 2, a brief introduction of mathematical morphology, which is crucial to understand the techniques employed in subsequent chapters, is given in an easy-to-understand manner. In Chapters 4 through 8, several data sets have been used to demonstrate numerous techniques. The specifications of those data sets have been provided in Chapter 3.

Pattern Retrieval (Chapter 4)

Original algorithms for the *retrieval* of unique geomorphologic networks, landforms, and threshold elevation regions (Sagar et al. 2000, Chockalingam and Sagar 2003, Sagar et al. 2003b, Sathymoorthy et al. 2007, Lim and Sagar 2008, Lim et al. 2009) for efficient characterization have been detailed. In contrast to other recent works, which have focused on the extraction of channel networks via algorithms that fail to precisely extract networks from tidal regions, the algorithms that Sagar et al. (2000) proposed can be generalized to both fluvial and tidal terrains. This piece of work helps to solve basic problems that all algorithms meant for the extraction of unique terrestrial connectivity networks have faced for over three decades. These algorithms are for unique feature retrieval from digital elevation maps. These algorithms grasped the importance of curvature concerning the framework to extract multiscale geomorphologic networks via systematically decomposing elevation surfaces and/or decomposed threshold elevation regions into their abstract structures that lead to valley and ridge connectivity networks. Approaches—which can be implemented on several geophysical and geomorphologic fields (e.g., digital elevation models (DEMs), clouds, and binary fractals) to segment them into regions of varied topological significance—have been demonstrated on cloud fields derived from MODIS data to better segment the regions within the cloud fields that have different compaction properties with varied cloud properties—that could be derived directly from elevation field, to quantify the spatial complexity that have relationships with conventional quantitative geomorphometric quantities of topological relevance—solved a basic problem by preserving the spatial variability which could not be achieved by planimetric-based measures.

Pattern Analysis (Chapters 5 through 8)

The techniques and methodology developed for geomorphologic pattern analyses provided/captured were demonstrated throughout Chapters 5 through 8. Studies related to quantitative characterization of spatiotemporal terrestrial complexity via scale-invariant measures that explain the commonly sharing physical mechanisms involved in terrestrial phenomena and processes were shown (Sagar et al. 1995a,b, 1998a, 1999, Sagar 1996, 1999a, Radhakrishnan et al. 2004, Teo et al. 2004, Chockalingam and Sagar 2005, Lian and Sagar 2005, 2006, Tay et al. 2005a,b, 2007, Lim and Sagar 2008b, Lim et al. 2011). The relationships derived serve to demonstrate the evidence of (1) self-organization via scaling laws in networks, hierarchically decomposed subwatersheds, and water bodies and their zones of influence (Sagar and Rao 1995b, Sagar 2000a, Sagar 2001c, Sagar et al. 2002, Sagar and Chockalingam 2004, Sagar and Tien 2004); (2) different universality classes of different terrestrial features; (3) relationships with laws such as Horton's laws, Hurst exponents, Hack's exponent, and other power-laws given in non-geoscientific context (Sagar and Chockalingam 2004, Sagar and Tien 2004, Tay et al. 2006, Sagar 2007); (4) limited utility of universal scaling laws in exploring possibilities to relate them with geomorphological processes; and (5) the need for alternative methods that provide scale-invariant but shape-dependent indexes for characterization of hillslopes and terrestrial surfaces (Sagar and Chockalingam 2004, Chockalingam and Sagar 2005, Tay et al. 2005b, 2007). In Chapters 4 and 5, a large number of surface water bodies (irrigation tanks), situated in the floodplain region of certain east-flowing rivers of India, which are retrieved from multi-date remotely sensed data, are analyzed in two-dimensional (2-D) space. Analysis was done primarily from the point of their size and shape distributions. In addition to this, basic measures of these water bodies were employed to show fractal length–area–perimeter relationships. Further investigations were carried out to include computations of fractal dimensions of skeletal networks of planar fractals and simulations of channel networks within fractal basins (Sagar et al. 1998, 2001). An F-SCN model has been simulated by employing nonlinear morphological transformations to construct other classes of network models, which can exhibit various empirical features that the random model cannot (Sagar and Murthy 2000, Sagar et al. 2001). In this model, it has been demonstrated how homogeneous and heterogeneous channel networks can be constructed. Applications of mathematical morphology transformations are shown to decompose fractal basins into nonoverlapping disks (NODs) of various shapes and sizes further to derive fractal power laws based on number–radius relationship (Sagar and Chockalingam 2004, Chockalingam and Sagar 2005). These networks facilitate to segment fractal DEM into sub-basins ranging from first to highest order. Host of allometric power-law relationships were drawn that were in good accord with other established network models and realistic networks. Entire framework was based on discrete rules and morphological transformations. Topologically, water bodies are the first-level topographic regions that get flooded, and as the flood level gets higher, adjacent water bodies merge. The looplike network that forms along all these merging points represents zones of influence of each water body (Sagar 2001, 2005, 2007). The geometric organizations of these two phenomena are respectively sensitive and insensitive to perturbation due to exogenic processes. Such interdependent phenomena follow the universal scaling laws found in other geophysical and biological contexts. In this work, universal scaling relationships among basic measures such as area, length, diameter, volume, and information about networks are exhibited by several natural phenomena to further retrieve and understand the common principles underlying the organization of these phenomena. In this study, a host of universal scaling laws in surface water bodies and their zones of influence that have similarities with several of these relationships encountered in various fields have been shown. Varied degrees of topographically convex regions within a catchment basin represent varied degrees of hillslopes. The nonnetwork space is akin to the space that is achieved by subtracting channelized portions contributed due to concave regions from the watershed space. This nonnetwork space is akin to non-channelized convex region within a catchment basin. An alternative shape-dependent quantity akin to fractal dimension to characterize this nonnetwork space has been proposed. Toward this goal, nonnetwork space is decomposed, in 2-D discrete space, into simple NODs of various sizes by employing mathematical morphological transformations and certain logical operations. Furthermore, the number of NODs of lesser than threshold radius is plotted against the radius, and the shape-dependent fractal dimension of nonnetwork space is computed. This study was extended to derive shape-dependent scaling laws as the laws derived from network measurements are shape independent. The relationship between the number of NODs and the radius of the disk provides an alternative fractal-like dimension that is shape dependent (Radhakrishnan et al. 2004, Sagar and Chockalingam 2004, Chockalingam and Sagar 2005). This was done with an aim to relate shape-dependent power laws with geomorphic processes such as hillslope processes, erosion, etc. Martian and terrestrial DEMs are analyzed by following granulometry and pattern spectrum concepts to derive shape–size complexity measures that provide new indexes to understand the Martian/terrestrial surfaces further to relate with several geomorphic processes. Simulating geodesic flow fields within a basin consisting of spatially distributed elevation regions, further to compute a geodesic spectrum, provides a unique quantitative (one-dimensional geometric support) geomorphologic indicator. Geodesic spectrum—that outperforms the conventional width function-based approach that is usually derived from planar forms of basin and its networks—construction involves basin as a random elevation field (e.g., DEM), and all threshold elevation regions are decomposed from DEM for understanding the shape–function relationship much better than that of width function.

Modeling (Chapter 9)

Computer simulations and modeling techniques demonstrated in this book provide insights to better understand certain geomorphologic and geophysical systems with the ultimate goal of developing cogent models in discrete space. The work is a fusion of computer simulations and spatial information theory and is closely related to the fields of mathematical geophysics and spatial informatics. The basic inputs required to understand the spatio-dynamical behavior of certain terrestrial phenomena will be drawn from multiscale/multitemporal satellite remotely sensed data. The three complex systems that are explained include the channelization process, surface water bodies, and elevation structures. Simulations allow us to gain a significantly good understanding of these complex systems in a way that is not possible with lab experiments. In regard to simulation of several possible behaviors of sand dunes and symmetrical fold, a first-order nonlinear difference equation that has the physical basis—to simulate all possible behaviors of these distinct phenomena—was considered. In these studies, critical inter-slipface angles for sand dune dynamics and inter limb angles for symmetrical fold under dynamics simulated under varied control parameters were proposed and shown via bifurcation phenomena.

Laws of geomorphic structures under the perturbations are provided and shown, through an interplay between numerical simulations and graphic analysis as to how systems traverse through various behavioral phases (Sagar et al. 1998, Sagar 2001). The discrete simulations are shown for the varied dynamical behavioral phases of certain geo(morpho)logic processes (e.g., water bodies (Sagar 2005), ductile symmetric folds (Sagar 1998), sand dunes (Sagar 1999, 2000, Sagar and Venu 2001, Sagar et al. 2003), and landscapes) under nonlinear perturbations that are caused due to *endogenic* and *exogenic* nature of forces.

Models for certain geomorphological processes in discrete space have been developed by simulating perturbations caused due to flood and drought (water body dynamical behavior), uplift and erosion (landscape dynamical behavior), shortening and amplification (fold dynamical behavior), and protruding and flattening (sand dune dynamical behavior) in a nonlinear fashion mimicking the realistic situations. Areal extents of a brackish water lagoon, Chilka Lake, are computed from the multi-date remotely sensed data, and the areal extent changes are modeled through logistic maps (Sagar and Rao 1995a–c). Spatiotemporal patterns of SWBs under the influence of temporally varied stream flow discharge behaviors are simulated in discrete space by employing geomorphologically realistic expansion and contraction transformations. Expansions and contractions of SWBs to various degrees, which are obvious due to fluctuations in stream flow discharge pattern, simulate the effects respectively owing to stream flow discharge that is greater or lesser than mean stream flow discharge. The cascades of expansion–contraction are systematically performed by synchronizing the stream flow discharge, which is represented as a template with definite characteristic information, as the basis to model the spatiotemporal organization of randomly situated surface water bodies of various sizes and shapes. The interplay between numerical simulations and graphic analysis has been shown to understand how these geomorphologically significant systems traverse through various behavioral phases.

Mathematical Morphology in GISci (Chapters 10 through 14)

Methods developed for spatial interpolation, visualization, and quantitative spatial reasoning (Sagar 2010, Sagar and Serra 2010, Rajashekara et al. 2012, Sagar et al. 2013) have been demonstrated. In an approach for spatial interpolation, Hausdorff dilation and Hausdorff erosion distances have been employed for the categorization of time-varying thematic maps depicting geomorphologic phenomenon and for the visualization of spatiotemporal behavior of such phenomenon by recursive generation of median elements. Spatial interpolation, which was earlier seen as a global transform, is extended by introducing *bijection* to deal with even connected components. This aspect solves problems of global nature in spatial–temporal GIS. Besides, a mathematical morphology-based algorithm has been proposed to generate contiguous maps from point data for better visualization. The use of thematic maps in time-sequential mode to visualize the spatiotemporal behavior of a phenomenon is demonstrated. Various other algorithms based on mathematical morphology that have been proposed and demonstrated are of use in quantitative spatial reasoning studies.

References

Chockalingam, L. and B. S. D. Sagar, 2003, Automatic generation of sub-watershed map from Digital Elevation Model: A morphological approach, *International Journal of Pattern Recognition and Artificial Intelligence*, 17(2), 269–274.

Chockalingam, L. and B. S. D. Sagar, 2005, Morphometry of networks and non-network spaces, *Journal of Geophysical Research-Solid Earth (American Geophysical Union)*, 110, B08203, doi:10.1029/2005JB003641.

Lian, T. L. and B. S. D. Sagar, 2005, Reconstruction of pore space from pore connectivity network via morphological transformations, *Journal of Microscopy (Oxford)*, 219(Pt 2), 76–85.

Lian, T. L. and B. S. D. Sagar, 2006, Modeling, characterization of pore-channel, throat and body, *Discrete Dynamics in Nature and Society*, 2006, 1–24, Article ID 89280.

Lim, S. L., V. C. Koo, and B. S. D. Sagar, 2009, Computation of complexity measures of morphologically significant zones decomposed from binary fractal sets via multiscale convexity analysis, *Chaos, Solitons & Fractals*, 41(3), 1253–1262.

Lim, S. L. and B. S. D. Sagar, 2008a, Cloud field segmentation via multiscale convexity analysis, *Journal Geophysical Research-Atmospheres*, 113, D13208, 17, doi:10.1029/2007JD009369.

Lim, S. L. and B. S. D. Sagar, 2008b, Derivation of geodesic flow fields and spectrum in digital topographic basins, *Discrete Dynamics in Nature and Society*, 2008, 26, Article ID 312870, doi:10.1155/2008/312870.

Lim, S. L., B. S. D. Sagar, V. C. Koo, and L. T. Tay, 2011, Morphological convexity measures for terrestrial basins derived from Digital Elevation Models, *Computers & Geosciences*, 37, 1285–1294.

Matheron, G., 1975, *Random Sets and Integral Geometry*, John Wiley & Sons, New York.

Najman, L. and H. Talbot, eds., 2010, *Mathematical Morphology: From Theory to Applications*, John Wiley & Sons, New York.

Pullar, D., 2001, MapScript: A map algebra programming language incorporating neighborhood analysis, *Geoinformatica*, 5, 145–163.

Radhakrishnan, P., B. S. D. Sagar, and L. L. Teo, 2004, Estimation of fractal dimension through morphological decomposition, *Chaos Solitons & Fractals* (an International Journal from Elsevier), 21(3), 563–572.

Rajashekara, H. M., P. Vardhan, and B. S. D. Sagar, 2012, Generation of zonal map from point data via weighted skeletonization by influence zone, *IEEE Geoscience and Remote Sensing Letters*, 9(3), 403–407.

Sagar, B. S. D., 1996, Fractal relations of a morphological skeleton, *Chaos, Solitons & Fractals*, 7(11), 1871–1879.

Sagar, B. S. D., 1998, Numerical simulations through first order nonlinear difference equation to study highly ductile symmetric fold (HDSF) dynamics: A conceptual study, *Discrete Dynamics in Nature and Society*, 2(4), 281–298.

Sagar, B. S. D., 1999a, Estimation of number-area-frequency dimensions of surface water bodies, *International Journal of Remote Sensing*, 20(13), 2491–2496.

Sagar, B. S. D., 1999b, Morphological evolution of a pyramidal sandpile through bifurcation theory: A qualitative model, *Chaos, Solitons & Fractals*, 10(9), 1559–1566.

Sagar, B. S. D., 2000a, Fractal relation of medial axis length to the water body area, *Discrete Dynamics in Nature and Society*, 4(1), 97.

Sagar, B. S. D., 2000b, Multi-fractal-interslipface angle curves of a morphologically simulated sand dune, *Discrete Dynamics in Nature and Society*, 5(2), 71–74.

Sagar, B. S. D., 2001a, Generation of self organized critical connectivity network map (SOCCNM) of randomly situated surface water bodies, letters to editor, *Discrete Dynamics in Nature and Society*, 6(3), 225–228.

Sagar, B. S. D., 2001b, Hypothetical laws while dealing with effect by cause in discrete space, letter to the editor, *Discrete Dynamics in Nature and Society*, 6(1), 67–68.

Sagar, B. S. D., 2001c, Quantitative spatial analysis of randomly situated surface water bodies through f–α spectra, *Discrete Dynamics in Nature and Society*, 6(3), 213–217.

Sagar, B. S. D., 2005, Discrete simulations of spatio-temporal dynamics of small water bodies under varied stream flow discharges, (invited paper), *Nonlinear Processes in Geophysics (American Geophysical Union)*, 12, 31–40.

Sagar, B. S. D., 2007, Universal scaling laws in surface water bodies and their zones of influence, *Water Resources Research*, 43(2), W02416.

Sagar, B. S. D., 2010, Visualization of spatiotemporal behavior of discrete maps via generation of recursive median elements, *IEEE Transactions on Pattern Analysis and Machine Intelligence*, 32(2), 378–384.

Sagar, B. S. D. and L. Chockalingam, 2004, Fractal dimension of non-network space of a catchment basin, *Geophysical Research Letters* (American Geophysical Union), 31(12), L12502.

Sagar, B. S. D., G. Gandhi, and B. S. P. Rao, 1995b, Applications of mathematical morphology on water body studies, *International Journal of Remote Sensing*, 16(8), 1495–1502.

Sagar, B. S. D. and K. S. R. Murthy, 2000, Generation of fractal landscape using nonlinear mathematical morphological transformations, *Fractals*, 8(3), 267–272.

Sagar, B. S. D., M. B. R. Murthy, and P. Radhakrishnan, 2003a, Avalanches in numerically simulated sand dune dynamics, *Fractals*, 11(2), 183–193.

Sagar, B. S. D., M. B. R. Murthy, C. B. Rao, and B. Raj, 2003b, Morphological approach to extract ridge-valley connectivity networks from Digital Elevation Models (DEMs), *International Journal of Remote Sensing*, 24(3), 573–581.

Sagar, B. S. D., C. Omoregie, and B. S. P. Rao, 1998, Morphometric relations of fractal-skeletal based channel network model, *Discrete Dynamics in Nature and Society*, 2(2), 77–92.

Sagar, B. S. D., N. Rajesh, S. A. Vardhan, and P. Vardhan, 2013, Metric based on morphological dilation for the detection of spatially significant zones, *IEEE Geoscience and Remote Sensing Letters*, 10(3), 500–504.

Sagar, B. S. D. and B. S. P. Rao, 1995a, Computation of strength of nonlinearity in lakes, letter to the editor, *Computers & Geosciences*, 21(3), 445.

Sagar, B. S. D. and B. S. P. Rao, 1995b, Fractal relation on perimeter to the water body area, *Current Science*, 68(11), 1129–1130.

Sagar, B. S. D. and B. S. P. Rao, 1995c, Possibility on usage of return maps to study dynamics of lakes: Hypothetical approach, *Current Science*, 68(9), 950–954.

Sagar, B. S. D. and B. S. P. Rao, 1995d, Ranking of lakes: Logistic maps, *International Journal of Remote Sensing*, 16(2), 368–371.

Sagar, B. S. D., C. B. Rao, and B. Raj, 2002, Is the spatial organization of larger water bodies heterogeneous? *International Journal of Remote Sensing*, 23(3), 503–509.

Sagar, B. S. D. and J. Serra, 2010, Spatial information retrieval, analysis, reasoning and modelling, *International Journal of Remote Sensing*, 31(22), 5747–5750.

Sagar, B. S. D., D. Srinivas, and B. S. P. Rao, 2001, Fractal skeletal based channel networks in a triangular initiator basin, *Fractals*, 9(4), 429–437.

Sagar, B. S. D. and T. L. Tien, 2004, Allometric power-law relationships in a Hortonian Fractal DEM, *Geophysical Research Letters (American Geophysical Union)*, 31(6), L06501.

Sagar, B. S. D. and M. Venu, 2001, Phase space maps of a simulated sand dune: A scope, *Discrete Dynamics in Nature and Society*, 6(1), 63–65.

Sagar, B. S. D., M. Venu, G. Gandhi, and D. Srinivas, 1998, Morphological description and interrelationship between force and structure: A scope to geomorphic evolution process modeling, *International Journal of Remote Sensing*, 19(7), 1341–1358.

Sagar, B. S. D., M. Venu, and K. S. R. Murthy, 1999, Do skeletal network derived from water bodies follow Horton's laws? *Journal Mathematical Geology*, 31(2), 143–154.

Sagar, B. S. D., M. Venu, and B. S. P. Rao, 1995a, Distributions of surface water bodies, *International Journal of Remote Sensing*, 16(16), 3059–3067.

Sagar, B. S. D., M. Venu, and D. Srinivas, 2000, Morphological operators to extract channel networks from Digital Elevation Models, *International Journal of Remote Sensing*, 21(1), 21–30.

Sathymoorthy. D., P. Radhakrishnan, and B. S. D. Sagar, 2007, Morphological segmentation of physiographic features from DEM, *International Journal of Remote Sensing*, 28(15), 3379–3394.

Serra, J., 1982, *Image Analysis and Mathematical Morphology*, Academic Press, London, U.K.

Soille, P., 1999, *Morphological Image Analysis: Principles and Applications*, Springer-Verlag, Heidelberg, Germany.

Stell, J. G., 2007, Relations in mathematical morphology with applications to graphs and rough sets, *Lecture Notes in Computer Science—Spatial Information Theory Book series*, DOI: 10.1007/978-3-540-74788-8, pp. 438–454.

Tay, L. T., B. S. D. Sagar, and H. T. Chuah, 2005a, Analysis of geophysical networks derived from multiscale digital elevation models: A morphological approach, *IEEE Geoscience and Remote Sensing Letters*, 2(4), 399–403.

Tay, L. T., B. S. D. Sagar, and H. T. Chuah, 2005b, Derivation of terrain roughness indicators via Granulometries, *International Journal of Remote Sensing*, 26(18), 3901–3910.

Tay, L. T., B. S. D. Sagar, and H. T. Chuah, 2006, Allometric relationships between travel-time channel networks, convex hulls, and convexity measures, *Water Resources Research (American Geophysical Union)*, 42(2), W06502,10.1029/2005WR004092.

Tay, L. T., B. S. D. Sagar, and H. T. Chuah, 2007, Granulometric analysis of basin-wise DEMs: A comparative study, *International Journal of Remote Sensing*, 28(15), 3363–3378.

Teo, L. L., P. Radhakrishnan, and B. S. D. Sagar, 2004, Morphological decomposition of sandstone pore-space: Fractal power-laws, *Chaos Solitons & Fractals*, 19(2), 339–346.

Tomlin, C. D., 1983, A map algebra, *Proceedings of Harvard Computer Graphics Conference*, Cambridge, MA, pp. 127–150.

2

Mathematical Morphology:
An Introduction

Birth of Mathematical Morphology

The first concepts of mathematical morphology were introduced by Georges Matheron (1975) as a part of his studies to find out the relationships between the geometry of porous media and their permeabilities in 1964, and later these studies were extensively developed by Jean Serra (1982) and followed by the scientists at Centre for Geostatistics and Mathematical Morphology (CGMM), Paris. This subject is mostly developed with having applications in stereology, microscopy, metallurgy, and in the fields of remote sensing, pattern recognition, and medical image processing. Most of the subject was developed at CGMM. Some others like Sternberg (1986) have introduced some of the pipeline transformations that are highly useful for grayscale functions. Even though mathematical morphology started around 1964, the work done was only on the binary images. The theory of mathematical morphology was introduced by J. Serra in 1975 and then developed by Lantuejoul (1978), Meyer (1980), and Beaucher (1990).

Mathematical morphology is originally based on set theory where sets represent objects in an image (Serra 1982). Mathematical morphology is a language like English. In this language, the basic operations like AND, OR, UNION, INTERSECTION, SUB, XOR, etc., are the characters. Using these characters, one can derive words like hit or miss transformation (HMT), erosion, dilation, etc. And one can also form sentences with the words (short/long) like opening, closing, rolling ball transformation, grassfire transformation (GFT), and cascade operations like cascade of erosion–dilation and cascade of dilation–erosion. Paragraphs can be formed to perform watershed, skiz, or thinning. In the successive sections, basics of set theory, binary morphology, mathematical representation of morphological processes, concept of structuring element, multiscale operations, skeletonization process, HMT and GFT and grayscale morphological operations are described with a diagrammatic representation.

Elements of Set Theory and Logical Operations

Logical operations are very helpful in understanding the morphologic concepts, and hence an elementary outline of the main facets of set theory is given here. Intersection, union, inclusion, and complement are some of the set operators. Mathematical morphology is based on these set operators. The main characteristics of the images will be preserved even after the transformation, implying a loss of information. A set of measurements can be computed to carry out quantitative analysis process provided the image is simplified.

Logical operations are shown illustratively on two binary images (X and Y) in Figure 2.1. These logical operations that are essential to understand mathematical morphological transformations have been illustrated in Figure 2.1a through e.

Grid Utilized for Morphological Transformations

Morphological transformations are carried out in discrete binary space. There are four basic units in the process of implementing morphological transformations: an image on a grid with a finite length and breadth, a subimage that is smaller than the image with a chosen grid that convolves (tessellate) over the image, a definite grammar that is looked for to obtain a similarity or non-similarity conditions, and finally an "action" initiated to generate the output.

In general, the rectangular grid is chosen to carry out the operations in discrete space. Operations on the square grid generate abruptness at the diagonal sites, whereas such abruptness is avoided in the usage of a hexagonal grid. The chosen grid to demonstrate various case studies in this book is a rectangle grid. Programming for a hexagonal grid is more complex in that each hexagonal location in terms of square matrix needs definition. The pixel arrangement in visual display monitor usually is on a rectangular grid, and hence pixel addressing is more flexible. The notion of a disk has to be considered since an image consists of a network of points dispatched on a discrete grid. In an octagonal grid, a point has eight neighbors, and a disk in this grid will be a square. Another type of grid is a hexagonal grid. The choice of the number of neighbors is more straightforward as a point has six neighbors, all at the same distance, and the disk is a hexagon. In a square grid, a point has four neighbors and a disk is a kind of a diamond. For a better understanding, the rectangular and hexagonal grids are shown in Figure 2.2.

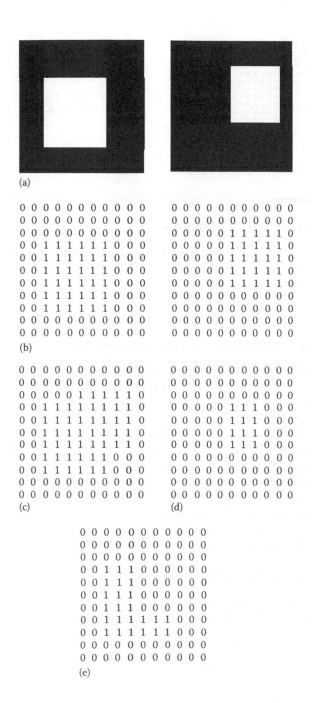

FIGURE 2.1
Illustrations of logical operations. A nonempty compact set and its representations: (a) in the form of a shape (foreground) shown in white shade and its complement (background) shown in black shade and (b) the discrete representation of the binary shape shown in (a), (c) $X \cup Y$, (d) $X \cap Y$, and (e) $X \backslash Y$.

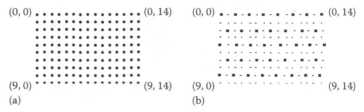

FIGURE 2.2
Grids: (a) square grid and (b) hexagonal grid.

Theory of Structuring Elements

Structuring element is a microstructure of the set with which transformations are to be performed. The role of structuring element is to unravel the hidden morphological properties of the set (X) that is transformed by structuring element (B) according to a particular rule. This functions as an interface between objective and subjective.

Characteristic Information of Structuring Element

Structuring element (B)—that possesses various characteristic information such as shape, size, orientation, and origin (Figure 2.3)—is used as a probing rule to perform the morphologic transformations on set X (Serra 1982). A symmetric template that performs various morphological transformations such as binary erosion, dilation, opening, and closing (Serra 1982) at various phases of this book is defined as follows: $B^s = [-b : b \in B]$, where B^s is obtained by rotating B by 180° on the plane. Broadly, the structuring elements (B) are categorized as symmetric and asymmetric types (Figure 2.4). We consider B that is symmetric with respect to the origin, circle in shape (on eight-connectivity grid), and of primitive size 3 × 3. The transpose of structuring element is shown in Figure 2.4. B_x will be the structuring element centered in x and B, the symmetric of B relative to its center. S_h will be translated set by vector h as shown in Figure 2.4.

Decomposition of Structuring Element

These structuring elements can be defined at will to unravel hidden properties of image under investigation. This choice is based on the type of result to

$$
\begin{array}{ccc}
1 & 1 & 1 \\
1 & \textcircled{1} & 1 \\
1 & 1 & 1
\end{array}
\qquad
\begin{array}{ccc}
0 & 1 & 0 \\
1 & \textcircled{1} & 1 \\
0 & 1 & 0
\end{array}
$$

FIGURE 2.3
Characteristics of flat structuring element: square in shape, size of 3 × 3, and symmetric about the origin.

```
1    1    1      1   1    1

1   (1)   1      1   1    1

1    1    1      1   1   (1)
    (a)                 (b)
```

FIGURE 2.4
Symmetrical and asymmetrical structuring elements. (a) Structuring element that is symmetric about the origin (in parenthesis) and (b) structuring element that is asymmetric about the origin (in parenthesis).

get the purpose of the transformation. Prior to understanding the impact of other types of structuring elements on the structure, it should be noted that there are several types of line structuring elements or one-dimensional (1-D) structuring elements. The line structuring elements with directions 0°, 45°, 90°, 135°, 180°, 225°, 270°, and 315° that are available on a square grid and 0°, 60°, 120°, 180°, 240°, and 300° that are on a hexagonal grid are shown in Figure 2.5.

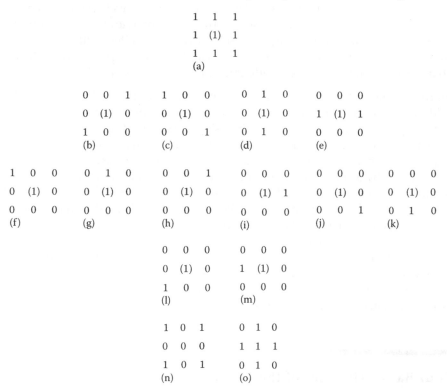

```
                    1   1   1
                    1  (1)  1
                    1   1   1
                       (a)

   0   0   1      1   0   0     0   1   0     0   0   0
   0  (1)  0      0  (1)  0     0  (1)  0     1  (1)  1
   1   0   0      0   0   1     0   1   0     0   0   0
      (b)            (c)           (d)           (e)

1  0  0    0  1  0    0  0  1    0  0  0    0  0  0    0  0  0
0 (1) 0    0 (1) 0    0 (1) 0    0 (1) 1    0 (1) 0    0 (1) 0
0  0  0    0  0  0    0  0  0    0  0  0    0  0  1    0  1  0
 (f)        (g)        (h)        (i)        (j)        (k)

                0   0   0      0   0   0
                0  (1)  0      1  (1)  0
                1   0   0      0   0   0
                   (l)           (m)

                1   0   1      0   1   0
                0   0   0      1   1   1
                1   0   1      0   1   0
                   (n)           (o)
```

FIGURE 2.5
Decomposition of circular structuring elements of eight-connectivity grid into 1-D structuring elements. (a) Flat symmetric structuring element, (b–e) one-dimensional symmetric structuring elements, (f–m) bi-point structuring elements, (n) structuring element B^1, and (o) structuring element B^2.

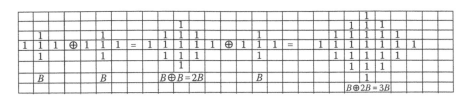

FIGURE 2.6
Minkowski sum of structuring elements $B \oplus B = 2B$.

These 1-D structuring elements can be used to shrink or expand the size of the objects in the given direction and may discard them. These line structuring elements are centered on the right side for erosion and on the left side for dilation. Many other objects can be generated with the composition of 1-D structuring elements, as shown in Figure 2.6.

Property of Iteration

To generate a large size erosion or dilation, the dilation as well as erosion can be iterated. Instead of using a larger structuring element, with the use of smaller structuring element repeatedly, one will get the same effect, although not all dilations with a large structuring element can be so decomposed. The nth size structuring element, denoted as nB, can be represented as $\underbrace{B \oplus B \oplus \cdots \oplus B}_{n\text{-times}} = nB$.

Figure 2.7 shows the process of generating larger size structuring elements (B) with structuring element of primitive size. As an example, we show in Figure 2.7 how we get $2B$ by adding $1B$ with $1B$. By using nth size B, nB, multiscale morphological transformations can be implemented. According to Matheron's (1975) approach, each image object is assumed to contain its boundary and thus can be represented by a closed subset of Euclidean space. In addition, many structuring templates are represented by a compact subset of E, so that constraints that correspond to the four principles of the theory of mathematical morphology ("Four Basic Principles of the Theory of Mathematical Morphology" section) such as invariance under translation, compatibility with change of scale, local knowledge, and upper semicontinuity (which are detailed just after the basic binary mathematical morphological transformations) will be imposed on morphological set transformations (erosion, dilation, opening, and closing) for precise extraction of topological information from the geomorphologic features.

Four Basic Principles of the Theory of Mathematical Morphology

Morphological transformations of an image object are said to be quantitative only if it satisfies four basic principles of the theory of mathematical morphology (Serra 1982, Maragos and Schafer 1986).

```
                                    1  1  1  1  1  1  1
                                    1  1  1  1  1  1  1
                    1  1  1  1  1   1  1  1  1  1  1  1
                    1  1  1  1  1   1  1  1  1  1  1  1
          1  1  1   1  1  1  1  1   1  1  1  1  1  1  1
          1  1  1   1  1  1  1  1   1  1  1  1  1  1  1
          1  1  1   1  1  1  1  1   1  1  1  1  1  1  1
          (a)          (b)            (c)
```

```
                                          1
                                       1  1  1
                          1         1  1  1  1  1
                       1  1  1   1  1  1  1  1  1  1
             1         1  1  1  1  1   1  1  1  1  1
          1  1  1         1  1  1      1  1  1  1
             1              1            1
          (d)          (e)            (f)
```

```
                          1  1  1
                       1  1  1  1  1
          1  1  1   1  1  1  1  1  1  1
       1  1  1  1  1  1  1  1  1  1  1
       1  1  1  1  1  1  1  1  1  1  1
       1  1  1  1  1  1  1  1  1  1
          1  1  1      1  1  1
          (g)          (h)
```

FIGURE 2.7
Scale–size aspects of structuring elements. (a–c) Square structuring element (which is also treated as a circle in eight-connectivity grid) of sizes B, 2B, and 3B, (d–f) rhombic structuring element (which is also treated as a circle in four-connectivity grid) of sizes B, 2B, and 3B, and (g and h) octagonal structuring element, which can be obtained by taking the Minkowski sum of symmetrical square and rhombic structuring elements of sizes 3 × 3, of B and 2B.

Invariance Under Translation

For any vector z in E, we have $X_z \oplus B = (X \oplus B)_z$. Besides, erosion or dilation by a single point is just a translation, i.e., $X \ominus \{b\} = X \oplus \{b\} = X_b$.

Erosion and Dilation

Erosion and dilation of X by B are increasing transformations with respect to X: $X_1 \subseteq X_2 \Rightarrow X_1 \oplus B \subseteq X_2 \oplus B$. However, it is decreasing with respect to B, i.e., $B_1 \subseteq B_2 \Rightarrow X \ominus B_2 \subseteq X \ominus B_1$. It can be inferred from these properties that if B contains the origin, then the erosion operation is an anti-extensive transformation, whereas the dilation is extensive, i.e., $X \ominus B \subseteq X \subseteq X \oplus B$.

Parallel Composition

The operations dilations and erosions distribute over set union and set intersection, respectively.

$$(X \cup Y) \oplus B = (X \oplus B) \cup (Y \oplus B)$$

$$(X \cap Y) \ominus B = (X \ominus B) \cap (Y \ominus B)$$

$$X \ominus (A \cup B) = (X \ominus A) \cap (X \ominus B)$$

X, M, Y, A, and $S \cdots$ = Subsets of E, x, m, y, a, and $s \cdots$ = elements or vectors of E.

Serial Composition

Successive erosions and dilations of a set X first by A and then by B equivalent to erosion and dilation, respectively, of X by their Minkowski sum $(A \oplus B)$ are given as follows:

$$(X \oplus A) \oplus B = X \oplus (A \oplus B)$$

$$(X \ominus A) \ominus B = X \ominus (A \oplus B)$$

Local Knowledge

Let M be a bounded analysis frame and X an image object that may exceed the mask M. Inside the mask M, we can know without error only the masked set $X \cap M$ and its transformed versions. However, erosions or dilations can be obtained from the original unmasked set M by a structuring element B without error inside a new mask $M^* = M \ominus B^s$.

Mathematical morphology is useful both in the transformation process and in the specific measurements later. One can use morphologic methods to transform the images to an appropriate condition, and then one can use the other transformations to calculate the various parameters like area, perimeter, length of an object, etc. In this book, only the binary mathematical morphological transformations have been detailed with respect to their usefulness in the contexts of geomorphology and GISci. In binary images, the sets (or objects) in question are represented as white (or black, depending on convention) pixels at the (x, y) coordinates in the image, defined in two-dimensional (2-D) integer space Z^2. However, morphological operations can also be extended to grayscale images that are represented as functions (Maragos 1989). Here, at the (x, y) coordinates of a pixel, it is assigned a value corresponding to its associated discrete gray-level value, $f(x, y)$. In this book, we also focus on morphological algorithms based on grayscale images that are more efficient and elegant than applying on binary images.

Binary Mathematical Morphological Operations

Certain notations that are used in the study are listed separately in the list of symbols. Most of the mathematical formalism and notations are adopted from Serra (1982). To understand this procedure, certain basic mathematical morphological transformations are detailed along with the list of symbols and notations. Mathematical morphology based on set theoretic concepts is a particular approach to the analysis of geometric properties of different structures. From geometrical point of view, morphological dilations and erosions are defined as set transformations that expand and contract a set. The morphological operators can be visualized as working with two images. The image being processed is referred to as the sets and other image as a structuring template. The main objective is to study the geometrical properties of a natural feature represented as a binary image by investigating its microstructures by means of "structuring templates," following Serra's concept (Serra 1982). It aims to extract information about the geometrical structure of an object (e.g., water body, basin, channel networks, and section of water bodies) by mathematical morphological concepts. In this book, specific geomorphological features are subjected to transformations by means of structuring element. The main characteristics of the structuring template are shape, size, origin, and orientation. The topological characteristics of water body such as spatial distribution, morphology, connectivity, convexity, smoothness, and orientation can be characterized by different structuring templates.

This section is devoted to give basic introduction on binary morphology. Basically, morphological transformations are of two types: (1) the basic operations including erosion, dilation, opening, and closing and (2) the homotopic operations linked to the skeleton including thinning, thickening, and HMT.

To perform certain operations on binary image used in this book, logical operations are of use. For example, in a binary image, pixels with 1s and 0s, respectively, denote pixels for set and set complement. Boolean operations link each logical operation. In a binary image, X, all pixels with a value 1 belong to the set X (foreground), and all 0 pixels to the complement set of X^c or the background. A spatial region is a connected, homogeneously 2-D cell.

Its formal definition is based on point-set topology with open and closed sets. Spatial sets referred here are defined as subsets of a metric space such as a Euclidean space. The discrete binary image, X, is defined as a finite subset of Euclidean 2-D space, R^2. The geometrical properties of a binary image possessing set (X) and set complement (X^c) are subjected to the morphological functions.

Minkowski Operations and Morphological Operations

Morphological transformations are based on Minkowski set addition and subtraction (Serra 1982, Maragos and Schafer 1986). The Minkowski set addition of two sets, X and B, is shown in Equation 2.1:

$$X \oplus B = (x + b : x \in X, b \in B) = \bigcup_{b \in B} X_b \tag{2.1}$$

X and B consist of all points c, which can be expressed as an algebraic vector addition $c = x + b$, where the vectors x and b, respectively, belong to X and B. The Minkowski set subtraction of B from X is denoted as Equation 2.2:

$$X \ominus B = (X^c \subseteq B)^c = \bigcap_{b \in B} X_{-b} \tag{2.2}$$

Morphological dilation and erosion are fundamental morphological operations (Serra 1982, Maragos 1989, 2005) that can be performed on any set (or map in binary form) on 2-D discrete space. Dilation and erosion are basic mathematical morphologic operators (Serra 1982, Maragos and Schafer 1986). These operations can be performed by employing the Boolean AND the Boolean OR operations (Maragos 2005) on any object, represented by the set X and its background by the set complement X^c (e.g., a map in binary form), of the 2-D Euclidean discrete space Z^2 by means of a (window) set B. The principle of a morphological transformation is based on the concept of structuring element denoted by B (Figure 2.4). This B will be used to compare the image under investigation. This comparison can be achieved by convoluting B such that its center hits all the points of the image X. For every position of B, the inclusion or intersection properties will be verified with the elements of the image.

Dilation

Morphological dilation of a set (X), on the 2-D Euclidean discrete space Z^2, is one of the important morphological operators (Serra 1982). Dilation combines two sets using vector addition of set elements. [X and B are sets in Euclidean space with elements x and b, respectively, $x = (x_1, x_2, ..., x_N)$ and $b = (b_1, b_2, ..., b_N)$ being N-tuples of element coordinates.] The dilation of X by B (structuring template) is the set of all possible vector sums of pairs of elements, one coming from X and the other from B. The dilation of a set, X, with structuring template, B, is defined as the set of all points such that B_x intersects X as shown in Equation 2.3. The dilation of X by B is defined as the set of all the points x that the translated B_x intersects X and is equivalent to the union of all the translates, mathematically denoted as Equation 2.3:

$$X \oplus \hat{B} = \{x : B_x \cap X \neq \emptyset\} = \bigcup_{b \in B} X_b \qquad (2.3)$$

where

X_b denotes the *translation* of X along the vector b, $X_b = \{x + b \mid x \in X\}$
$\hat{B} = \{x : -x \in B\}$ is the *symmetric* of B with respect to origin

Illustrative example explaining morphological dilation is shown in Figure 2.8. Of late, through Boolean OR transformation, it was shown that Minkowski set addition and the morphological dilation are the same (Maragos 2005). In the modern view of mathematical morphology, based on the adjunction property, dilation and Minkowski addition are equivalent. The Boolean OR transformation of X by B is equivalent to the Minkowski set addition $\oplus$ of X by B. This operation that expands image object is dilation of X by B: $X \oplus B \triangleq \{z : (B^s)_{+z} \cap X \neq \emptyset\} = \bigcup_{y \in B} X_{+y}$, where X_{+y} denotes the *translation* of X along the vector y, $X_{+y} \triangleq \{x + y \mid x \in X\}$, and $B^s \triangleq \{x : -x \in B\}$ is the *symmetric* of B with respect to origin. This operation enlarges the objects, and neighboring particles will be connected. The small holes inside the image will be filled and gulfs on the boundary by this dilation transformation.

The translates involved in dilation (2.3) of a set (Figure 2.8a) containing five elements by symmetric B of primitive size 3×3 and of circle in eight connectivity in shape are shown in Figure 2.8b through f. Here, while matching

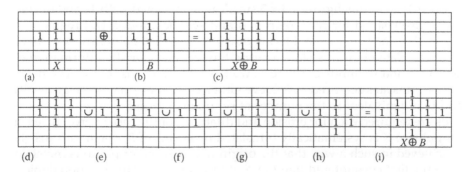

FIGURE 2.8
Dilation of set X by symmetric structuring element B of primitive size 3×3 square (Figure 2.4a). The involved five translates are also shown: (a) a set X with five foreground elements shown with 1s, (b) a structuring element B of size 3×3 and symmetric about the origin at center, (c) dilation of X by B, (d–h) five translates of each element of X by B for dilation, and (i) dilation of X by B obtained by taking the union of five translates shown in (d–h).

the first encountered set point at location (2, 1) with reference to center point of *B*, we check for exact overlap with all points in *B* with all set points. As for the first encountered set point, we see that there is a mismatch. Then the points of *B* that are not exactly matched with set points would be placed at locations beyond the set points. This can be better comprehended from the first translate shown in Figure 2.8d. Similarly, the second and further translates are shown. As at the third encountered set point the matching is exactly identified by means of *B*, there is no change observed in the corresponding translate. The union of all these translates produces dilated version of *X* by *B* as illustrated in matrix form (Figure 2.8i).

Erosion

Structuring element *B* will be moved from top to bottom and from left to right by applying the criterion of erosion principle to achieve shrinking. When the rectangle, *B*, is centered on one point of the frame of the image *X*, then it will be truncated, and only its intersection with the shape is kept. *Erosion* transformation, of *X* by *B* expressed in Equation 2.4, *denoted by* $\ominus$, is defined as the set of points *x* such that the translated B_x is contained in the original set *X* and is equivalent to the intersection of all the translates:

$$X \ominus B = \{x : B_x \subseteq X\} = \bigcap_{b \in B} X_{-b} \qquad (2.4)$$

The Boolean AND transformation of *X* by B^s is equivalent to the Minkowski set subtraction, of *X* by *B*, $X \ominus B \triangleq \{z : B_{+z} \subseteq X\} = \bigcap_{y \in B} X_{-y}$. This operation shrinks the input image object.

The rule followed to translate the set elements to further achieve erosion is slightly different from the rule followed in dilation process. For better understanding, this transformation is illustrated in matrix form (Figure 2.9a). In this figure, a 3×3 size *X* is represented with 1s and 0s that respectively stand for set foreground and set background (X^c) regions. In Figure 2.9a and b, five set points are obvious. These set points are systematically translated in terms of symmetric *B* with characteristic information of size 3×3 and rhombus in shape as well as with center as origin. The number of translates required to achieve either erosion or dilation (Figure 2.9a and b) is equivalent to the number of set points present. Hence, five translates are required each for erosion and dilation. For the case of erosion, each set point in *X* is systematically translated by means of *B*. The first translate is achieved in such a way that the origin in *B* (i.e., center point) is matched with the first encountered point of *X* at location (2, 1). This location depicts the second column of first scan line of *X*. Then we observe that *B* is not exactly overlapped with all the neighborhood set points. Hence, we consider this as a "mismatch," and the first encountered set point is transformed into set background point. This is shown in the first translate involved in the erosion process. Similar translation is done for the second encountered

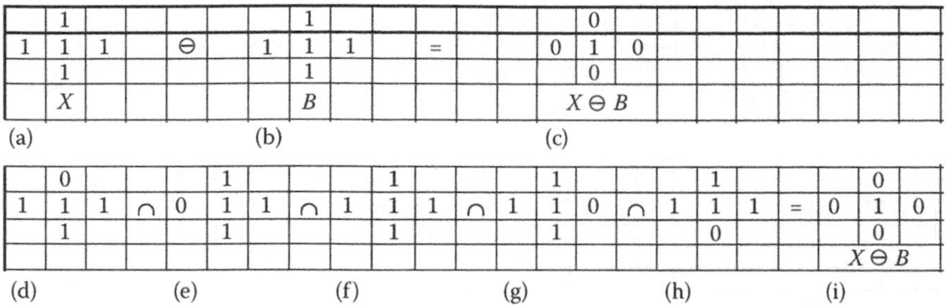

	1				1					0	
1	1	1	⊖	1	1	1	=	0	1	0	
	1				1				0		
	X				*B*				*X* ⊖ *B*		

(a)	(b)	(c)

	0				1				1				1				1				0	
1	1	1	∩	0	1	1	∩	1	1	1	∩	1	1	0	∩	1	1	1	=	0	1	0
	1				1				1				1				0				0	
																				X ⊖ *B*		

(d)	(e)	(f)	(g)	(h)	(i)

FIGURE 2.9
Diagrammatic representation of morphological erosion process. The involved five translates are also shown: (a) a set *X* with five foreground elements shown with 1s, (b) a structuring element *B* of size 3 × 3 and symmetric about the origin at center, (c) erosion of *X* by *B*, (d–h) five translates of each element of *X* by *B* for erosion, and (i) erosion of *X* by *B* obtained by taking the union of five translates shown in (d–h).

set point located at (1, 2) to check whether it exactly matches with *B*. As this second set point also mismatches with reference to the origin of *B*, the second translate for set point at location (1, 2) is transformed into set background point. Similar exercise provides five translates as shown in Figure 2.9d through h. It is obvious that the translate achieved for the third encountered set point at location (2, 2) exactly matches with *B*. Hence, no change is observed in the corresponding translate. Further, the intersection of all the translates provides eroded version of *X* by *B* (Figure 2.9i). This operation shrinks image objects. Isolated points and the small particles will be removed by this operation. It shrinks the other particles, discards peak on the boundary of the object, and disconnects. The dilation with an elementary structuring template expands the set with a uniform layer of elements, while the erosion operator eliminates a layer from the set. To avoid confusion, $(X \ominus B)$ and $(X \oplus B)$ are simply referred to as erosion and dilation. It is worth mentioning here that Minkowski addition and subtraction are akin to the morphological dilation and erosion as long as the structuring template (*B*) is of symmetric type. Hereafter, the dilation and erosion of *X* by *B* are denoted as $(X \oplus B)$ and $(X \ominus B)$, respectively. See Serra (1982), Maragos and Schafer (1986), and Maragos (2005) for detailed explanations and implementations of these fundamental morphologic transformations along with their algebraic properties.

Opening and Closing

By employing erosion and dilation of *X* by *B*, opening and closing transformations, *respectively denoted with symbols* o *and* •, could be defined. Cascade of erosion–dilation is called opening transformation (Equation 2.5). The dilation followed by erosion is called closing transformation (Equation 2.6):

$$X \circ B = ((X \ominus B) \oplus B) \tag{2.5}$$

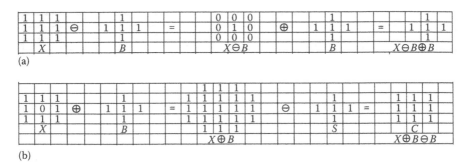

(a)

(b)

FIGURE 2.10
Diagrammatic representation of basic cascade morphological transformations: (a) opening and (b) closing.

$$X \bullet B = ((X \oplus B) \ominus B) \tag{2.6}$$

These transformations are illustrated in Figure 2.10a and b, where cascade of erosion followed by dilation of X of size 3×3 with nine set points by means of B is shown. To perform erosion first on the nine set points, nine translates are required. Then the resultant eroded version would be dilated to achieve the opened version of X by B as shown in Figure 2.10a. Similarly, to achieve closed version of X by B (Figure 2.10b), we first perform dilation on X of size 3×3 with nine set points by means of B followed by erosion on the resultant dilated version. To perform these transformations shown in Figure 2.10, by changing the scale of B, one requires taking the addition of B by B to a desired level.

These cascade transformations are idempotent (Serra 1982). The opening and closing operations are idempotent (Serra 1982, Maragos 1989) as shown in Equations 2.7 and 2.8:

$$(((X \ominus B) \oplus B) \ominus B \oplus B) = (X \ominus B) \oplus B = X \circ B \tag{2.7}$$

$$(((X \oplus B) \ominus B) \oplus B \ominus B) = (X \oplus B) \ominus B = X \bullet B \tag{2.8}$$

However, these transformations (Equations 2.7 and 2.8) can be carried out according to the multiscale approach (Maragos 1989). In the multiscale approach, the size of the structuring template will be increased from iteration to iteration. But a variation will be identified while performing either opening or closing as multiscale operations/cycles according to Equations 2.9 and 2.10:

$$(X \ominus (B \oplus B) \oplus (B \oplus B)) = (((X \ominus B) \ominus B) \oplus B \oplus B) = (X \ominus 2B) \oplus 2B = X \circ 2B \tag{2.9}$$

$$(X \oplus (B \oplus B) \ominus (B \oplus B)) = (((X \oplus B) \oplus B) \ominus B \ominus B) = (X \oplus 2B) \ominus 2B = X \bullet 2B \tag{2.10}$$

Theoretically, the aforementioned expression is true. Another way of performing opening is the right-hand-side notation.

Multiscale Morphological Operations

Multiscale dilation and erosion can be performed by varying the size of *structuring element nB*, where $n = 0, 1, 2, ..., N$. Dilations and erosions can also be performed iteratively, as follows:

$$(X \oplus nB) = (X \oplus B) \oplus B \oplus \cdots \oplus B \qquad (2.11)$$

$$(X \ominus nB) = (X \ominus B) \ominus B \ominus \cdots \ominus B \qquad (2.12)$$

where $n = 0, 1, 2, ..., N$. Figures 2.11 and 2.12 show effects of iterative dilations and erosions. In this section, opening and closing operations are performed on the basis of cycles. As shown in Equations 2.9 and 2.10, the size of structuring element will be changed as follows; if B is of size 3×3 pixels, this means that instead of using a larger structuring element, it is often possible to use a smaller one repeatedly to get the same effect. We employ recursive erosions and dilations to perform multiscale opening and closing transformations in Equations 2.13 and 2.14:

$$(X \circ nB) = ((X \ominus nB) \oplus nB) \qquad (2.13)$$

$$(X \bullet nB) = ((X \oplus nB) \ominus nB) \qquad (2.14)$$

where n (homothetic parameter) is the number of times the transformations are repeated. Illustrative examples of multiscale opening and closing

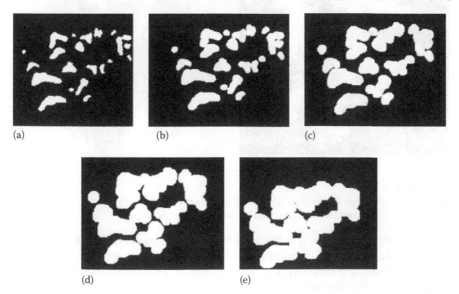

(a)　　　　　　　　(b)　　　　　　　　(c)

(d)　　　　　　　　(e)

FIGURE 2.11

Iterative dilations and their effects. (a) Set showing various objects, (b–e) set of objects obtained after first, second, third, and fourth dilation cycles, respectively.

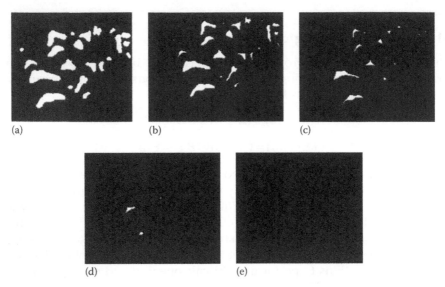

(a) (b) (c)

(d) (e)

FIGURE 2.12
Iterative erosions and their effects. (a) Set showing various objects, (b–e) set of objects obtained after first, second, third, and fourth erosion cycles, respectively.

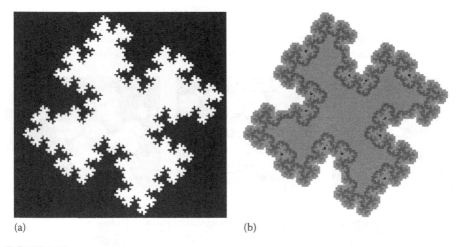

(a) (b)

FIGURE 2.13
Multiscale morphological opening and closing. (a) Binary Koch quadric fractal and (b) dilated, original, opened, and eroded fractal coded by different shades of gray.

transformations are shown in Figure 2.13. See *Matheron* (1975) and *Serra* (1982) for morphological transformations and their numerous applications. These transformations are employed systematically as explained in the equations in entire book. See Serra (1982), Maragos and Schafer (1986), Maragos (1989), and Maragos (2005) for a more detailed exposition of these fundamental transformations together with its algebraic properties.

Homotopic Operations Based on Basic Binary Morphological Transformations

The transformations from the field of mathematical morphology such as erosion, dilation, and opening discussed so far are used to extract morphological skeletal network (MSN).

Morphological Skeleton

Morphological skeleton is a one-pixel-wide caricature that summarizes the overall shape, size, orientation, and association of a geometric structure from which inferences can be drawn. The general term structure is to connote "the expression of external morphology of the objects" (e.g., water body). Components of such structures are traditional characteristics of shape, in two dimensions, and outline textural details. Highly symmetrical objects have the skeletons with symmetry. The more irregular is the object, the more irregular is its skeleton.

A connectivity preserving way of erosion called skeletonization is described by Blum (1973). The resulting skeleton is one picture element (pixel) thick objects, which have the same connectivity as the original object. Skeletons are of special interest because they reflect the structure of the original objects in their end pixels and vertices. The concept of skeletonization is developed by mathematical morphologists (Lantujoul 1978, Serra 1982, Maragos and Schafer 1986). The skeleton or medial axis of a set is the line made up of those points for which the distance to the boundary of the set is reached by at least two points. The skeleton of a geometric structure (Figure 2.14a) viewed as a subset of R^2 (Euclidean space) is defined as the set of the centers of the maximal disks inscribable inside the structure. A disk is maximal if it is not properly contained in any other disk totally included in the structure. Hence, a maximal disk must touch the boundary of the structure at least at two different points. The combination of centers of the maximal disks inscribable is a skeleton. This concept is being extensively applied in several fields such as biological shape description (Blum 1973), pattern recognition (Margos and Schafer 1986, Maragos 1989), and metallography with highly promising results. Some examples can be seen in Maragos and Schafer (1986). Figure 2.14a through j shows the skeleton extraction process.

This skeletonization concept is developed by mathematical morphologists (Lantuejoul 1978). The skeleton of a geometric structure can be mathematically defined as Equations 2.15 and 2.16:

$$Sk_n(X) = ((X \ominus nB) \backslash (X \ominus nB) \circ B) \tag{2.15}$$

```
1 1 1 1 1      0 0 0 0 0      0 0 0 0 0      0 1 1 1 0
1 1 1 1 1      0 1 1 1 0      0 0 0 0 0      1 1 1 1 1
1 1 1 1 1      0 1 1 1 0      0 0 1 0 0      1 1 1 1 1
1 1 1 1 1      0 1 1 1 0      0 0 0 0 0      1 1 1 1 1
1 1 1 1 1      0 0 0 0 0      0 0 0 0 0      0 1 1 1 0
(a)            (b)            (c)            (d)

0 0 0 0 0      0 0 0 0 0      1 0 0 0 1      0 0 0 0 0
0 0 1 0 0      0 0 0 0 0      0 0 0 0 0      0 1 0 1 0
0 1 1 1 0      0 0 0 0 0      0 0 0 0 0      0 0 0 0 0
0 0 1 0 0      0 0 0 0 0      0 0 0 0 0      0 1 0 1 0
0 0 0 0 0      0 0 0 0 0      1 0 0 0 1      0 0 0 0 0
(e)            (f)            (g)            (h)

               0 0 0 0 0      1 0 0 0 1
               0 0 0 0 0      0 1 0 1 0
               0 0 1 0 0      0 0 1 0 0
               0 0 0 0 0      0 1 0 1 0
               0 0 0 0 0      1 0 0 0 1
               (i)            (j)
```

FIGURE 2.14
(a) Set X (also treated as zeroth eroded version), (b) first eroded X, (c) erosion of eroded X, (d) opening of zeroth eroded X, (e) opening of first eroded version of X, (f) opening of second eroded version of X, (g) skeletal subsets of order zero, (h) skeletal subsets of first order, (i) skeletal subsets of second eroded version of X, and (j) skeletal network of a set shown in (a).

$n = 0, 1, 2, \ldots, N$

$$Sk(X) = \bigcup_{n=0}^{N} Sk_n(X) \tag{2.16}$$

where $Sk_n(X)$ denotes the nth skeletal subset of set (X). In the aforementioned expression, subtracting from the eroded versions of X, their opening by B retains only the angular points. The union of all such possible points produces skeletal network.

Hit or Miss Transformation

HMT is another important morphological operation. Let B be composed of the two disjoint sets B^1 and B^2; then the HMT of X by B is defined as the set of all points where B_x^1 is included in X and B_x^2 is included in X^c. The set X^c is the accompaniment of X and B_x^i, $i = 1,2$ denotes the translation of B^1 by x. This HMT is expressed as Equation 2.17:

$$(X \ominus B^1) = \left\{ x : B_x^1 \subseteq X; B_{x^c}^2 \subseteq X^c \right\} \tag{2.17}$$

0	0	0	0	0	0	0	0	0	0	0	0	0	0	0	0	0	0
0	0	0	1	0	0	0	0	0	0	0	0	0	0	0	0	0	0
0	0	1	1	1	0	0	1	0	0	0	0	1	0	0	0	0	0
0	1	1	1	1	1	1	1	1	1	1	1	1	1	1	0	0	0
0	0	1	1	1	0	0	1	0	0	1	0	0	0	0	0	0	0
0	0	0	1	0	0	0	0	0	0	0	0	0	0	0	0	0	0
0	0	0	0	0	0	0	0	0	0	0	0	0	0	0	0	0	0

(a)

B_1^1 =	0	1	0		B_2^1 =	1	0	1		$B_1^1 \cup B_2^1$ =	1	1	1
	1	1	1			0	0	0			1	1	1
	0	1	0			1	0	1			1	1	1

(b) (c) (d)

1	1	0	1	0	1	1	1	1	1	1	1	1	1	1	1	1	1
1	0	0	0	0	0	0	1	0	1	1	1	0	1	0	1	1	
0	0	0	1	0	0	0	0	0	0	0	0	0	0	0	0	0	0
1	0	1	1	1	0	0	1	0	0	1	0	0	1	0	1	1	
0	0	0	1	0	0	0	0	0	0	0	0	0	0	0	0	0	0
0	0	0	0	0	0	0	0	0	0	0	0	0	0	0	0	0	0
1	0	0	0	0	0	0	*1*	0	0	1	0	1	1	1	1	1	
1	1	0	1	0	1	1	1	1	1	1	1	1	1	1	1	1	1

(e)

0	0	0	0	0	0	0	0	0	0	0	0	0	0	0	0	0	0
0	0	0	0	0	0	0	0	0	0	0	0	0	0	0	0	0	0
0	0	0	0	0	0	0	0	0	0	0	0	0	0	0	0	0	0
0	0	0	0	0	0	0	1	0	0	0	0	0	0	0	0	0	0
0	0	0	0	0	0	0	0	0	0	0	0	0	0	0	0	0	0
0	0	0	0	0	0	0	0	0	0	0	0	0	0	0	0	0	0
0	0	0	0	0	0	0	0	0	0	0	0	0	0	0	0	0	0
0	0	0	0	0	0	0	0	0	0	0	0	0	0	0	0	0	0

(f)

FIGURE 2.15

HMT. (a) Numerals 1s and 0s, respectively, represent X and X^c; (b) B_1^k; (c) B_2^k; (d) $B = B_1^k \cup B_2^k$, in (b–d) the origin is the center of the 3 × 3 square; (e) erosion of X by B_1^k is shown in bold (**1**) and erosion of X^c by B_2^k is shown in italic (*1*). To obtain the eroded version, each network element (complement) is translated with respect to $B_1^k(B_2^k)$ of size one centered on $X_i(X_i^c)$ to check whether all the elements in $X(X^c)$ overlaps with the neighboring elements of $X_i(X_i^c)$. If exact overlap occurs, there would be no change required in the translate; otherwise, the centered position in the image would be removed. Similarly, all other elements, $X_i, i = 0, 1, 2, ..., n$ $(X_i^c, i = 0, 1, 2, ..., n)$ are translated by changing the nonoverlapping properties with respect to $B_1^k(B_2^k)$. This erosion transformation is required to understand $X \ominus B_1^k$ and $X^c \ominus B_2^k$. (f) In one of the grids at (7, 4), the eroded version of X by B_1^k is intersected with the eroded version of X^c by B_2^k. Such intersecting portion results from the HMT; in other words, $(X * \{B\})$.

Suppose B^2 is chosen as the kernel complement of B^1, the expression (2.17) can be rewritten as Equation 2.18:

$$X \otimes \{B\} = (X \ominus B^1) \cap (X^c \ominus B^2) \tag{2.18}$$

where W is the kernel with finite support. The HMT can be used to detect the occurrence of the exact pattern (B^1) in the image X. This transformation is to compute the area occupied by an object in the binary form (Figure 2.15).

Grassfire Transformation

The principle is that if we assume the center of the object possesses the wet grass, the remaining part of the object contains the dry grass. If the fire is lit along the boundary points of the dry grass, and the fire is allowed to propagate toward the wet grass at uniform speed, the dry grass will be burned leaving the wet grass unburned. This transformation can be achieved through performing the consecutive erosions. The boundaries of different degrees of eroded sets are termed as the fire frontlines. These successive fire frontlines are the boundaries of the successive eroded sets.

Convex Hull of Sets

A Euclidean set X is convex if and only if the line segments joining any two pair of points lie entirely within the set. Consequently, a convex hull, $CH(X)$ (Figure 2.16b), is defined as the smallest convex polygon containing all points x in the set X. It can be easily visualized by imagining an elastic band stretched open to enclose the given object. When the elastic band is released, it will assume the shape of the required convex hull. Soille (1998) proposed the idea, based on morphological transformations, of generating convex hull of a set by intersecting all half-plane closings encompassing the set. For a given angle θ, there are two half-planes (denoted by π_θ^+ and π_θ^-) which correspond to this orientation, the second one being the complement of the first, i.e., $(\pi_\theta^+)^c = \pi_\theta^-$. The intersections of the half-plane closings obtained for all possible half-plane orientations result in the convex hull of the set. It is mathematically denoted as Equation 2.19:

$$CH(A) = \bigcap_\theta \left[\phi_{\pi_\theta^+}(A) \cap \phi_{\pi_\theta^-}(A) \right] \tag{2.19}$$

where

$\phi_{\pi_\theta^+}$ represents the half-plane closings at π_θ orientation
$\phi_{\pi_\theta^-}$ symbolizes the closings with the complement of the corresponding half-plane

See Soille (1998) for more details about the construction of convex hull of sets.

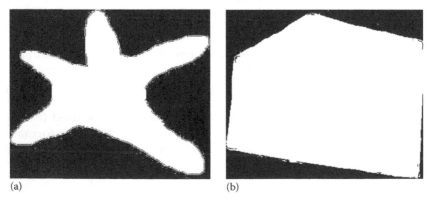

(a)   (b)

FIGURE 2.16
Convex hull of a set. (a) Concave set and (b) convex hull of set shown in (a).

Grayscale Morphological Operations

Grayscale image (e.g., raster digital elevation map) is denoted as a function (e.g., Figure 2.17a) represented by a nonnegative 2-D sequence $f(m, n)$, which assumed $J + 1$ possible intensity values: $j = 0, 1, 2, ..., J$. As we deal with 8 bit/pixel digital topographic data, $J = 255$. The function, f, we deal with is discrete, defined on a (rectangular) subset of the discrete plane Z^2. For the following discussion, we deal with digital image functions of the form $f(x, y)$ and structuring element B. Again, $f(x, y)$ is a grayscale input image defined as a finite subset in Z^2, while B is a binary pattern.

Grayscale Dilation and Erosion

The erosion (dilation) of f by B replaces the value of f at a pixel (x, y) by the *minima (maxima)* of the values of f over a structuring template B. We represent these gray-level morphological transformations as Equations 2.20 and 2.21.

Grayscale erosion is defined as Equation 2.20:

$$(f \ominus B)(x, y) = \min_{(i,j) \in B} \{f(x+i, y+j)\} \qquad (2.20)$$

Morphological grayscale dilation is defined as Equation 2.21:

$$(f \oplus B)(x, y) = \max_{(i,j) \in B} \{f(x-i, y-j)\} \qquad (2.21)$$

where B is a discrete binary template (e.g., Figure 2.4a). $(f \ominus B)$ and $(f \oplus B)$ can be obtained by computing *minima* and *maxima*, respectively over a moving template B. From Equations 2.20 and 2.21, it is obvious that erosion is the

5	6	2	0	1	9	7
9	3	1	0	5	4	5
6	4	3	6	9	5	5
1	7	5	7	4	2	4
4	5	3	8	5	7	3
8	3	6	1	2	0	8
3	8	7	3	6	8	9

(a)

.	.	.	.	.	.	.
.	1	0	0	0	1	.
.	1	0	0	0	2	.
.	1	3	3	2	2	.
.	1	1	1	0	0	.
.	3	1	1	0	0	.
.	.	.	.	.	.	.

(b)

.	.	.	.	.	.	.
.	9	6	9	9	9	.
.	9	7	9	9	9	.
.	7	8	9	9	9	.
.	8	8	8	8	8	.
.	8	8	8	8	9	.
.	.	.	.	.	.	.

(c)

FIGURE 2.17

Grayscale erosion and dilation: morphological and logical transformations. (a) Grayscale function (f) of size 7 × 7, (b) erosion ($f \ominus B$), and (c) dilation ($f \oplus B$) transformed by means of a flat structuring element of size 3 × 3, which is symmetric about the origin and square in shape (Figure 2.4a).

duality of dilation because eroding the foreground pixels is equivalent to dilating the background pixels (Serra 1982). Dilation will expand an object in question, while erosion will make it shrink. Figure 2.17b and c illustrates grayscale erosions and dilations.

Grayscale Opening and Closing

The cascades of dilation and erosion operations result in opening and closing operations, which are used for smoothing purposes. Erosion is the dual of dilation as eroding the foreground pixels is equivalent to dilating the background pixels. Opening and closing are both based on the basic morphological transformations. Opening of f by B is achieved by first eroding f followed by dilating with respect to B and is mathematically shown as Equation 2.22:

$$(f \circ B) = ((f \ominus B) \oplus B) \tag{2.22}$$

where $\circ$ denotes the symbol for opening.

The definition of closing is the reverse of opening, where dilation of f by B is performed first, followed by erosion with respect to B. Closing of f by B is defined as the dilation of f by B followed by erosion with respect to B, which is mathematically represented as Equation 2.23:

$$(f \bullet B) = ((f \oplus B) \ominus B) \tag{2.23}$$

where $\bullet$ denotes the symbol for closing.

Opening eliminates specific image details smaller than B, removes noise, and smoothens the boundaries from the inside, whereas closing fills holes in objects, connects close objects or small breaks, and smoothens the boundaries from the outside. Figure 2.18 illustrates the grayscale opening and closing operations on a synthetic function.

.	.	.	.	.	.	.
.	1	1	0	2	2	.
.	3	3	3	3	2	.
.	3	3	3	3	2	.
.	3	3	3	3	2	.
.	3	3	1	1	0	.
.	.	.	.	.	.	.

(a)

.	.	.	.	.	.	.
.	6	6	6	9	9	.
.	6	6	6	9	9	.
.	7	7	7	8	8	.
.	7	7	8	8	8	.
.	8	8	8	8	8	.
.	.	.	.	.	.	.

(b)

FIGURE 2.18
Grayscale morphological and logical transformations. (a) Opening and (b) closing of grayscale function shown in Figure 2.17a, transformed by means of a flat structuring element of size 3 × 3, which is symmetric about the origin and square in shape (Figure 2.4a).

Multiscale Grayscale Morphological Operations

Multiscale opening of scale n is defined as erosion of the image by B for n-times followed by dilation with the same B for n-times (Equation 2.24). By duality, multiscale closing of scale n is defined as dilation of f by B for n-times followed by erosion by B for n-times (Equation 2.25). These multiscale opening and closing transformations are mathematically represented as $(f \circ nB) = ((f \ominus nB) \oplus nB)$ and $(f \bullet nB) = ((f \oplus nB) \ominus nB)$, respectively, where the scaling factor, $n = 0, 1, 2, \ldots, N$:

$$(f \circ nB) = ((f \ominus nB) \oplus nB) \tag{2.24}$$

$$(f \bullet nB) = ((f \oplus nB) \ominus nB) \tag{2.25}$$

These multiscale openings and closings of f by B are represented as (1) $((f \ominus nB) \oplus nB) = (((f \ominus B) \ominus B \ominus \cdots \ominus B) \oplus B \oplus B \oplus \cdots \oplus B) = (f \circ nB)$ and (2) $((f \oplus nB) \ominus nB) = (((f \oplus B) \oplus B \oplus \cdots \oplus B) \ominus B \ominus B \ominus \cdots \ominus B) = (f \bullet nB)$ at scale $n = 0, 1, 2, \ldots, N$, respectively. Performing opening and closing iteratively by increasing the size of B transforms a grayscale image (e.g., DEM) into respective lower resolutions. Multiscale opening and closing of DEM by nB affect spatially distributed elevation regions in the form of smoothing of contours to various degrees. The shape and size of B control the shape of smoothing and the scale, respectively. Figure 2.19 illustrates the effects of multiscale grayscale opening and closing transformations.

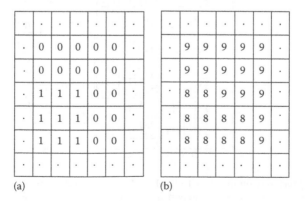

(a)   (b)

FIGURE 2.19
Recursive application of grayscale (a) opening and (b) closing transformations. $(f \circ B)$ and $(f \bullet B)$ are shown in Figure 2.18a and b. $(f \circ 2B) = (f \circ (B \oplus B))$ and $(f \bullet 2B) = (f \bullet (B \oplus B))$ denote opening and closing of two cycles.

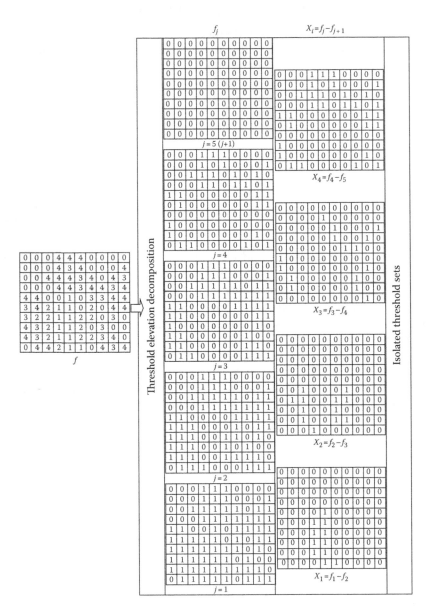

FIGURE 2.20
Original image f has maximum intensity level $I = 4$. Threshold-decomposed zones f_i with $i =$ 1, 2, 3, 4, and 5 ($I + 1$) are respectively shown, along with the isolated sets with index i ranging from 1, 2, …, I. The sets X_i are isolated by $f_j - f_{j+1}$.

Threshold Decomposition of a Function

By thresholding f at all possible intensity levels (e.g., topographic elevations of DEM), $0 \leq j \leq J$, we obtain threshold-decomposed binary images according to Equation 2.26:

$$f_j(m,n) = \begin{cases} 1, & f(m,n) \geq j \\ 0, & f(m,n) < j \end{cases} \tag{2.26}$$

Thresholded sets, decomposed from f, take values 0 and 1 (the pixels with 1 and 0 represented with white and black colors denote respectively sets and their complements). The sets (f_j) form a sequence of sets that characterize f entirely and are such that for any threshold elevations j and $j + 1$ with $(j+1) \geq (j) \Rightarrow (f_{j+1}) \subseteq (f_j)$, for j ranging between 1 and J—as illustrated in Figure 2.20. A synthetic function consists of nine zones (or sets) with designated-set orders ranging from 1 to 9. We express this through Figure 2.20. The union of these sets (f_j) satisfies the inclusion relationship (Maragos and Ziff 1990) as shown in Equation 2.27:

$$f = \bigcup_{i=1}^{I} f_i \tag{2.27}$$

References

Beucher, S., 1990, Segmentation d-images et morphologic mathematique, These Docteur en Morphologic Mathematique, Ecole des Mines de Paris, Paris, France.

Blum, H., 1973, Biological shape and visual sciences (Part I), 1, *Theoretical Biology*, 38, 205–287.

Lantuejoul, C., 1978, La sequelettisation et son application aux mesures topologiques des mosaiques polycristallines, These de Docteur-Ingnieur, School of Mines, Paris, France.

Maragos, P. A., 1989, Pattern spectrum and shape representation, *IEEE Transactions on Pattern Analysis and Machine Intelligence*, 11, 701–716.

Maragos, P., 2005, Morphological filtering for image enhancement and feature detection. In: *The Image and Video Processing Handbook*, ed. A. C. Bovik, Elsevier Academic Press, Amsterdam, the Netherlands, pp. 135–156.

Maragos, P. A. and R. W. Schafer, 1986, Morphological skeleton representation and coding of binary images, *IEEE Transactions on Acoustics, Speech and Signal Processing*, ASSP-34(5), 1228–1244.

Maragos, P. and R. D. Ziff, 1990, Threshold superposition in morphological image analysis systems, *IEEE Transactions on Pattern Analysis and Machine Intelligence*, 12(5), 498–504.

Matheron, G., 1975, *Random Sets and Integral Geometry*, Wiley, New York.

Meyer, F., 1980, Feature extraction by mathematical morphology in the field of quantitative cytology, Technical report of Ecole nationale superiere des mines de Paris, Fountainbleau, France.

Serra, J., 1982, *Image Analysis and Mathematical Morphology*, Academic Press, New York, p. 610.

Soille, P., 1998, Grey scale convex hulls: Definition, implementation and applications. In: *Proceedings ISMM'98*, Vancouver, British Columbia, Canada, pp. 83–90.

Sternberg, S. R., 1986, Greyscale morphology, *Computer Vision, Graphics, and Image Processing*, 35, 333–355.

3

Simulated, Realistic Digital Elevation Models, Digital Bathymetric Maps, Remotely Sensed Data, and Thematic Maps

This chapter provides briefly the details and specifications of various data sets that have been subsequently employed to demonstrate various approaches and algorithms provided in the chapters that follow. These data sets include simulated and realistic digital elevation models (DEMs), digital bathymetric maps (DBMs), remotely sensed satellite data, and various thematic maps. This chapter is segregated into five sections that respectively provide details on the following aspects: (1) numerical array as a spatial function, (2) generation of planar fractal basins (sets), (3) generation of fractal landscapes and fractal DEMs (F-DEMs) (functions), (4) realistic DEMs and DBMs, and (5) remotely sensed satellite data.

Numerical Array as a Spatial Function

As an example, a simulated DEM is shown in Figure 3.1 with three spatially distributed elevation regions numerically represented as 1s, 2s, and 3s (Chockalingam and Sagar 2003). Typical channel and ridge connectivity networks can be extracted from such a function.

Generation of Planar Fractal Basins (Sets)

To generate a model that conforms to the natural river basin, at least in statistical sense, it is essential to have the broad outline of the basin in the form of a polygon (i.e., an initiator) and the generating mechanism that transforms the initiator as a fractal basin. Generating mechanism needs to be designed by considering the following conditions:

- Area of the basin should be constant under succession of change in scale.
- The basin outline should possess increasing number of crenulations with increasing number of iterations.

1	1	1	1	1	1	1	1	1	1	1
1	2	2	2	2	2	2	2	2	2	1
1	2	2	2	2	2	2	2	2	2	1
1	2	2	3	3	3	3	3	2	2	1
1	2	2	3	3	3	3	3	2	2	1
1	2	2	3	3	3	3	3	2	2	1
1	2	2	3	3	3	3	3	2	2	1
1	2	2	3	3	3	3	3	2	2	1
1	2	2	2	2	2	2	2	2	2	1
1	2	2	2	2	2	2	2	2	2	1
1	1	1	1	1	1	1	1	1	1	1

FIGURE 3.1
Simulated DEM with three spatially distributed elevation regions represented numerically as 1s, 2s, and 3s. (From Chockalingam, L. and Sagar, B.S.D., *J. Pattern Recogn.*, 17(2), 269, 2003.)

- With iterative process to simulate basin outlines, the basin outlines should not self-intersect.
- The length of river network should increase with increasing iteration. The generating mechanism plays an important role while transforming the initiator as a fractal basin. Homogeneous and heterogeneous channel network patterns result respectively from a symmetric generator with nonrandom rule and either a symmetric or asymmetric generator with random rule. Also, the characteristics of the network depend on the overall shape of the initiator–basin. Asymmetric fractal basins arise due to asymmetric outline of the initiator and due to the generating mechanism as well as the adopted rule to transform the initiator as fractal basin. This model has two sequential phases.

Fractal Basin Generation

To generate fractal basins with fractal dimensions ranging from 1 to 2 in two-dimensional (2-D) space, one begins with two shapes: (1) broad outline of the basin as polygon, an initiator–basin (Figure 3.2a), and (2) a generator (Figure 3.2b). The latter is an oriented broken line made up of N equal sides of length r. Each stage of the construction begins with a broken line and consists in replacing each straight interval with a copy of the generator, reduced and displaced to have the same end points as those of the interval being replaced. In all cases, $D = \text{Log}N/\text{Log}(1/r)$. Step 1 is to draw the segment of length $(0,1)$, which is one side length in the

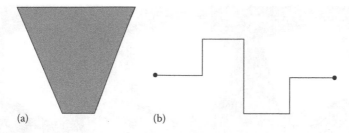

FIGURE 3.2
(a) A triangular–initiator basin and (b) generating mechanism. (From Sagar, B.S.D. et al., *Fractals*, 9(4), 429, 2001.)

initiator–basin (Figure 3.2a). Step 2 is to draw the kinked curves each made up of N intervals superposable upon the segment. Step 3 is to replace each of the N segments used in step 2 by a kinked curve obtained by reducing the curve of step 1 in the ratio $r(N) = 1 = r$. One obtains altogether N^2 segments of length $1/(r)^2$. Iterating this process adds further details. This process of generating fractal basin is based on the principle involved in the generation of Koch curves by considering the bounded initiator–basin. The boundary of the fractal basin possesses many V- and Λ-shaped crenulations. These crenulations in the outline of the fractal basin and in the successive erosion frontlines determine the whole channel network pattern. By following this process for two iterations, a fractal basin (Figure 3.3) of size 400 × 400 pixels is generated where a triangle set (Figure 3.2a) and a generating rule (Figure 3.2b) act as an initiator and a generator (Mandelbrot 1982). By using the generator (Figure 3.2b), and the five other initiators that include square, pentagon, hexagon, heptagon, and octagon (Figure 3.4a through e), five fractal basins have also been generated (Figure 3.5a through e; Sagar 1996, Sagar et al. 1998, 2001, Radhakrishnan et al. 2004, Sagar and Tien 2004, Lim et al. 2009).

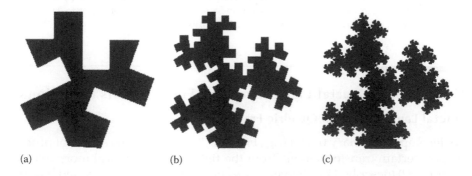

FIGURE 3.3
(a–c) First-, second-, and third-order fractal basins generated by the generator shown in Figure 3.2b. (From Sagar, B.S.D. et al., *Fractals*, 9(4), 429, 2001.)

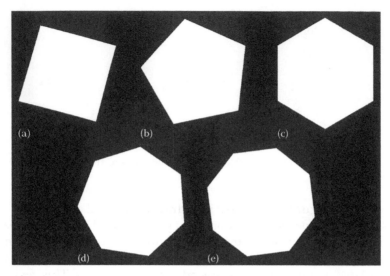

FIGURE 3.4
Initiator–basins of (a) four sides, (b) five sides, (c) six sides, (d) seven sides, and (e) eight sides. (From Sagar, B.S.D., *Chaos Soliton. Fract.*, 7(11), 1871, 1996; Sagar, B.S.D. et al., *Discrete Dyn. Nat. Soc.*, 2(2), 77, 1998.)

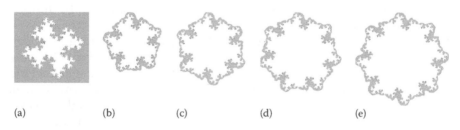

FIGURE 3.5
Third-order Koch quadric binary fractal basins from (a) four sided, (b) five sided, (c) six sided, (d) seven sided, and (e) eight sided initiators. (From Sagar, B.S.D., *Chaos Soliton. Fract.*, 7(11), 1871, 1996; Sagar, B.S.D. et al., *Discrete Dyn. Nat. Soc.*, 2(2), 77, 1998.)

Generation of Fractal Landscapes and Fractal DEMs (Functions)

Fractal Landscape from Quadric Fractal Basin

To decompose a binary fractal (e.g., Figure 3.5a) into several regions of prominence, certain transformations from the field of mathematical morphology (Serra 1982) (described in Chapter 2) are considered. The decomposed binary fractal subsets will be dilated by a specific structuring template to find out the various regions of prominence. In the following, how a binary fractal is decomposed into various regions of prominence is detailed. A binary fractal basin (Figure 3.5a) is considered. The flow direction network (FDN) (Figure 3.6) is extracted according to Equations 2.15 and 2.16. By implementing this procedure, the decomposed FDN subsets of this binary fractal are dilated to the same degree in order to decompose the binary fractal into its regions of prominence. Figure 3.7 shows the simulated DEM with various

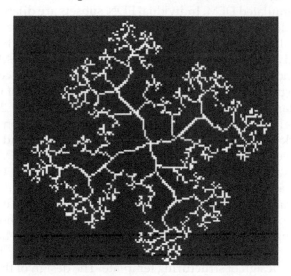

FIGURE 3.6
Fluid FDN extracted from binary fractal basin. (From Sagar, B.S.D. and Murthy, K.S.R., *Fractals*, 8(3), 267, 2000.)

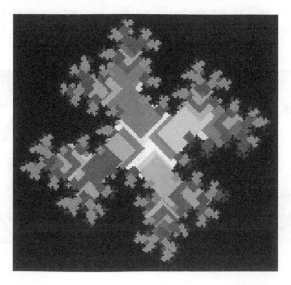

FIGURE 3.7
Binary fractal basin after decomposition into TPRs. (From Sagar, B.S.D. and Murthy, K.S.R., *Fractals*, 8(3), 267, 2000.)

regions of topological prominence. A square structuring template (*B*) is considered for a similar decomposition. However, other types of structuring templates unravel other topological characteristics of the landscape. In this sample study, the union of dilated and coded FDN subsets starts from $n = N$ to $n = 0$. Various regions indicated by different shades represent various elevation levels in simulated DEM. Individual FDN subsets are dilated to the same degree and coded with respective shades by following the sequential steps shown as (a) dilation of FDN subsets as $FDN_n \oplus nB = TPR_n$, (b) gray-coding of each dilated FDN subset, producing a transcendental DEM (Figure 3.7) from binary fractal (Figure 3.5a). The binary fractal basin is decomposed into various topologically prominent regions (TPRs), the surface of which is akin to the fractal landscape (Sagar et al. 2000, Sagar and Murthy 2000). Each of the shaded regions is treated as a specific region of elevation in the DEM. Light and dark regions are assumed to represent higher and lower elevations, respectively. The 3-D surfaces are plotted for this DEM (Figure 3.7) with vertical exaggerations 5 (Figure 3.8a) and 7 (Figure 3.8b). The variations in the fractal landscape topography are subjected to change in the shape and other characteristic information of the structuring template. It is worthwhile to mention that the morphology of regions of prominence (extracted by decomposing the binary fractal using the procedures detailed sequentially) is liable to vary with changing structuring templates. The precision depends on the design of structuring template. The design of the structuring template can be made by taking into consideration the morphological characteristics of each elevation level and interrelationships among all the spatially distributed elevation levels from a morphological standpoint, and an asymmetric structuring template (*B*), where *B* is not equal to the transpose of *B*, can also be considered to have more realistic landscapes. The structural

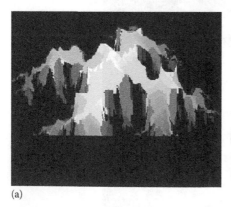 

(a)   (b)

FIGURE 3.8
Fractal landscape generated from Figure 3.5a. Light and dark regions of DEM are visualized as high and low elevations, respectively (vertical exaggeration: (a) 5 and (b) 7). (From Sagar, B.S.D. and Murthy, K.S.R., *Fractals*, 8(3), 267, 2000.)

variation in the surface topography determines the formation of dendrites which is a natural phenomenon. The topological description of the binary fractal provides a basis for the classification of the internal region that is topologically important. This study may be useful to show some meaningful inferences with elevation characteristics. This study is of practical interest to geomorphologists, as the simulated landscape and FDNs are akin to the natural landscape possessing alluvial fans.

Fractal Landscape from Triadic Fractal Basin

In a similar fashion, another case has been considered. A triangular initiator–basin is transformed as a fractal basin (Figure 3.3; Sagar et al. 2001) by following the principle involved in Koch curve generation. This binary fractal basin has been decomposed into TPRs. These TPRs have been assigned gray shades assuming that the TPRs of specific gray level represent a spatially distributed region of a specific elevation. A detailed procedure to simulate a fractal basin may be seen in Sagar and Murthy (2000). This simulated F-DEM thus generated is shown in Figure 3.9. We define the Hortonian F-DEM of a fluvial basin as a finite subset of 2-D Euclidean space that can have values between 0 and 255, each representing spatially distributed elevation region. We simulate this DEM by considering a binary fractal basin (X) (Figure 3.3) that possesses 1s and 0s, respectively, representing

FIGURE 3.9
Simulated fractal DEM achieved through morphological decomposition procedure. (From Sagar, B.S.D. and Tien, T.L., *Geophys. Res. Lett. (Am. Geophys. Union)*, 31(6), L06501, 2004.)

topological space of the basin and its complement. We consider a specific generating mechanism to simulate boundaries of binary fractal basin at different scales by considering two postulates: (1) the area of the basin is constant under the succession of scale changes, and (2) the length of the channel network should be varied under the succession of scale change to make the basin Hortonian. We decompose this binary fractal basin into TPRs by employing morphological erosions, dilations, and logical difference and union operations to simulate F-DEM. The simulation of internal topology of the basin within a defined geometric boundary is referred to as gray-level F-DEM (Figure 3.9).

Realistic DEMs and DBMs

DEM is a basic discrete representation of terrestrial surface. In a raster grid, each grid cell possesses spatial coordinates (x, y) and a number representing elevation values at spatial coordinates (x, y). DEM is denoted as a function represented by a nonnegative 2-D sequence $f(m, n)$, which assumed $I + 1$ possible intensity (elevation) values: $i = 0, 1, 2, ..., I$. The data range in synthetic data is within the interval from 0 to 255 ($I = 255$) elevations. The function, f, is discrete, defined on a (rectangular) subset of the discrete plane Z^2. The higher the intensity value, the higher is the topographic elevation, and vice versa.

The availability of multitemporal and multiscale DEMs, essentially derived from remotely sensed data, has changed the scenario of terrain modeling or quantitative geomorphometric studies. The techniques applied (that have emerged) to (1) extract automatically the features of geophysical relevance, (2) analyze them in spatiotemporal mode, (3) model and simulate various surficial processes in discrete space, and (4) characterize terrestrial surfaces for catchment classification by using DEMs are briefly reviewed in this chapter. As it stands, stand-alone techniques by making use of rather mixed mathematical techniques are employed to deal with these four points. In recent years, several mathematical morphological–based algorithms have been employed to perform various forms of DEM analysis, ranging from feature extraction to modeling and simulation of various terrestrial processes. This review also discusses on how mathematical morphology (a field most appropriate to conduct research on terrain modeling studies) has been applied to (1) generate DEMs at multiple scales, (2) extract multiscale networks, (3) derive shape–size descriptors, (4) simulate processes mimicking natural events/episodes, (5) segment/classify subbasins of a region, and (6) derive a host of new geomorphic indicators/descriptors. Some of the results achieved by the application of mathematical morphology to address the topic of terrain analysis are provided illustratively. Despite the advantages of these morphological-based algorithms over conventional methodologies, there exist several open problems that are briefly discussed in the concluding part.

Generally, raw elevation data in the form of stereophotographs or field surveys and the equipment necessary to process these data are not readily available to potential users of a DEM. Most users are therefore forced to rely on DEMs published by government agencies. The most common forms of DEMs available are those produced by digitizing the contours on existing topographic maps, known as cartometric DEMs. Existing plates used for printing maps are scanned. The resulting raster is vectorized and edited. The contours are tagged with elevations, and additional elevation data are created from the hydrograph layer such as shorelines, which provide additional contours.

Since the 1980s, there has been an increasing move toward using automated digital correlation techniques to generate what is known as photogrammetric DEMs directly from stereoscopic imagery, especially where contour data are not readily available or are not accurate enough. Photogrammetry can be done manually or automatically. In the manual method, an operator looks at a pair of stereophotos through a stereoplotter and must move two dots together until they appear to be one lying just at the surface of the ground. In the automatic method, an instrument calculates the parallax displacement of a large number of points. For example, for the United States Geological Survey (USGS) 7.5 min quadrangles, the Gestalt Photo Mapper II correlates 500,000 points. The extraction of elevation from photographs is confused by flat areas, especially lakes, and wherever the ground surface is obscured by objects such as buildings and trees.

Since the 1990s, DEMs are being consistently used in most of the studies related to watersheds. The DEMs derived directly through automated DEM generation that take advantage of stereo viewing capability of satellite images are preferred to those derived indirectly from digitized topographic maps. These automated maps are comparable to, or better than, those obtained through topographic maps in rugged or mountainous regions. The present case study illustrates a simple and elegant methodology for extracting drainage networks from DEMs, in general, by successfully implementing it on a transcendentally generated elevation model (Figures 3.8 and 3.9) obtained by considering a third-order Koch quadric and triadic fractal basins (Figures 3.3 and 3.5a).

- A contour-based DEM of Gunung Ledang region (Figure 3.10a) along with one of its subwatershed (Figure 3.10b) is generated by considering the information from surveyed topographic maps (Chockalingam and Sagar 2003, 2005).
- A small area of Yellowstone DEM (Figure 3.11) is also used, the details of which can be seen at http://edcwww.cr.usgs.gov/glis/hyper/guide/usgs_dem (last accessed November 12, 1999) (Sagar et al. 2003).

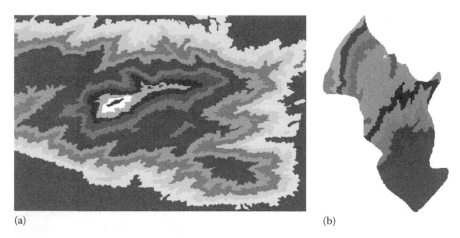

(a)   (b)

FIGURE 3.10
(a) Contour-based DEM of a part of Gunung Ledang region (b) a subwatershed from (a). (From Chockalingam, L. and Sagar, B.S.D., *J. Geophys. Res. Solid Earth (Am. Geophys. Union)*, 110, B08203, 2005.)

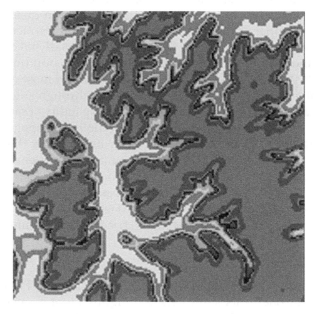

FIGURE 3.11
Sample DEM of size 200 × 200 pixels of a small part of the United States (downloaded from the Internet). (From Sagar, B.S.D. et al., *Int. J. Remote Sens.*, 24(3), 573, 2003.)

- The Malaysian government, under the coordination of the Malaysia Center for Remote Sensing (MACRES), participated in the Airborne Synthetic Aperture Radar/Topographic Synthetic Aperture Radar (AIRSAR/TOPSAR) PACRIM program jointly organized by the National Aeronautics and Space Administration and the Commonwealth Scientific and Industrial Research Organization. Polarimetric AIRSAR and interferometric TOPSAR data are used for terrain-related analysis. In this book, we analyze the interferometrically derived TOPSAR DEM of Cameron Highlands region of Malaysia (Figure 3.12). This region comprises a series of mountain stations at altitudes between 500 and 1300 m (Tay et al. 2005a,b, 2007, Lim et al. 2011). The real-world DEMs correspond to the topographic synthetic aperture radar (TOPSAR) DEMs of Cameron Highlands (Figure 3.12a) and Petaling regions (Figure 3.12b) of Malaysia from Tay et al. (2007). The Cameron

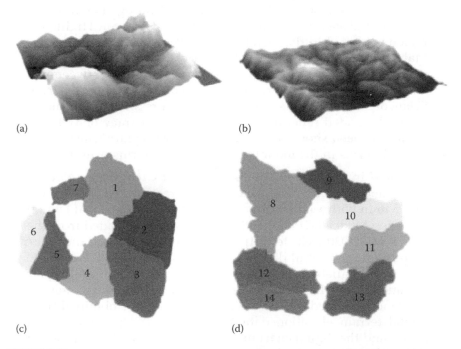

(a)   (b)

(c)   (d)

FIGURE 3.12
(a) Three-dimensional shaded relief image of TOPSAR DEM of Cameron Highlands, Malaysia; (b) 3-D shaded relief image of TOPSAR DEM of Petaling, Malaysia; (c) seven delineated subbasins in different colors of Cameron Highlands DEM; and (d) seven delineated subbasins in different colors of Petaling DEM. (From Tay, L.T. et al., *Int. J. Remote Sens.*, 26(18), 3901, 2005a; Tay, L.T. et al., *IEEE Geosci. Remote Sens. Lett.*, 2(4), 399, 2005b; Tay, L.T. et al., *Int. J. Remote Sens.*, 28(15), 3363, 2007.)

Highlands study area is enclosed by latitudes 4°31′–4°36′N and longitudes 101°15′–101°20′E, while the Petaling region is enclosed by latitudes 2°59′–3°02′N and longitudes 101°37′–101°40′E. Cameron Highlands is a highland region situated in the state of Pahang, Malaysia. It has hilly terrain with elevation range in between 400 and 1800 m. The Petaling region is comparatively flat with an altitude not more than 215 m. Cameron Highlands DEM covers an area of 900 × 900 pixels with 10 m resolution, while Petaling DEM covers a region of 750 × 800 pixels with 5 m resolution. Fourteen subbasins were demarcated from DEMs of Cameron Highlands and Petaling regions (Figure 3.12a and b). Each of the 14 subbasins (Figure 3.12c and d) has different value of I, depending on its maximum altitude. The Cameron Highlands subbasins are high-altitude basins whereas the Petaling subbasins have relatively lower altitudes, hence the value I for Cameron Highlands subbasins is generally greater than that of Petaling subbasins. For instance, basin 1 (of Cameron Highlands region) has value I as 1280 m while basin 8 (of Petaling region) has a maximum elevation (and thus I) of 208 m. Cameron Highlands region is located in the eastern part of Perak state in Peninsular Malaysia. The physical relief of this area is rough where it comprises a series of mountainous forest at altitudes between 400 and 1800 m. The Petaling region is located in the southern part of Selangor state in Peninsular Malaysia. Figure 3.12a and b shows their 3-D shaded relief images. The height accuracy of TOPSAR DEMs has been shown to be 1 m root mean square (rms) in flat areas, 3 m rms in the mountain areas, and 2 m rms overall.

- The DEM in Figure 3.13 shows the area of Great Basin, Nevada (Sathymoorthy et al. 2007). The DEM was rectified and resampled to 925 m in both x and y directions. The DEM is a Global Digital Elevation Model (GTOPO30 DEM) and was downloaded from the USGS GTOPO30 website (http://edcwww.cr.usgs.gov/landdaac/gtopo30/gtopo30.html [last accessed January 3, 2005]). GTOPO30 DEMs are available at a global scale, providing a digital representation of the Earth's surface at a 30 arc-s sampling interval. The land data used to derive GTOPO30 DEMs are obtained from digital terrain elevation data (DTED), the 1° DEM for United States, and the digital chart of the world (DCW). The accuracy of GTOPO30 DEMs varies by location according to the source data. The DTED and the 1u dataset have a vertical accuracy of ±30 m, while the absolute accuracy of the DCW vector dataset is ±2000 m horizontal error and ±650 vertical error. The DEM of Great Basin has a mean gradient of 4.94.

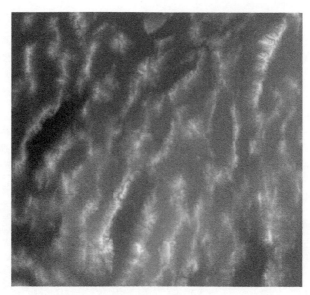

FIGURE 3.13
GTOPO30 DEM of Great Basin, Nevada. The elevation values of the terrain (minimum 1005 m and maximum 3651 m) are rescaled to the 0–255 interval (the brightest pixel has the highest elevation). The scale is approximately 1:3,900,000. (From Sathymoorthy, D. et al., *Int. J. Remote Sens.*, 28(15), 3379, 2007.)

Synthetic Basins and DBMs

The three synthetic cases (see Figure 13.14) with varied internal topographic regions that replicate the (1) flat, (2) undulated without channels, and (3) with channels conspicuous in bottom topography area considered. In reality, these three cases are synthesized forms of bottom topography of shallow water regimes (e.g., shallow lakes with flat bottom topography), bays and estuaries, and basins of floodplains and tidal environments.

- *Case 1*: Single inlet from which the water propagates uniformly within the mask-set (Figure 13.14a). With this assumption, oscillations in tidal levels and forcing influence the whole tidal basin that is assumed to be flat.

- *Case 2*: Single inlet from which the water would first flow into channelized regions followed by inland water. Here, channelized set and inlets (Figure 13.14b) are with different elevations. Nevertheless, in contrast to Case 1, flow fields in channelized sets maintain orthogonality with the flow fields in non-channelized

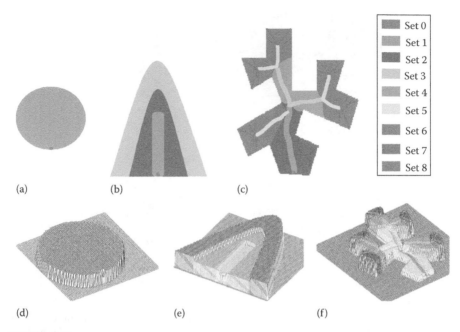

FIGURE 3.14
Tidal basins with different assumptions: (a) flat tidal basin, (b) tidal basin with channelized and non-channelized zones (multiple sets of topological significance), and (c) tidal basin with multiple sets, sets indexed with even and odd indexes, respectively, refer to channelized and non-channelized zones. (d–f) Three-dimensional mesh representations of three synthetic tidal basins shown in (a–c). (From Lim, S.L. and Sagar, B.S.D., *Discrete Dyn. Nat. Soc.*, 2008(312870), 26, 2008b.)

sets. This is both physically and intuitively justified due to the fact that flow propagation in channelized zones precedes flow propagation in non-channelized regions.

- *Case 3*: Single inlet and water flows alternatively into channel region and into inland until the propagating waterfronts reach the basin boundary (Figure 13.14c). Criteria followed to simulate flow fields—satisfying the fact that channelized and non-channelized regions are relatively with different mean elevations—include the following: Let X be the set of one channel, and it makes an arborescence from the inlet point from which the water flows into channels and their inlands, which is therefore the base of the trunk of the tree. This tree is connected, by definition, because we work on one channel set, where the water flow is coming up uniquely. Each branch is assimilated to a segment (if not, we subdivide the branch into a short succession of segments based on the following criteria: (1) the mean elevation, (2) width of segments, (3) direction of flow, and (4) the depths by taking the structure of an ascending tree). Then, disconnect each branch, by removing its first point (that of the subdivision with the upstream branch).

- Besides these synthetic cases, DBM of parts of Central San Francisco bay and DEM of Coastal Santa Cruz regions (Figure 3.15) are considered. Central San Francisco Bay bathymetry data, acquired with multibeam system, have been utilized here with permission from USGS. The coastal San Francisco Bay's bathymetry has been acquired through Multibeam Sonar System, collected in 1997 using a Simrad EM 1000 multibeam swath mapping system (http://sfbay.wr.usgs. gov/highlight_archives/new1998.html [last accessed October 9, 2007] and http://terraweb.wr.usgs.gov/projects/SFBaySonar/ [last accessed October 9, 2007]). The region of interest in San Francisco Bay area, of size 512 × 480 pixels, encompasses approximately from 37°48′41″N to 37°51′34″N, and from 122°26′2″W to 122°29′28″W.

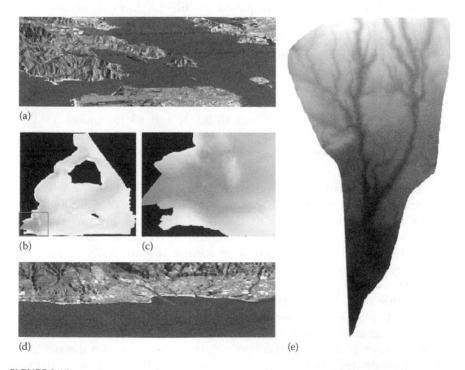

(a)

(b)   (c)

(d)   (e)

FIGURE 3.15
(a) Three-dimensional view of remote sensing data of Central SF Bay, (b) bathymetry of Central SF Bay, (c) bathymetry of inset of (c), (d) 3-D view of Santa Cruz, and (e) digital elevation map of Santa Cruz. (From Lim, S.L. and Sagar, B.S.D., *Discrete Dyn. Nat. Soc.*, 2008(312870), 26, 2008b.)

- *Santa Cruz digital elevation model*: The 10 m grid spacing digital elevation map (DEM) of Santa Cruz is downloaded from San Francisco Bay Area Regional Database (BARD) homepage (http://bard.wr.usgs.gov/htmldir/dem_html/index.html [last accessed October 9, 2007]), provided by USGS. At a scale of 1:24,000, it is available as 7.5 min standard DEM format on Universal Transverse Mercator (UTM) projection on North American Datum of 1927 (NAD 27), in the unit of meters for elevation relative to the National Geodetic Vertical Datum of 1929 (NGVD 29) (http://rockyweb.cr.usgs.gov/nmpstds/acrodocs/dem/1DEM0897.PDF [last accessed October 9, 2007]). The region of interest in Santa Cruz, of size 346 × 654 pixels, covers approximately from 36°56′35″N to 37°00′00″N, and from122°03′56″W to 122°05′38″W.

Remotely Sensed Satellite Data

Remotely sensed satellite data acquired by IRS-1A LISS III sensors and MODIS channels have been considered throughout Chapters 4 through 8. The general specifications, such as the geographical coordinates and spatial and spectral resolutions, are briefly described in the following:

- IRS-1C remotely sensed data (Figure 3.16) of regions between the geographical coordinates (a) 18°00′ and 18°30′ N and 83°15′ and 83°45′ E, and (b) 18°00′ and 18°30′ N and 83°15′ and 83°45′ E (Sagar et al. 1995a, 1995b, Sagar 1999, Sagar et al. 2002, Sagar 2007) are used as source data to demarcate a large number of surface water bodies.

- The data set containing large number of water bodies in 528 km^2 region is taken for this study. This sample consists of a number of surface water bodies larger than 32.5 m^2. The limit 36.25 m^2 represents the smallest water body that could be traced accurately from IRS-1A (LISS II) data in geocoded format (Figure 3.17). It lies in between the geographical coordinates of 18°15′ and 18°30′N and 83°30′ and 83°45′E belonging to the 65N/11 Survey of India (SOI) topographic map. These were extracted from IRS-1A remotely sensed data of a region situated between the geographical coordinates 18°15′ and 18°30′N and 83°30′ and 83°45′E belonging to the 65N/11 SOI topographic map that covers a part of Vizianagaram district of Andhra Pradesh, India. Since the resolution of IRS-1A (LISS II) data is 36.25 m by 36.25 m, the minimum limit considered was 36.25 m^2 to trace the water bodies for this analysis. The sample consists of a number of water bodies larger than 36.25 m by 36.25 m n where n

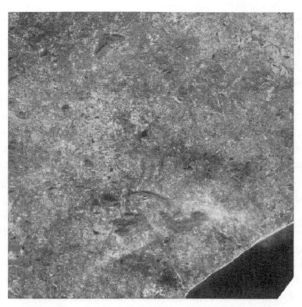

FIGURE 3.16
Geocoded IRS-1A (LISS II) data of a region situated between the geographical coordinates of 18°00′ and 18°30′N and 83°15′ and 83°45′E 36.25 m by 36.25 m resolution acquired on August 3, 1993.

FIGURE 3.17
Geocoded IRS-1C (LISS II) data situated between the geographical coordinates of 18°15′ and 18°30′N and 83°30′ and 83°45′E, belonging to the 65 N/II SOI topographic map, of 36.25 m by 36.25 m resolution acquired on August 3, 1993. SOI 65 N/11.

should be more than 20 pixels. The limit 36.25 m × 36.25 m represents the smallest water body that could be traced accurately from IRS-1A (LISS II) data in geocoded format (Figure 3.17). Since the resolution of IRS-1A (LISS II) data is 36.25 m by 36.25 m and the minimum limit considered is 36.25 m by 36.25 m n, the water bodies not eliminated on the basis of this criterion are traced. To carry out water body distribution studies automatically, all traced surface water bodies are digitized by a digital Pulnix camera. The water bodies, which are present in the major part of the image, are kept in the file of size 480 by 480 pixels (368.6 km by 368.6 km). By giving a specific threshold value, the entire data are kept in the form of water body and no-water body regions.

- Other region considered is consisting of a large number of semiartificial irrigation tanks of various sizes and shapes, of a floodplain region of Gosthani River (one of the east-flowing rivers of India) situated between 18°00′ and 18°15′N latitudes and 83°15 and 83°30′E longitudes (Figure 3.18). These water bodies are controlled by topography, and at one side, minor bunds are constructed in order to store the water. The general spatial patterns of these water bodies are uniquely determined by general river flow patterns within a floodplain region. This floodplain region in general is with <2° slope, in which the topographic undulations are not too rough to

FIGURE 3.18
IRS-1A LISS III image situated between 18°00′ and 18°15′N latitudes and 83°15′ and 83°30′E longitudes.

discard the assumption that flood propagation is rather isotropic. The water bodies, from Landsat and System 10 pour Observation de la Terre (SPOT) Geocoded visual (paper) products acquired in 1986 and 1990, are not very conspicuous. This is because of the fact that the water bodies are rather (1) shallow and (2) polluted with acute siltation problem. For decades, no attempts have been made to de-silt these water bodies. For these two reasons, extraction/ tracing of these water bodies directly from remotely sensed data is an arduous task.

- The mathematical morphological techniques were applied on surface water bodies extracted from the digital data sets of small area (5 km × 5 km), acquired through SPOT (PLA mode) digital data of December 1990 (Figure 3.19), to demonstrate the two algorithms respectively to compute basic measures and directional relationships of water bodies (Sagar et al. 1995b). The sample consisted of nine water bodies each with a minimum area of 100 m² as SPOT (PLA) digital data possess 10 m × 10 m resolution (Figure 3.19).

- To test whether the water body follows Horton's laws, an image of the Nizamsagar reservoir situated in Andhra Pradesh, India, acquired on April 13, 1989 from IRS-1A LISS III sensor with a resolution of 72 m (Figure 3.20; Sagar et al. 1999) is considered.

FIGURE 3.19
IRS LISS III digital data.

FIGURE 3.20
Extracted Nizamsagar reservoir from IRS-1A LISS III, 72 m resolution. (From Sagar, B.S.D. et al., *J. Math. Geol.*, 31(2), 143, 1999.)

- Moderate Resolution Imaging Spectroradiometer (MODIS) data are the ideal data source for resource and environmental remote sensing monitoring on the regional scale. We consider two cloud images (Figure 3.21a and c), belonging to regions situated between the spatial coordinates 60°–74′E, 20°–28′N and 150°–162′°W, 2°–10.5′N, respectively, acquired through MODIS (http://modis-atmos.gsfc.nasa.gov/IMAGES/index.html) (Lim and Sagar 2008a). Figure 3.21a was acquired on June 30, 2006 (Day 181), at 1510 UTC with a size of 851 × 621 pixels, while Figure 3.21c was obtained on May 28, 2007, at 2355 UTC with a size of 797 × 512 pixels. The original images are in true color (bands 1, 4, and 3 for red, green, and blue channels, respectively). The images are then converted to grayscale functions $f(x, y)$ (multilevel signals) with 8 bits/pixel. Thus, the values for $f(x, y)$ vary from 0 to 255, with value 0 representing black and value 255 representing pure white (an ideal case); values in between 0 and 255 (i.e., 1–256) denote shades of gray that range sequentially from dark gray to lighter gray shades. A true color image is a 24-bit image consisting of 8-bit images, for red (R), green (G), and blue (B) planes, respectively. As processing of a 24-bit image is computationally highly expensive, a 24-bit true color image is converted into an 8-bit grayscale image. Processing an 8-bit image is

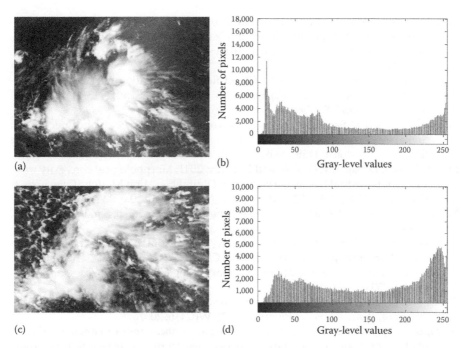

FIGURE 3.21
(a) Isolated Moderate Resolution Imaging Spectroradiometer (MODIS) cloud (cloud-1), (b) histogram of Figure 3.21a, (c) isolated MODIS cloud (cloud-2), and (d) histogram of Figure 3.21c. Refer to text for the spatial and time details of clouds acquired by MODIS satellites. (From Lim, S.L. and Sagar, B.S.D., *J. Geophys. Res. Atmos.*, 113, D13208, 2008a.)

straightforward and is acceptable in remotely sensed data analysis studies. The corresponding histograms of Figure 3.21a and c are shown in Figure 3.21b and d, respectively. It is observed that most of the pixel values are distributed either toward dark gray to black or toward light gray to white, with a peak value at 255 (white) in Figure 3.21b. This can be explained using Figure 3.21a such that a large patch of seemingly thick white cloud constitutes the middle of the image.

References

Chockalingam, L. and B. S. D. Sagar, 2003, Automatic generation of sub-watershed map from Digital Elevation Model: A morphological approach, *International Journal of Pattern Recognition and Artificial Intelligence*, 17(2), 269–274.

Chockalingam, L. and B. S. D. Sagar, 2005, Morphometry of networks and non-network spaces, *Journal of Geophysical Research-Solid Earth (American Geophysical Union)*, 110, B08203, doi:10.1029/2005JB003641.

Lim, S. L., V. C. Koo, and B. S. D. Sagar, 2009, Computation of complexity measures of morphologically significant zones decomposed from binary fractal sets via multiscale convexity analysis, *Chaos, Solitons & Fractals*, 41(3), 1253–1262.

Lim, S. L. and B. S. D. Sagar, 2008a, Cloud field segmentation via multiscale convexity analysis, *Journal Geophysical Research-Atmospheres*, 113, D13208, 17, doi:10.1029/2007JD009369.

Lim, S. L. and B. S. D. Sagar, 2008b, Derivation of geodesic flow fields and spectrum in digital topographic basins, *Discrete Dynamics in Nature and Society*, 2008, Article ID 312870, 26, doi:10.1155/2008/312870.

Lim, S. L., B. S. D. Sagar, V. C. Koo, and L. T. Tay, 2011, Morphological convexity measures for terrestrial basins derived from Digital Elevation Models, *Computers & Geosciences*, 37(9), 1285–1294, doi:10.1016/j.cageo.2010.10.002.

Mandelbrot, B. B., 1982, *Fractal Geometry of Nature*, W. H. Freeman, San Francisco, CA, p. 468.

Radhakrishnan, P., B. S. D. Sagar, and L. L. Teo, 2004, Estimation of fractal dimension through morphological decomposition, *Chaos Solitons & Fractals*, 21(3), 563–572.

Sagar, B. S. D., 1996, Fractal relations of a morphological skeleton, *Chaos, Solitons & Fractals*, 7(11), 1871–1879.

Sagar, B. S. D., 2007, Universal scaling laws in surface water bodies and their zones of influence, *Water Resources Research*, 43(2), W02416, 2007.

Sagar, B. S. D., G. Gandhi, and B. S. P. Rao, 1995a, Applications of mathematical morphology on water body studies, *International Journal of Remote Sensing*, 16(8), 1495–1502.

Sagar, B. S. D. and K. S. R. Murthy, 2000, Generation of fractal landscape using nonlinear mathematical morphological transformations, *Fractals*, 8(3), 267–272.

Sagar, B. S. D., M. B. R. Murthy, C. B. Rao, and B. Raj, 2003, Morphological approach to extract ridge-valley connectivity networks from Digital Elevation Models (DEMs), *International Journal of Remote Sensing*, 24(3), 573–581.

Sagar, B. S. D., C. Omoregie, and B. S. P. Rao, 1998, Morphometric relations of fractal-skeletal based channel network model, *Discrete Dynamics in Nature and Society*, 2(2), 77–92.

Sagar, B. S. D., C. B. Rao, and B. Raj, 2002, Is the spatial organization of larger water bodies heterogeneous? *International Journal of Remote Sensing*, 23(3), 503–509.

Sagar, B. S. D. and D. Srinivas, 1999, Estimation of number-area-frequency dimensions of surface water bodies, *International Journal of Remote Sensing*, 20(13), 2491–2496.

Sagar, B. S. D., D. Srinivas, and B. S. P. Rao, 2001, Fractal skeletal based channel networks in a triangular initiator basin, *Fractals*, 9(4), 429–437.

Sagar, B. S. D. and T. L. Tien, 2004, Allometric power-law relationships in a Hortonian Fractal DEM, *Geophysical Research Letters, (American Geophysical Union)*, 31(6), L06501.

Sagar, B. S. D., M. Venu, and K. S. R. Murthy, 1999, Do skeletal network derived from water bodies follow Horton's laws? *Journal Mathematical Geology*, 31(2), 143–154.

Sagar, B. S. D., M. Venu, and B. S. P. Rao, 1995b, Distributions of surface water bodies, *International Journal of Remote Sensing*, 16(16), 3059–3067.

Sagar, B. S. D., M. Venu, and D. Srinivas, 2000, Morphological operators to extract channel networks from Digital Elevation Models, *International Journal of Remote Sensing*, 21(1), 21–30.

Sathymoorthy, D., P. Radhakrishnan, and B. S. D. Sagar, 2007, Morphological segmentation of physiographic features from DEM, *International Journal of Remote Sensing*, 28(15), 3379–3394.

Serra, J., 1982, *Image Analysis and Mathematical Morphology*, Academic Press, New York, p. 610.

Tay, L. T., B. S. D. Sagar, and H. T. Chuah, 2005a, Analysis of geophysical networks derived from multiscale digital elevation models: A morphological approach, *I.E.E.E Geoscience and Remote Sensing Letters*, 2(4), 399–403.

Tay, L. T., B. S. D. Sagar, and H. T. Chuah, 2005b, Derivation of terrain roughness indicators via Granulometries, *International Journal of Remote Sensing*, 26(18), 3901–3910.

Tay, L. T., B. S. D. Sagar, and H. T. Chuah, 2007, Granulometric analysis of basinwise DEMs: A comparative study, *International Journal of Remote Sensing*, 28(15), 3363–3378.

United States Geological Survey (USGS). http://bard.wr.usgs.gov/htmldir/dem_html/index.html (last accessed October 9, 2007).

United States Geological Survey (USGS). http://rockyweb.cr.usgs.gov/nmpstds/acrodocs/dem/1DEM0897.PDF (last accessed October 9, 2007).

United States Geological Survey (USGS). http://sfbay.wr.usgs.gov/highlight_archives/new1998.html (last accessed October 9, 2007).

United States Geological Survey (USGS). http://terraweb.wr.usgs.gov/projects/SFBaySonar (last accessed October 9, 2007).

4

Feature Extraction

Terrestrial surfaces are composed of regions with varied degrees of concavity and convexity. Physiographic and geomorphologic processes can be better explained by quantitative descriptions of these concavities and convexities. Valley and ridge connectivity networks are the abstract structures of concave and convex zones of terrestrial surfaces. The extraction of these abstract structures has been explained in this chapter. These abstract structures are important from the point of terrestrial surface characterization quantitatively via morphometry (Horton 1945), hypsometry (Strahler 1957), allometry (Maritan et al. 1996, Rodriguez-Iturbe and Rinaldo 1997, Sagar and Tien 2004), morphological decompositions (Sagar and Chockalingam 2004), and granulometry (Tay et al. 2005b, 2007).

Processing of remotely sensed data in both spatial and frequency domains has received wide attention. Remotely sensed data are available for the characterization of various phenomena related to terrestrial, lunar, and planetary surfaces and atmospheric phenomena such as clouds in spatio-temporal mode. The characterization of such data is primarily of interest to hydrologists, geomorphologists, geologists, geophysicists, ecologists, climatologists, and environmentalists. The application of remote sensing in various fields is greatly realized in the last three decades. One of the data derivable from remotely sensed data is a digital elevation model (DEM) that provides rich clues about the physiographic constitution of Earth and Earth-like planetary surfaces. From such data, a lot of useful thematic information could be extracted. Such thematic information serves various purposes of significance in studies such as flood mapping, military mapping, city planning, and so on.

This chapter addresses extraction of features of terrestrial significance. The features include unique ridge and channel networks, physiographic features such as mountain objects, hierarchically decomposed subwatersheds, and varied topologically significant regions of cloud fields. This chapter is broadly categorized into three sections: the "Unique Feature Retrieval via Binary Skeletonization" section, the "Retrieval of Physiographic Features from DEMs via Morphological Segmentation" section, and the "Extraction of Morphologically Significant Zones" section.

Unique Feature Retrieval via Binary Skeletonization

Unique terrestrial features such as channel networks and ridge networks are conspicuous from DEMs. DEMs can be obtained from the remotely sensed satellite data having stereo viewing capability. The application of morphological operators provides interesting approaches to retrieve unique terrestrial features from DEMs. An algorithm is developed to extract unique terrestrial features from DEMs and has been demonstrated on simulated DEMs and realistic DEMs. This algorithm isolates V-shaped and Λ-shaped crenulations from elevation contours of DEMs. These crenulations testify the presence of ridge and channel connectivity network subsets. Further, the algorithm is based on the principle involved in morphological skeleton.

This section of the chapter addresses the extraction process in two ways: (i) by considering only binary morphological transformation, which is time consuming, computationally expensive, but yields precise results, and (ii) by considering grayscale morphological transformation, which involves lesser time, but has limitations that could be seen in the retrieval of features in terms of lack of proper connectivity.

Some Background Studies of Unique Feature Extraction

Channel network is an interesting terrestrial feature. Besides surveyed topographic maps where one can see these channel networks as blue lines, aerial photographs were employed to map terrestrial features relevant for the quantitative description of drainage basins. Feature retrieval from remotely sensed data has several advantages compared to those taken from topographic map sources. The superiority of aerial photographs over topographic maps was shown by Bunik and Turner (1971). Of late, much more advanced and inexpensive data acquired through remote sensing satellite sensors have provided the required data at multiple spatial and temporal scales. To deal with the quantitative description of terrestrial complexity, unique features retrieved from multiscale and multitemporal satellite data are of immense use. One of such source data includes DEMs acquired at varied spatiotemporal scales. The DEM is an array of numbers representing the spatial distribution of elevation heights over a terrestrial surface, and such DEMs could be derived indirectly from digitized topographic maps or directly through photogrammetric processing of medium-scale, black and white metric aerial photographs (Franklin 1990).

Any terrestrial surface characterization involves three sequential steps: (i) processing of stereo remote sensing data, (ii) DEM generation from remotely sensed satellite data, and (iii) retrieval of significant terrestrial features from DEMs of varied spatiotemporal scales. Pioneering researchers have addressed these steps (O'Callaghan and Mark 1984, Yuan and Vanderpool 1986, Jenson 1987, Jenson and Domingue 1988, Mark 1988,

Martz and de Jong 1988, Morris and Heerdegen 1988, Qian et al. 1990, Fairfield and Leymarie 1991, Freeman 1991, Chorowicz et al. 1992, Tribe 1992). Elegant approaches to extract valley networks from DEMs have been proposed by O'Callaghan and Mark (1984) and Tarboton et al. (1991). These valley networks, in other words stream lines, are conspicuous as blue lines in topographic maps, but no topographic map provides ridge networks. Many algorithms to extract these unique topological networks from DEMs have limitations when the DEM possesses non-hilly regions.

In spite of the availability of numerous algorithms that have addressed the problems of the extraction of unique networks from DEMs (Peucker and Douglas 1975, Mark et al. 1982, Mark 1983a,b, O'Callaghan and Mark 1984, Jenson 1985, Band 1986, Franklin 1990), these extraction procedures have several limitations when DEM possesses non-hilly regions. Network features extracted from DEMs of tidal regions are erroneous and, in turn, warrant new approaches that can take curvature into consideration (Sagar et al. 2000, 2001, 2003).

What Do Angular Points in DEM Represent?

Angular points are conspicuous in any DEM representing spatially distributed elevation regions in gray levels. These angular points are of two types: V-shaped and Λ-shaped portions of the elevation contours in two-dimensional space. These angular points are referred to as crenulations that indicate the presence of unique topological networks (Morisawa 1957). The complex organization of channel networks is determined by the spatial positions of the crenulations in the sequential elevation contours (e.g., Drfiffwood, Nashville, Little Tijunga maps of U.S. Geological Survey, Strahler 1964). The overall structure of the spatial distribution of elevation regions determines the branching organization of channel and ridge networks.

Widths of crenulations of the lower-elevation contours are more than that of higher-elevation contours. The number of crenulations in the higher-elevation contours is more than that of lower-elevation contours. Crenulations are flow paths of streams. In turn, the streams that flow through the crenulations of higher-elevation contours are narrow and are more in number than that of streams flowing through lower-elevation contours. Two streams flow through the crenulations of successive higher-elevation contours and bifurcate at some point, which is another crenulation that is wider. The stream at that bifurcation point is wider than the previous level. If one follows Horton–Strahler ordering scheme (that is explained in Chapter 5), all stream segments flowing through higher-elevation contours are designated with order-1 streams, and whenever the order-1 segments join together, they form the order-2 segments, the widths of which are more than that of lower-order streams. This indicates that the number of lower-order stream segments is more than that of higher orders. This description explains the importance of isolating all possible crenulations, which are flow paths of stream lines, from contours of DEM. The union of all crenulations isolated from DEM would provide the valley connectivity network. Now the task is to provide an algorithm to isolate all possible crenulations from a given DEM.

Streams flow between the ridges. Takayasu (1990) mentioned that ridges are Brownian motions that act as barriers. In geomorphologic terms, an elevation contour possesses crenulations of "V" and "Λ" shapes, which can be termed respectively as the paths of ridge and valley connectivity networks. In mathematical terms, these networks are *suprema* and *infima*.

Valley Connectivity Network Extraction from DEM Using Binary Morphological Operations

Streams (water flow) flow through valley (channel) networks. Hence, the extracted valley connectivity network could be treated as stream network. The extraction of valley connectivity network from DEM, a function denoted as $f(x, y)$, is addressed in the following section.

$f(x, y) = Z$, where Z ranges between 0 and 255 as long as the DEM considered is stored in 8 bits per pixel. Values 0 and 255, respectively, represent the lowest and highest elevations shown in a normalized scale. The spatially distributed elevation regions with values 0 and 255, respectively, are represented as black and white shades, and the intermediary elevation zones occupy the other shades between black (0) and white (255).

The procedure to extract valley connectivity networks from $f(x, y)$, in other words DEM, involves two broad steps: threshold decomposition and morphological skeletonization.

Threshold decomposition of $f(x, y)$: When the dynamic range is wide such that DEM possesses all elevations ranging from 0 to 255, such a DEM can be decomposed into a maximum of 255 sets. Each set represents a region in planar form above certain threshold level. This threshold decomposition is shown in Equation 4.1:

$$f^t = \begin{cases} 1, & f(x,y) \geq t \\ 0, & f(x,y) < t \end{cases}$$
(4.1)

where $0 \leq f(x, y) \leq 255$, and t can be a chosen value(s) from 0 to 255. For example, refer to Figure 4.1a and treat it as a synthetic DEM consisting of three spatially distributed elevation regions. By following threshold decomposition procedure shown in Equation 4.1, this synthetic DEM could be decomposed into three threshold elevation regions (Figure 4.1b through d). These threshold decomposed elevation regions follow inclusion relationship (e.g., Figure 4.2) as $f^3 \subseteq f^2 \subset f^1$. The stack of these threshold elevation regions f^t, for t = 1, 2, and 3, forms the original DEM. Henceforth we refer the threshold elevation regions f^t as set X^t for simplicity. The crenulations in the boundary of X^t for each threshold value indicate the presence of

(a)

1	1	1	1	1	1	1	1	1	1	1
1	2	2	2	2	2	2	2	2	2	1
1	2	2	2	2	2	2	2	2	2	1
1	2	2	3	3	3	3	3	2	2	1
1	2	2	3	3	3	3	3	2	2	1
1	2	2	3	3	3	3	3	2	2	1
1	2	2	3	3	3	3	3	2	2	1
1	2	2	2	2	2	2	2	2	2	1
1	2	2	2	2	2	2	2	2	2	1
1	2	2	2	2	2	2	2	2	2	1
1	1	1	1	1	1	1	1	1	1	1

(b)

1	1	1	1	1	1	1	1	1	1	1
1	1	1	1	1	1	1	1	1	1	1
1	1	1	1	1	1	1	1	1	1	1
1	1	1	1	1	1	1	1	1	1	1
1	1	1	1	1	1	1	1	1	1	1
1	1	1	1	1	1	1	1	1	1	1
1	1	1	1	1	1	1	1	1	1	1
1	1	1	1	1	1	1	1	1	1	1
1	1	1	1	1	1	1	1	1	1	1
1	1	1	1	1	1	1	1	1	1	1

(c)

0	0	0	0	0	0	0	0	0	0	0	0
0	1	1	1	1	1	1	1	1	1	1	0
0	1	1	1	1	1	1	1	1	1	1	0
0	1	1	1	1	1	1	1	1	1	1	0
0	1	1	1	1	1	1	1	1	1	1	0
0	1	1	1	1	1	1	1	1	1	1	0
0	1	1	1	1	1	1	1	1	1	1	0
0	1	1	1	1	1	1	1	1	1	1	0
0	1	1	1	1	1	1	1	1	1	1	0
0	1	1	1	1	1	1	1	1	1	1	0
0	0	0	0	0	0	0	0	0	0	0	0

(d)

0	0	0	0	0	0	0	0	0	0	0	0
0	0	0	0	0	0	0	0	0	0	0	0
0	0	0	0	0	0	0	0	0	0	0	0
0	0	0	1	1	1	1	1	0	0	0	0
0	0	0	1	1	1	1	1	0	0	0	0
0	0	0	1	1	1	1	1	0	0	0	0
0	0	0	1	1	1	1	1	0	0	0	0
0	0	0	0	0	0	0	0	0	0	0	0
0	0	0	0	0	0	0	0	0	0	0	0
0	0	0	0	0	0	0	0	0	0	0	0

FIGURE 4.1
Simulated DEM with three spatially distributed elevation regions represented numerically as
1s, 2s, and 3s. (a) A simulated DEM and (b–d) threshold decomposed elevation regions with t
greater than or equal to 1, 2, and 3.

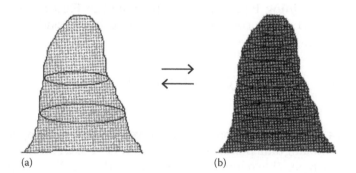

(a)   (b)

FIGURE 4.2
Threshold decomposition of a DEM: (a) function and mapping and (b) pile of sets. (From Sagar,
B.S.D. et al., *Int. J. Remote Sens.*, 21(1), 21, 2000.)

valley connectivity network subsets. The procedure to extract subsets
of valley connectivity network of entire considered DEM is as follows.
Understanding this procedure requires basic knowledge of mathematical
morphological operations (see Chapter 2).

We know that $X^t \subseteq X^{t-1}$. Let $f(x, y)$ consist of three spatially distributed
elevation regions, and $f(x, y)$ could be decomposed into threshold elevation
regions as X^1, X^2, and X^3 (e.g., Figure 4.1b through d):

$$X^3 \subseteq X^2 \subset X^1 \tag{4.2}$$

These decomposed elevation regions consist of 1s (set) and 0s (set comple-ment). To determine the possible valley connectivity network of $f(x,y)$, X^t needs to be transformed into its abstract network, in other words morpho-logical skeleton $SK(X^t)$.

Morphological Skeleton of X^t

Computation of morphological skeleton of X^t involves two steps: (i) compu-tation of skeletal subsets of order "n," where "n" ranges from zero to N, and (ii) union of all nth-order skeletal subsets. Equations 4.3 and 4.4 explain these two steps.

$$SK_n(X^t) = ((X^t \ominus nB) \setminus (X^t \ominus nB)) \circ B, \quad \text{where } n = 0,1,2,\dots,N \qquad (4.3)$$

where
X^t is an elevation region obtained with a threshold value "t"
nB is a structuring element B of size (radius) n
$\ominus$ and $\circ$, respectively, denote symbols for morphological erosion and mor-phological opening, and reverse solidus indicates set subtraction

The process explaining Equation 4.3 is illustrated in Figure 4.3. Once the skeletal subsets of X^t are computed, according to the second step, skeletal network of X^t, $SK(X^t)$ can be derived as

$$SK(X^t) = \bigcup_{n=0}^{N} SK_n(X^t) \qquad (4.4)$$

In Equation 4.3, subtracting from the eroded version of X^t their opening by B retains only the angular points, which are points of skeletal network. The union of all such possible points produces a network from X^t. The application of this procedure on all possible threshold decomposed elevation regions yields t number of skeletal networks like $SK(X^{t+1})$, $SK(X^{t+2})$,…

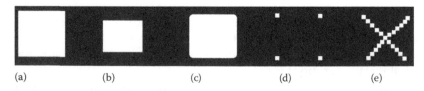

(a)	(b)	(c)	(d)	(e)

FIGURE 4.3
Steps involved in the computation of morphological skeleton of a set X. (a) Set X, (b) $X \ominus B$, (c) $(X \ominus B) \circ B$, (d) $SK_1(X) = (X \ominus B) \setminus (X \ominus B) \circ B$, and (e) $SK(X) = \bigcup_{n=0}^{N} SK_n(X)$. For more details, refer Figure 2.14.

The computation of possible valley connectivity network of DEM further involves the following steps:

Step 1: $(SK(X^t) \cup X^{t+1}) \setminus X^{t+1}$

Step 2: $(SK(X^{t+1}) \cup X^{t+2}) \setminus X^{t+2}$

$$(4.5)$$

For the example of synthetic DEM, $f(x,y)$, and the three decomposed threshold elevation regions X^1, X^2, and X^3 (Figure 4.1b through d), the earlier sequential steps are shown illustratively (Figure 4.4). Corresponding figure-wise equations are shown as follows:

$$SK(X^1) = \bigcup_{n=0}^{N} ((X^1 \ominus nB) \setminus (X^1 \ominus nB) \circ B) \ \text{(Figure 4.4a)} \qquad (4.6)$$

$$SK(X^2) = \bigcup_{n=0}^{N} ((X^2 \ominus nB) \setminus (X^2 \ominus nB) \circ B) \ \text{(Figure 4.4b)} \qquad (4.7)$$

$$SK(X^3) = \bigcup_{n=0}^{N} ((X^3 \ominus nB) \setminus (X^3 \ominus nB) \circ B) \ \text{(Figure 4.4c)} \qquad (4.8)$$

$$VCN(X^1) = (SK(X^1) \cup X^2) \setminus X^2 \qquad (4.9)$$

$$VCN(X^2) = (SK(X^2) \cup X^3) \setminus X^3 \qquad (4.10)$$

$$VCN(X^3) = (SK(X^3) \cup X^4) \setminus X^4, \quad \text{where } X^4 = \varnothing \qquad (4.11)$$

Figure 4.4a through f has been generated by employing Equations 4.6 through 4.11. Finally, the valley connectivity network of DEM considered can be obtained by Equation 4.12:

$$VCN(f) = VCN(X^1) \cup VCN(X^2) \cup VCN(X^3) \qquad (4.12)$$

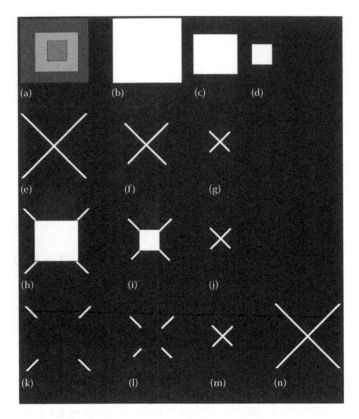

FIGURE 4.4
Steps involved in the extraction of valley connectivity networks from DEM. (a) Synthetic DEM shown with three spatially distributed elevation regions (in three different shades), (b–d) threshold decomposed elevation regions, (e–g) skeletal networks of threshold elevation regions, (h–j) union of skeletal networks with threshold elevation regions of next level, (k–m) isolated skeletal networks within the threshold elevation regions, and (n) final skeletal network (valley connectivity network) of synthetic DEM shown in Figure 4.4a.

This approach has been demonstrated on a transcendentally generated fractal DEM (Figure 3.7). The details about the generation of fractal DEM can be seen from Chapter 3. The DEM shown in Figure 3.7 is used to demonstrate the sequential steps explained earlier to extract valley connectivity network. Fifteen elevation regions could be decomposed from the fractal DEM (Figure 3.7) by following the threshold decomposition procedure according to Equation 4.1. Each of those 15 threshold decomposed regions $(X^1, X^2, ..., X^{15})$, respectively, is obtained by choosing threshold values ranging from 1 to 15. In this DEM, 1s and 15s, respectively, denote lowest and highest elevations. Each threshold decomposed region is transformed into its morphological skeleton network, according to Equations 4.3 and 4.4. Further, according to Equation 4.5, the possible valley connectivity network between the two successive threshold decomposed elevation regions is computed. Such valley connectivity networks

1										1
	1								1	
		1						1		
			1				1			
				1		1				
					1					
				1		1				
			1				1			
		1						1		
	1								1	
1										1

FIGURE 4.5
Drainage network in the simulated DEM shown in Figure 4.1.

that are further superposed on one another according to Equations 4.6 through 4.12 are also explained via illustrations (Figure 4.5). The final valley connectivity network for the fractal DEM is shown in Figures 3.7 and 4.5.

DEM could be inverted such that higher-elevation regions, depicted, for instance, with value 15 as in the DEM considered, are replaced with lower-elevation regions, depicted with value 1. Such an inverted DEM could be employed to implement the entire approach to obtain ridge connectivity network. To demonstrate how DEM could be employed to retrieve both valley and ridge connectivity networks, another simulated DEM (Figure 3.9) is considered. This DEM and its inverse were considered, and the approach explained and demonstrated on DEM (Figure 3.7) was applied to retrieve possible valley and ridge connectivity networks.

This approach makes it clear that geophysical networks from realistic DEMs, generated from high-resolution remotely sensed data, could be precisely extracted. The precision and accuracy of such retrieved networks heavily depend on the accuracy of DEMs. This approach is highly sensitive to the noise in the DEM. Another limitation of this approach is that it is computationally expensive. But this approach could be generalized to both fluvial and tidal regions.

Superficially, the reader would agree that simple binary morphological operations—such as erosion, dilation, and cascade of erosion–dilation, and certain logical operations like set subtraction, intersection, and union operations and their systematic use—could extract valley and ridge connectivity networks from DEMs. The answer for a question on how accurate is the retrieved network is that as far as DEM is precise and noise-free, the network is very precise and accurate. The main limitation of this approach is that this entire process is time consuming. Is there a more straightforward approach, whereby one can apply such an approach directly on grayscale DEM? To answer this nontrivial question, a set of grayscale morphological operations has been employed to extract similar networks from DEMs directly. This study is presented in the next section.

Ridge and Valley Connectivity Networks via Grayscale Skeletonization

Valley and ridge connectivity networks are two singular networks of topological interest that could be extracted from DEM directly by employing grayscale morphological operations. The details of these grayscale morphological operations and their implementations are given in Chapter 2.

The extraction of these two singular networks, without involving threshold decomposition processes and binary morphological operations employed in the previous section, could be done by employing grayscale erosion and grayscale dilation operations by means of a flat structuring element B, simple algebraic subtraction, and computing *suprema*. Further, this extraction approach simply follows grayscale skeletonization involving the earlier-mentioned operations.

Given a DEM, denoted for simplicity as f, the extraction of valley connectivity network follows Equations 4.13 and 4.14:

$$VCN_n(f) = ((f \ominus nB) - (f \ominus nB)) \circ B, \quad \text{where } n = 0, 1, 2, \cdots, N \quad (4.13)$$

$$VCN(f) = \bigvee_{\forall n}(VCN_n(f)) \quad (4.14)$$

By employing similar steps on inverted DEM, denoted as f^{-1}, ridge connectivity network could be obtained as per Equations 4.15 and 4.16:

$$RID_n(f) = ((f^{-1} \ominus nB) - (f^{-1} \ominus nB)) \circ B, \quad \text{where } n = 0, 1, 2, \ldots, N \quad (4.15)$$

$$RID(f) = \bigvee_{\forall n}(RID_n(f)) \quad (4.16)$$

One can convert these two singular networks, $VCN(f)$ and $RID(f)$, into binary forms by choosing threshold value 1 as shown in Equations 4.17 and 4.18:

$$VCN = \begin{cases} 1, & VCN(f) \geq 1 \\ 0, & VCN < 1 \end{cases}, \quad (4.17)$$

and

$$RID = \begin{cases} 1, & RID(f) \geq 1 \\ 0, & RID < 1 \end{cases} \quad (4.18)$$

With the implementation of aforementioned sequential steps on a given DEM, one can superpose these two singular networks on one another as $VCN \cup RID$. From such a superposed version, it could be seen that segments

FIGURE 4.6
Subwatershed isolated from contour-based DEM
of a part of Gunung Ledang region.

belonging to valley connectivity network are embedded between the segments of ridge connectivity networks. In other words, ridge connectivity network acts as a barrier for valley connectivity network.

A part of DEM of Gunung Ledang region (Figure 4.6) prepared by digitizing the elevation contours traced from topographic maps is considered to implement grayscale skeletonization to extract valley connectivity network and ridge connectivity network (Figure 4.7). For a better visual appeal, gray shades are assigned for those regions and valley connectivity network segments embedded between the ridge lines (Figure 4.8).

DEMs generated from remotely sensed satellite data are available at multiple spatial scales. The higher the spatial resolution, the higher is the intricacy of the network extracted, and vice versa. The lower spatial resolution of DEM could be used to extract sparser network. These statements further support the idea of verifying scaling behavior of the networks extracted from multiple spatial scales. To demonstrate what has been described, a DEM (Figure 3.11) is considered and has been smoothened recursively by applying Gaussian Blurring technique (Table 4.1). Those transformed DEMs are shown in Figure 4.9a through e. It is obvious that the spatial resolution of DEM is coarsened. On these five DEMs that are with decreasing spatial resolutions, grayscale skeletonization technique has been applied with respect to a square-shaped symmetric flat structuring element. The resultant valley and ridge connectivity networks from respective DEMs are shown

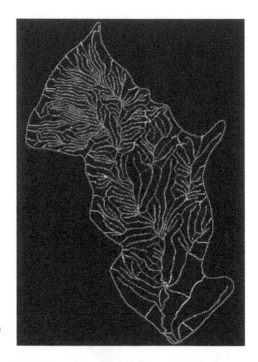

FIGURE 4.7
Automatically extracted channel (branched) and ridge (looplike) networks.

FIGURE 4.8
Automatically generated subwatershed map.

TABLE 4.1

Box-Counting Dimension of Multiscale
Valley Connectivity Networks (VCNs)
and Ridge Connectivity Networks
(RCNs) Extracted from a Sample DEM

Gaussian Blurring Level (in Pixels(s) Radius)	VCNs	RCNs
1	1.7895	1.7819
2	1.7295	1.7320
3	1.6364	1.6520
4	1.5792	1.5653
5	1.5157	1.5152

Source: Sagar, B.S.D. et al., *Int. J. Remote Sens.*, 24(3), 573, 2003.

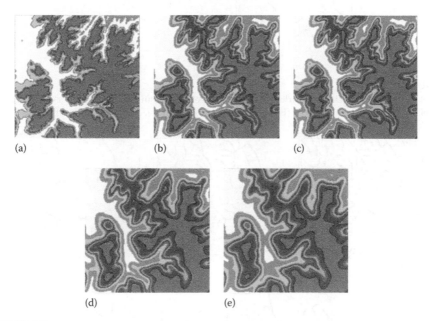

(a)　　　　(b)　　　　(c)

(d)　　　　(e)

FIGURE 4.9
DEM after Gaussian blurring (a) with a 1-pixel radius, (b) with a 2-pixel radius, (c) with a 3-pixel radius, (d) with a 4-pixel radius, and (e) with a 5-pixel radius. (From Sagar, B.S.D. et al., *Int. J. Remote Sens.*, 24(3), 573, 2003.)

in Figures 4.10a through e and 4.11a through e. From these networks, it is apparent that the intricacy of the network is reduced with increasing degree of smoothening (Gaussian blur levels). Such networks obtained from multiscale DEMs would further facilitate other approaches to understand the scale-invariant properties of terrestrial surfaces. This aspect is addressed in Chapter 5.

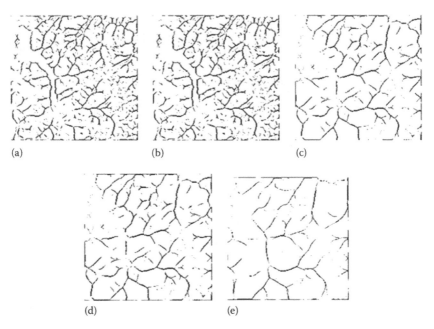

FIGURE 4.10

(a–e) Valley connectivity networks extracted respectively from DEM after Gaussian blurring levels shown in Figure 4.9 (a–e). (From Sagar, B.S.D. et al., *Int. J. Remote Sens.*, 24(3), 573, 2003.)

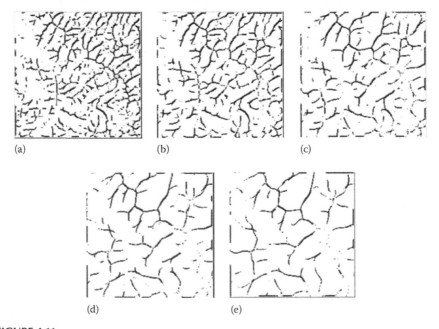

FIGURE 4.11

(a–e) Ridge connectivity networks extracted respectively from DEM after Gaussian blurring levels shown in Figure 4.9 (a–e). (From Sagar, B.S.D. et al., *Int. J. Remote Sens.*, 24(3), 573, 2003.)

This algorithm has also been applied on a simulated DEM (Figure 3.9). The elevation contours and the valley and ridge connectivity networks extracted from this simulated DEM are shown in Figure 4.12a through c. This straightforward approach that does not involve threshold decomposition, benefits from the advantage of crenulations, which are present in each successive erosion frontline of all the spatially distributed elevation regions. This approach may fail to extract precise networks from DEMs of lesser rugged terrains (e.g., tidal regions). Another major limitation of this approach is that sometimes discontinuities in the networks extracted—from DEMs with coarser resolution—are observed. To fix this problem, a combination of one-dimensional (1-D) structuring elements of different directions could be employed instead of a single primitive structuring element considered in this section.

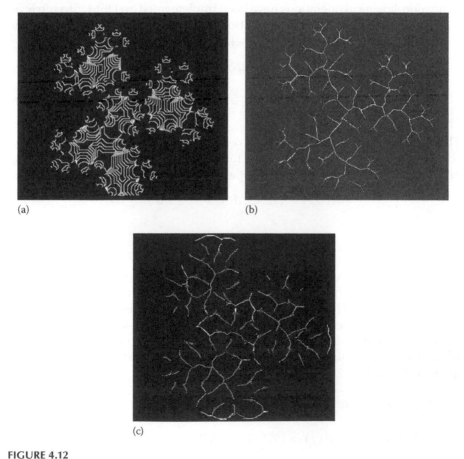

(a)   (b)

(c)

FIGURE 4.12

(a) Elevation contours of a simulated fractal DEM, (b) Valley, and (c) ridge connectivity networks. (From Sagar, B.S.D. et al., *Int. J. Remote Sens.*, 24(3), 573, 2003.)

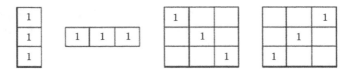

FIGURE 4.13
Line segment structure elements in four directions.

A flat symmetric structuring element (*B*) of primitive size 3 × 3 considered in the extraction of two singular networks explained in the "Ridge and Valley Connectivity Networks via Grayscale Skeletonization" section is of square shape (Figure 2.4a). Such a structuring element could be decomposed into four 1-D structuring elements (Figure 4.13). These four directional structuring elements (*B^i*) are denoted as B^1, B^2, B^3, and B^4 such that $\bigcup_{i=1}^{4} B^i = B$. By treating center element of each of 1-D structuring element as an origin, grayscale morphological operations have been implemented on DEM, *f*, to retrieve connectivity network. Thus, the equations involving (*B^i*) to extract networks from DEM take the form of Equations 4.19 and 4.20:

$$RID_n^i(f) = ((f \ominus nB^i) - (f \ominus nB^i) \circ B^i) \tag{4.19}$$

$$RID(f) = \bigvee_{\substack{n=0 \\ \forall i}}^{N} RID_n^i(f) \tag{4.20}$$

The earlier two steps are performed on inverted-DEM, f^{-1}, to extract valley connectivity network by employing four directional structuring elements, and the corresponding Equations 4.21 and 4.22 are given as follows:

$$VCN_n^i(f) = ((f^{-1} \ominus nB^i) - (f^{-1} \ominus nB^i) \circ B^i) \tag{4.21}$$

$$VCN(f) = \bigvee_{\substack{n=0 \\ \forall i}}^{N} VCN_n^i(f) \tag{4.22}$$

The symbol "∨" in Equations 4.20 and 4.22 denote the supremum of all outputs. The valley connectivity network could be treated as dual of ridge connectivity network. Hence, the dual operators of those employed to extract ridge connectivity network directly from DEM. Morphological dilation and closing are dual operators of morphological erosion and opening, respectively. Hence, the equation to compute valley connectivity networks is also given as Equation 4.23:

$$VCN_n^i(f) = ((f \oplus nB^i) - (f \oplus nB^i) \bullet B^i)$$

$$VCN(f) = \bigvee_{\substack{n=0 \\ \forall i}}^{N} VCN_n^i(f) \Rightarrow VCN^1(f) \vee VCN^2(f) \vee VCN^3(f) \vee VCN^4(f)$$

$$\tag{4.23}$$

Note that the opening and closing operations in Equations 4.19, 4.21, and 4.23 are performed with respect to the structuring element of primitive size and of specific direction (i). Thus, the four directional structuring elements (Figure 4.13) are considered and not for the exotic directions of larger-sized squares. Since we employed grayscale morphological operations and suprema operation to obtain these networks, the gray-level values depicting these networks are multilevel. By using thresholding approach, the extracted networks could be converted into binary form, where all nonzero pixels would be assigned as 1s and 0 pixels remain its value.

This approach—where grayscale morphological operations have been implemented with respect to four directional structuring elements decomposed from a square structuring element—has been implemented on a DEM (Figure 3.12a) and also on multiscale DEMs (Figure 4.14a through c) generated by nonlinear smoothening filtering process. The unique connectivity networks from those multiscale DEMs are extracted by following Equations 4.19 through 4.23. The results are shown in Figure 4.15a through f. Morphological thinning has been applied on the networks obtained to keep the final networks (Figure 4.15a through f) as one-pixel-wide caricatures. It is obvious that the intricacy in these networks changes with increasing degree of smoothening. This technique yields networks with lesser discontinuities compared to that of the technique explained in the previous section. These multiscale networks offer new insights for terrestrial characterization studies. This aspect has been dealt with in Chapter 6.

The morphology-based frameworks and their applications shown in the "Unique Feature Retrieval via Binary Skeletonization" section are useful to extract the networks from both fluvial and tidal systems.

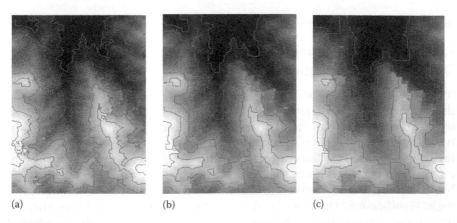

(a) (b) (c)

FIGURE 4.14
Cameron DEM at multiscales. (a) $k = 1$, (b) $k = 5$, and (c) $k = 10$. (From Tay, L.T. et al., *IEEE Geosci. Remote Sens. Lett.*, 2(4), 399, 2005a.)

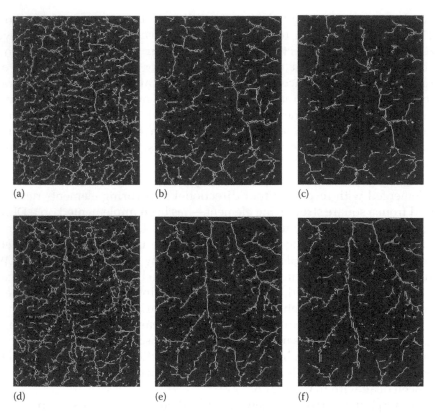

FIGURE 4.15
Multiscale ridge networks, (a) $k = 1$, (b) $k = 5$, and (c) $k = 10$, and multiscale channel networks, (d) $k = 1$, (e) $k = 5$, and (f) $k = 10$, extracted from the corresponding multiscale DEMs shown in Figure 4.14a through c. (From Tay, L.T. et al., *IEEE Geosci. Remote Sens. Lett.*, 2(4), 399, 2005a.)

Retrieval of Physiographic Features from DEMs via Morphological Segmentation

In the "Unique Feature Retrieval via Binary Skeletonization" section, three morphological approaches to extract unique topological networks of significance to characterize further the terrestrial surfaces in quantitative terms have been shown with demonstrations. These networks facilitate ways to employ characterization techniques such as morphometry, allometry, granulometry, and morphological shape decomposition. Besides these source networks, there exist other features of geomorphologic significance. This section deals with extraction of features such as mountains, basins, and piedmont slopes. Toward this, morphology-based segmentation approach has been developed and has demonstrated its utility on DEMs.

Physical features and attributes of terrestrial surface are important to be retrieved. These features provide basic characteristics of land surface. In the classification of landforms of a terrain, the first phase includes proper detection of the physiographic features broadly categorized into three predominant features such as mountains, basins, and piedmont slopes. Until recent past, mapping the features has been done through fieldwork and visual interpretation of topographies, which proved to be tedious.

The following are as per the classic literature:

- Mountains are the portions of terrains that are sufficiently elevated above the surrounding land (between the elevation of 300 and 600 m) and have comparatively steep sides. The two important distinctive parts of a mountain include (i) the highest point, in other words the peak, and (ii) the mountain side, i.e., the part of a mountain between the peak and the foot (Bates and Jackson 1987) (>6° slope).

- Basins are topographic regions from which drainage networks receive runoff through flow and groundwater flow. All the surface land from the highest point of land down to the stream bottom is considered as part of the drainage networks' basin. Basins are generated through the receival of tributaries carried by drainage networks in land slope regions (Monkhouse 1965) (<3° slope).

- Piedmont slopes are the parts of the terrain that are not classified as mountains or basins. Piedmont slopes form either narrow rings surrounding mountain ranges or gently sloping plains in between a basin and a mountain or eroded mountain remnants surrounded by basins (Miliaresis and Argialas 1999) (between 3° and 6° slope).

In what follows in this section, a mathematical morphological approach to segment a given DEM, (f), into these three predominant physiographic features is included.

Figure 4.16a is a flowchart showing all the basic and sequential steps involved in the morphology-based segmentation approach to map three predominant regions from a given DEM. The gradient (slope) values are higher for steep valley sides or cliff sides, and these values are at their minimal at the pits and peaks. Due to these conditions, simple thresholding is not an appropriate choice to map the three predominant features from DEM. Morphology-based algorithm developed and demonstrated on DEM involves the following important steps:

- Ultimate erosion (Figure 4.16b) to extract peaks and pits
- Conditional dilation on the extracted peaks for mountain extraction
- Conditional dilation on the extracted pits for basin extraction

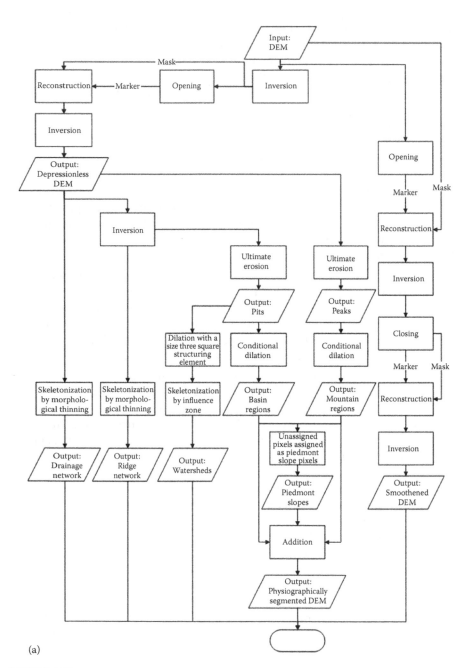

(a)

FIGURE 4.16
(a) Proposed physiographic segmentation algorithm. (From Sathymoorthy, D. et al., *Int. J. Remote Sens.*, 28(15), 3379, 2007.)

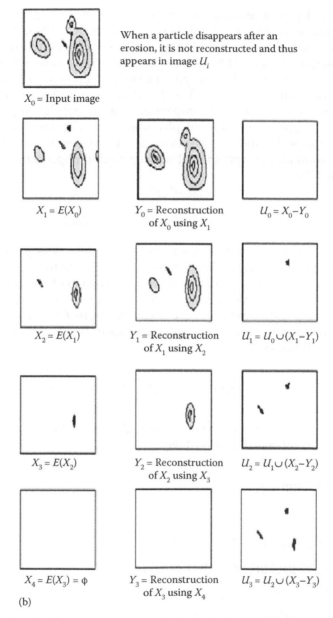

X_0 = Input image

When a particle disappears after an erosion, it is not reconstructed and thus appears in image U_i

$X_1 = E(X_0)$

Y_0 = Reconstruction of X_0 using X_1

$U_0 = X_0 - Y_0$

$X_2 = E(X_1)$

Y_1 = Reconstruction of X_1 using X_2

$U_1 = U_0 \cup (X_1 - Y_1)$

$X_3 = E(X_2)$

Y_2 = Reconstruction of X_2 using X_3

$U_2 = U_1 \cup (X_2 - Y_2)$

$X_4 = E(X_3) = \phi$

Y_3 = Reconstruction of X_3 using X_4

$U_3 = U_2 \cup (X_3 - Y_3)$

(b)

FIGURE 4.16 (continued)
(b) An example of the ultimate erosion operation. Ultimate erosion is implemented through the iterative erosion of the image until all objects vanish (images X_i) and the reconstruction of each eroded image using the eroded image, $E(X_i)$, as the mask and the erosion of smaller size as the marker. The reconstructed images (images Y_i) are subtracted from the corresponding eroded images to form the eroded sets (images U_i). The final resultant image is known as the ultimate erode set. (Reproduced from Duchene, P. and Lewis, D., *Visilog 5 Documentation*, Noesis Vision Inc., Quebec, Canada, 1996. With permission.)

The peaks and pits of terrain, respectively, refer to the highest points of the mountains and the lowest points of the basin. In DEMs, these peaks and pits are respectively surrounded by regions of lower and higher elevations. The segmentation of DEMs into predominant regions should begin with the extraction of peaks and pits. Peaks and pits could be extracted by employing ultimate erosion. The DEM (Figure 3.13) of Great Basin, Nevada, has been considered to implement the sequential steps to partition DEM into three predominant physiographic features. Treating this DEM as f, first peaks and pits are extracted by systematically following ultimate erosion and morphological reconstruction approaches as shown in Equation 4.24:

$$f \ominus B; \quad (f \ominus B) \ominus B; \cdots ; \quad (f \ominus B) \ominus B \ominus \cdots \ominus B \qquad (4.24)$$

In Equation 4.24, a symmetric flat structuring element B of primitive size 3×3 is employed to erode f and is denoted as $(f \ominus B)$. On this eroded image, another erosion B is performed to obtain eroded version of f by two cycles of erosion. This version of eroded image is denoted by $(f \ominus B) \ominus B$, which could be written as $(f \ominus 2B)$ since $(f \ominus B) \ominus B$ is equivalent to $f \ominus (B \oplus B)$. This process is continued until f is eroded completely such that it satisfies the property shown in Equation 4.25:

$$(f \ominus NB) = \varnothing | (f \ominus (N-1)B) \neq \varnothing \qquad (4.25)$$

The sequence of eroded versions of f involved in this ultimate erosion process is shown as follows (Equation 4.26):

$$
\begin{aligned}
&(f \ominus 0B) = f, \\
&(f \ominus B) \\
&(f \ominus B)B = (f \ominus 2B) = f \ominus (B \oplus B) \\
&((f \ominus B) \ominus B) \ominus B = f \ominus (B \oplus B \oplus B) = (f \ominus 3B) \\
&\vdots \\
&((f \ominus B) \ominus B) \ominus \cdots \ominus B = f \ominus (B \oplus B \oplus \cdots \oplus B) = (f \ominus NB)
\end{aligned}
\qquad (4.26)
$$

In Equation 4.26, there is a sequence of $(f \ominus NB)$, where NB is large enough to transform f into empty set (dark region). By taking the morphological reconstruction process—that involves computing point-wise minima between successive eroded versions—a set of other sequential images is obtained as shown in Equations 4.27:

$$(f \ominus 0B) \wedge (f \ominus B) = g_0$$
$$(f \ominus B) \wedge (f \ominus 2B) = g_1$$
$$\vdots \qquad\qquad (4.27)$$
$$(f \ominus (N-2)B) \wedge (f \ominus (N-1)B) = g_{N-2}$$
$$(f \ominus (N-1)B) \wedge (f \ominus BN) = g_{N-1}$$

In Equation 4.27, $g_0, g_1, \cdots, g_{N-2}, g_{N-1}$ represent reconstructed versions. Simple algebraic subtraction of these reconstructed versions from corresponding eroded versions is shown as Equation 4.28:

$$(f \ominus 0B) - g_0 = U_0$$
$$(f \ominus B) - g_1 = U_1$$
$$\vdots \qquad\qquad (4.28)$$
$$(f \ominus (N-2)B) - g_{N-2} = U_{N-2}$$
$$(f \ominus (N-1)B) - g_{N-1} = U_{N-1}$$

Peaks of DEM, f, could be obtained by the sequential steps shown in Equation 4.29:

$$U_1 = U_0 \cup ((f \ominus B) - g_1)$$
$$U_2 = U_1 \cup ((f \ominus 2B) - g_2)$$
$$\vdots \qquad\qquad (4.29)$$
$$U_{N-2} = U_{N-3} \cup ((f \ominus (N-2)B) - g_{N-2})$$
$$U_{N-1} = U_{N-2} \cup ((f \ominus (N-1)B) - g_{N-1})$$

Small peaks in the DEM would disappear by applying grayscale erosion $(f \ominus nB)$. The grayscale erosion removes bright areas (higher elevation regions in DEM). While bright areas are removed, valley regions that are relatively darker regions in DEM get expanded. By repeatedly applying grayscale erosions on the eroded versions of DEM, one can achieve ultimate eroded erosion of DEM, denoted as $(f \ominus NB)$.

Each level of grayscale eroded version of DEM would be used, and point-wise minima between the successive eroded versions would yield images that are referred as morphological reconstructed DEMs. Such a morphological reconstruction isolated certain features within an image. The idea of morphological reconstruction stems from geodesic transformations.

Grayscale morphological reconstruction maintains peak removal effect of erosion by avoiding the valley widening effect. Subtraction of reconstructed eroded image from the original image yields peaks that were removed by grayscale erosion.

The extraction of peaks from a DEM could be done by performing ultimate erosion on DEM and also by performing grayscale morphological reconstruction on each eroded image into the preceding eroded version of DEM (Duchene and Lewis 1996). Final resultant image yields peaks of the DEM. Applying similar process on the inverted DEM yields pits of the DEM. Such extracted peaks and pits act as important markers (seeds) for segmenting the DEM into three predominant physiographic features.

The mean gradient computed for DEM of Great Basin of Nevada is 4.94. The pixels in the gradient range of 0°–57.12° of the DEM are rescaled to an interval of 0–255. As per the gradient details of this DEM, 37.46% (34,248), 21.32% (19,488), and 39.92% (36,491) pixels fall respectively in the gradient categories >6°, 3°–6°, and <3°. Simple thresholding of this grayscale DEM yields segmented details shown in Figure 4.17a. The details in this figure further support the argument that simple thresholding does not yield proper results as it fails to classify the peaks and mountain tops of the DEM as mountain pixels (the pixels in white shade).

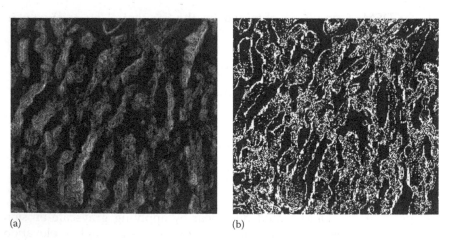

(a)　　　　　　　　　　　　　　　　　　　　(b)

FIGURE 4.17
Gradient analysis of the DEM of Great Basin, Nevada. (a) The pixels of the DEM (in the gradient range of 0°–57°12′) rescaled to the 0–255 interval (the brightest pixel has the highest gradient). (b) Gradient thresholding of the DEM. The pixels in white have a gradient higher than 6°, the pixels in gray have a gradient between 3° and 6°, and the pixels in black have a gradient less than 3°. (From Sathymoorthy, D. et al., *Int. J. Remote Sens.*, 28(15), 3379, 2007.)

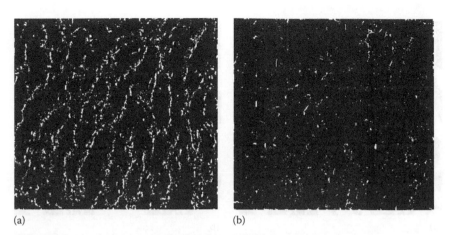

(a)   (b)

FIGURE 4.18

Extraction of (a) peaks and (b) pits from the DEM of Great Basin, Nevada. (From Sathymoorthy, D. et al., *Int. J. Remote Sens.*, 28(15), 3379, 2007.)

Peaks and pits of this DEM are extracted by performing ultimate erosion transformation combined with grayscale morphological reconstruction respectively on DEM and on inverted DEM. The numbers of peaks and pits extracted respectively from DEM and inverted DEM are 1315 (Figure 4.18a) and 559 (Figure 4.18b). The peaks (Figure 4.18a) and pits (Figure 4.18b) extracted from the DEM are required as markers (seeds) to map mountains and basin regions, respectively.

Mountain Extraction

Conditional Dilation of the Peaks of the DEM

Conditional dilation is performed on the peaks (Figure 4.18a) extracted from the DEM by means of a symmetric flat square structuring element of size 3 × 3. This conditional dilation was recursively applied until no further changes are produced (Figure 4.19a). After each cycle of conditional dilation, the boundary pixels of the dilated peaks that have gradient lesser than 6° are deleted. The foreground (in white pixels) and background (in black pixels) in the image produced from this step (Figure 4.19a), respectively, represent mountain and non-mountain pixels. However, some small islands of non-mountain pixels surrounded by mountain pixels are observed on mountain tops. These islands have been emerged due to spurious peaks.

Removal of Small Islands of Mountain Pixels Observed in Flat Areas

Like the small islands of non-mountain pixels surrounded by mountain pixels observed, small islands of mountain pixels surrounded by

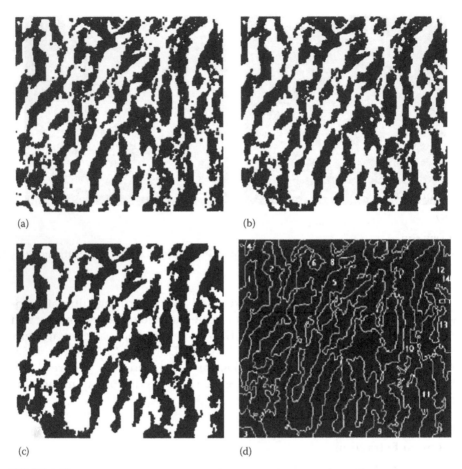

(a)   (b)

(c)   (d)

FIGURE 4.19
Mountain extraction. (a) The mountain pixels (the pixels in white) of the DEM. The black pixels are non-mountain pixels. (b) The mountain pixels after the removal of erroneous non-mountain regions enclosed by mountain pixels. (c) The mountain pixels after removal of erroneous mountain pixels. (d) The identification of the individual mountain objects. (From Sathymoorthy, D. et al., *Int. J. Remote Sens.*, 28(15), 3379, 2007.)

non-mountain pixels are also observed. These small islands of non-mountain pixels or mountain pixels appeared due to noise in the DEM. The noise in the DEM yields spurious peaks and pits along with genuine peaks and pits. These spurious peaks and pits cannot form larger mountain regions. These small islands of mountain regions with sizes lesser than 180 pixels are converted into non-mountain pixels (Figure 4.19b). Similarly, the mountain pixels after the removal of erroneous mountain pixels are shown in Figure 4.19c. Fourteen distinct mountain objects could be extracted and labeled (Figure 4.19d). Mountain regions have occupied a total of 42,168 (46.13%) pixels. Based on the characteristics such as size

TABLE 4.2

Numerical Description of the Mountain Objects Extracted from the DEM of Great Basin

Object ID	Area (Pixels)	Perimeter Length (Pixels)	Maximum Elevation (Grayscale Value)	Mean Gradient (°)
1	1,249	202	232	11.80
2	5,653	1118	255	11.44
3	682	249	227	9.64
4	1,274	106	223	8.40
5	274	68	229	10.26
6	3,387	762	248	10.16
7	201	80	185	6.47
8	12,326	2693	216	10.06
9	430	111	216	10.92
10	6,336	1020	184	9.61
11	299	160	216	10.96
12	1,663	417	179	11.14
13	1,044	319	198	8.28
14	82	57	174	6.24
15	62	43	186	7.01
16	76	38	201	7.52
17	3,596	814	186	12.65
18	2,816	485	210	9.83
19	507	116	156	7.16
20	211	68	170	8.54

Source: Sathymoorthy, D. et al., *Int. J. Remote Sens.*, 28(15), 3379, 2007.

perimeter, maximum elevation, and mean gradient, each mountain object extracted is described (Table 4.2).

Basin Extraction

Once pits are extracted by performing ultimate erosion combined with systematic grayscale morphological reconstruction on inverted DEM, from DEM, the basin could be mapped according to the following three steps.

Conditional Dilation of Pits of the DEM

Pits will be dilated by means of square-shaped symmetric structuring element of size 3 × 3. The boundary pixels of the dilated pits that have gradient higher than >3° would be deleted. Until no further changes are produced, the conditional dilation of the pits would be repeated. The foreground pixels and the background pixels in the image produced from this step, respectively, are basin and non-basin pixels.

Removal of Small Islands of Non-Basin Pixels Enclosed within Basin Regions

Those pixels that are not classified as pits, and those pixels, in the image produced out of the previous step, that are flagged as basin pixels due to their gradient being more than 3° would be removed by assigning them as basin pixels.

Removal of Small Islands of Basin Pixels Observed in Non-Basin Areas

Due to spurious pits, erroneous basin regions would be formed. Spurious pits do not form larger basin regions, as there are large gradient values in their surroundings. Such erroneous basin pixels would be removed by converting them into non-basin pixels. If the DEM is completely free from noise such that there are no spurious pits and peaks, then the removal of small islands will not come into picture.

The pits (Figure 4.18b) extracted from DEM (Figure 3.13) are conditionally dilated until convergence (Figure 4.20a). The small islands of non-basin pixels surrounded by basin pixels are assigned as basin pixels (Figure 4.20b). Those basin regions with size lesser than 180 pixels are selected and removed by converting them into non-basin pixels (Figure 4.20c). Classified basin pixels are found to be about 36,642 (40.21%) pixels.

Extraction of Piedmont Slopes

Those pixels that are neither classified as mountain pixels nor as basin pixels would be classified as piedmont slope pixels. Mountains (Figure 4.19) and basins (Figure 4.20) segmented from the DEM (Figure 3.13) are shown. Those pixels that are neither classified as mountains nor as basins are isolated (Figure 4.21). Those isolated non-mountain and non-basin pixels are nothing but piedmont slope pixels. Piedmont slopes occupy 11,037 (12.11%) pixels in the DEM considered.

The combination of the mountains (Figure 4.19), basins (Figure 4.20), and piedmont regions (Figure 4.21) extracted from the DEM form the physiographically segmented DEM. These results are compared with results obtained via seed-ridge and seed-valley growing approaches, proposed by Miliaresis and Argialas (1999), to map mountains, basins, and piedmont regions. The seed-ridge and seed-valley approaches are extracted by the following runoff simulation process. These seed-ridge and seed-valley are like peaks and pits that are extracted via ultimate erosion and grayscale morphological reconstructions. Seed-growing is similar to conditional dilation employed in our approach. The results obtained via these two approaches are interesting, and a few comparisons have been made later. Figure 4.22a and b, respectively, shows physiographic segmentations of DEM obtained by performing morphology-based algorithm and seed-growing approach. The application of the latter approach results in 40,419 (43.50%) pixels being classified as mountain pixels, whereas basin

FIGURE 4.20

Basin extraction. (a) The basin pixels (the pixels in white) of the DEM. The black pixels are non-basin pixels. (b) The basin pixels after the removal of erroneous non-basin regions enclosed by basin pixels. (c) The basin pixels after the removal of erroneous basin pixels. (From Sathymoorthy, D. et al., *Int. J. Remote Sens.*, 28(15), 3379, 2007.)

and piedmont regions, respectively, have occupied 26,835 (29%) and 25,574 (27.51%) pixels. Thirty-six distinct mountain regions have been observed. It is observed that mountain objects in Figure 4.22b are narrower than those observed in Figure 4.22a. A number of mountain objects seen as broken objects (Figure 4.22b) appeared as clustered mountain objects (Figure 4.22a). This difference is due to the fact that seed-ridge pixel image contains a number of peaks extracted via morphology-based algorithm. Region-growing approach is unable to retrieve all the mountain regions of the DEM, in particular mountaintop regions. Due to the reason that seed-valley pixel image does not contain a number of pits extracted by morphology-based algorithm, the basin regions in Figure 4.22b are smaller

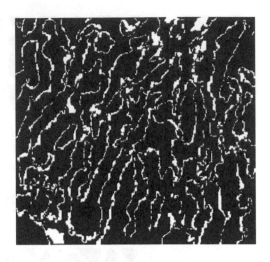

FIGURE 4.21
Piedmont slope regions (the pixels in white). (From Sathymoorthy, D. et al., *Int. J. Remote Sens.*, 28(15), 3379, 2007.)

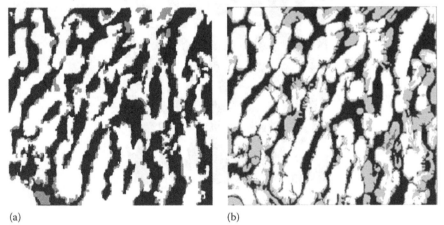

(a)   (b)

FIGURE 4.22
Mountain pixels are the pixels in white, the piedmont pixels are the pixels in gray, and the basin pixels are the pixels in black. (a) The results obtained using the developed algorithm. (b) The results obtained in Miliaresis and Argialas (1999). (From Sathymoorthy, D. et al., *Int. J. Remote Sens.*, 28(15), 3379, 2007.)

than those of Figure 4.22a. In view of this, region-growing approach could not extract all the basin regions of the DEM.

Runoff simulation approach may not extract seed-ridge and seed-valley pixels, which are important in the physiographic segmentation of DEM, from relatively flat regions of DEM. This limitation results in errors in seed-ridge and seed-valley, further causing errors in extracted mountain and basin objects. However, morphology-based algorithm has the potential to efficiently operate on even flat regions of DEM.

To summarize this section on physiographic segmentation of DEMs into mountains, basins, and piedmont regions, ultimate erosions is used as a first step followed by grayscale morphological reconstruction essentially to extract peaks and pits from the DEM. Further the peaks and pits are conditionally dilated to obtain mountains and basins. Those pixels that do not belong to either mountains or basins are classified as piedmont regions.

Extraction of Morphologically Significant Zones

Many terrestrial phenomena are available as fields across spatial scales and temporal scales. Now remote sensing satellites sense the terrestrial surfaces via various sensing mechanisms and acquire terrestrial data in frequent intervals. Such data provide information in terms of fields such as elevation fields, soil moisture fields, rainfall field, temperature field, etc. Characterizations of such fields are important from the point of understanding the structure and process relationships.

Various themes representing natural features such as lakes, water bodies, threshold elevation regions, rainfall spread, temperature spread, and porous medium of rocks are sets that show theme and no-theme regions, in other words set and set complement. Such themes are in binary forms. Spatial fields such as rainfall, landscape, temperatures, clouds, vegetation, and elevations are spatially heterogeneous, to varied degrees, in their spatial and/ or temporal organizations. These spatial fields possess varied degrees of spatiotemporal complexities. Each spatial field represents a phenomenon possessing varied spatial complexities from one locale to another locale within a spatial field. Such spatial fields can be decomposed into threshold sets. Decomposing either a spatial field or a threshold set decomposed from a spatial field into morphologically significant regions is an important study. The morphologically significant regions possess varied degrees of spatial complexities. This section deals with the following aspects:

1. Decomposition of a binary fractal into morphologically significant regions
2. Segmentation of a cloud field into morphologically significant regions

These set-like fractal and spatial field–like clouds are the source data to explain the application of multiscale binary and grayscale morphological opening transformations.

Decomposition of Morphologically Significant Zones from a Binary Fractal

Box counting method (Feder 1988) and cube counting method (Douketis et al. 1995, Robertson et al. 1995, Zahn and Zösch 1999) are elegant ways for computing fractal dimensions of fractal sets and fractal functions.

Based on mathematical morphological transformations, attempts have been made to compute fractal dimensions of spatial objects and spatial fields. These morphology-based approaches include boundary dilation method (Flook 1978) and Minkowski–Bouligand dimension (Schroeder 1991). Generalized dimension computation proposed by Halsey et al. (1986) has also been proved powerful in computing spatial complexities of sets and spatial fields. Sagar (1996) proposed an approach that requires skeletal network of spatial set to establish a relationship between the ratio of the number of skeletal segments of successive orders and the ratio of mean lengths of skeletal segments of successive orders. This approach has been demonstrated on a fractal Koch quadric shape, and the full-length details are given in Chapter 5. Another approach based on morphological shape decomposition computes a number–radius relation (Lian et al. 2004, Radhakrishnan et al. 2004, Sagar and Chockalingam 2004, Chockalingam and Sagar 2005). Multiscale networks' length versus scale (radius of structuring element) provides a graphical relationship and a fractal dimension–like quantity (Sagar et al. 2003, Tay et al. 2005). Multiscale convexity analysis has been employed to segment cloud fields (Lim and Sagar 2008). Allometry-based analysis also provides a host of power-law relationships (Sagar 2007).

A Koch quadric binary fractal object (Figure 3.5a) is considered to demonstrate multiscale convexity analysis-based segmentation. This segmentation yields morphologically significant regions (zones) of fractal. This approach involves the following five steps:

1. Generation of multiscale fractal
2. Construction of convex hulls of multiscale fractals
3. Estimation of convexity measures of multiscale fractals
4. Derivation of morphologically significant threshold scales and convexity measures
5. Partition of fractal into morphologically significant regions

Multiscale opening has been performed on fractal until Nth level such that the opening of fractal by Nth size B yields an empty set as shown in Equation 4.30. Let X be a Koch quadric fractal, and B be a structuring element of primitive size 3×3:

$$(X \ominus B) \oplus B = X \circ B$$

$$((X \ominus B) \ominus B) \oplus B \oplus B = (X \ominus (B \oplus B)) \oplus (B \oplus B) = (X \ominus 2B) \oplus 2B = X \circ 2B$$

$$\vdots \tag{4.30}$$

$$X \circ (N-1)B \neq \varnothing$$

$$X \circ NB = \varnothing$$

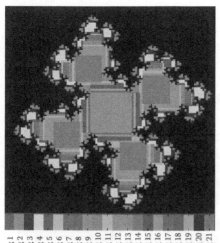

Opening 1 Opening 2 Opening 3 Opening 4 Opening 5 Opening 6 Opening 7 Opening 8 Opening 9 Opening 10 Opening 11 Opening 12 Opening 13 Opening 14 Opening 15 Opening 16 Opening 17 Opening 18 Opening 19 Opening 20 Opening 21

FIGURE 4.23
Opened fractal at multiscales. (From Lim, S.L. et al., *Chaos Solitan. Fract.*, 41(3), 1253, 2009.)

The Nth size B that made X, in the fractal case, empty set is 22 after opening shown as $(X \circ 22B) = \varnothing$. The outputs obtained after each cycle of opening (up to 21 cycles of opening) are gray shaded, and those gray-shaded opened versions of fractal are superposed (Figure 4.23). With increasing cycle of opening, the area of fractal gets reduced, and application of multiscale opening follows the property as Equation 4.31:

$$(X \circ NB) \subseteq (X \circ (N-1)B) \subseteq \cdots \subseteq (X \circ 2B) \subseteq (X \circ B) \subseteq X \qquad (4.31)$$

Fractal after each cycle of opening is considered, and convex hull is constructed according to the approach described in the following section.

Binary Convex Hull Construction

Convex hull construction is done according to half-plane-based closing, which is due to Soille (1998). A set X is convex if and only if the line segments joining any two pair of points lie entirely within the set. The fractal set (Figure 3.5a) considered here is not a convex set. Convex hull, $CH(X)$, is defined as the smallest convex polygon containing all points x in the set X. Convex hulls can be constructed for points spread randomly over a geographical space, for spatial objects (Figure 3.5a) and for spatial fields (Figure 3.21a). An approach that was followed to construct convex hull is based on half-plane closings (Soille 1998). This approach generates convex hull of a set (Figure 4.24a) by intersecting all half-planes encompassing the set. Half-planes are denoted by π_θ^+ and π_θ^- for a given angle θ. For every angle θ, there would be two half-planes (e.g., left–right, right–left, top–bottom, bottom–top, and so on). Convex hull of a set would be produced by intersecting the half-plane closings for all possible orientations and is mathematically denoted as Equation 4.32:

$$CH(X) = \cap \left(\phi_{\pi_\theta^+}(X) \cap \phi_{\pi_\theta^-}(X) \right) \tag{4.32}$$

where π_θ^+ and π_θ^-, respectively, denote closings of half-planes at π_θ^+ and π_θ^- orientations (π_θ^- is the complement of the half-plane with orientation π_θ^+).

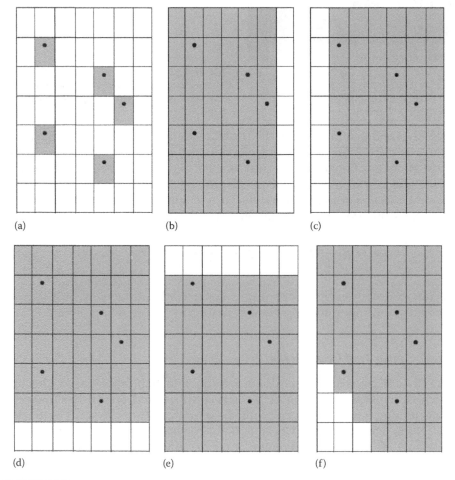

(a)   (b)   (c)

(d)   (e)   (f)

FIGURE 4.24
Example showing the steps resulting in the convex hull of a set that consists of five isolated points. Convex hull construction of binary point data via half-plane closings: (a) a set that consists of five points, (b) closing of X by the right-vertical half-plane, (c) closing of X by the left-vertical half-plane, (d) closing of X by the lower-horizontal half-plane, (e) closing of X by the upper-horizontal half-plane, (f) closing of X by $3\pi/4$ left half-plane.

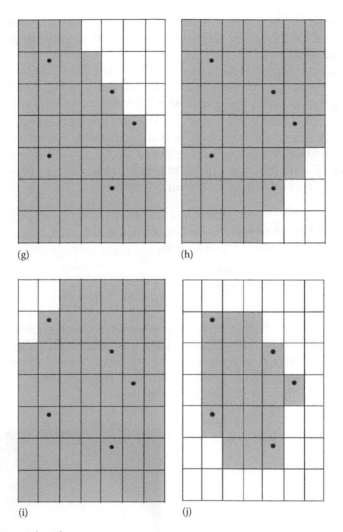

FIGURE 4.24 (continued)
Example showing the steps resulting in the convex hull of a set that consists of five isolated points. Convex hull construction of binary point data via half-plane closings: (g) closing of X by 3π/4 right half-plane, (h) closing of X by π/4 right half-plane, (i) closing of X by π/4 left half-plane, and (j) intersection of closings (b) through (i). (From Lim, S.L. et al., *Chaos Solitan. Fract.*, 41(3), 1253, 2009.)

In the example, convex hull construction for set X representing five points is shown (Figure 4.24a). Closing of this set X by respective half-planes is shown with shades (Figure 4.24b through i). Eight half-plane closed versions shown are respectively obtained by right-vertical, left-vertical, lower-horizontal, upper-horizontal, 3(π/4) left, 3(π/4) right, (π/4) right, and (π/4) left. Finally, these eight half-plane closed versions (Figure 4.24b through i) are intersected to obtain the convex hull (Figure 4.24j). It is obvious that this convex hull is the closed set that encloses the five points in the original set X (Figure 4.24a).

This binary convex hull construction approach is employed to construct convex hulls for the 21 opened versions of fractal set shown in Figure 4.23. The corresponding convex hulls are also gray shaded and superposed on one another (Figure 4.25). $A[\cdot]$ is cardinality of set $[\cdot]$. The areas of each opened version of the set and its corresponding convex hull are computed. The areas of $(X \circ nB)$ and $CH(X \circ nB)$ are denoted by $A(X \circ nB)$ and $A(CH(X \circ nB))$. It is worth mentioning that $(X \circ nB) \subseteq (X \circ (n-1)B)$, and hence $CH(X \circ nB) \subseteq CH(X \circ (n-1)B)$. In turn, their areas satisfy the following properties:

1. $A(X \circ (n+1)B) \le A(X \circ nB)$
2. $A(CH(X \circ (n+1)B)) \le A(CH(X \circ nB))$
3. $(X \circ nB) \subseteq CH(X \circ nB)$ and
4. $A(X \circ nB) \le A(CH(X \circ nB))$

For the fractal set X (Figure 3.5a), 21 opened versions (Figure 4.23) and 21 corresponding convex hulls (Figure 4.25) are generated. Graphical plots between the nth size B and the areas of nth-level opened version and its convex hull are shown (Figure 4.26a). It is obvious from this graph (Figure 4.26a) that the areas decrease with increasing size (n) of B.

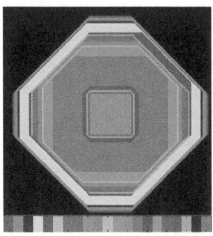

FIGURE 4.25
Convex hull of opened fractal at multiscales.
(From Lim, S.L. et al., *Chaos Solitan. Fract.*, 41(3), 1253, 2009.)

Convex hull 1
Convex hull 2
Convex hull 3
Convex hull 4
Convex hull 5
Convex hull 6
Convex hull 7
Convex hull 8
Convex hull 9
Convex hull 10
Convex hull 11
Convex hull 12
Convex hull 13
Convex hull 14
Convex hull 15
Convex hull 16
Convex hull 17
Convex hull 18
Convex hull 19
Convex hull 20
Convex hull 21

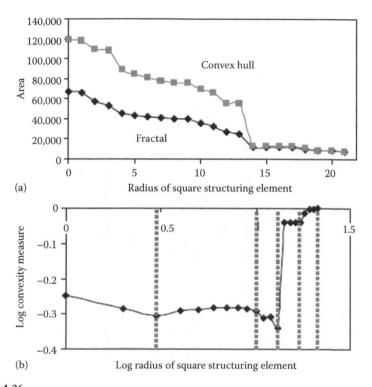

(a)

(b)

FIGURE 4.26

(a) Area of fractal and its convex hull, at increasing size of square structuring element, and (b) convexity measure at increasing size of square structuring element in logarithmic representation. (From Lim, S.L. et al., *Chaos Solitan. Fract.*, 41(3), 1253, 2009.)

Convexity measures for all 21 opened versions of fractal sets are computed according to Equation 4.33:

$$CM(X \circ nB) = \frac{A(X \circ nB)}{A(CH(X \circ nB))}$$

(4.33)

This convexity measure ranges between 0 and 1, since $A(X \circ nB) \leq A(CH(X \circ nB))$. This convexity measure would be 1 if and only if $(X \circ nB) \equiv CH(X \circ nB)$. The convexity measures of multiscale fractals, generated through multiscale morphological opening transformation, are plotted as functions of the scale, i.e., nB (Figure 4.26b). This graphical relationship provides a basis to determine crossover scales further to determine the transition zones between the morphological phases. These transition levels are in fact the threshold levels of opening cycle number at which the morphological constitution of fractal shape shows significant (sudden) change (Table 4.3). These threshold

TABLE 4.3

Complexity Measures of Morphologically Significant Zones Decomposed from Various Fractal Shapes (Islands)

Fractal Type	Zone 1		Zone 2		Zone 3		Zone 4		Zone 5		Zone 6		
	CM	NCM	CM	NCM	CM	NCM	CM	NCM	CM	NCM	CM	NCM	FD
Quadric	1.05	0.35	2.21	0.31	1.47	0.49	0	0	1.52	0.38	0	0	1.74
Random quadric	1.92	0.48	1.53	0.51	1.78	0.36	1.10	0.27	0.93	0.47	0	0	1.69
Triadic	1.81	0.36	2.42	0.19	0.81	0.27	2.00	0.29	1.65	0.06	0	0	1.8
Random triadic	2.20	0.44	1.88	0.47	2.34	0.29	2.16	0.27	1.75	0.19	0	0	1.76

Source: Lim, S.L. et al., *Chaos Solitan. Fract.*, 41(3), 1253, 2009.

CM, complexity measure; NCM, normalized complexity measure; FD, fractal dimension computed via box counting method.

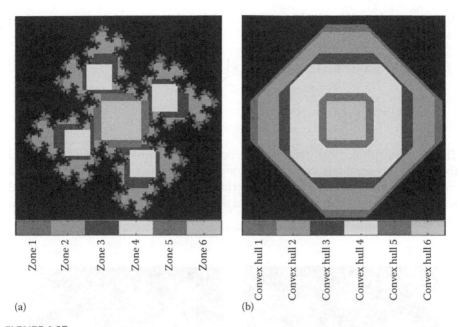

FIGURE 4.27
(a) Opened fractal at crossover scales and (b) convex hulls of crossover scale opened fractals. (From Lim, S.L. et al., *Chaos Solitan. Fract.*, 41(3), 1253, 2009.)

opening levels, for the fractal evolved through multiscale opening transformation, are opening cycles of 0, 3, 10, 13, 17, and 21 (Figure 4.27a). Figure 4.27a shows that fractal at these threshold opening levels are gray shaded and are superposed. Corresponding gray-shaded convex hulls are also superposed (Figure 4.27b). The evolving fractal set at these threshold opening levels are denoted as $(X \circ 0B)$, $(X \circ 3B)$, $(X \circ 10B)$, $(X \circ 13B)$, $(X \circ 17B)$, and $(X \circ 21B)$. Morphologically significant zones (X_i) are extracted by simple logical subtraction as shown in Equation 4.34:

$$X_1 = (X \circ 0B) \backslash (X \circ 3B)$$

$$X_2 = (X \circ 3B) \backslash (X \circ 10B)$$

$$X_3 = (X \circ 10B) \backslash (X \circ 13B)$$

$$X_4 = (X \circ 13B) \backslash (X \circ 17B) \qquad (4.34)$$

$$X_5 = (X \circ 17B) \backslash (X \circ 21B)$$

$$X_6 = (X \circ 21B) \backslash (X \circ 22B)$$

The morphologically significant zones isolated according to Equation 4.34 satisfy the following properties:

1. $X = X_1 \cup X_2 \cup X_3 \cup X_4 \cup X_5 \cup X_6 = \bigcup_{i=1}^{6} X_i$

2. $A(X) = A\left(\bigcup_{i=1}^{6} X_i\right)$

These six morphologically significant zones isolated are shown in Figure 4.28a through f. In a similar fashion, morphologically significant zones are extracted from third-order deterministic Koch triadic and random Koch triadic fractal sets, and random Koch quadric fractal sets (Peitgen et al. 2004) (Figure 4.29a through c).

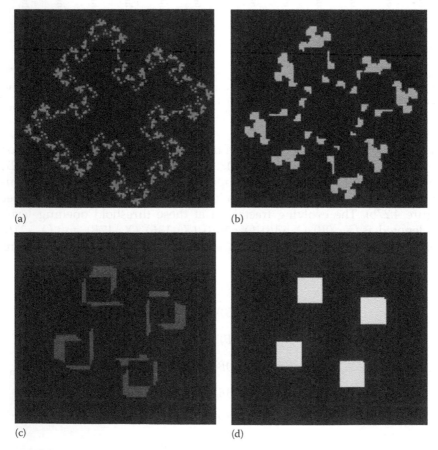

(a)   (b)

(c)   (d)

FIGURE 4.28
Zones segmented from quadric fractal object. (a) X_1, (b) X_2, (c) X_3, (d) X_4.

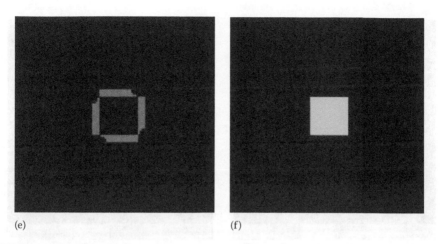

(e)　　　　　　　　　　　　　(f)

FIGURE 4.28 (continued)
Zones segmented from quadric fractal object. (e) X_5, and (f) X_6. (From Lim, S.L. et al., *Chaos Solitan. Fract.*, 41(3), 1253, 2009.)

How a given set (thematic map) could be converted to morphologically significant zones is shown. Multiscale morphological binary opening transformation, half-plane closing to construct convex hulls, and convexity measures have been collectively employed to achieve the objective. Morphological phases, in terms of crossover scales obtained from a relationship between scale and convexity measure, are taken as the bases to partition binary objects.

How this entire approach can be extended to partition spatial fields that involve multivalues? One crude approach is by converting spatial fields into threshold sets. But in the section that follows, multiscale grayscale morphological opening transformation, grayscale convex hull construction, and convexity measures have been employed to segment spatial fields (e.g., cloud field) into different regions of prominence.

Cloud Field Segmentation via Multiscale Convexity Analysis

Besides many, cloud is one of the good examples of spatial field. MODIS (moderate resolution imaging spectroradiometer) provides cloud field data in very short time intervals. Such a cloud field data resembles function $f(x, y)$ depicting spectral values at each spatial position (x, y). A cloud field can be segmented through simple thresholding technique. But the results derived via thresholding technique may not provide any structurally significant details of cloud. To decompose cloud fields, $f(x, y)$, a multiscale convexity analysis-based approach has been proposed and has been demonstrated on cloud fields (Figure 3.21a and c) isolated from MODIS data. This approach involves the following:

1. Generation of cloud field at multiple coarser scales
2. Construction of grayscale convex hulls of multiscale cloud fields
3. Computation of convexity measures across multiscales

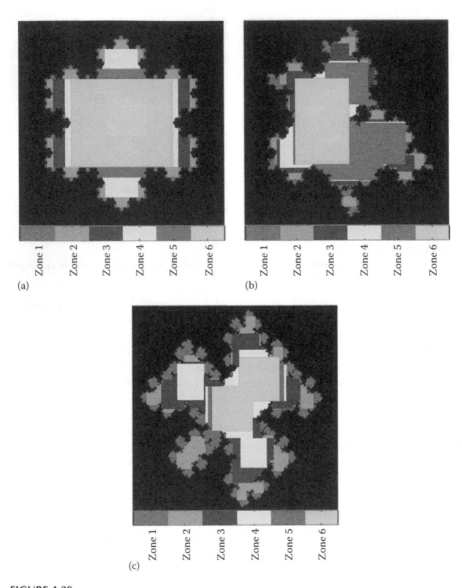

FIGURE 4.29
Morphologically significant zones decomposed from (a) Koch triadic fractal island, (b) random
Koch triadic fractal island, and (c) random Koch quadric fractal island. (From Lim, S.L. et al.,
Chaos Solitan. Fract., 41(3), 1253, 2009.)

Clouds that exist in various shapes and sizes are formed through condensation
and deposition of fine water droplets and ice crystals. With the advent of satellite remote sensing and computer-assisted mapping techniques, understanding spatiotemporal characteristics of cloud fields has been greatly enhanced.
Many researchers provided elegant approaches to derive macroscale and

microscale atmospheric fields such as cloud top pressure, aerosol concentration, and cloud particle effective radius from remotely sensed satellite data (Ackerman et al. 1998, and references therein). MODIS channels provide different data of land, sea, and atmosphere fields. Through seminal studies, characterization of these fields has received wide attention (Inoue 1987, Rossow 1989, Gao and Goetz 1991, King et al. 1992, 1996, Frey et al. 1995, Hutchison and Hardy 1995, Ackerman 1997). Through spatial variability tests, several researchers have addressed the topic of retrieval of significant zones from cloud fields possessing both naturally and anthropogenically generated aerosols with varied concentrations and from cloud particle effective radius maps.

Geophysical fields are spatially heterogeneous to varied degrees. These fields, to name a few, include landscapes, rainfall fields, cloud fields, and fields depicting various macroscale atmospheric fields. Many fractal-based characteristics provide scale-invariant measures to characterize these geophysical fields (Mandelbrot 1982).

A cloud possesses surface, the structure of which is highly time dependent. Characteristics of cloud fields could be better understood by segmenting the cloud fields. Cloud fraction, cloud top pressure, cloud optical depth, column water vapor, and cloud particle effective radius are some of the macroscale atmospheric fields. Spatial patterns of these fields, as observed from MODIS data, are analyzed by Mote and Frey (2006).

Satellite cloud scenes are classified into distinct regions via K-means clustering algorithm (Gordon et al. 2005). The characteristics of radiation transport in inhomogeneous clouds are studied using three-dimensional (3-D) simulations of radioactive transport and the independent pixel approximation (Zinner et al. 2006).

Cloud fields isolated from satellite data possess brightness values that are distributed heterogeneously. This heterogeneity is due to the presence of cloud ice, cloud water, and aerosols. Simple thresholding technique can be employed to segment cloud fields if the brightness values are distributed homogeneously across all the spatial coordinates. The brightness values are heterogeneously distributed in all realistic clouds. Shape-based segmentation procedure is an appropriate one to segment cloud field into morphologically significant zones. The segmentation approach based on multiscale convexity analysis to partition cloud field into morphologically significant regions is explained in the "Computation of Convexity Measure for Spatial Fields" section.

Generation of Cloud Field at Multiple Coarser Spatial Scales

Multiscale grayscale morphological opening transformation has been applied on two cloud fields (Figure 3.21a and c) to generate coarsened version of the cloud fields. A cloud field is an aggregation of various sub-images (cloud subfields). The increasing degree of grayscale opening filters out the subfield of increasing sizes.

Let f be a cloud field. The application of multiscale opening on the cloud fields is according to Equation 4.35:

$$(f \ominus nB) \oplus nB = f \circ nB, \quad \text{where } n = 0, 1, 2, \cdots, N \qquad (4.35)$$

With increasing n, the cloud field's spatial resolution will be reduced. The cloud field gets flattened, mimicking the generation of cloud field at coarser resolutions, under the influence of increasing degree of opening. Selected levels of cloud function transformed via multiscale opening are shown in Figure 4.30. These selected levels are taken from the total 100 opened cloud images using multiscale opening transformation according to Equation 4.35. The n values, ranging from 1 to 100, have been employed in generating 100 levels of opened versions, and square type of B, that is flat, symmetric, and primitive size of 3×3, is employed. The area of cloud field f is computed according to Equation 4.36:

$$A(f) = \sum_{x,y} f(x, y) \qquad (4.36)$$

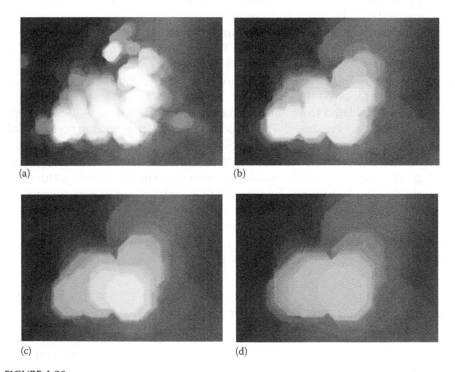

(a) (b)

(c) (d)

FIGURE 4.30
(a–d) 25 Cycles, 50 cycles, 75 cycles, and 100 cycles of opened versions of cloud function shown in Figure 3.21a.

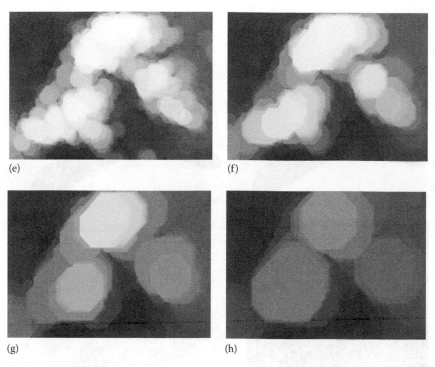

(e)

(f)

(g)

(h)

FIGURE 4.30 (continued)
(e–h) 25 cycles, 50 cycles, 75 cycles, and 100 cycles of opened versions of cloud function shown in Figure 3.21c. (From Lim, S.L. and Sagar, B.S.D., *J. Geophys. Res.*, 113, D13208, 2008, doi:10.1029/2007JD009369.)

The area of $(f \ominus nB)$ is greater than $(f \ominus (n+1)B)$. The increasing levels of opened versions of cloud fields satisfy the following property:

$$A(f \circ NB) \leq A(f \circ (N-1)B) \leq \cdots \leq A(f \circ 2B) \leq A(f \circ B) \leq A(f)$$

Construction of Grayscale Convex Hull

Computation of convex hull of a function, a multivalued field, could be done via half-plane closings. Typically, the convex hull of a spatial object, in other words a set (Figure 4.31a), looks like Figure 4.31b, whereas the convex hull of a synthetic cloud field (Figure 4.31c) is as shown in Figure 4.31e. The 3-D views of Figure 4.31c and e are respectively shown in Figure 4.31d and f. The mathematical explanation of convex hull construction of a multivalued field (Figure 4.32a) is given later. Convex hull construction of a function, f (e.g., Figure 4.32a), requires generation of half-plane closings. With half-plane closing, holes in the function would be filled, small breaches would be connected, and overall spatially complex field would be converted into rather smooth field.

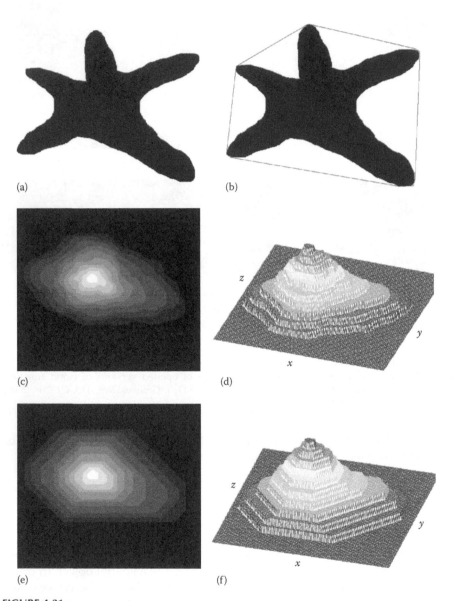

FIGURE 4.31

(a) Threshold set decomposed from a synthetic cloud function, (b) convex hull of a threshold set shown in (a), (c) a synthetic cloud function consists of 10 gray levels—which can be decomposed maximum into 10 threshold sets, (d) 3-D representation of synthetic cloud function—shown in (c)—x, y depict spatial coordinates and z represents corresponding gray levels at respective x, y spatial coordinates, (e) convex hull of synthetic cloud function shown in (c), and (f) 3-D representation of convex hull shown in (e). (From Lim, S.L. and Sagar, B.S.D., *J. Geophys. Res.*, 113, D13208, 2008, doi:10.1029/2007JD009369.)

Half-plane closing of subset of *f*

19	25	21	30	25
14	17	16	222	20
8	12	240	254	208
9	209	250	255	254
15	208	240	253	252

⟶ Direction of translation

(a)

Previous value = 0 (init)
Maximum along line = 19
Current value = max (0,19)

19	25	21	30	25
14	17	16	222	20
8	12	240	254	208
9	209	250	255	254
15	208	240	253	252

(b)

Previous value = 19
Maximum along line = 209
Current value = max (19,209)

19	25	21	30	25
19	17	16	222	20
19	12	240	254	208
19	209	250	255	254
19	208	240	253	252

(c) Second translation

Previous value = 209
Maximum along line = 250
Current value = max (209,250)

19	209	21	30	25
19	209	16	222	20
19	209	240	254	208
19	209	250	255	254
19	209	240	253	252

(d) Third translation

Previous value = 250
Maximum along line = 255
Current value = max (250,255)

19	209	250	30	25
19	209	250	222	20
19	209	250	254	208
19	209	250	255	254
19	209	250	253	252

(e) Fourth translation

Previous value = 255
Maximum along line = 254
Current value = max (255,254)

19	209	250	255	25
19	209	250	255	20
19	209	250	255	208
19	209	250	255	254
19	209	250	255	252

(f) Fifth translation

Final result

19	209	250	255	255
19	209	250	255	255
19	209	250	255	255
19	209	250	255	255
19	209	250	255	255

(g)

FIGURE 4.32
(a–g) Sequential steps involved in obtaining successive five translates (b–f) of a function of size
5 × 5 shown in (a)—via left-vertical half-plane to achieve half-plane closing of the function,
and (g) half-plane closing obtained by left-vertical half-plane of a function shown in (a). (From
Lim, S.L. and Sagar, B.S.D., *J. Geophys. Res.*, 113, D13208, 2008, doi:10.1029/2007JD009369.)

Figure 4.32a through g illustrates the half-plane closings by left-vertical half-plane. A sub-image of Figure 4.33a that is of array size 7 × 7, of size 5 × 5, is considered to explain half-plane closing by means of left-vertical half-plane. Left-vertical half-plane is moved to the first column of the function. The gray values (or brightness values) in the first column include (from top to bottom) 19, 14, 8, 9, and 15, with a value 19 being the maximum. First translation of the half-plane closing by left-vertical half-plane is shown in Figure 4.32b. The first translation involves replacing all the values in that column with a maximum value if such a value is not lesser than the value in the previous translation. This process of replacing the value with the

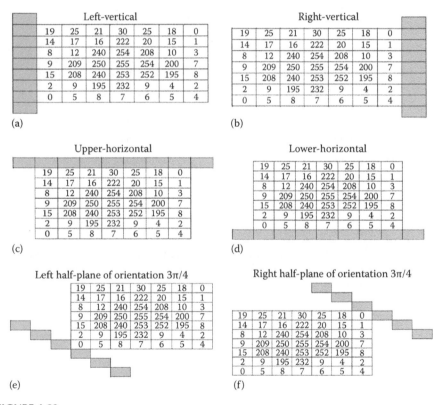

(a) Left-vertical   (b) Right-vertical

(c) Upper-horizontal   (d) Lower-horizontal

(e) Left half-plane of orientation 3π/4   (f) Right half-plane of orientation 3π/4

FIGURE 4.33
Half-plane closing of grayscale function *f* using eight directions. Different half-planes of eight directions are considered to obtain eight half-plane closings. (a–h) Function with half-planes of specific directions, (i) all eight half-planes with the function, (j and k) half-plane closings, obtained by an approach explained in Figure 4.3d through j, according to corresponding direction of half-planes shown in figures (l–q), and (r) point-wise minima of all half-plane closings shown in (j and k) yields convex hull of original function. (From Lim, S.L. and Sagar, B.S.D., *J. Geophys. Res.*, 113, D13208, 2008, doi:10.1029/2007JD009369.)

Left half-plane of orientation π/4

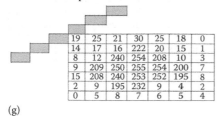

19	25	21	30	25	18	0
14	17	16	222	20	15	1
8	12	240	254	208	10	3
9	209	250	255	254	200	7
15	208	240	253	252	195	8
2	9	195	232	9	4	2
0	5	8	7	6	5	4

(g)

Right half-plane of orientation π/4

19	25	21	30	25	18	0
14	17	16	222	20	15	1
8	12	240	254	208	10	3
9	209	250	255	254	200	7
15	208	240	253	252	195	8
2	9	195	232	9	4	2
0	5	8	7	6	5	4

(h)

A gray-scale function *f* of 7 rows by 7 columns

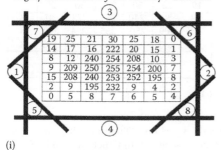

19	25	21	30	25	18	0
14	17	16	222	20	15	1
8	12	240	254	208	10	3
9	209	250	255	254	200	7
15	208	240	253	252	195	8
2	9	195	232	9	4	2
0	5	8	7	6	5	4

(i)

Closing by left-vertical half-plane

19	209	250	255	255	255	255
19	209	250	255	255	255	255
19	209	250	255	255	255	255
19	209	250	255	255	255	255
19	209	250	255	255	255	255
19	209	250	255	255	255	255
19	209	250	255	255	255	255

(j)

Closing by right-vertical half-plane

255	255	255	255	254	200	8
255	255	255	255	254	200	8
255	255	255	255	254	200	8
255	255	255	255	254	200	8
255	255	255	255	254	200	8
255	255	255	255	254	200	8
255	255	255	255	254	200	8

(k)

Closing by upper-horizontal half-plane

30	30	30	30	30	30	30
222	222	222	222	222	222	222
254	254	254	254	254	254	254
255	255	255	255	255	255	255
255	255	255	255	255	255	255
255	255	255	255	255	255	255
255	255	255	255	255	255	255

(l)

Closing by lower-horizontal half-plane

255	255	255	255	255	255	255
255	255	255	255	255	255	255
255	255	255	255	255	255	255
255	255	255	255	255	255	255
253	253	253	253	253	253	253
232	232	232	232	232	232	232
8	8	8	8	8	8	8

(m)

Closing by left half-plane of orientation 3π/4

255	255	255	255	255	255	255
253	255	255	255	255	255	255
240	253	255	255	255	255	255
208	240	253	255	255	255	255
15	208	240	253	255	255	255
5	15	208	240	253	255	255
0	5	15	208	240	253	255

(n)

FIGURE 4.33 (continued)
Half-plane closing of grayscale function *f* using eight directions. Different half-planes of eight directions are considered to obtain eight half-plane closings. (a–h) Function with half-planes of specific directions, (i) all eight half-planes with the function, (j and k) half-plane closings, obtained by an approach explained in Figure 4.3d through j, according to corresponding direction of half-planes shown in figures (l–q), and (r) point-wise minima of all half-plane closings shown in (j and k) yields convex hull of original function. (From Lim, S.L. and Sagar, B.S.D., *J. Geophys. Res.*, 113, D13208, 2008, doi:10.1029/2007JD009369.)

(continued)

Closing by right half-plane of orientation 3π/4

255	254	222	30	25	18	0
255	255	254	222	30	25	18
255	255	255	254	222	30	25
255	255	255	255	254	222	30
255	255	255	255	255	254	222
255	255	255	255	255	255	254
255	255	255	255	255	255	255

(o)

Closing by left half-plane of orientation π/4

19	25	25	30	240	254	255
25	25	30	240	254	255	255
25	30	240	254	255	255	255
30	240	254	255	255	255	255
240	254	255	255	255	255	255
254	255	255	255	255	255	255
255	255	255	255	255	255	255

(p)

Closing by right half-plane of orientation π/4

255	255	255	255	255	255	255
255	255	255	255	255	255	254
255	255	255	255	255	254	252
255	255	255	255	254	252	195
255	255	255	254	252	195	8
255	255	254	252	195	8	5
255	254	252	195	8	5	4

(q)

19	25	25	30	25	18	0
19	25	30	222	30	25	8
19	30	240	254	222	30	8
19	209	250	255	254	200	8
15	208	240	253	252	195	8
5	15	208	232	195	8	5
0	5	8	8	8	5	4

(r) $$CH(f) = \Lambda_{\theta}[\phi_{\pi_{\theta}^+}(f)\Lambda\phi_{\pi_{\theta}^-}(f)]$$

FIGURE 4.33 (continued)
Half-plane closing of grayscale function *f* using eight directions. Different half-planes of eight directions are considered to obtain eight half-plane closings. (a–h) Function with half-planes of specific directions, (i) all eight half-planes with the function, (j and k) half-plane closings, obtained by an approach explained in Figure 4.3d through j, according to corresponding direction of half-planes shown in figures (l–q), and (r) point-wise minima of all half-plane closings shown in (j and k) yields convex hull of original function. (From Lim, S.L. and Sagar, B.S.D., *J. Geophys. Res.*, 113, D13208, 2008, doi:10.1029/2007JD009369.)

maximum value in that column, as long as that maximum value is not less than the value in the previous translation, would be repeated until the last column in that direction (i.e., left to right). It is called closing of function by left-vertical half-plane, once all the columns of the function in left–right direction are translated via left–right vertical plane. This closing of function by left-vertical half-plane is denoted by $\phi_{\pi^+}(f)$. Since there are five columns in the function (Figure 4.32a), there are 5 + 1 translations, last translation being the closing of the function by means of left-vertical half-plane. In a similar fashion, if the process of replacing the values from rightmost column until the leftmost column, then it would be called closing of the function by right-vertical half-plane, which is denoted as $\phi_{\pi_{\theta}^-}(f)$. Closings of the function by other half-planes (e.g., top-horizontal and bottom-horizontal) could be generated by changing the directions. To have a clearer understanding of convex hull construction, an array of size 7 × 7 (Figure 4.33a) showing multivalued function is considered. Closings by half-planes of eight directions are shown in Figure 4.33a through q. The point-wise minimum of all the closings of the function by all directions yields convex hull (Figure 4.33r).

A general equation to construct convex hull of the function (f) is shown in Equation 4.37:

$$CH(f) = \bigwedge_{\theta} (\phi_{\pi_{\theta}^{+}}(f) \wedge \phi_{\pi_{\theta}^{-}}(f))$$ (4.37)

where

$(\pi_{\theta}^{+})^{c} = (\pi_{\theta}^{-})$ denotes two half-planes at orientation θ

$\phi(f)$ denotes the closing of a grayscale image (f)

By following this approach (Soille 1998), convex hulls for all 100 multiscale opened versions of the fields are constructed. Figure 4.34a through i shows eight closings of a cloud field obtained respectively by eight different half-planes. Figure 4.35a through h shows convex hulls of corresponding multiscale opened versions of a cloud field shown in Figure 4.30a through h.

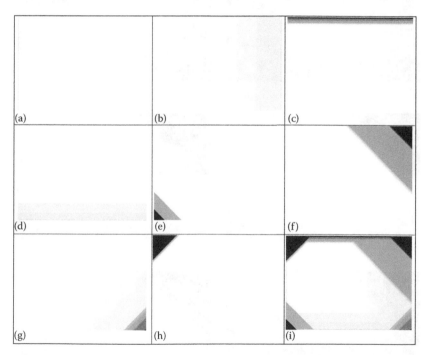

FIGURE 4.34
Convex hull generation of cloud function (Figure 3.21a) by half-planes—due to Soille (1998): (a) left-vertical half-plane, (b) right-vertical half-plane, (c) upper-horizontal half-plane, (d) lower-horizontal half-plane, (e) left half-plane of orientation $3\pi/4$, (f) right half-plane of orientation $3\pi/4$, (g) right half-plane of orientation $\pi/4$, (h) left half-plane of orientation $\pi/4$, and (i) intersection of all half-plane closings from (a) to (h) results in grayscale convex hull of cloud function shown in Figure 3.21a. (From Lim, S.L. and Sagar, B.S.D., *J. Geophys. Res.*, 113, D13208, 2008, doi:10.1029/2007JD009369.)

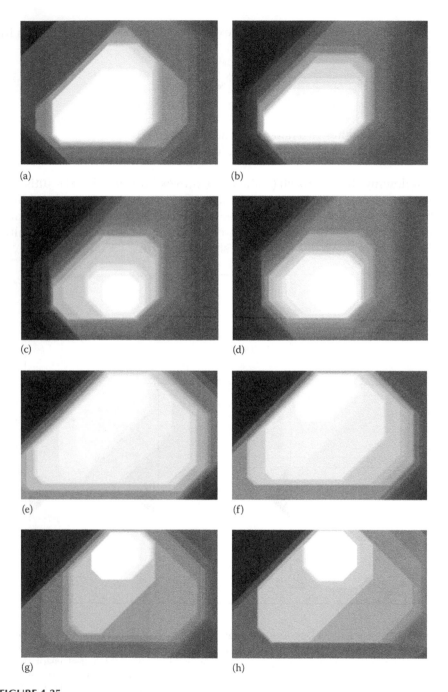

FIGURE 4.35
Convex hulls (a–d) of 25th, 50th, 75th, and 100th opened versions of cloud 1 and (e–h) of 25th, 50th, 75th, and 100th opened versions of cloud 2. (From Lim, S.L. and Sagar, B.S.D., *J. Geophys. Res.*, 113, D13208, 2008, doi:10.1029/2007JD009369.)

Areas of Multiscale Clouds and Their Convex Hulls

Sum of all the grayscale values (or brightness values) of cloud field (f) over all the spatial coordinates (x, y) is the area of (f). Similarly, sum of the grayscale values of convex hull of (f), $CH(f)$, over all the spatial coordinates (x, y) is the area of the convex hull. They are represented as

$$A(f) = \sum_{x,y} f(x,y), \quad A(CH(f)) = \sum_{x,y} CH(f(x,y))$$

These areas for all the 100 multiscale opening versions of the two cloud fields (Figure 3.21a and c) and their corresponding convex hulls are computed. They satisfy the following properties:

1. $A(f) \geq A(f \circ B) \geq \cdots \geq A(f \circ (N-1)B) \geq A(f \circ NB)$
2. $A(CH(f)) \geq A(CH(f) \circ B) \geq \cdots \geq A(CH(f) \circ (N-1)B) \geq A(CH(f) \circ NB)$
3. $A(f) \leq A(CH(f))$, and $A(f \circ nB) \leq A(CH(f \circ nB))$

Figure 4.36a and b shows graphical relationships for the two cloud functions, between the areas of cloud and its convex hull. These areas of opened versions of cloud functions and their corresponding convex hulls are plotted as functions of size of B, in other words scale "n" (Figure 4.36a and b). It is obvious that with increasing size, these areas are decreasing. Areas of convex hulls of corresponding opened versions of both the cloud fields have been plotted as functions of areas of opened versions of the clouds (Figure 4.36c and d). According to Equations 4.38 and 4.39, probability distributions of two cloud functions and their corresponding convex hulls across the scales (opening versions) have been computed and plotted as functions of scale (Figure 4.36e and f):

$$P_f(n, B) = \frac{A(f \circ nB) - A(f \circ (n+1)B)}{A(f)} \tag{4.38}$$

$$P_{CH(f)}(n, B) = \frac{A(CH(f \circ nB)) - A(CH(f \circ (n+1)B))}{A(CH(f))} \tag{4.39}$$

where $A(f \circ nB)$ and $A(CH(f \circ nB))$ represent the areas of (f) opened by nth size B and corresponding convex hull. $A(f)$ and $A(CH(f))$ are the areas of original cloud function and its corresponding convex hull, respectively. The probability distribution values computed according to Equations 4.38 and 4.39 satisfy the following properties: (i) $\sum_{n=0}^{N} P_f(n, B) = 1$ and (ii) $\sum_{n=0}^{N} P_{CH(f)}(n, B) = 1$.

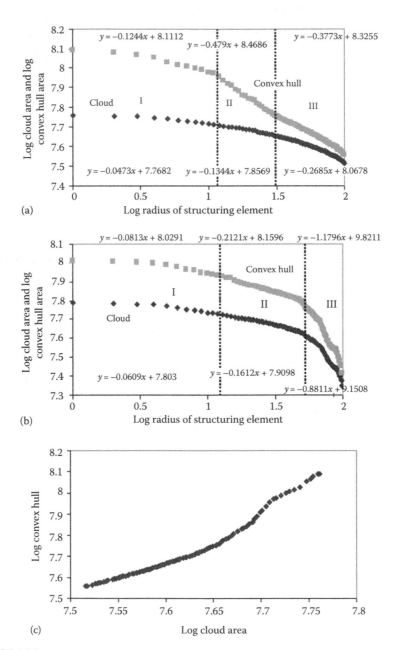

(a)

(b)

(c)

FIGURE 4.36
(a) Log–log graph between cloud area and convex hull versus corresponding radius of structuring element for cloud 1, (b) log–log graph between cloud area and convex hull versus corresponding radius of structuring element for cloud-2, (c) log–log graph of convex hull versus cloud area for cloud 1.

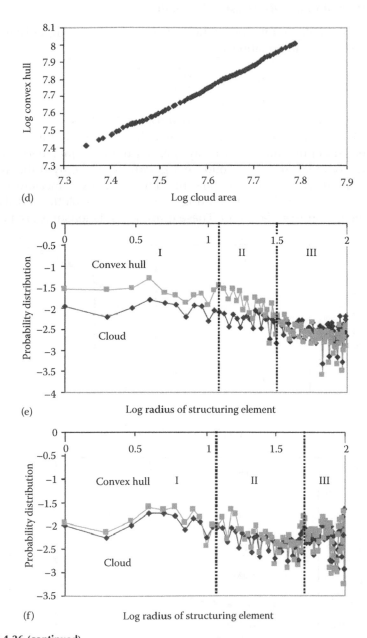

FIGURE 4.36 (continued)
(d) Log–log graph of convex hull versus cloud area for cloud 2, (e) log–log graph between the radii of structuring templates and corresponding probability distribution values for cloud 1, and (f) log–log graph between the radii of structuring templates and corresponding probability distribution values for cloud 2. (From Lim, S.L. and Sagar, B.S.D., *J. Geophys. Res.*, 113, D13208, 2008, doi:10.1029/2007JD009369.)

Computation of Convexity Measure for Spatial Fields

Areas of cloud field and its convex hull are employed to compute convexity measure according to Equation 4.40:

$$CM(f) = \frac{A(f \circ nB)}{A(CH(f \circ nB))} \tag{4.40}$$

This measure characterizes spatial heterogeneity. This measure ranges between 0 and 1. The upper bound of convexity measure is 1 as the area of the convex hull of the function is always greater than or equal to its function. The convexity measure 1 of a cloud field is valid if and only if the areas of the cloud and its convex hull are the same. These convexity measures of two cloud functions and their corresponding convex hulls are plotted as functions of the scale n (Figure 4.37a and b). These graphical relationships are taken as

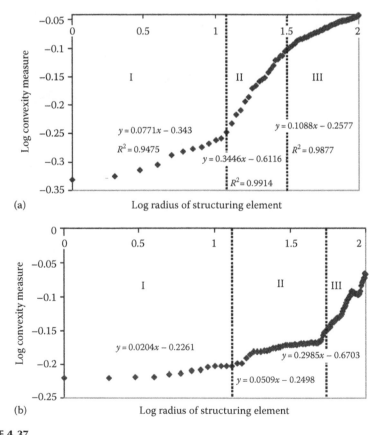

FIGURE 4.37
(a) Log–log graph of convexity measures with increasing radius of structuring element for cloud 1, and (b) log–log graph of convexity measures with increasing radius of structuring element for cloud 2. (From Lim, S.L. and Sagar, B.S.D., *J. Geophys. Res.*, 113, D13208, 2008, doi:10.1029/2007JD009369.)

the bases to mark the crossover scales further to determine the morphological regimes. The transition lines between the morphological regimes are demarcated (as vertical lines in Figure 4.37a and b). The basis to classify the morphological regimes of a cloud field appears valid in terms of the relationships between scale factor n and convexity measure relationship (e.g., Figure 4.37a and b). It is obvious from these plots that they do not possess universal scaling relationships. The convexity measure pattern across scales is divided into three groups (Figure 4.37). Groups I, II, and III are for the convexity measures corresponding to structuring elements from $n = 1$ to 11, $n = 12$ to 31, $n = 32$ to 100, respectively. In group I, the graph shows a rather flat curve with a slope value of 0.077, whereas the slope values are 0.3446 and 0.1088 for groups II and III, respectively. Similar graphical analysis is followed to find out morphological regimes for second cloud function (Figure 4.37b).

Figure 4.38a and c is generated on the basis of segregated phases shown in Figure 4.37a and b. It is conspicuous from the sequence of convexity

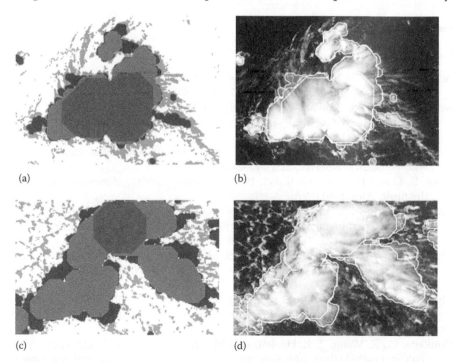

(a)　　　　　　　　　　　(b)

(c)　　　　　　　　　　　(d)

FIGURE 4.38
(a) Superposed gray-shaded binarized (by choosing threshold gray-level value 128) cloud 1 images at threshold-opening cycles, (b) boundaries of 12th, 32nd, and 100th opened cloud 1 images superimposed on the original cloud image, (c) superposed gray-shaded binarized (by choosing threshold gray-level value 110) cloud 2 images at threshold-opening cycles, and (d) boundaries of 12th, 49th, and 100th opened cloud-2 images superimposed on the original cloud image. Different regions—that are categorized broadly as inner, middle, and outer regions—depict zones with different spatial heterogeneities. (From Lim, S.L. and Sagar, B.S.D., *J. Geophys. Res.*, 113, D13208, 2008, doi:10.1029/2007JD009369.)

measures of opened versions of cloud 1 that there is a sudden change at radii 12, 32 (crossover scales). The cloud images at the 12th cycle, 32nd cycles, and 100th cycle of opening have been converted into binary images by choosing a common threshold value. Each of those three binary images is appropriately gray shaded and superposed on one another (Figure 4.38a). The boundaries of these binary images have also been superposed on the original cloud (Figure 4.38b). It is convincing through visual inspection that the regions within these boundaries have different degrees of spatial homogeneities. In a similar fashion, cloud 2 has also been partitioned into three morphologically significant regions (Figure 4.38c and d). This entire framework—where grayscale convex hull construction, multiscale morphological opening transformation, and convexity measure computations play vital roles—can be extended to segment DEMs into morphologically significant regions.

References

Ackerman, S. A., 1997, Remote sensing aerosols from satellite infrared observations, *Journal of Geophysical Research*, 102, 17069–17079.

Ackerman, S. A., K. I. Strabala, W. P. Menzel, R. A. Frey, C. C. Moeller, and L. E. Gumley, 1998, Discriminating clear sky from clouds with MODIS, *Journal of Geophysical Research*, 103, 32141–32157.

Band, L. E., 1986, Topographic partition of watersheds with digital elevation models, *Water Resources Research*, 22(1), 15–24.

Bates, R. L. and J. A. Jackson (Eds.), 1987, *Glossary of Geology*, American Geological Institute, Alexandria, VA.

Bunik, H. F. and K. A. Turner, 1971, Remote sensing applications to the quantitative analysis of drainage networks, *American Society of Photogrammetry Proceedings*, Fall meeting, pp. 71–313.

Chockalingam, L. and B. S. D. Sagar, 2005, Morphometry of network and non-network space of basins, *Journal of Geophysical Research-Solid Earth*, 110, B08203, 15, doi:10.1029/2005JB003641.

Chorowicz, J., C. Ichoku, S. Riazanoff, Y. J. Kim, and B. Cervelle, 1992, A combined algorithm for automated drainage network extraction, *Water Resources Research*, 28, 1293–1302.

Douketis, C., Z. Wang, T. L. Haslett, and M. Moskovits, 1995, Fractal character of cold-deposited silver films determined by low-temperature scanning tunneling microscopy, *Physical Review B*, 51(16), 11022–11031.

Duchene, P. and D. Lewis, 1996, *Visilog 5 Documentation*, Noesis Vision Inc., Quebec, Canada.

Fairfield, J. and P. Leymarie, 1991, Drainage networks from grid Digital Elevation Models, *Water Resource Research*, 27, 709–717.

Feder, J., 1988, *Fractals*, Plenum Press, New York.

Flook, A. G., 1978, The use of dilation logic on the quantimet to achieve fractal dimension characterization of textured and structured profiles, *Powder Technology*, 21, 295–298.

Franklin, S. E., 1990, Topographic context of satellite spectral response, *Computers & Geosciences*, 16, 1003–1010.

Freeman, T. G., 1991, Calculating catchment area with divergent flow based on a regular grid, *Computers & Geosciences*, 17, 413–422.

Frey, R. A., S. A. Ackerman, and B. J. Soden, 1995, Climate parameters from satellite spectral measurements, I, Collocated AVHRR and HIRS/2 observations of the spectral greenhouse parameter, *Journal of Climate*, 9, 327–344.

Gao, B. C. and A. F. H. Goetz, 1991, Cloud area determination from AVIRIS data using water vapor channels near 1 mm, *Journal of Geophysical Research*, 96, 2857–2864.

Gordon, N. D., J. R. Norris, C. P. Weaver, and S. A. Klein, 2005, Cluster analysis of cloud regimes and characteristic dynamics of midlatitude synoptic systems in observations and a model, *Journal of Geophysical Research*, 110, D15S17, doi:10.1029/2004JD005027.

Halsey, T. C., M. H. Jensen, L. P. Kadanoff, I. Procacia, and B. I. Shraiman, 1986, Fractal measures and their singularities: The characterization of strange sets, *Physical Review A*, 33, 1141–1151.

Horton, R. E., 1945, Erosional development of stream and their drainage basin: Hydrological approach to quantitative morphology, *Geophysical Society of America Bulletin*, 56, 275–370.

Hutchison, K. D. and K. R. Hardy, 1995, Threshold functions for automated cloud analyses of global meteorological satellite imagery, *International Journal of Remote Sensing*, 16, 3665–3680.

Inoue, T., 1987, A cloud type classification with NOAA 7 split window measurements, *Journal of Geophysical Research*, 92, 3991–4000.

Jenson, S. K., 1985, Automated derivation of hydrologic basin characteristics from digital elevation models, *Proceedings of Auto-Carto 7, Digital Representations of Spatial Knowledge*, American Society of Photogrammetry and American Society on Surveying and Mapping, Washington, DC, pp. 301–310.

Jenson, S. K., 1987, Methods and applications in surface depression analysis, Paper presented at *Auto-Carto 8*, Baltimore, MD.

Jenson, S. K. and J. O. Domingue, 1988, Extracting topographic structure from digital elevation data for geographic information system analysis, *Photogrammetric Engineering and Remote Sensing*, 54, 1593–1600.

King, M. D. et al., 1996, Airborne scanning spectrometer for remote sensing of cloud, aerosol, water vapor and surface properties, *Journal of Atmospheric and Oceanic Technology*, 13, 777–794.

King, M. D., Y. J. Kaufman, W. P. Menzel, and D. Tanre, 1992, Remote sensing of cloud, aerosol, and water vapor properties from the Moderate Resolution Imaging Spectrometer (MODIS), *IEEE Transactions on Geoscience and Remote Sensing*, 30, 2–27.

Lian, T. L., P. Radhakrishnan, and B. S. D. Sagar, 2004, Morphological decomposition of sandstone pore-space: Fractal power-laws, *Chaos, Solitons & Fractals*, 19(2), 339–346.

Lim, S. L., V. C. Koo, and B. S. D. Sagar, 2009, Computation of complexity measures of morphologically significant zones decomposed from binary fractal sets via multiscale convexity analysis, *Chaos, Solitons & Fractals*, 41(3), 1253–1262.

Lim, S. L. and B. S. D. Sagar, 2008, Cloud field segmentation via multiscale convexity analysis, *Journal of Geophysical Research*, 113, D13208, doi:10.1029/2007JD009369.

Mandelbrot, B., 1982, *Fractal Geometry of Nature*, Freeman, San Francisco, CA, p. 468.

Maritan, A., F. Coloairi, A. Flammini, M. Cieplak, and J. R. Banavar, 1996, Universality classes of optimal channel networks, *Science*, 272, 984.

Mark, D. M., 1983a, Automated detection of drainage networks from digital elevation models, Auto-arto VI, *Proceedings of the Sixth International Symposium on Automated Cartography*, The steering committee for the sixth International symposium on automated Cartography, Ottawa, Ontario, Canada, pp. 288–289.

Mark, D. M., 1983b, Relations between field-surveyed channel networks and map-based geomorphometric measures, Inez, Kentucky, *Annals of American Association of Geography*, 73, 358–372.

Mark, D. M., 1988, Network models in geomorphology, Chapter 4. In: *Modelling Geomorphological Systems*, ed. M. G. Anderson, John Wiley & Sons Ltd., Chichester, U.K., pp. 73–97.

Mark, D., J. Dozier, and J. Frew, 1982, Automated basin delineation from digital terrain data, *NASA Technical Memorandum*, 84984, 21.

Martz, L. W. and E. de Jong, 1988, Catch: A Fortran program for measuring the catchment area from digital elevation models, *Computers & Geosciences*, 14, 627–640.

Miliaresis, G. C. and D. P. Argialas, 1999, Segmentation of physiographic features from Global Digital Elevation Model/GTOPO30, *Computers & Geosciences*, 25, 715–728.

Monkhouse, F. J., 1965, *Principles of Physical Geography*, University of London Press Ltd., New York.

Morisawa, M. E., 1957, Accuracy of determination of stream lengths from topographic maps, *American Geophysical Union Transactions*, 38, 86–88.

Morris, D. G. and R. G. Heerdegen, 1988, Automatically derived catchment boundaries and channel networks and their hydrological applications, *Geomorphology*, 1, 131–141.

Mote, P. W. and R. Frey, 2006, Variability of clouds and water vapor in low latitudes: View from Moderate Resolution Imaging Spectroradiometer (MODIS), *Journal of Geophysical Research*, 111, D16101, doi:10.1029/2005JD006791.

O'Callaghan, J. F. and D. M. Mark, 1984, The extraction of drainage networks from digital elevation data, *Computer Vision Graphics Image Processing*, 28, 323–344.

Peitgen, H. O., H. Jürgens, and D. Saupe, 2004, *Chaos and Fractals: New Frontiers of Science*, Springer, New York.

Peucker, T. K., and D. H. Douglas, 1975, Detection of surface specific points by local parallel processing of discrete terrain elevation data, *Computer Graphics and Image Processing*, 4, 375–387.

Qian, J., R. W. Ehrich, and J. B. Campbell, 1990, DNESYS-An expert system for automatic extraction of drainage networks from digital elevation data, *IEEE Transactions on Geoscience and Remote Sensing*, 28, 29–44.

Radhakrishnan, P., L. L. Teo, and B. S. D. Sagar, 2004, Estimation of fractal dimension through morphological decomposition, *Chaos, Solitons & Fractals*, 21(3), 563–572.

Robertson, M. C., C. G. Sammis, M. Sahimi, and A. J. Martin, 1995, Fractal analysis of three-dimensional spatial distribution of earthquakes with a percolation interpretation, *Journal of Geophysical Research*, 100(B1), 609–620.

Rodriguez-Iturbe, I. and A. Rinaldo, 1997, *Fractal River Basins: Chance and Self-Organization*, Cambridge University Press, Cambridge, U.K.

Rossow, W. B., 1989, Measuring cloud properties from space: A review, *Journal of Climate*, 2(3), 201–213.

Sagar, B. S. D., 1996, Fractal relations of a morphological skeleton. *Chaos, Solitons & Fractals*, 7, 1871–1879.

Sagar, B. S. D., 2007, Universal scaling laws in surface water bodies and their zones of influence, *Water Resources Research*, 43(2), W02416–W06502, doi:10.1029/2006WR005075.

Sagar, B. S. D. and L. Chockalingam, 2004, Fractal dimension of nonnetwork space of a catchment basin, *Geophysical Research Letters*, 31, L12502, doi:10.1029/2004GL019749.

Sagar, B. S. D., M. B. R. Murthy, C. B. Rao, and B. Raj, 2003, Morphological approach to extract ridge-valley connectivity networks from Digital Elevation Models (DEMs), *International Journal of Remote Sensing*, 24(3), 573–581.

Sagar, B. S. D., D. Srinivas, and B. S. P. Rao, 2001, Fractal skeletal based channel networks in a triangular initiator basin, *Fractals*, 9(4), 429–437.

Sagar, B. S. D. and T. L. Tien, 2004, Allometric power-law relationships of Hortonian fractal digital elevation model, *Geophysical Research Letters*, 31, L06501, doi:10.1029/2003GL019093.

Sagar, B. S. D., M. Venu, and D. Srinivas, 2000, Morphological operators to extract channel networks from digital elevation models, *International Journal of Remote Sensing*, 21(1), 21–30.

Sathymoorthy, D., P. Radhakrishnan, and B. S. D. Sagar, 2007, Morphological segmentation of physiographic features from DEM, *International Journal of Remote Sensing*, 28(15), 3379–3394.

Schroeder, M., 1991, *Fractals, Chaos, Power Laws: Minutes from an Infinite Paradise*, W.H. Freeman, New York, pp. 41–45.

Soille, P., 1998, Grey scale convex hulls: Definition, implementation and applications, in *Proceedings of the Fourth International Symposium on Mathematical Morphology and Its Applications to Image and Signal Processing*, Amsterdam, the Netherlands, pp. 83–90, Springer, New York.

Strahler, A. N., 1957, *Handbook of Applied Hydrology*, ed. V. T. Chow, McGraw-Hill, New York.

Strahler, A. H., 1964, Quantitative geomorphology of drainage basins and channel networks. In: *Handbook of Applied Hydrology*, ed. V. T. Chow, McGraw-Hill, New York.

Takayasu, H., 1990, *Fractals in Physical Sciences*, Manchester University Press, New York, p. 170.

Tarboton, D. G., R. L. Bras, and I. Rodriguez-Iturbe, 1991, On the extraction of channel networks from digital elevation data, *Hydrological Processes*, 5(1), 81–100.

Tay, L. T., B. S. D. Sagar, and H. T. Chuah, 2005a, Analysis of geophysical networks derived from multiscale digital elevation models: A morphological approach, *IEEE Geoscience and Remote Sensing Letters*, 2(4), 399–403.

Tay, L. T., B. S. D. Sagar, and H. T. Chuah, 2005b, Derivation of terrain roughness indicators via granulometries, *International Journal of Remote Sensing*, 26(18), 3901–3910.

Tay, L. T., B. S. D. Sagar, and H. T. Chuah, 2007, Granulometric analysis of basin-wise DEMs: A comparative study, *International Journal of Remote Sensing*, 28(15), 3363–3378.

Tribe, A. J., 1992, Problems in automated recognition of valley features from digital elevation models and a new method toward their resolution, *Earth Surface Processes Landforms*, 17, 437–454.

Yuan, L. P. and N. L. Vanderpool, 1986, Drainage network simulation, *Computers & Geosciences*, 12, 653–665.

Zahn, W. and A. Zösch, 1999, The dependence of fractal dimension on measuring conditions of scanning probe microscopy, *Fresenius Journal of Analytical Chemistry*, 365(1–3), 168–172.

Zinner, T., B. Mayer, and M. Schroder, 2006, Determination of three dimensional cloud structures from high-resolution radiance data, *Journal of Geophysical Research*, 111, D08204, doi:10.1029/2005JD006062.

5

Terrestrial Surface Characterization:
A Quantitative Perspective

Quantitative characterization of terrestrial surface is important to understand terrestrial processes. Quantitative characterization can be carried out by computing morphometric parameters for the unique networks derived from digital elevation models (DEMs). Stream network is a basic input, from which various basic measures can be computed. Those basic measures pave a way to carry out morphometric analysis.

This chapter provides basic details to carry out morphometric analysis of treelike networks. Treelike networks are predominant in physiographic, biological, geological, and sociological domains. The networks, which are essentially in branched patterns, in other words loop less networks, that we considered to demonstrate conventional morphometric analysis include:

- Networks derived from fractal basins of both random and deterministic
- Networks derived from DEMs (both synthetic and realistic)
- Networks derived from planar features such as lakes, water bodies

The morphometric analysis of networks provides features such as lakes and water bodies further to understand the spatial complexity of a phenomenon from which the networks extracted. Two ways of characterization of a phenomenon are characterization of phenomenon itself and characterization of abstract structure, which is like a branched network, of the phenomenon. In this chapter, the latter aspect is demonstrated.

Network Morphometry: A Valuable Tool to Characterize Surficial Phenomena: A Review

Arboreal networks, like branched trees, have been characterized via fractal description (Shlesinger and West 1991). Other networks that have been characterized via similar descriptors include lung morphogenesis (Nelson and

Manchester 1988), stream networks (Horton 1945, Strahler 1964, Shreve 1967, Mandelbrot 1982, La Barbera and Rosso 1987, 1989, Tarbotan et al. 1990, Marani et al. 1991, Rosso et al. 1991, Masek and Turcotte 1993), physical networks, and social networks. Derivation of several morphometric (topological) quantities to characterize numerous realistic and synthetic networks of geophysical importance has been addressed (Howard 1990, Takayasu 1990, Rinaldo et al. 1993, Maritan et al. 1996a,b, Rodriguez-Iturbe and Rinaldo 1997, Sagar et al. 1998b, Dodds and Rothman 1999, Sagar et al. 2001, Maritan et al. 2002, Sagar and Tien 2004).

Many of these works employ classical morphometric analysis founded by Horton (1945) and Strahler (1957). The first step in the network morphometric analysis includes designation of branched network segments with ordering. This order designation mechanism is due to Horton and Strahler, which is popularly known as Horton–Strahler order designation mechanism.

Horton–Strahler Order Designation

The order of the network ranges from 1 to n (any finite number). All open-ended segments are designated as first-order segments. Second-order segment begins from the point where two first-order segments meet. Similarly, when two second-order segments meet, a third-order segment begins. If any lower-order segment joins a higher-order segment, then the higher-order continues until it meets another higher-order segment for the beginning of still higher-order segment. This order designation continues until the whole network's segments are designated with orders ranging from $\omega - 1$ to Ω, where ω would begin from 2. In short, when two streams of order i and j merge, a stream of order ω is formed, and Equation 5.1 explains this ordering scheme mathematically.

$$\omega = \max\left\{i, j, Int\left[1 + \left(\frac{1}{2}\right)(i + j)\right]\right\} \qquad (5.1)$$

where function $Int[\cdot]$ denotes the integer part of the argument. Figure 5.1 illustrates this ordering scheme. The order of the network shown in Figure 5.1 is termed as $\Omega = 4$. This ordering scheme is a basic prerequisite to compute morphometric parameters, popularly known as Horton's laws of networks. The Horton laws include law of numbers, law of lengths, and law of areas. From these laws, bifurcation ratio (R_B), stream length ratio (R_L), and stream area ratio (R_A) are defined (Schumm 1956).

Ratio of the number of stream segments of a given order $N(\omega, \Omega)$ to the number of stream segments with the immediate higher order $N(\omega + 1, \Omega)$—this definition is expressed mathematically as Equation 5.2.

$$R_B = \frac{N(\omega - 1, \Omega)}{N(\omega, \Omega)}, \quad \omega = 2, 3, \ldots, \Omega \qquad (5.2)$$

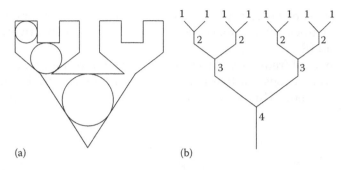

(a)   (b)

FIGURE 5.1
(a) The fractal structure and (b) the morphological skeleton after designating Strahler's ordering. (From Sagar, B.S.D. et al., *Discrete Dyn. Nat. Soc.*, 2, 77, 1998.)

Ratio of mean length of segments of order ω, $\bar{L}(\omega, \Omega)$, and mean length of segments of the immediate lower order, $\bar{L}(\omega-1, \Omega)$, yields stream length ratio R_L. This is mathematically shown as Equation 5.3.

$$R_L = \frac{\bar{L}(\omega, \Omega)}{\bar{L}(\omega-1, \Omega)}, \quad \omega = 2, 3, \ldots, \Omega \tag{5.3}$$

where

$\bar{L}(\omega, \Omega) = \sum_{i=1}^{N} L_i(\omega, \Omega) \Big/ N(\omega, \Omega)$ is the mean length of stream order ω
(R_B) and (R_L) are Horton's laws of number and length, respectively

Stream area ratio (R_A) is the ratio of the mean stream area of order ω and the mean stream area of order ω − 1. The area $A(\omega, \Omega)$ is the area drained directly by the particular stream of order ω and also the area drained by tributaries of lower order ω − 1, joining the stream of order ω. This quantity proposed by Schumm (1956) is expressed as Equation 5.4.

$$R_A = \frac{\bar{A}(\omega, \Omega)}{\bar{A}(\omega-1, \Omega)}, \quad \omega = 2, 3, \ldots, \Omega \tag{5.4}$$

where $\bar{A}(\omega, \Omega)$ is obtained by dividing the total area drained $A(\omega, \Omega)$ by the number of stream segments of order ω.

For the homogeneous river basins, these empirical laws of stream numbers, lengths (Horton 1945), and stream areas (Schumm 1956) are constant. These three topological quantities—(R_B), (R_L), and (R_A)—can also be estimated by plotting the logarithm values of stream numbers, mean lengths, and mean areas of order ω of Ω as functions of order ω. Slope values of those best-fit lines for numbers, lengths, and areas respectively denote (R_B), (R_L), and (R_A).

Ratio of channel frequency (F) and the square of the channel density (ρ^2) yields a universal constant that is of use to test how closely Horton basin obeys Melton's law. This ratio, also called Melton's law, is expressed in Equation 5.5.

$$\frac{F}{\rho^2} \tag{5.5}$$

where (F) is estimated as a ratio of the total number of channel segments of all orders ranging from $\omega - 1 = 1$ to $\omega = \Omega$, and area of basin of order Ω, and ρ is estimated as a ratio of total length of stream network of basin of order Ω and the total area of the basin of order Ω. These two parameters are shown as Equations 5.6 and 5.7.

$$F = \frac{\sum_{\omega=1}^{\Omega} N(\omega, \Omega)}{A} \tag{5.6}$$

$$\rho = \frac{L(\Omega)}{A(\Omega)} \tag{5.7}$$

This law provides a value of about 0.69, through which a basin can be decided whether it is of Hortonian type or non-Hortonian type.

Based on these Horton's laws of number, length, and area, as well as Melton's law, a host of topological quantities such as (R_B), (R_L), (R_A), (F), and ρ for various networks have been defined. The networks are extracted from synthetic planar basins—such as triadic fractal, quadric fractal, fractal basins with five-, six-, seven-, and eight-sided—DEMs of eight subbasins belonging to Gunung Ledang region of Malaysian peninsular, and Nizamsagar reservoir.

The considered source data—with details on how they have been simulated, generated, or created—that include synthetic planar basins, DEMs of eight subbasins, and Nizamsagar reservoir are shown in Figures 3.5a, 3.7, 3.9, and 3.20. The procedure followed to extract networks from these source data is morphological skeletonization explained in detail in Chapter 4, in particular in the "Some Background Studies of Unique Feature Extraction," "Valley Connectivity Network Extraction from DEM Using Binary Morphological Operations," and "Ridge and Valley Connectivity Networks via Grayscale Skeletonization" sections. The corresponding networks, extracted from these source data, are shown in

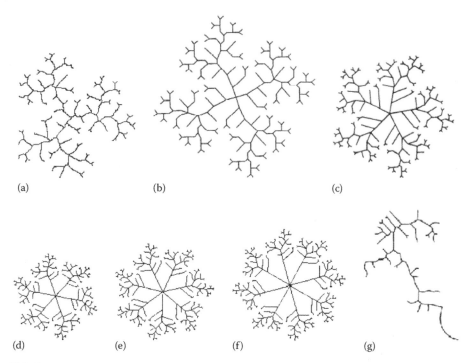

FIGURE 5.2
Networks in (a) three-sided fractal basin, (b) four-sided fractal basin, (c) five-sided fractal basin, (d) six-sided fractal basin, (e) seven-sided fractal basin, (f) eight-sided fractal basin, and (g) Nizamsagar reservoir. (From Sagar, B.S.D. et al., *Discrete Dyn. Nat. Soc.*, 2, 77, 1998; Sagar, B.S.D. et al., *Fractals*, 9, 429, 2001; Sagar, B.S.D. et al., *J. Math. Geol.*, 31(2), 143, 1999.)

Figures 5.2 and 5.3. All these networks are abstract structures in branched form summarizing the overall structures, orientations of the corresponding phenomena. Each of these branched networks is designated with orders by following Horton–Strahler ordering scheme (Equation 5.1).

Order-wise segments' lengths, number are computed. These are basic measures required to compute topological quantities such as bifurcation and length ratios. Other associated parameters such as density and frequency of the networks could be computed based on these basic measures. These topological quantities and other power-law relationships derived for all the aforementioned networks belonging to varied phenomena of terrestrial importance are tabulated in Tables 5.1 through 5.4. These quantities have been related with other power-laws drawn based on allometry, fractal analysis, etc. These scaling (fractal) and allometric relationships are described in the next chapter.

Morphometric quantities computed for these networks extracted from synthetic planar fractal basins are shown in Table 5.4. It is obvious that

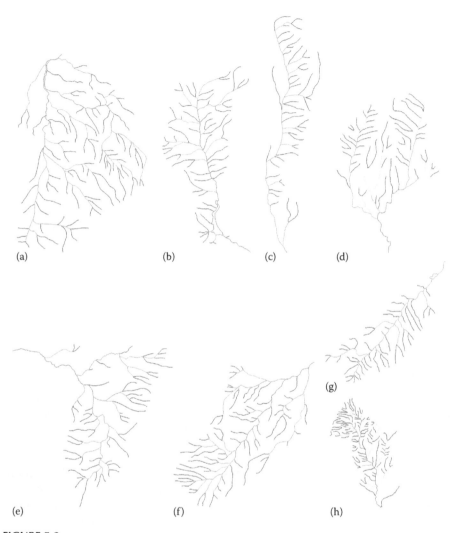

FIGURE 5.3
(a–h) Networks after order designation of eight basins of Gunung Ledang region. (From Chockalingam, L. and Sagar, B.S.D., *J. Geophys. Res.*, 110, B08203, 2005.)

there are networks with three trees, four trees, five trees, six trees, seven trees, and eight trees (Figure 5.2a through f). A generation mechanism employed to generate planar fractal basins from initiators ranging from three-sided triangle to eight-sided octagon is with eight sides that possess fractal dimension of $\log 8/\log 4 = 1.5$. The networks from these six planar fractal basins are extracted by following morphological skeletonization (Equations 4.3 and 4.4).

TABLE 5.1

Fractal Dimensions for the Morphological Skeleton Network:
Comparison between Length–Area Measures and the Estimated
Values from Order Ratios

Parameter	Estimated Values	Equation No.
Order ratio		
Bifurcation ratio	2.33	5.2
Skeleton length ratio	1.725	5.3
Skeleton area ratio	2.385	5.4
Total skeleton length vs. area		
Exponent β	0.98	5.16
Fractal dimension $(D_{TS}) = 2\beta$	1.96	
Main skeleton length vs. area		
Exponent α	0.612	5.13
Fractal dimension $(d) = 2\alpha$	1.224	
Estimation of fractal dimensions from order ratios		
$D = \dfrac{\text{Log } R_B}{\text{Log } R_L}$	1.56	5.8 and 5.9
$d = 2\dfrac{\text{Log } R_L}{\text{Log } R_A}$	1.25	5.11
$D_{TS} = \dfrac{\text{Log } R_B}{\text{Log } R_A}$	1.92	5.12
$D_{TS} = \dfrac{\text{Log } R_L}{\text{Log } R_B}$	1.23	

Source: Sagar, B.S.D., *Chaos Soliton Fract.*, 7(11), 1871, 1996.

TABLE 5.2

Fractal Dimensions of the Structure and Its
Morphological Skeleton Length: Comparison
between Box Counting Measures and the
Estimated Values from Morphometric Orders

Measured Fractal Dimensions through the Box Counting Method		
D (Generated fractal)	D_{TS} (Morphological skeleton)	d (Main skeleton length)
1.50	1.56	1.23

Source: Sagar, B.S.D., *Chaos Soliton Fract.*, 7(11), 1871, 1996.

TABLE 5.3

Basic Measures of Morphological Skeletons of the Second-Order Fractal Basins

Initiator	A	P	l	No. of Orders				Main Length of Individual Order (L)				Mean Areas of Individual Order (A)				Total Length
				1	2	3	4	1	2	3	4	1	2	3	4	
Five sided	110	160	16	130	50	15	5	0.56	1.1	1.67	7.5	0.313	1.21	2.78	14.06	190.25
Six sided	166.32	192	18	156	60	18	6	0.46	1.05	1.5	7.25	0.213	1.102	2.26	13.33	205.5
Seven sided	232.63	224	22	182	70	21	7	0.65	1.3	1.67	10	0.42	1.7	2.78	28.57	313.6
Eight sided	309.1	256	26	208	80	24	8	0.54	1.26	1.67	11	0.29	1.6	2.78	30.37	341

Source: Sagar, B.S.D. et al., *Discrete Dyn. Nat. Soc.*, 2, 77, 1998.

TABLE 5.4

Certain Order Ratios of Morphological Skeletons of the Second-Order Fractal Basins

	Bifurcation Ratio				Length Ratio				Area Ratio						
	2	3	4		2	3	4		2	3	4				
Initiator	N_1/N_2	N_2/N_3	N_3/N_4	R_B	$\bar{L}_2/\bar{L}_1$	$\bar{L}_3/\bar{L}_2$	$\bar{L}_4/\bar{L}_3$	R_L	$\bar{A}_2/\bar{A}_1$	$\bar{A}_2/\bar{A}_1$	$\bar{A}_2/\bar{A}_1$	R_A	R_C	ρ	F
Five sided	2.6	3.33	3	2.98	1.96	1.52	4.5	2.66	3.87	2.3	5.06	3.74	0.86	1.73	1.82
Six sided	2.6	3.33	3	2.98	2.28	1.43	4.83	2.85	5.17	2.05	5.9	4.37	0.91	1.24	1.45
Seven sided	2.6	3.33	3	2.98	2	1.29	5.99	3.09	4.05	1.63	10.3	5.32	0.93	1.35	1.21
Eight sided	2.6	3.33	3	2.98	2.33	1.33	6.6	3.41	5.51	1.74	10.9	6.06	0.95	1.1	1.04

Source: Sagar, B.S.D. et al., *Discrete Dyn. Nat. Soc.*, 2, 77, 1998.

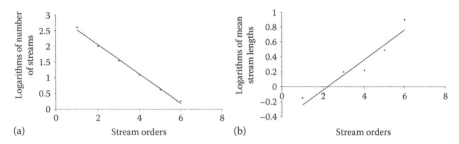

(a)   Stream orders   (b)   Stream orders

FIGURE 5.4
Statistical results of F-SCNs from triangular initiator–basin. (a) The log of the number of channel segments of a given order plotted against that order, and (b) the log of the average length of
channel segments of a given order plotted against that order. Horton's laws state that a natural
drainage basin will yield a linear relation on each graph. (From Sagar, B.S.D. et al., *Fractals*, 9,
429, 2001.)

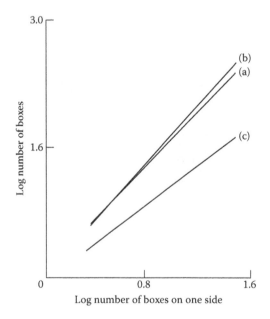

FIGURE 5.5
Fractal plots of (a) fractal structure; (b) total morphological skeleton length; and (c) main
skeleton length through the box counting method. (From Sagar, B.S.D. et al., *Discrete Dyn.
Nat. Soc.*, 2, 77, 1998.)

Graphical plots show relationships between the order of network segments and the logarithms of number and lengths of corresponding order
(Figures 5.4 through 5.10). The antilogs of slope values computed for these
relationships yield bifurcation ratio and length ratio respectively for the networks under study. Fractal dimensions of the basins and their corresponding networks as well as their fractal-length-area-permeter relationships are
shown in Tables 5.5 through 5.7. Similar morphometric quantities computed

for the networks (Figure 5.10) derived from DEMs of realistic basins and for the network (Figure 5.9) extracted from planar Nizamsagar reservoir have been shown in Tables 5.8 and 5.10.

Mandelbrot (1982) confirmed that the fractal dimension of river network has a relationship with Horton's laws of number and length. This relation is shown in Equation 5.8.

$$D = \frac{\text{Log } R_B}{\text{Log } R_L} \tag{5.8}$$

where D is fractal dimension that can be computed by conventional box counting method (Feder 1988). A series of studies have been published (La Barbera and Rosso 1987, 1989, Feder 1988, Tarbotan et al. 1990, Rosso et al. 1991, Stark 1991), where the relationships between fractal dimensions

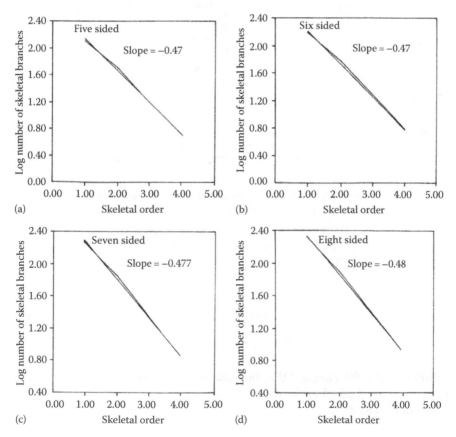

FIGURE 5.6
Graphs of (a–d) stream order number versus the logarithm of number of skeleton branches.

(continued)

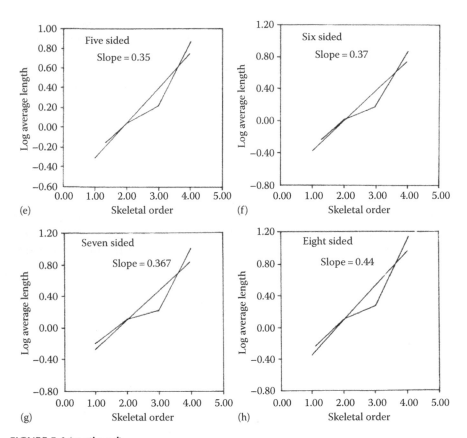

FIGURE 5.6 (continued)
Graphs of (e–h) stream order number versus the logarithm of mean length for five-, six-, seven-, and eight-sided fractal networks. (From Sagar, B.S.D. et al., *Discrete Dyn. Nat. Soc.*, 2, 77, 1998.)

computed through various approaches and the morphological quantities have been shown. These relationships, shown as Equation 5.9, include

$$D_1 = \frac{\text{Log } R_B}{\text{Log } R_L}, \quad R_B \geq R_L$$

$$D_1 = 1, \quad R_B < R_L$$

(5.9)

Tarbotan et al. (1990) argued that the ratio of logarithmic values of R_B and R_L needs to be multiplied with fractal dimension of main stream length (d), as shown in Equation 5.10.

$$D_2 = d \left(\frac{\text{Log } R_B}{\text{Log } R_L} \right)$$

(5.10)

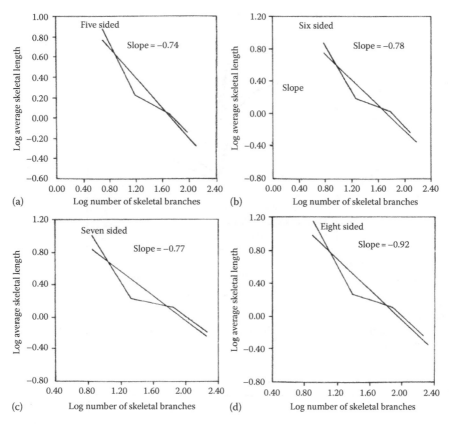

FIGURE 5.7
(a–d) Graphs showing the logarithm of number of skeletons versus the logarithm of average length for five-, six-, seven-, and eight-sided fractal networks. (From Sagar, B.S.D. et al., *Discrete Dyn. Nat. Soc.*, 2, 77, 1998.)

Feder (1988) has provided relationships, by involving not only R_B and R_L, but also R_A, as (Equation 5.11)

$$D_3 = 2\left(\frac{\text{Log } R_L}{\text{Log } R_B}\right), \quad R_B \geq R_A$$

$$D_3 = 2\left(\frac{\text{Log } R_L}{\text{Log } R_A}\right), \quad R_B < R_A$$

(5.11)

Another relation shown by Rosso et al. (1991) was (Equation 5.12)

$$D_4 = 2\left(\frac{\text{Log } R_B}{\text{Log } R_A}\right), \quad R_B \geq R_A$$

$$D_4 = 2, \quad R_B < R_A$$

(5.12)

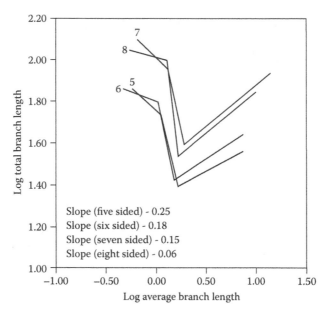

FIGURE 5.8
Graph showing the log (average branch length) versus log (total branch length). (From Sagar, B.S.D. et al., *Discrete Dyn. Nat. Soc.*, 2, 77, 1998.)

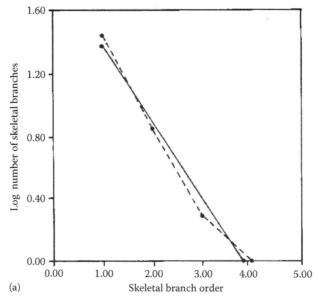

(a)

FIGURE 5.9
Statistical results of morphological skeleton network of Nizamsagar reservoir. (a) The log of the number of skeletal segments of a given order plotted against that order.

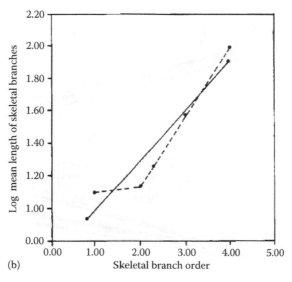

(b)

FIGURE 5.9 (continued)
Statistical results of morphological skeleton network of Nizamsagar reservoir. (b) The log of the average length of skeletal segments of a given order plotted against that order. Horton's laws state that a natural river network yields a linear relation on each graph. (From Sagar, B.S.D. et al., *J. Math. Geol.*, 31(2), 143, 1999.)

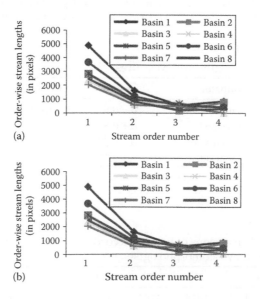

(a)

(b)

FIGURE 5.10
Graphs showing the log (average branch length) versus log (total branch length) for networks of eight basins of Gunung Ledang region shown in Figure 5.3. (a) Stream order versus number of streams, (b) stream order versus stream lengths.

(continued)

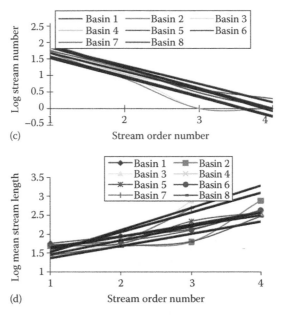

FIGURE 5.10 (continued)
Graphs showing the log (average branch length) versus log (total branch length) for networks of eight basins of Gunung Ledang region shown in Figure 5.3. (c) Stream order versus logarithm of number of streams, and (d) stream order versus logarithm of mean stream lengths. (From Chockalingam, L. and Sagar, B.S.D., *J. Geophys. Res.*, 110, B08203, 2005.)

TABLE 5.5

Length–Area Measures

Initiator	$l \sim A^{\alpha}$	$L \sim A^{\beta}$
Five sided	$\alpha = 0.59$	$\beta = 1.112$
Six sided	$\alpha = 0.56$	$\beta = 1.04$
Seven sided	$\alpha = 0.57$	$\beta = 1.055$
Eight sided	$\alpha = 0.57$	$\beta = 1.0132$

Source: Sagar, B.S.D. et al., *Discrete Dyn. Nat. Soc.*, 2, 77, 1998.

TABLE 5.6

Fractal Dimensions of F-SCNs according to the Derivations Proposed by Geomorphologists

Initiator	D_1	D_2	D_3	D_4	D_5	D_6
Five sided	1.116	1.305	1.48	2	1.74	1.25
Six sided	1.042	1.183	1.422	2	1.78	1.18
Seven sided	1	1.09	1.36	2	1.77	1.15
Eight sided	1	1.03	1.36	2	1.92	1.06

Source: Sagar, B.S.D. et al., *Discrete Dyn. Nat. Soc.*, 2, 77, 1998.

TABLE 5.7

Fractal Dimensions of Fractal Basins, Morphological Skeleton, and Main Skeletal Length

Initiator	Measured Fractal Dimensions through Box Counting Method		
	D-Fractal Basin	DTS-Skeleton	d-Mail Channel Length
Five sided	1.72	1.63	1.16
Six sided	1.77	1.66	1.13
Seven sided	1.81	1.70	1.13
Eight sided	1.85	1.77	1.14

Source: Sagar, B.S.D. et al., *Discrete Dyn. Nat. Soc.*, 2, 77, 1998.

TABLE 5.8

Basic Measures of Morphological Skeletal Network, Certain Morphometric Order Ratios, and Dissection Properties of Nizamsagar Reservoir

No. of Skeletal Orders				Length of Skeletal Orders				R_B	R_L	D_2	D
1	2	3	4	1	2	3	4				
28	7	2	1	356	96	76	200	3.33	2.16	1.912	1.92

Source: Sagar, B.S.D. et al., *J. Math. Geol.*, 31(2), 143, 1999.

If a Hortonian system implies an area that tends to infinity as the order tends to infinity, then D_4 proposed by Rosso et al. (1991) has no hydrological relevance (Stark 1991). It is true that as the resolution is refined, the area of the basin (Hortonian basins) does not change, but the length of the network does.

Order versus Number/Mean Length of Network

Order-wise segment numbers and mean lengths are plotted as functions of the number of orders for the different considered networks of synthetic networks, realistic DEMs, water bodies, and pore connectivity networks (Figures 5.5 through 5.10). The slope values for the two plots, viz., (1) order versus number and (2) order versus mean lengths, that are shown for all the considered networks are shown in Tables 5.4 and 5.8.

Mean Length versus Number

Similarly, order-wise mean lengths of segments of all the considered networks (ranging from synthetic networks, realistic stream networks, and abstract networks of water bodies) are plotted as functions of corresponding segment number (Figures 5.5 through 5.10). The slope values computed for the best-fit lines are shown in Tables 5.4 and 5.8. The values obtained by subtracting those slope values from 1 yield fractal dimensions D_s. Such fractal dimensional values for the networks considered are also shown in Tables 5.3, 5.5 through 5.8.

Without following any ordering scheme for the networks extracted from the basins, and water bodies, the following measures, also known as fractal dimensions of basin, total network and main (longitudinal) networks, are computed by following box counting method (Feder 1988). D, D_{TS}, and d, respectively, denote fractal dimensions of main structure (e.g., basin, water body), fractal dimension of total network, and fractal dimension of main stream segment. Based on these D, D_{TS}, and d, several other morphometric relationships are proposed for the networks considered. These relationships and the numerical results obtained for all the considered networks are shown in Tables 5.2, 5.7, and 5.8.

Interestingly, these morphometric quantities have relations with other popular power-laws, scale-invariant properties, fractal dimensions that could be computed—by taking the basic measures such as longitudinal length ($L_{\parallel}$), transverse length ($L_{\perp}$), area (A), perimeter (P), and total network length (L)—for various terrestrial phenomena. These phenomena include networks, basins, water bodies, zones of influence, and subbasins.

In what follows, standard allometric and fractal power-law relationships that explain self-organization characteristics of various phenomena have been explained.

Length–area relationship: Through length–area relationship, power-laws (e.g., α, β) could be derived. For instance, from geomorphology point of view, basin area (A) and main stream length (l_{mc}) of the stream network provided the following relationship (Equation 5.13):

$$l \sim A^{\alpha} \tag{5.13}$$

where α is a power-law, popularly known as Hack's law (Hack 1957). If a basin is perfectly circular in shape, and also the main stream length is equivalent to the diameter of the circular basin, then the relationship shown in Equation 5.13 takes the form of Equation 5.14.

$$l = \sqrt{A} = A^{0.5} \tag{5.14}$$

But α is always greater than 0.5 for the realistic basins and water bodies. Hack (1957) found that this α for a river basin in Virginia and Maryland had a value of about 0.6. Mandelbrot (1982) described that Equation 5.15 provides fractal dimension (d) of main stream length as (Equation 5.15)

$$d = 2\alpha \tag{5.15}$$

The area (A) and the total stream length (L) are related according to Equation 5.16

$$L \sim A^{\beta} \tag{5.16}$$

where β is a fitted exponent.

Fractal Relationship of Medial Axis Length to the Water Body Area

Hack (1957) proposed a power-law between the length and the area of basin as $l \sim A^{h}$, which is like Equation 5.13, where l and A are main stream length and area of basin, respectively. In standard dimensional analysis, the power-law as h is 0.5. But in realistic basins, this h is larger than 0.5. For a large number of water bodies extracted from IRS-1A LISS IIII remotely sensed data, this length–area relationship has been verified. Mandelbrot (1982) demonstrated that the power value in the relation $(l \sim (1/A^{h}))$ is not 2 for nonstandard shapes such as basins and water bodies. Length of a medial axis of a nonstandard shapes is longer than its longitudinal length. Fractal dimension that could be computed for such a medial axis is denoted as d. Then the relationship between the length of medial axis and the area of nonstandard shape is taken in the form of Equation 5.17.

$$l \sim A^{d/2} \tag{5.17}$$

About 160 surface water bodies have been traced from remotely sensed satellite data (Figure 3.16) situated between the geographical coordinates of 18°00′–18°30′N latitudes and 83°15′–83°45′E longitudes. Medial axes of these 160 water bodies have been extracted, and their corresponding lengths have been computed. Fractal dimensions of these lengths have been found to be about 1.113 ± 0.01. Logarithms of these lengths have been plotted as a function of logarithms of their corresponding areas. Slope value computed for a line that is best fit for this graphical relationship is found to be 0.556, which is precisely 1.113/2 = 0.5565.

Fractal Relation of Perimeter to the Water Body Area

A fractal relation of perimeter to the area of water body, which is in a nonstandard shape, is given in the form of Equation 5.18 (Lovejoy 1982, Mandelbrot et al. 1984).

$$A \sim P^{\alpha} \tag{5.18}$$

where α for classical Euclidean shapes would be 2 and is lesser than 2 for nonstandard shapes like water bodies and basins. According to Euclidean law, the area–perimeter relationship is $A \sim P^{2}$, further supporting that the water body has the fractal dimension of exactly 1, which indicates that water body has a smooth circular outline. For shapes that possess the fractal dimension of outline (D) of 1, the power-law of 2 is satisfied as $2/D = 2/1 = 2$. This D for realistic surface water bodies is more than 1, and hence the power-law would be less than 2. Mandelbrot (1967) computed the fractal dimension of West Coast of Britain as 1.26. In a way, the boundaries of water bodies are also like coastlines. Since the water bodies are in the non-fractal shapes, the power-law (α) in the area–perimeter relationship is $2/D$. To verify this, about 200 water bodies have been traced from remotely sensed satellite data for a region situated between 18°00′–18°30′N latitudes and 83°15′–83°45′ E longitudes (Figure 3.16). Out of these 200 water bodies, 4 water bodies have been selected, and their fractal dimensions have been computed as 1.51, 1.49, 1.49, and 1.46.

Logarithms of areas of 200 water bodies have been plotted as functions of logarithms of corresponding perimeters (Figure 5.11). The sizes of water bodies range from 0.05 to 0.8 km². The data are well fitted by the power-law $A \sim P^{1.30}$ for the 200 water bodies, and the fractal dimension of perimeter could be computed according to Equation 5.18, as $2/D = \alpha$, i.e., $D = 2/\alpha \Rightarrow 2/1.3 = 1.53$. This fractal dimension of perimeter of 1.53 is observed very close to the fractal dimension of boundaries of four selected water bodies from the data computed as 1.51, 1.49, 1.49, and 1.46. This value of 1.53 is also close to that of the Brownian mountain lakes, for which $D = 1.50$ (Mandelbrot 1982, Schroeder 1991). This area–perimeter fractal relationship of water bodies is one of the important geomorphologic characteristics (Sagar and Rao 1995).

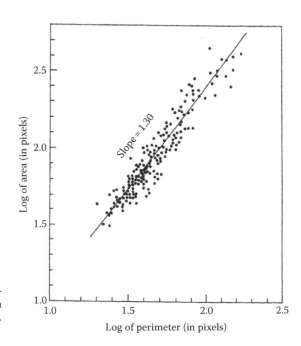

FIGURE 5.11
Logarithm of water body area versus logarithm of perimeter. (From Sagar, B.S.D. and Rao, B.S.P., *Curr. Sci.*, 68, 1129, 1995.)

Allometric Scaling Relationships in Hortonian Fractal Digital Elevation Model

Scaling laws shown for water bodies in previous sections include length–area and perimeter–area relationships. A host of allometric power-law relationships have also been shown for Hortonian fractal-DEM (F-DEM). It has been found that the F-DEM is geomorphologically realistic from the viewpoint of its Hortonity and scaling laws.

An F-DEM (Figures 3.7 and 3.9)—generated via decomposition of a fractal binary basin into topologically significant regions and gray-shading schemes—that follows Hortonity has been considered. The channel and ridge connectivity networks from this Hortonian F-DEM extracted by following network extraction algorithm explained in Chapter 4 have been employed to hierarchically decompose this DEM into subbasins of several orders. In turn, DEM of sixth order could be decomposed into subbasins of lower orders.

Drainage basin of the fluvial systems on Earth could be better described by self-affine properties (Tarboton et al. 1988, Rodriguez-Iturbe and Rinaldo 1997). Within a drainage basin belonging to a landscape, the structural organization can be better determined by the two unique topological connectivity networks that include loopless valley connectivity network (VCN) and looplike ridge connectivity network. Popular Horton's laws of number and length have been proposed by considering the loopless VCNs. The VCN segments embedded between the ridge connectivity networks, which are looplike networks, exist between ridges that are Brownian motion–like (Takayasu 1990). Several researches have shown that various types of networks follow allometric scaling relationships (Maritan et al. 1996a,b, Rodriguez-Iturbe and Rinaldo 1997, Banavar et al. 1999, Veitzer and Gupta 2000, Banavar et al. 2002, Maritan et al. 2002).

The valley and ridge connectivity networks (Figure 5.12a) extracted from Hortonian F-DEM (Figure 3.9) have been employed to verify the allometric power-laws. Since the VCN (Figure 5.12a) is extracted from F-DEM, this network has also been referred to as fractal-skeletal-based channel network (F-SCN) model (Figure 5.2a) and has been following Horton's laws (Sagar et al. 1998b, 2001, 2003, Sagar and Murthy 2000). See Chapter 4 for the morphological equations to extract these connectivity networks. The union of the two unique networks, VCN and RID, is shown in Figure 5.12a, and its mathematical representation is $RID \cup VCN$. The extracted VCN of F-DEM is designated with Horton–Strahler ordering scheme, and the network yields sixth order. The fractal dimension estimated by employing R_B and R_L values computed according to the approach detailed in the previous section yields 1.76, which is in good agreement with realistic geomorphologic networks.

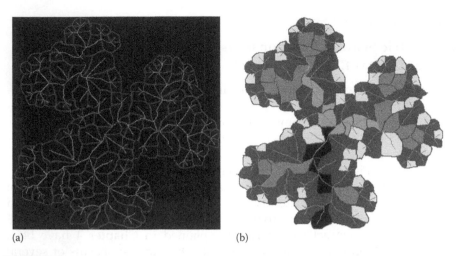

(a)   (b)

FIGURE 5.12
(a) Loop-like ridge connectivity and loopless channel connectivity networks, and (b) subbasins of sixth-order basin. (From Sagar, B.S.D. and Tien, T.L., *Geophys. Res. Lett.*, 31, L06501, 2004.)

From this sixth-order F-DEM, 2 fifth-order, 5 fourth-order, 10 third-order, 36 second-order, and 81 first-order subbasins could be decomposed hierarchically. These decomposed subbasins have been shown in Figure 5.12b.

To show relationships (scaling), the basic measures required include area, main length, perimeter, longitudinal length, and transverse length of all subbasins decomposed from F-DEM. The definitions of these basic measures are given briefly.

Basic measures: For a given basin-like F-DEM, the organization of total VCN is shown in Equation 5.19.

$$\bigcup_{\omega-1}^{\Omega}\left(VCN\left(\omega-1_{i=1}^{N},\Omega\right)\right) \tag{5.19}$$

where
   ω is designated order of network segments
   i is index of the network segment belonging to order ω
   Ω is order of the basin

For each network segment with index i, the order ω, there will be a contributing area that is precisely computed as the area embedded between the ridges that surround a network segment i of order ω. This measure is shown in Equation 5.20.

$$A(VCN(\omega-1_{i},\Omega)) \tag{5.20}$$

Main length of the basin in longitudinal direction (L_{mc}) is the length of the main channel from the extremity to the outlet of the basin of order ω. This is denoted as Equation 5.21.

$$L_{mc}(VCN(\omega_i, \Omega)) \tag{5.21}$$

The total contributing area of all network segments of all orders and the total length of all segments of all orders are computed according to Equations 5.22 and 5.23.

$$\sum_{\omega_{i=1}^n=1}^{\Omega} A(VCN(\omega_i, \Omega)) \tag{5.22}$$

and

$$\sum_{\omega_{i=1}^n=1}^{\Omega} L(VCN(\omega_i, \Omega)) \tag{5.23}$$

Perimeter (P) for subbasins of each network segment with index i of order ω is the boundary length of such a subbasin. Total perimeter length of all network segments of all orders will be computed according to Equation 5.24.

$$\sum_{\omega_{i=1}^n=1}^{\Omega} P(VCN(\omega_i, \Omega)) \tag{5.24}$$

Total perimeter length is equivalent to the total length of all ridges of F-DEM. Besides these basic measures that rely mostly on network organization, other basic measures such as longitudinal length ($L_{\parallel}$) and transverse length ($L_{\perp}$) have also been computed for all segments ranging from 1 to n, for all subbasins ranging from order 1 to Ω. These basic measures for all the subbasins decomposed from F-DEM have been tabulated (Table 5.9). Based on these basic measures, several allometric scaling relationships have been derived.

Scaling Laws in F-DEM

Several allometric relationships have been derived between the basic measures (Table 5.10). These allometric relationships yield power-law values found to be universal scaling laws (Table 5.10). These allometric relationships shown are for A and L_{mc}, A and P, $L_{\perp}$ and $L_{\parallel}$, and $L_{\perp}$ and L_{mc} for F-DEM.

TABLE 5.9

Basic Measures of All Subbasins Hierarchically Decomposed from F-DEM

Basin Order	Area (Pixels)	Perimeter (Pixels)	Longitudinal Length (Pixels)	Transverse Length (Pixels)	Main Channel Length (Pixels)
6	64,447	1969	338	330	471
5	18,713	757	176	160	245
5	21,594	840	184	180	217
4	5,237	428	111	90	118
4	6,903	478	114	114	142
4	6,709	445	105	94	124
4	7,770	458	116	103	131
4	9,309	550	132	119	156
3	2,133	216	54	49	72
3	1,719	193	56	40	69
3	1,958	217	56	52	67
3	1,495	172	48	40	57
3	1,850	207	53	51	64
3	1,667	184	46	45	59
3	2,002	243	65	57	61
3	2,172	243	65	53	75
3	2,162	232	58	58	65
3	2,392	261	69	59	86
3	1,865	214	55	51	63
3	6,517	453	122	81	121
2	1,282	182	55	38	64
2	1,182	161	47	33	60
2	1,254	177	53	34	62
2	793	139	39	32	36
2	1,269	184	47	39	51
2	2,089	246	63	62	75
2	774	134	35	34	37
2	673	128	32	32	33
2	1,271	177	55	35	64
2	1,042	151	44	32	55
2	1,408	185	46	43	68
2	1,582	195	50	43	61
2	1,726	227	59	49	61
2	625	112	34	24	30
2	638	137	41	31	34
2	1,009	193	48	39	36
2	889	140	41	31	35
2	797	142	40	33	36
2	811	140	41	28	36
2	562	110	35	22	27
2	632	114	30	29	39
2	760	163	40	39	30

TABLE 5.9 (continued)

Basic Measures of All Subbasins Hierarchically Decomposed from F-DEM

Basin Order	Area (Pixels)	Perimeter (Pixels)	Longitudinal Length (Pixels)	Transverse Length (Pixels)	Main Channel Length (Pixels)
2	885	153	38	35	41
2	789	130	38	29	42
2	789	125	35	30	36
2	692	126	36	29	38
2	660	120	34	26	45
2	1,000	168	47	38	58
2	747	123	31	30	36
2	1,307	175	46	42	56
2	1,304	195	52	44	56
2	780	130	37	31	39
2	419	98	33	19	29
2	540	106	38	25	36
2	565	114	33	25	30
2	1,282	203	59	40	62
1	804	128	35	31	32
1	162	58	17	14	10
1	156	65	18	17	11
1	218	60	17	15	13
1	364	87	24	22	16
1	212	65	20	15	14
1	191	61	20	12	14
1	173	58	16	14	13
1	284	75	21	19	18
1	364	93	26	22	16
1	407	93	26	21	14
1	298	79	23	18	15
1	127	56	22	9	18
1	295	81	21	21	19
1	360	88	30	16	12
1	248	69	19	18	14
1	837	139	34	34	35
1	327	82	24	20	15
1	282	88	28	14	16
1	167	57	17	13	12
1	1,060	154	40	39	48
1	462	94	33	22	22
1	158	55	18	11	16
1	398	92	29	19	20
1	239	64	18	16	14
1	1,507	190	51	44	63
1	188	66	19	17	13

(*continued*)

TABLE 5.9 (continued)

Basic Measures of All Subbasins Hierarchically Decomposed from F-DEM

Basin Order	Area (Pixels)	Perimeter (Pixels)	Longitudinal Length (Pixels)	Transverse Length (Pixels)	Main Channel Length (Pixels)
1	152	62	17	16	12
1	287	68	18	18	14
1	274	70	20	17	17
1	1,295	185	53	35	53
1	171	57	16	14	14
1	330	79	21	21	22
1	296	72	21	17	8
1	298	88	24	22	18
1	251	77	21	19	21
1	470	100	26	25	18
1	754	132	40	28	33
1	248	65	19	15	23
1	145	53	17	12	16
1	138	52	15	14	13
1	108	44	13	11	8
1	435	94	28	21	18
1	185	57	16	15	13
1	274	76	23	16	11
1	315	78	22	19	22
1	182	60	18	17	11
1	260	75	20	18	18
1	241	66	18	17	15
1	156	69	17	15	9
1	259	75	19	18	18
1	365	82	23	20	22
1	266	68	18	17	14
1	301	84	25	19	22
1	238	71	21	17	14
1	257	66	19	16	21
1	784	129	43	24	29
1	223	64	18	16	15
1	221	63	20	14	16
1	340	87	23	22	12
1	307	76	22	18	17
1	395	92	24	23	22
1	491	99	26	25	24
1	1,275	155	41	36	43
1	309	82	21	21	23
1	184	58	18	13	15
1	284	74	20	19	15
1	243	70	20	17	17
1	263	68	19	17	13

TABLE 5.9 (continued)

Basic Measures of All Subbasins Hierarchically Decomposed from F-DEM

Basin Order	Area (Pixels)	Perimeter (Pixels)	Longitudinal Length (Pixels)	Transverse Length (Pixels)	Main Channel Length (Pixels)
1	263	73	20	18	14
1	177	62	18	15	10
1	510	99	26	26	20
1	308	75	21	18	20
1	170	61	17	16	19
1	295	78	22	19	19
1	284	75	22	16	14
1	338	79	21	20	19
1	211	66	17	16	19
1	291	84	25	20	26
1	262	74	22	17	13
1	230	60	18	14	20
1	135	50	14	13	12
1	225	62	17	16	14
1	314	86	24	20	16
1	315	80	23	19	17
1	265	77	21	20	19
1	304	74	23	17	14

Source: Sagar, B.S.D. and Tien, T.L., *Geophys. Res. Lett.*, 31, L06501, 2004.

Figure 5.13 shows the double logarithmic plots, for different variables that exhibit universal scaling relationships. A popular relationship between A and L_{mc}, also known as Hack's law, is shown in Equation 5.25.

$$L_{mc} \sim A^h \tag{5.25}$$

where $h \geq 0.5$, and it was reported that this h ranges between 0.56 and 0.6 for realistic basins (Hack 1957, Maritan et al. 1996b). By considering lengths and contributing areas of each network segment with index i for subbasins of order 1, order-wise power-laws, similar to Hack's law, have been computed (Table 5.10). Power-law values, for this relationship, for subbasins of orders ranging from 1 to 6 respectively include 0.502, 0.56, 0.56, 0.55, 0.55, and 0.56. Power-law relationship between the variables $L_\perp$ and $L_\sim$ yields Hurst exponent (H) according to Equation 5.26.

$$L_\perp \sim L_\parallel^H \tag{5.26}$$

A basin is said to be self-similar if this (H) is exactly 1. If this $H < 1$, then the corresponding basin is referred to be self-affine. For all the subbasins of order (ω), double logarithmic graphs are plotted between $L_\perp$ and $L_\sim$, and Hurst exponents have been computed for basins of all orders ranging from 1 to 6 respectively as 0.94, 0.94, 0.96, 0.98, 0.94, and 0.98. These (H) values

TABLE 5.10

Power-Law Values among Allometric Measures of F-DEM

Relations	Notations	For All Orders	Basin's Order					
			1	2	3	4	5	6
A and L_{mc}	h	0.55	0.502	0.56	0.56	0.55	0.55	0.56
A and P	α	1.35	1.31	1.36	1.41	1.44	1.48	1.46
P and L_{mc}	β	1.39	1.51	1.32	1.28	1.26	1.23	1.23
L_{mc} and $L_{\parallel}$	—	0.97	0.92	1.01	1.04	1.03	0.94	0.95
$L_{\perp}$ and $L_{\parallel}$	H	0.95	0.94	0.94	0.96	0.98	0.94	0.98
α and h	—	0.39	0.38	0.41	0.39	0.38	0.37	0.38
H and h	—	1.80	1.87	1.70	1.74	1.77	1.80	1.80
β and h	—	1.34	1.30	1.36	1.41	1.44	1.48	1.46
β and α	—	0.52	0.50	0.55	0.55	0.55	0.55	0.53
$2h$	D_{Lmc}	1.06	1.00	1.11	1.11	1.10	1.10	1.12
$2/\alpha$	D_P	1.48	1.53	1.47	1.42	1.39	1.35	1.37
D_{Lmc} and D_P	D_P	0.70	0.65	0.75	0.78	0.80	0.81	0.81
$1+\dfrac{D_{Lmc}}{1+H}$	—	1.55	1.52	1.57	1.59	1.56	1.57	1.57

Source: Sagar, B.S.D. and Tien, T.L., *Geophys. Res. Lett.*, 31, L06501, 2004.

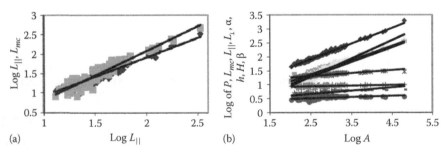

FIGURE 5.13

Allometric relationships among basic measures. (a) Squares show that the relationship between L_k and L_{mc}, and triangles indicate relationship between L_k and L_{mc}. Note that the former relationship enables the self-affinity of the basin and its subbasins. (b) Triangles show the area–transverse length relation. The crosses indicate the area–main channel length relationship. The area–perimeter relationship is shown with diamonds, and squares show the relationship between area and longitudinal relationship. The area, perimeter, and mean length are in units of pixels, and relationships between the logarithm of area of the basin and (with open stars), h (with plus), H (with solid stars), and β (with minus). (From Sagar, B.S.D. and Tien, T.L., *Geophys. Res. Lett.*, 31, L06501, 2004.)

further testify that these subbasins are self-affine. Power-law values denoted by h, α, β, and H are respectively derived for the variables A and L_{mc}, A and P, L_{mc} and P, and $L_{\perp}$ and $L_{\vee}$. These power-law values derived by considering all subbasins of all orders are respectively 0.53, 1.35, 1.38, and 0.95.

Many studies related to allometric relationships provided power-law values for the basins of similar or higher-order Hortonian basins. From such studies, it is difficult to understand the extent of deviations across order-wise subbasins (i.e., lowest-order subbasin to highest-order basins). To understand the extent of deviations in these power-law values across order-wise subbasins, these relationships among basic measures have been shown not only for higher-order basins but also for subbasins within the basin with order Ω. Fractal dimension of main channel length ($D_{L_{mc}}$) is equivalent to $2h$ and is estimated for F-DEM to be 1.06. However, it is interesting to note that significant deviations have been observed in these relationships for order-wise subbasins of lower order (Table 5.10). It is interesting to note significant deviations in the scaling laws from the lower-bound $1+(D_{L_{mc}}/(1+H)) = 3/2$. For those networks possessing $h = 0.5$ (e.g., topological random networks), this lower-bound $1+(D_{L_{mc}}/(1+H)) = 3/2$ (Veitzer and Gupta 2000). Along with other allometric power-law relationships, the estimates for $1+(D_{L_{mc}}/(1+H))$ for order-wise subbasins have been tabulated in Table 5.10. Decomposed subbasins of various orders and their corresponding main lengths of F-DEM are illustrated (Figure 5.14). The allometric relationships and power-law values for other popular network models—such as Scheideggar networks, Peano networks, optimal channel networks (OCNs), realistic networks, and F-SCNs—have been shown in Table 5.11. The estimates derived for F-DEM and the subbasins decomposed from F-DEM are geomorphologically realistic as OCNs and realistic river networks.

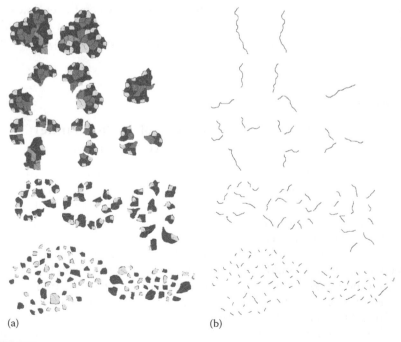

(a)   (b)

FIGURE 5.14
(a) Subbasins decomposed from a Hortonian F-DEM and (b) corresponding main lengths. (From Sagar, B.S.D. and Tien, T.L., *Geophys. Res. Lett.*, 31, L06501, 2004.)

TABLE 5.11

Scaling Exponents for Several Networks

Network	D_{Lmc}	h	H	$1 + \dfrac{D_{Lmc}}{1 + H}$
Scheideggar	1	2/3	1	5/3
Peano	1	1/2	1	3/2
OCN (fractal)	1.05	0.56	0.88	1.56
F-SCN	1.06	0.55	0.95	1.54
Tirso (IT)	1.05	0.53	0.94	1.54

Source: Sagar, B.S.D. and Tien, T.L., *Geophys. Res. Lett.*, 31, L06501, 2004.

These power-law relationships provide insights to understand landscape organization and commonly sharing physical mechanisms between the basins. Overall structure of the network determines the geomorphologic processes and functions. Most of the allometric power-laws derived for subbasins of various orders rely heavily on planimetric measures. Little emphasis was laid on the network organization. To show the impact of small network geometric organization in understanding basin processes and functions, a host of new power-law relationships have been proposed. These new power-law relationships rely on travel time networks, corresponding convex hulls, and convexity measures. In what follows, allometric relationships between travel time channel networks, convex hulls, and convexity measures have been provided by highlighting their importance in understanding basin structure in a better way.

Allometric Relationships between Travel Time Channel Networks, Convex Hulls, and Convexity Measures

Convex hull of a non-convex branched (loopless) network can be treated as a basin, although some basins are similar to close hulls. It is known that a branched network possesses open-ended network segments, lower-order to highest-order network segments, and an outlet. Travel time networks of a network with an outlet could be generated by recursively removing open-ended points of network until the travel time network reaches outlet. A new topological quantity that has not been noted thus far could also be derived by computing convex hulls of these travel time networks. Convexity measures could be computed by taking the ratio of lengths of travel time networks and the areas of corresponding convex hulls of travel time networks. The three significant scaling relationships derived among lengths of travel time networks, areas of corresponding convex hulls, and convexity measures include (1) $L(X_n) \sim A(C(X_n))^{\alpha}$, (2) $CM(X_n) \sim 1/L(X_n)^{\beta}$, and (3) $CM(X_n) \sim 1/A(C(X_n))^{\gamma}$, where α, β, and γ are power-law values.

Shape is an important factor in geomorphologic analysis. In the previous sections of this chapter, importance of topological geometry of networks in understanding various geomorphologic processes has been highlighted. Various quantitative characteristics could be derived from such networks via morphometric, fractal, and allometric scaling analyses (Horton 1945, Rigon et al. 1996, Rodriguez-Iturbe and Rinaldo 1997, Turcotte 1997, Rinaldo et al. 1998, Sagar et al. 1998a,b, 2001, Maritan et al. 2002, Sagar and Chockalingam 2004, Sagar and Tien 2004, Chockalingam and Sagar 2005). By defining topological aggregation, structure and length of network pattern, elongation, and general shape of basin, a large number of scaling coefficients could be defined (Rinaldo et al. 1998). Some of these scaling coefficients, to name a few, include h, ε, H, and β. The ranges of these coefficients for realistic networks are respectively 0.53–0.60, 0.65–0.90, 0.70–1.00, and 0.41–0.46 (Rinaldo et al. 1998). Geomorphologic width functions and random cascade models that are based on network links and contributing areas would provide ways to characterize basin processes (Marani et al. 1991, Gupta and Waymire 1993, Marani et al. 1994, Veneziano et al. 2000). This type of characterization does not involve general geometric organization and the diverging angles between network segments, which have a major role in understanding the processes in geomorphologic basins. However, allometric relationships derived based on travel time networks, convex hulls, and convexity measures involve these two important features.

An example of non-convex set-like branched networks (Figure 5.15a; Turcotte et al. 1998), and its corresponding convex hull (Figure 5.15b) are shown. They are respectively denoted by X and $C(X)$. A boundary along the path taken by a tight rubber ring when it is warped around the non-convex set (e.g., Figure 5.15a) is treated as a convex hull (e.g., Figure 5.15b).

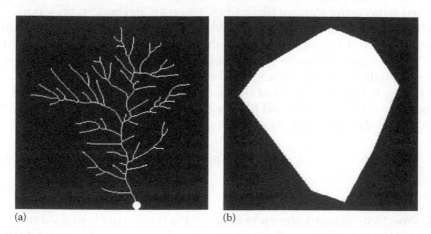

(a)   (b)

FIGURE 5.15
(a) An example of fourth-order channel network (non-convex set) and (b) its convex hull. A stationary outlet is shown as a round dot in (a). (From Tay, L.T. et al., *Water Resour. Res.*, 42, W06502, 2006.)

The derivation of a host of new allometric power-laws requires following three steps:

1. Generation of travel time network sequence
2. Convex hull construction for corresponding travel time networks
3. Computation of three basic measures

The implementations of the earlier three steps have been demonstrated on a model network with a stationary outlet (Figure 5.15a).

By treating a treelike network (e.g., branched river networks, Figure 5.15a) with an outlet as a dry-tree, a fire is lit at all the extremities of the network and allow that fire to propagate at uniform speed toward outlet. As a result, the network is progressively burned out as time progresses. The network that is burned across time intervals visualized could be treated as travel time networks. This process is mathematically explained—by denoting the network and the sequence of travel time networks respectively as X and X_n, where n is 0, 1, 2, ..., N—as follows. Morphological pruning of branched network iteratively yields travel time networks.

Let X and B denote non-convex VCN and structuring element possessing certain characteristic information such as shape, size, orientation, and origin (Chockalingam and Sagar 2005). Structuring element B could be decomposed into various ways. The set X is a loopless network of one-pixel-wide caricature. Such an X is a composition of N network subsets, Nth level subset(s) being the outlet(s). The first-level open-ended network subsets are extremities of the network, which are also termed as source points. The two transformations essentially required to generate travel time networks include hit-or-miss transformation (HMT) of X by disjointed structuring element (B^1) and (B^2), and an algebraic subtraction.

HMT of X by B is denoted by $\left(X * \{B\}\right)$, where $\{B\} = \left(B_1^1 \cup B_2^1\right)$ (Figure 5.16). Erosions of X of B_1^1 and X^c by B_2^1 are shown in Figure 5.16 (Jang and Chin 1990). This figure illustrates X (network elements) and X^c respectively represented with 1s and 0s. B_1^1 and B_2^1 are two disjointed sets, and the union of them is B. Logical union and intersection of eroded versions of X and X^c obtained with respect to B_1^1 and B_2^1 have been shown in Figure 2.15. The HMT of X by B is shown in Equation 5.27.

$$X * \{B\} = \left(X \ominus B_1^1\right) \cap \left(X^c \ominus B_2^1\right) \tag{5.27}$$

where $\{B\} = \left(B_1^1 \cup B_2^1\right)$.

Connectivity network shown in Figure 5.15a is recursively pruned by performing the following set of transformations (Equations 5.28 through 5.31).

$$X * \{B\} = \left(X \ominus B_1^k\right) \cap \left(X^c \ominus B_2^k\right) \tag{5.28}$$

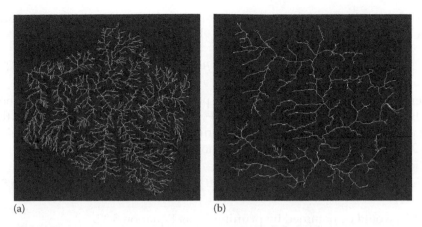

FIGURE 5.16
Channel networks derived for 14 subbasins of (a) Cameron Highlands and (b) Petaling Jaya regions of Malaysia. (From Tay, L.T. et al., *Int. J. Remote Sens.*, 28(15), 3363, 2007.)

$$
B_1^1 = \begin{matrix} 1 & 0 & 0 \\ 0 & 1 & 0 \\ 0 & 0 & 0 \end{matrix} \quad
B_1^2 = \begin{matrix} 0 & 0 & 1 \\ 0 & 1 & 0 \\ 0 & 0 & 0 \end{matrix} \quad
B_1^3 = \begin{matrix} 0 & 0 & 0 \\ 0 & 1 & 0 \\ 0 & 0 & 1 \end{matrix} \quad
B_1^4 = \begin{matrix} 0 & 0 & 0 \\ 0 & 1 & 0 \\ 1 & 0 & 0 \end{matrix}
$$

$$
B_1^5 = \begin{matrix} \times & 1 & \times \\ 0 & 1 & 0 \\ 0 & 0 & 0 \end{matrix} \quad
B_1^6 = \begin{matrix} 0 & 0 & \times \\ 0 & 1 & 1 \\ 0 & 0 & \times \end{matrix} \quad
B_1^7 = \begin{matrix} 0 & 0 & 0 \\ 0 & 1 & 0 \\ \times & 1 & \times \end{matrix} \quad
B_1^8 = \begin{matrix} \times & 0 & 0 \\ 1 & 1 & 0 \\ \times & 0 & 0 \end{matrix}
$$

$$
B_2^1 = \begin{matrix} 0 & 1 & 1 \\ 1 & 0 & 1 \\ 1 & 1 & 1 \end{matrix} \quad
B_2^2 = \begin{matrix} 1 & 1 & 0 \\ 1 & 0 & 1 \\ 1 & 1 & 1 \end{matrix} \quad
B_2^3 = \begin{matrix} 1 & 1 & 1 \\ 1 & 0 & 1 \\ 1 & 1 & 0 \end{matrix} \quad
B_2^4 = \begin{matrix} 1 & 1 & 1 \\ 1 & 0 & 1 \\ 0 & 1 & 1 \end{matrix}
$$

$$
B_2^5 = \begin{matrix} \times & 0 & \times \\ 1 & 0 & 1 \\ 1 & 1 & 1 \end{matrix} \quad
B_2^6 = \begin{matrix} 1 & 1 & \times \\ 1 & 0 & 0 \\ 1 & 1 & \times \end{matrix} \quad
B_2^7 = \begin{matrix} 1 & 1 & 1 \\ 1 & 0 & 1 \\ \times & 0 & \times \end{matrix} \quad
B_2^8 = \begin{matrix} \times & 1 & 1 \\ 0 & 0 & 1 \\ \times & 1 & 1 \end{matrix}
$$

FIGURE 5.17
Disjointed structuring templates in eight directions. (From Tay, L.T. et al., *Water Resour. Res.*, 42, W06502, 2006.)

where B_1^k and B_2^k are disjointed structuring elements of $\{B\}$, with $k = 1, 2, \ldots, 8$. Figure 5.17 shows disjointed structuring elements in eight directions.

A pruned version of X could be obtained by subtracting $(X * \{B\})$ from X as shown in Equation 5.29.

$$X \otimes \{B\} = X - (X * \{B\}) \tag{5.29}$$

where $X \otimes \{B\}$ is the first pruned version, and it is denoted as X_1, $X * \{B\} = (X \ominus B_1^k) \cap (X^c \ominus B_2^k)$, and $(X \ominus B_1^k) = \{x | (B_1^k)_x \subseteq X\}$, and $\{B\} = \{(B_1^1, B_1^2, \ldots, B_1^8), (B_2^1, B_2^2, \ldots, B_2^8)\}$ (Figure 5.17). The "×s" in Figure 5.17 signify the "don't care" condition, in the sense that it does not matter whether the pixel in that location has a value of 0 or 1. Equation 5.30 elaborates it further.

$$X \otimes \{B\} = ((\cdots((X \otimes B^1) \otimes B^2)\cdots) \otimes B^8) \qquad (5.30)$$

where $X \otimes \{B\}$ is an output obtained by peeling of X in one pass with B^1, then peeling of the result in one pass with B^2, and so on until X is spurred in the last pass with B^8. By repeating this pruning process on X_1, which is nothing but $X \otimes \{B\}$, X_2 would be obtained as shown in Equation 5.31.

$$X_2 = X_1 - (X_1 * \{B\}) = X_1 \otimes \{B\} \qquad (5.31)$$

And X_3 would be obtained by pruning X_2 as Equation 5.32.

$$X_3 = X_2 - (X_2 * \{B\}) = X_2 \otimes \{B\} \qquad (5.32)$$

This iteration of pruning goes on until the last point, which is an outlet of branched network (X) reaches as explained in Equation 5.33.

$$X_N = X_{N-1} \otimes \{B\} \qquad (5.33)$$

where X_N is the outlet.

Each level of pruned network X_n, for n ranging from 1 to N, is gray shaded and superposed on one another as shown in Equation 5.34.

$$X = \bigcup_{\forall n} X_n \qquad (5.34)$$

where X_n is the gray-shaded pruned network. This is shown in Figure 5.18. Snapshots from a sequence of travel time networks generated for a model network (Figure 5.15a) are shown in Figure 5.19. Some properties of these travel time networks ranging from $X_1, X_2, \ldots, X_N$ satisfy the following relations:

1. $X = \bigcup_{n=0}^{N} (X_n - X_{n+1})$
2. $X_N \subset X_{N-1} \subset \cdots \subset X_2 \subset X_1 \subset X$
3. $X, X_1, X_2, \ldots, X_N$ are obtained by iterative pruning

The logic behind following morphological pruning recursively is to peel off the extremities of loopless branched network (X) iteratively based on a postulate that computes time required for the particle (e.g., in a fluid flow-like river stream) to reach the outlet (X_N). Bifurcation points in the network would be encountered during this iterative pruning process. In realistic

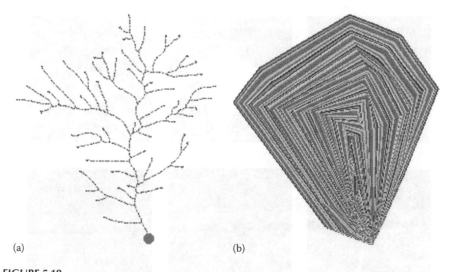

(a)

(b)

FIGURE 5.18
(a) Gray-shaded travel time network being pruned iteratively till it reaches the outlet, and (b) gray-shaded union of convex hulls of networks pruned to different degrees. (From Tay, L.T. et al., *Water Resour. Res.*, 42, W06502, 2006.)

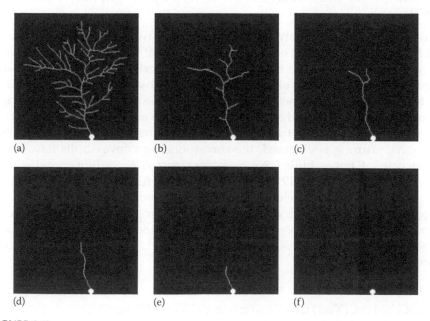

(a)   (b)   (c)

(d)   (e)   (f)

FIGURE 5.19
Snapshots of pruned versions obtained at iterations of 1, 50, 100, 150, 200, and 246, sequentially generated (a–f) travel time networks.

(continued)

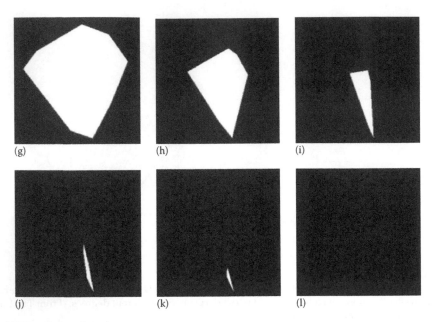

FIGURE 5.19 (continued)
Snapshots of pruned versions obtained at iterations of 1, 50, 100, 150, 200, and 246, sequentially generated (g–l) their corresponding convex hulls. (From Tay, L.T. et al., *Water Resour. Res.*, 42, W06502, 2006.)

network case (e.g., river network), a path between two points is not the shortest Pythagorean distance, but has some amount of sinuosity. Time required for a particle to reach the outlet in a flow network that is in straight path would be lesser than that of the tortuous flow network. The network would be treated as symmetric network if the travel time required for all particles from all source points of the network to reach the outlet is the same.

Once the sequence of travel time networks that could be obtained by recursive pruning is obtained, the corresponding convex hulls have been constructed. Convex hulls have been constructed according to half-plane closing approach (Soille 1998) that is explained in Chapter 4. Snapshots of the convex hulls of the selected travel time networks have been shown in Figure 5.19. Each corresponding convex hull is properly gray shaded, and a superposed version has been shown in Figure 5.18c. Convex hull of a travel time network of order n, X_n, is denoted by $C(X_n)$. These $C(X_n)$ possess the following properties:

1. $X_n \subseteq C(X_n)$
2. $C(X_N) \subset C(X_{N-1}) \subset C(X_{N-2}) \subset \cdots \subset C(X_2) \subset C(X_1) \subset C(X)$
3. $A(C(X_N)) \leq A(C(X_{N-1})) \leq A(C(X_{N-2})) \leq \cdots \leq A(C(X_2)) \leq A(C(X_1))$
   $\leq A(C(X))$
4. $L(X_n) \leq A(C(X_n))$

$L(X_n)$ and $A(C(X_n))$, respectively, denote the length of travel time network and the area of convex hull of travel time network. Ratio between $L(X_n)$ and $A(C(X_n))$ yields convexity measure (Heijmans and Tuzikov 1998, Zunic and Rosin 2004). This convexity measure of travel time network, denoted by $CM(X_n)$ (Equation 5.35), ranges between 0 and 1. $CM(X_n)$ would be 1, if and only if X_n is convex. The rate of change in the areas of $C(X_n)$ is relatively faster than that of the length of X_n. Hence, the convexity measures of decreasing $L(X_n)$ and $A(C(X_n))$ converge.

$$CM(X_n) = \frac{L(X_n)}{A(C(X_n))} \tag{5.35}$$

There are three novel allometric relationships between the lengths of the sequential travel time networks $(L(X_n))$, the area of convex hull, $A(C(X_n))$, and the corresponding convexity measure $CM(X_n)$. Those relationships include Equations 5.36 through 5.38.

$$L(X_n) \sim (A(C(X_n)))^\alpha \tag{5.36}$$

$$CM(X_n) \sim \frac{1}{L(X_n)^\beta} \tag{5.37}$$

$$CM(X_n) \sim \frac{1}{A(C(X_n))^\gamma} \tag{5.38}$$

The model network (Turcotte et al. 1998) considered to show the three new allometric relationships is of size 256 × 256 pixels (Figure 5.15a). Morphological pruning of this network took 246 iterations to reach the outlet. These 246 travel time networks and their corresponding convex hulls (Figure 5.18) are gray shaded as per Equation 5.39.

$$X = \bigcup_{n=0}^{N-1} \bigcup_{i=n}^{255} (X_n - X_{n+1})^i \tag{5.39}$$

and

$$C(X) = \bigcup_{n=0}^{N-1} \bigcup_{i=n}^{255} (C(X_n) - C(X_{n+1}))^i \tag{5.40}$$

It is obvious that a network that is asymmetric possesses an asymmetric convex hull. It is defined that a network is symmetric if there exists a geometric similarity between all possible convex hulls of corresponding travel time network sequence. If the length of the travel time network $(L(X_n))$ equals the area of the corresponding convex hull, then the upper limit of convexity measure, i.e., 1, would be attained. An example network is a network that fills the space satisfying the space-filling characteristic. Peano-curve-like network is one of such examples, which has Hausdorff dimension of 2 in two-dimensional Euclidean space.

The convexity measure is similar to drainage density. A relationship between width function of a basin and the convexity measure that is explained here is worth exploring. Figure 5.20a shows the allometric relationships between $L(X_n)$ and $A(C(X_n))$, and their convexity measures $CM(X_n)$. These relationships that are statistically significant have been shown for the model network as shown in Equations 5.41 through 5.43.

$$L(X_n) \sim (A(C(X_n)))^{0.57} \tag{5.41}$$

$$CM(X_n) \sim \frac{1}{L(X_n)^{0.7}} \tag{5.42}$$

$$CM(X_n) \sim \frac{1}{A(C(X_n))^{0.43}} \tag{5.43}$$

Lengths and convexity measures of travel time networks have been plotted as functions of areas of convex hulls and lengths of travel time networks (Figure 5.20). Fractal nature of the topological and geometric organization of the network could be better understood from these allometric relationships. Similar process of generating a sequence of travel time networks and constructing convex hulls has been performed on the realistic channel networks extracted from the DEM of Cameron Highlands of Malaysian Peninsular situated between the geographical coordinates 101°15′–101°20′E and 4°31′–4°36′N (Figure 3.12). These networks partitioned into seven subbasins have been shown in Figures 5.16a and 5.21. For model network, a linear relationship is observed between $CM(X_n)$ and $CM(X_{n+1})$ (Figure 5.20b).

The allometric power-law relationships shown for the model network have also been computed for the travel time networks generated for the seven networks belonging to seven subbasins (Figure 5.21). Those results have been provided for those seven subbasins along with model network (Table 5.12; Figure 5.22). This table also shows the network morphometric quantitative (R_B) and (R_L) and other two popular power-laws, viz., Hack's and Hurst's exponents. By comparison, it could be seen that α and h values are similar, though the α values are slightly higher. However, Hurst exponents (H) have not shown any significant relationships with the new allometric power-laws shown for travel time networks. It has been

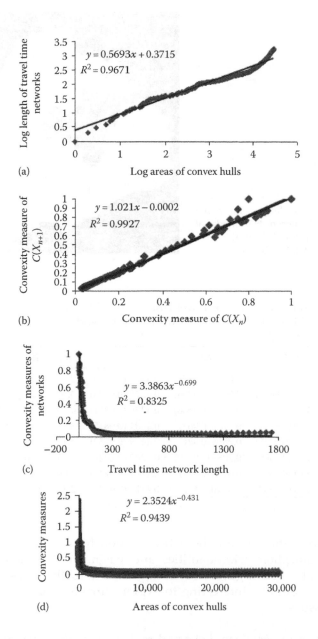

FIGURE 5.20
Cross-plots between (a) lengths of the sequential pruned networks and the corresponding areas of convex hulls in logarithm scale; (b) convexity measures at time n and at time $n + 1$; (c) lengths and convexity measures in logarithm scale; and (d) areas of convex hulls and convex measures in logarithm scale. (From Tay, L.T. et al., *Water Resour. Res.*, 42, W06502, 2006.)

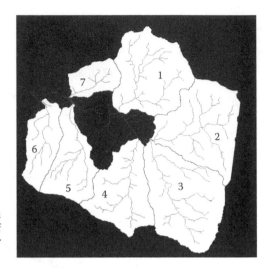

FIGURE 5.21

Channel networks derived for seven subbasins of Cameron Highlands of Malaysia. (From Tay, L.T. et al., *Water Resour. Res.*, 42, W06502, 2006.)

TABLE 5.12

Allometric Power-Laws between Travel Time Channel Networks, Convex Hulls, and Convexity Measures for Model Network and Networks of Seven Basins of Cameron Highlands

Network	α (R^2)	σ (R^2)	λ (R^2)	R_B	R_L	h	H
Model	0.5693 (0.9671)	0.6988 (0.8325)	0.4307 (0.9439)	3.84	1.66		
Basin 1	0.5777 (0.9883)	0.7109 (0.9358)	0.4223 (0.9783)	3.60	2.21	0.5414	0.9714
Basin 2	0.5774 (0.9925)	0.7189 (0.9586)	0.4226 (0.9861)	4.35	2.25	0.5561	1
Basin 3	0.5799 (0.9934)	0.7131 (0.963)	0.4201 (0.9875)	3.31	2.39	0.5612	0.9256
Basin 4	0.5521 (0.9835)	0.7814 (0.92)	0.4479 (0.9752)	4.47	3.18	0.5671	0.9506
Basin 5	0.5798 (0.9905)	0.7083 (0.9469)	0.4202 (0.982)	3.31	2.16	0.5766	0.9162
Basin 6	0.5819 (0.9865)	0.6955 (0.925)	0.4181 (0.9743)	4.00	2.64	0.5746	0.8597
Basin 7	0.5885 (0.9887)	0.68 (0.9348)	0.4115 (0.9772)	2.82	2.39	0.5548	0.8950

Source: Tay, L.T. et al., *Int. J. Remote Sens.*, 28(15), 3363, 2007.

postulated that these novel allometric relationships may be having relations with morphometric quantities and basin width functions. But it requires results on a large number of networks to further substantiate this postulate. It would be interesting to explore further work to address the following issues:

1. A large number of synthetic and realistic branched networks with different topologies to find out whether the novel allometric power-laws are of universality class.

2. Between the elongated basin and radial basin, what would be the differences in the ranges of these proposed power-law values derived from travel time networks and convexity measures?

3. Between the elongated basin and radial basin, what would be the differences in the rates of change in the convexity measures across travel time networks?

To properly address the aforementioned open problems, one needs to consider various networks bearing different geometries with stationary longitudinal length ($L_\parallel$) and varying transverse length ($L_\perp$) by maintaining ($L_\perp$) always lesser than stationary ($L_\parallel$). From such varied networks, the power-laws based on travel time networks need to be compared with popular Hack's and Hurst's exponents of the networks.

From this work, it is inferred that the rates of change in the length of travel time network and in the areas of corresponding convex hulls are related, and such a relationship exhibits scale-invariant characteristics. From these length–area measures, convexity measures derived for a dynamically shrinking travel time network propagating toward its outlet provide insights for geomorphologists to understand the characterization process. The new power-laws that are shown here complement with other popular scaling coefficients, which further help understanding the commonly shared physical mechanisms in different river basins. What is interesting most is to find out whether these power-law values of a network of higher order hold good for the networks of lower order decomposed from higher-order network. This interesting postulate can be addressed by hierarchically decomposing the higher-order basin into subbasins of lower order, which further facilitates to classify the subbasins. Such an exercise would provide insights to explore links to find out what makes a subbasin different from the other within a larger basin.

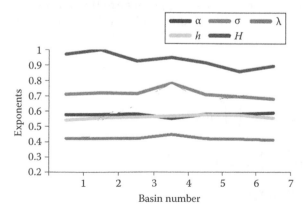

FIGURE 5.22
Basin-wise exponents that include new exponents computed based on travel time networks and their convex hulls, and other popular coefficients. (From Tay, L.T. et al., *Water Resour. Res.*, 42, W06502, 2006.)

Universal Scaling Laws in Water Bodies and Their Zones of Influence

Based on topologically significant loopless (channel), looplike (watershed boundaries) networks, and the spatial distribution of topographic depression, landscape organization could be better explained. Quantitative description of land surface via morphometric analysis, and allometry of networks has been detailed in the previous sections of this chapter. Spatially distributed surface water bodies that usually occur at the topographic depressions also provide clues to understand the spatiotemporal organization of land surface. Moreover, during the flood period, surface water bodies are first-level topographic regions that get flooded. Adjacent water bodies would get merged as the flood level increases. Zones of influence of water bodies would be looplike network that form along the merging points of floodwater frontlines propagating from the water bodies. Water bodies and their zones of influence are other two topologically interdependent phenomena that follow the universal scaling laws. These two phenomena are like valley and ridge connectivity networks. A host of universal scaling laws have been derived for water bodies and their zones of influence, and found that these two interdependent phenomena of varied shapes and sizes belong to different universality classes.

Markers and masks are like water bodies and their zones of influence. Other examples include seed and pulp, skeleton and body, river network and basin, and grain and pore. For a cluster of N number of water bodies within a biogeographic boundary, there would be N number of zones of water body influences. The floodwater frontlines that propagate from all N number of water bodies would get extinguished at meeting points. The lines formed at these extinguishing points are the boundary lines of zones of influence of water bodies. Water body and its corresponding zone of influence are denoted by X_i and $ZI(X_i)$, and the following properties would be satisfied:

1. $X_i \subseteq ZI(X_i)$

2. $\displaystyle\bigcup_{i=1}^{N} X_i = X$ and $\displaystyle\bigcup_{i=1}^{N} ZI(X_i)$

3. $\displaystyle\left(\bigcup_{i=1}^{N} X_i\right) \subseteq \left(\bigcup_{i=1}^{N} ZI(X_i)\right)$

4. $A(X_i) \leq A(ZI(X_i))$

Figure 5.23 shows the schematic of water bodies that are contained in their zones of influence. A large number of allometric scaling relationships have been proposed for model networks and for realistic networks (e.g., Horton 1945,

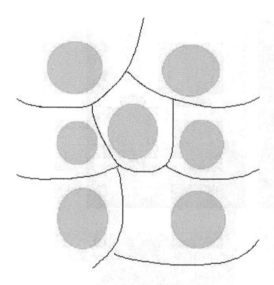

FIGURE 5.23
Schematic section diagram showing that the water bodies (circular objects) are smaller than the influence zones (regions within the black boundary). (From Sagar, B.S.D., *Water Resour. Res.*, 43(2), W02416, 2007.)

Langbein 1947, Hurst 1951, Hack 1957, Mandelbrot 1982, Mesa and Gupta 1987, Robert and Roy 1990, Rosso et al. 1991, Ijjasz-Vasquez et al. 1993, Sagar and Rao 1995, Maritan et al. 1996a,b, Rodriguez-Iturbe and Rinaldo 1997, Banavar et al. 1999, Dodds and Rothman 1999, Sagar 2000, Veitzer and Gupta 2000, Maritan et al. 2002, Sagar and Srinivas 2002, Sagar and Chockalingam 2004, Sagar and Tien 2004, Chockalingam and Sagar 2005, Tay et al. 2006). These universal scaling relationships have been derived between many allometric and morphometric parameters estimated on the basis of the analysis of geomorphologic data available for numerous models and realistic networks and basins.

Water bodies and their zones of influence do possess characteristics such as shape, size, lengths, and orientations. Zones of influence are the zones that are prone to flooding from their corresponding water bodies. Scaling relationships have been proposed for water bodies (Sagar and Rao 1995, Sagar 2000), but no attempt has been made until 2007 to verify scaling laws for the zones of influence. In what follows, the universal scaling laws, derived via allometric relationships and granulometric analysis, of both water bodies and their zones of influence are provided to further compare them to find out whether they belong to single universality class or not.

A region of a large number of semi-artificial irrigation tanks of various sizes and shape has been considered in Figure 5.24a. See Chapter 3 about this study area specification and other physiographic conditions.

Zones of Influence ($ZI(X_i)$)

Zones of influence for the large number of water bodies (Figure 5.24b) are constructed by following skeletonization of zone of influence (SKIZ), which is due to Lantuejoul (1978, 1980). From DEMs, catchments can be divided by various approaches, one of which is an elegant watershed transformation

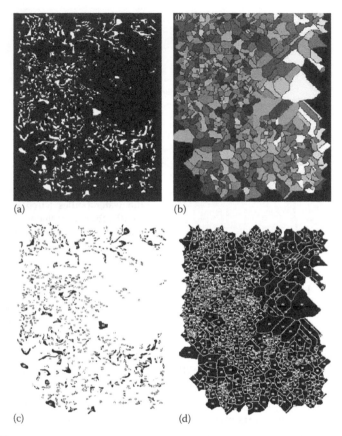

FIGURE 5.24
(a) A section consisting of a large number of small water bodies traced from floodplain region of Gosthani river, and (b) zones of influence of water bodies shown in Figure 5.24a. Different gray shades are used to distinguish the adjacent influence zones, and (c and d) the labels of water bodies and their corresponding zones of influence (see Table 5.13). (From Sagar, B.S.D., *Water Resour. Res.*, 43(2), W02416, 2007.)

(Beaucher 1990). Catchments are like zones of influence. Catchments have regional minima, in other words "markers." In the case of zones of influence map, water bodies act as markers as water bodies are situated in the regions <2° slope, and there would be no major topographic undulations in such regions. In view of this, the impact of topography is left out while deriving zones of influence of water bodies. Since the study region considered here is <2° slope, which is free from major topographic undulations (with the exception of a few isolated hummocks of rather gently sloped hills), it is considered that floodwater that propagates from water bodies (markers) would fill the no-water body region like a process similar to dilation that was adopted to generate zones of influence. Figure 5.24c and d respectively shows the indexes (labels) for water bodies and their corresponding zones of influence.

Allometric universal scaling laws have been computed for water bodies and their corresponding zones of influence. In what follows, an approach has been followed to generate zones of influences of water bodies.

Generation of Zones of Influence of Water Bodies

Recursively applying the dilation transformation on water bodies would give an effect of continuous flooding. Such a process of implementing iterative dilations of water bodies yields dilated portions of the number of water bodies reaching extinguishing points. The process of SKIZ is shown mathematically as six steps.

Step 1: Dilate X_i and X_js by nth size B, where $i \neq j$ and i, j range from 1 to N. Denote such dilated versions as $(X_i \oplus nB)$ and $\left(\bigcap_{\substack{j \neq i \\ i=1}}^{N} X_j \oplus nB \right)$.

Step 2: Subtract $\left(\bigcup_{\substack{j \neq i \\ i=1}}^{N} X_j \oplus nB \right)$ from $(X_i \oplus nB)$ and denote the result as $ZI_n(X_i)$.

Step 3: Repeat steps 1 and 2 by increasing n from 1 to N until convergence.

Step 4: Take the union of all the subtracted versions obtained for all n's of B to obtain $ZI_n(X_i)$ as shown in $ZI(X_i) = \bigcup_{n=0}^{N} (ZI_n(X_i))$.

Step 5: Repeat steps 1–4 by changing i, i.e., X_i, where i is ranging from 1 to I to obtain $ZI(X_i)$ for all i's.

Step 6: Take the union of all $ZI(X_i)$ for all i's shown mathematically as $ZI(X) = \bigcup_{i=1}^{I} (ZI(X_i))$.

The earlier six-step approach has been implemented on a section containing a large number of small water bodies (Figure 5.24a) to finally obtain their zones of influence (Figure 5.24b). These $ZI(X_i)$ satisfy the following properties:

1. $X_i \subseteq ZI(X_i) \Rightarrow \bigcup_{i=1}^{I} X_i \subseteq (ZI(X_i))$

2. $N(X_i) = N(ZI(X_i))$: The number of water bodies in the section would be equivalent to the number of zones of influence

3. $A(X_i) < A(ZI(X_i))$

4. $L_{\parallel}(X_i) < L_{\parallel}(ZI(X_i))$

5. $L_{\perp}(X_i) < L_{\perp}(ZI(X_i))$

Allometry-Based Scaling Laws

Basic measures such as area (A), longitudinal length $(L_{\parallel})$, and transverse length $(L_{\perp})$ of each water body and its corresponding zone of influence have been computed (Table 5.13). Several allometric relationships have been derived (Figures 5.25 and 5.26). These relationships are between A and $L_{\perp}$,

TABLE 5.13

Basic Measures of Water Bodies and Their
Zones of Influence

M	$Area_M$	$L_{\parallel M}$	$L_{\perp M}$	X	$Area_X$	$L_{\parallel X}$	$L_{\perp X}$
2	349	38	26	2	1,434	56	50
5	307	30	24	1	1,612	71	59
7	48	10	8	4	827	41	28
8	51	11	9	8	199	19	16
10	95	19	9	14	469	35	25
11	292	31	17	11	712	47	28
12	299	38	19	12	1,491	66	40
13	85	15	10	19	757	38	37
14	1822	61	60	9	7,473	101	100
17	70	13	7	5	1,148	63	44
18	200	24	13	21	1,003	56	56
20	48	10	9	20	601	41	25
21	73	11	9	6	2,042	62	56
23	244	25	24	15	2,040	60	50
24	86	12	12	26	514	31	25
25	40	8	6	24	519	37	31
26	517	30	28	25	3,234	75	74
27	131	22	8	17	1,373	60	34
28	141	17	16	3	2,649	84	69
29	129	20	16	10	1,213	59	50
30	192	24	13	16	1,912	74	47
31	171	23	16	18	1,295	53	37
32	136	27	10	34	1,847	50	50
33	101	17	9	27	895	56	34
35	157	16	13	22	1,817	56	56
36	390	37	18	40	1,699	53	46
37	155	16	14	37	393	31	22
38	73	10	10	35	580	32	26
39	186	21	21	36	1,255	56	54
42	85	12	9	41	358	28	20
44	621	50	24	29	2,920	75	56
45	9	3	3	38	189	18	12
46	86	13	9	28	1,386	50	47
47	160	18	16	44	642	41	28
50	264	43	18	48	1,055	50	38
51	39	11	5	49	237	23	19
52	210	20	19	42	592	37	35
53	83	13	11	33	2,756	65	60
54	433	38	18	47	3,512	88	79
57	168	23	11	51	693	41	37

TABLE 5.13 (continued)

Basic Measures of Water Bodies and Their
Zones of Influence

M	Area$_M$	$L_{\parallel M}$	$L_{\perp M}$	X	Area$_X$	$L_{\parallel X}$	$L_{\perp X}$
59	131	19	11	56	676	52	40
60	385	33	17	30	6,100	125	103
61	1476	86	58	52	4,768	112	78
62	189	17	16	50	635	37	28
63	179	24	11	46	1,621	57	56
64	540	44	36	58	2,729	81	59
65	691	78	54	62	2,601	75	69
69	106	19	12	59	744	53	38
71	384	44	26	57	2,480	78	69
72	275	30	13	13	7,361	178	134
74	34	8	7	63	761	40	32
75	302	30	25	69	881	46	42
76	63	11	10	66	548	31	26
78	559	47	26	54	3,691	100	63
79	206	22	22	55	1,847	72	62
82	674	51	22	76	2,123	78	50
83	1534	63	52	75	6,917	119	103
84	37	8	6	67	946	47	34
86	50	12	6	60	5,231	132	90
88	1293	57	38	65	7,372	160	140
89	166	21	11	83	658	37	22
91	262	38	14	71	5,383	94	90
92	100	22	8	78	1,762	59	53
94	93	16	8	80	848	47	34
95	1158	94	23	81	5,236	150	93
96	239	28	16	82	2,470	90	65
97	77	15	10	84	848	47	25
98	158	17	15	85	520	28	28
99	159	21	16	86	1,230	50	50
100	236	47	11	87	1,289	65	37
101	422	32	27	90	1,741	78	47
102	370	39	22	88	2,009	56	53
104	280	32	15	100	1,264	63	35
105	48	10	6	91	328	28	22
106	353	26	20	101	925	40	32
107	89	18	8	92	757	40	31
109	67	12	11	89	994	50	28
111	36	8	6	97	330	28	22
112	351	31	29	104	1,342	50	37
113	34	7	7	95	637	41	37

(continued)

TABLE 5.13 (continued)

Basic Measures of Water Bodies and Their
Zones of Influence

M	$Area_M$	$L_{\|M}$	$L_{\perp M}$	X	$Area_X$	$L_{\|X}$	$L_{\perp X}$
114	1143	96	61	98	5,055	122	84
116	103	14	13	99	3,316	79	78
117	62	9	8	94	761	38	35
118	796	52	48	96	5,663	91	85
120	56	11	7	108	729	35	28
121	147	24	15	106	800	41	34
122	252	34	13	105	1,187	58	37
125	350	36	19	109	3,370	75	72
126	68	13	7	111	423	31	18
127	137	20	15	102	1,917	93	78
128	83	20	7	107	2,662	78	66
129	171	30	16	110	2,925	85	68
130	129	24	10	117	1,142	63	31
131	252	23	19	115	1,417	65	43
132	448	42	18	120	1,648	50	50
133	130	18	14	116	3,009	75	59
134	573	37	28	114	2,508	75	62
135	796	73	34	119	12,177	169	157
136	161	25	13	123	1,010	41	40
137	152	22	10	124	2,705	91	50
138	165	27	9	122	8,189	140	100
139	145	17	15	112	2,872	97	90
140	52	12	6	125	2,374	69	53
141	428	59	13	127	3,747	95	78
142	184	21	14	113	1,633	85	81
143	238	27	17	128	1,699	79	65
145	245	23	20	118	1,571	84	78
146	507	35	33	129	2,498	71	66
148	102	14	12	137	4,586	92	87
149	457	36	32	132	3,820	91	56
150	314	30	18	126	11,046	156	103
151	816	43	36	133	2,644	91	66
152	87	14	13	140	3,806	103	100
153	138	15	14	135	2,025	62	60
155	134	14	14	130	2,860	72	62
156	82	15	13	134	3,451	72	72
157	205	21	16	139	1,283	53	53
158	453	36	30	136	4,188	96	76
159	155	22	16	142	1,198	51	47
160	30	7	6	141	994	50	32

TABLE 5.13 (continued)

Basic Measures of Water Bodies and Their
Zones of Influence

M	Area$_M$	$L_{\parallel M}$	$L_{\perp M}$	X	Area$_X$	$L_{\parallel X}$	$L_{\perp X}$
163	101	18	8	143	863	51	31
164	103	16	10	147	1,271	47	45
167	8	3	3	138	2,822	72	53
168	337	33	17	148	1,593	72	38
169	185	25	13	151	12,592	166	135
170	65	13	9	160	440	34	19
171	237	25	23	154	1,166	50	47
174	41	10	8	158	833	46	33
175	60	11	6	145	1,779	72	62
176	254	33	22	153	2,011	69	62
177	55	10	9	156	623	35	35
178	90	14	11	155	1,502	58	56
180	51	12	9	170	198	22	19
181	81	13	10	150	2,900	84	60
183	78	13	11	163	1,251	53	44
184	45	12	7	179	683	38	31
185	102	19	11	162	661	37	37
188	274	27	18	168	1,747	53	44
190	31	7	6	171	652	37	34
191	361	23	22	167	2,497	69	60
192	93	15	8	174	343	28	25
193	33	9	7	166	309	38	35
194	55	11	8	176	1,857	62	47
195	89	13	12	169	925	44	40
196	173	22	15	165	2,287	65	50
197	169	16	17	159	7,882	131	114
199	312	27	17	181	1,075	44	35
200	122	16	16	182	421	41	19
202	132	19	17	175	767	44	28
203	116	15	13	189	496	31	28
204	26	8	4	173	1,234	59	38
206	88	15	10	188	781	34	28
208	259	21	17	186	2,528	72	72
211	195	24	18	177	1,112	41	40
213	37	12	4	197	91	13	7
214	705	44	31	194	2,368	69	66
215	147	16	13	190	177	22	16
216	225	21	18	192	904	50	34
218	63	15	8	201	510	34	25

(continued)

TABLE 5.13 (continued)

Basic Measures of Water Bodies and Their
Zones of Influence

M	$Area_M$	$L_{\|M}$	$L_{\perp M}$	X	$Area_X$	$L_{\|X}$	$L_{\perp X}$
220	532	42	31	202	1,598	62	40
221	55	15	6	152	4,509	100	85
222	396	39	19	198	1,820	66	47
227	54	13	8	203	793	40	35
230	234	36	16	180	3,073	115	97
231	116	19	10	205	1,159	40	34
232	33	7	6	208	927	35	30
233	159	20	16	211	624	37	37
235	26	7	6	219	72	15	6
236	24	6	5	204	1,428	56	47
240	169	28	10	214	1,184	47	35
242	60	11	9	231	184	18	14
243	584	41	39	217	1,533	66	62
244	72	16	8	222	244	27	19
245	41	8	7	121	14,700	316	306
247	457	44	19	229	1,260	55	40
248	124	24	9	221	752	47	31
249	73	15	6	199	3,524	103	75
253	118	15	14	228	740	38	25
254	72	11	11	226	504	31	29
255	217	19	17	234	646	31	28
257	94	13	11	238	1,211	56	37
258	486	34	19	193	10,377	147	125
260	59	14	6	239	723	38	32
261	93	13	12	218	1,099	57	44
262	22	6	5	237	167	25	12
263	84	13	12	240	727	37	31
265	70	13	7	241	288	25	22
266	1290	91	59	242	3,461	103	79
267	47	10	6	225	3,227	93	78
269	54	11	8	243	710	33	31
270	331	28	21	244	1,094	44	34
271	92	12	9	250	424	31	31
273	247	22	17	246	899	38	31
275	114	20	9	224	11,416	157	137
276	34	8	6	210	7,018	132	125
277	1374	72	66	245	3,001	72	71
278	64	10	10	248	514	28	25
279	180	29	9	247	1,455	50	38
280	44	10	6	253	1,251	60	59

TABLE 5.13 (continued)

Basic Measures of Water Bodies and Their
Zones of Influence

M	Area$_M$	$L_{\parallel M}$	$L_{\perp M}$	X	Area$_X$	$L_{\parallel X}$	$L_{\perp X}$
281	431	27	25	251	2,126	65	50
282	33	10	5	255	201	18	18
284	107	20	10	258	592	28	25
285	48	10	7	260	292	22	19
286	237	29	16	256	1,173	50	37
288	361	36	17	262	1,542	52	44
289	132	15	15	264	1,193	53	44
290	44	10	7	254	1,508	59	47
292	112	20	18	267	1,340	56	34
293	54	11	7	266	443	28	28
294	179	30	9	261	1,150	56	44
295	55	13	6	259	11,414	147	119
296	18	6	4	271	144	24	16
297	60	12	11	257	755	47	41
298	40	7	7	272	488	28	22
299	104	17	14	274	343	28	22
300	128	14	14	270	772	41	38
301	141	21	9	263	2,399	69	66
302	71	12	11	268	670	37	37
303	134	15	15	235	37,839	290	247
304	82	15	8	279	280	29	20
305	92	15	10	276	910	44	37
306	96	20	14	277	1,139	41	34
308	123	19	13	273	1,524	63	47
309	67	12	9	280	1,280	53	37
310	223	18	17	275	1,410	47	41
311	58	12	7	281	454	34	33
312	116	15	10	282	592	31	31
314	85	16	6	285	866	44	31
315	390	30	24	283	1,669	72	47
316	258	25	15	289	1,394	60	41
318	161	27	8	284	2,188	66	66
320	57	11	9	269	6,733	137	134
321	78	13	9	290	595	44	34
322	52	9	8	297	784	37	31
323	92	14	11	294	556	31	31
324	61	14	5	296	392	25	19
326	76	15	8	301	997	40	37
327	199	19	16	293	879	50	28

(continued)

TABLE 5.13 (continued)

Basic Measures of Water Bodies and Their
Zones of Influence

M	$Area_M$	$L_{\parallel M}$	$L_{\perp M}$	X	$Area_X$	$L_{\parallel X}$	$L_{\perp X}$
330	85	11	11	299	736	44	44
331	176	18	12	298	730	40	32
332	194	21	20	287	2,384	65	60
334	31	7	6	302	364	22	22
335	85	14	11	306	686	35	28
337	548	50	32	303	1,846	66	56
338	141	31	7	307	1,061	53	34
339	55	12	6	311	3,661	100	73
340	144	22	15	305	1,316	50	50
342	102	14	11	304	801	41	35
344	387	27	21	314	1,921	60	50
345	276	29	15	308	1,665	53	50
347	116	18	11	309	1,024	47	40
349	602	47	26	315	2,292	79	47
350	219	27	14	313	1,352	53	40
351	137	19	17	320	971	42	31
352	108	15	12	317	1,244	44	44
354	306	31	20	319	3,364	99	66
356	457	35	33	322	2,104	69	50
357	110	16	10	323	877	50	34
358	1574	88	42	325	6,942	125	88
361	90	18	10	327	964	47	47
362	86	16	10	328	1,150	56	41
363	396	45	26	335	1,183	59	37
365	125	19	17	332	1,741	60	50
367	66	13	9	334	571	31	28
368	42	9	8	341	520	25	25
369	28	7	5	337	496	34	25
371	114	19	9	339	1,869	53	50
372	101	14	11	336	1,464	50	41
373	124	19	12	343	752	37	34
374	70	12	10	342	514	37	28
375	90	17	9	345	243	31	12
376	118	21	10	344	2,260	80	62
378	75	12	12	333	541	28	28
379	163	24	12	350	561	46	25
380	95	16	10	351	599	31	31
381	61	11	10	356	355	25	22
383	907	62	38	346	4,565	109	78
385	156	30	9	357	1,473	54	44

TABLE 5.13 (continued)

Basic Measures of Water Bodies and Their Zones of Influence

M	$Area_M$	$L_{\parallel M}$	$L_{\perp M}$	X	$Area_X$	$L_{\parallel X}$	$L_{\perp X}$
386	63	11	10	359	544	31	28
387	162	20	14	361	388	28	22
388	438	46	24	360	562	37	25
389	85	17	9	355	2,526	69	60
390	90	20	10	349	370	28	16
391	2718	69	61	340	19,290	225	139
392	200	20	17	362	1,222	50	43
393	76	13	13	370	1,484	59	50
394	4	3	2	367	355	27	22
395	399	33	28	368	1,569	50	50
397	169	18	18	364	1,575	66	31
399	55	13	8	366	2,367	63	53
400	186	17	16	376	1,794	62	50
402	341	34	19	379	824	41	37
403	109	21	8	372	1,024	61	59
404	130	14	13	378	1,153	47	34
405	170	33	12	371	2,838	81	57
407	127	21	19	374	2,671	69	66
408	450	35	26	383	1,640	68	41
409	126	16	16	381	1,814	60	59
410	639	56	20	386	2,220	72	60
411	693	44	42	382	2,566	75	66
412	463	37	19	385	1,965	56	50
413	45	8	8	391	271	28	28
414	86	14	13	348	12,538	178	150
420	88	17	14	354	2,060	135	131
421	176	19	17	401	610	32	28
422	50	9	9	393	1,179	53	49
425	472	41	25	394	2,093	85	56
426	46	11	6	404	622	31	28
427	958	86	41	389	4,852	153	72
428	642	47	23	396	3,152	85	78
429	87	16	15	406	741	41	34
430	65	11	10	408	301	25	25
432	55	9	8	398	1,613	49	44
434	359	37	36	397	4,686	112	71
435	132	20	17	409	1,105	43	38
436	298	23	23	410	1,227	51	44
437	483	36	25	417	2,133	63	62

(continued)

TABLE 5.13 (continued)

Basic Measures of Water Bodies and Their
Zones of Influence

M	Area$_M$	$L_{\parallel M}$	$L_{\perp M}$	X	Area$_X$	$L_{\parallel X}$	$L_{\perp X}$
438	136	17	14	403	1,255	48	34
439	271	30	19	413	1,656	50	44
440	156	27	18	415	1,118	53	37
441	46	11	6	416	748	41	28
442	87	15	7	411	1,678	75	47
443	60	14	7	414	920	52	25
444	237	25	16	363	13,890	169	156
446	908	56	32	423	2,016	75	44
447	534	37	25	419	2,410	62	56
448	74	15	10	420	460	38	27
449	272	26	21	424	1,208	63	53
451	82	14	12	421	736	38	29
452	1605	85	51	358	8,976	216	175
453	53	11	8	426	198	25	19
454	189	26	19	430	886	43	38
456	212	30	14	358	8,976	216	175
457	151	27	11	425	2,151	75	50
458	66	15	6	428	1,130	47	40
460	68	15	7	435	302	31	21
461	183	30	11	434	1,368	57	34
462	140	22	14	433	997	44	33
463	84	16	10	438	1,027	43	40
464	128	20	13	436	892	54	37
465	30	8	5	441	220	21	19
466	97	12	12	437	760	37	31
467	196	30	15	445	650	35	32
468	57	21	7	443	690	41	34
471	218	18	18	402	5,899	106	90
472	732	50	38	432	4,239	117	72
473	288	33	17	422	2,606	81	81
474	59	11	6	446	572	31	28
475	376	33	24	444	3,020	79	65
476	298	26	21	440	3,593	84	78
477	459	33	20	431	3,005	79	63
478	498	41	19	447	2,570	66	50
479	75	15	8	452	759	36	30
480	129	19	12	427	4,873	131	100
481	106	21	8	439	2,299	68	56
482	220	25	12	448	1,966	62	50

TABLE 5.13 (continued)

Basic Measures of Water Bodies and Their
Zonesof Influence

M	$Area_M$	$L_{\parallel M}$	$L_{\perp M}$	X	$Area_X$	$L_{\parallel X}$	$L_{\perp X}$
484	54	10	10	454	709	41	28
485	217	27	19	449	3,185	97	72
486	1399	72	53	450	11,918	159	109
487	37	11	4	451	1,732	62	47
488	67	13	8	455	778	34	31
489	344	36	13	460	1,454	69	34
491	425	46	18	458	1,186	59	41
492	134	22	13	459	1,010	41	31
493	304	28	22	461	2,394	65	62
494	952	69	57	453	3,317	103	78
495	111	19	17	457	800	35	31
496	66	14	7	465	699	44	29
497	358	41	27	466	1,078	53	44
498	155	23	11	462	769	40	28
499	64	10	9	469	1,247	41	40
501	97	14	11	463	2,572	66	58
502	474	41	29	468	2,107	66	62
504	110	13	12	464	1,320	41	41
505	56	6	1	471	183	21	18
506	1008	63	45	475	3,581	78	62
508	69	13	8	478	476	29	25
510	807	45	44	476	2,686	75	62
511	707	59	38	479	2,184	75	62
512	122	14	12	472	1,348	51	47
513	423	29	26	480	894	44	29
514	1258	58	38	474	4,062	81	78
515	218	18	18	481	1,850	61	42
516	363	29	21	485	832	40	34
519	196	19	15	486	1,223	47	41
520	871	49	32	487	2,643	97	91
521	142	17	15	489	1,253	50	35
522	480	40	20	488	2,143	69	63
523	132	16	13	482	10,475	193	91
524	44	8	7	491	560	31	25
525	428	39	22	484	2,521	97	50
527	966	48	38	493	4,983	87	79
528	267	35	21	498	1,337	53	47
529	64	11	7	496	249	22	21
530	630	63	47	503	5,202	97	91

(continued)

TABLE 5.13 (continued)

Basic Measures of Water Bodies and Their
Zones of Influence

M	Area$_M$	L$_{\parallel M}$	L$_{\perp M}$	X	Area$_X$	L$_{\parallel X}$	L$_{\perp X}$
531	102	14	9	502	1,724	75	40
532	154	18	12	504	485	41	19
533	115	15	10	500	9	3	3
534	290	31	14	499	2,388	69	63
535	69	13	7	501	918	42	35
536	646	57	30	507	1,456	78	42
537	89	13	9	490	2,352	88	75
538	135	15	14	511	1,329	62	62
539	73	13	7	494	1,559	66	34
540	59	10	7	495	2,012	63	38
541	140	19	11	509	1,384	60	40
543	1543	76	68	521	6,599	115	79
544	183	16	14	517	704	44	31
545	335	39	14	508	1,346	44	44
546	2018	108	68	516	7,658	169	85
547	192	28	18	512	1,490	56	47
549	66	10	9	515	376	31	25
550	238	34	18	514	3,359	94	50
551	154	23	13	519	1,651	57	44
553	107	14	14	518	2,121	72	53
554	194	21	14	522	2,079	59	59
555	157	21	13	524	2,713	90	63
556	181	25	16	523	1,219	50	41
558	1093	80	51	520	6,892	147	106
560	300	25	23	527	2,851	69	65
561	132	15	11	532	255	25	15
562	407	27	23	528	4,257	78	75
563	217	20	19	525	2,224	75	56
564	162	27	18	534	525	38	25
565	609	61	30	526	5,956	116	107
566	83	16	9	537	361	41	22
567	356	27	23	530	2,134	66	63
568	648	59	37	535	4,225	126	62
570	247	30	13	529	2,800	75	62
571	177	25	11	543	661	35	25
572	132	17	13	533	4,633	90	72
573	70	13	8	536	1,353	44	43
574	65	13	8	545	616	35	31
575	52	10	9	538	2,250	59	59
577	670	57	30	544	3,708	103	72

TABLE 5.13 (continued)

Basic Measures of Water Bodies and Their
Zones of Influence

M	Area$_M$	$L_{\parallel M}$	$L_{\perp M}$	X	Area$_X$	$L_{\parallel X}$	$L_{\perp X}$
578	122	26	14	551	1,514	50	50
579	92	14	14	539	1,496	66	53
582	97	15	10	541	2,047	66	47
584	47	10	7	550	1,102	46	41
587	465	35	32	560	2,189	60	57
588	57	13	7	540	1,646	60	59
589	1071	41	38	552	3,964	88	62
590	329	24	23	555	1,782	56	53
591	213	25	14	561	1,272	66	35
592	71	11	9	553	1,421	53	46
593	619	47	30	549	3,650	122	84
594	685	53	23	554	3,570	106	60
595	298	21	21	542	4,372	100	75
596	130	22	16	557	1,307	50	37
598	546	35	30	556	3,184	87	68
599	397	29	28	559	3,064	72	65
600	242	25	22	564	1,088	56	28
601	132	16	15	563	1,786	66	50
602	463	29	29	566	1,669	69	63
603	201	26	12	562	1,656	65	57
604	459	35	29	568	2,505	72	59
606	302	32	26	567	2,090	69	50
607	67	16	5	547	2,989	85	81
608	307	42	18	571	1,878	59	50
609	92	19	9	569	1,248	50	38
610	140	19	18	572	562	41	32
612	59	10	9	580	2,723	75	63
613	124	19	14	558	5,972	103	103
615	144	21	9	578	740	46	28
616	40	10	9	576	609	38	25
617	478	40	27	573	2,867	72	63
618	136	20	15	574	1,739	63	44
619	741	47	39	580	2,723	75	63
620	433	31	27	579	1,670	59	46
621	97	13	11	586	268	22	16
623	531	43	33	575	4,328	100	69
624	43	9	8	583	643	37	31
625	117	20	10	587	432	25	25
626	58	10	10	585	1,328	53	44

(continued)

TABLE 5.13 (continued)

Basic Measures of Water Bodies and Their
Zones of Influence

M	$Area_M$	$L_{\|M}$	$L_{\perp M}$	X	$Area_X$	$L_{\|X}$	$L_{\perp X}$
627	206	27	15	588	1,655	59	34
628	71	13	7	594	367	28	19
629	80	16	8	593	605	44	28
630	192	22	14	590	1,018	40	37
631	146	18	13	591	2,037	69	59
632	52	11	9	597	721	43	28
633	130	22	8	604	162	21	13
634	25	6	5	589	899	41	34
635	117	13	12	595	1,115	44	41
636	235	21	19	592	2,326	77	67
637	51	12	7	598	797	41	34
638	84	20	6	599	1,541	50	50
641	328	25	21	608	1,518	60	56
642	361	29	19	600	2,757	91	63
643	65	13	8	601	2,994	78	54
644	286	30	19	605	2,659	84	62
645	69	17	6	603	3,892	82	79
646	78	13	9	614	447	31	29
647	332	43	16	607	1,390	66	47
650	334	37	32	609	2,496	66	60
651	56	11	6	616	868	37	37
653	4	3	2	610	691	44	38
654	200	21	12	613	2,317	97	50
655	65	15	11	612	1,427	60	34
656	593	51	33	611	3,767	100	81
658	332	31	19	625	1,213	40	37
659	74	17	7	617	1,411	62	61
662	202	26	15	619	2,457	75	53
663	68	12	10	630	1,631	50	40
664	124	18	12	633	809	48	44
665	455	32	30	629	1,505	59	43
666	2523	78	61	632	7,153	116	100
667	122	16	14	622	1,543	69	55
668	41	11	7	623	1,912	75	50
669	246	20	17	626	5,957	106	81
670	483	36	28	634	4,708	141	122
671	191	19	16	638	695	35	32
672	125	16	13	627	3,437	71	69
673	131	18	11	628	1,693	78	31
675	348	27	23	631	3,707	75	62

TABLE 5.13 (continued)

Basic Measures of Water Bodies and Their
Zones of Influence

M	$Area_M$	$L_{\parallel M}$	$L_{\perp M}$	X	$Area_X$	$L_{\parallel X}$	$L_{\perp X}$
676	214	27	16	621	2,065	84	81
677	140	15	13	635	880	44	25
678	296	31	17	615	7,017	157	103
679	153	21	11	637	647	31	29
680	73	13	10	638	695	35	32
681	127	14	12	639	1,475	59	40
682	366	29	22	641	7,416	132	105
683	398	32	25	644	1,781	71	41
684	1354	59	56	641	7,416	132	105
685	1035	68	37	642	5,326	154	106

Source: Sagar, B.S.D., *Water Resour. Res.*, 43(2), W02416, 2007.

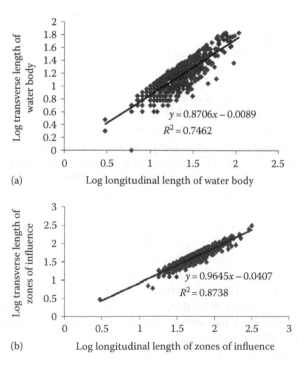

(a)

(b)

FIGURE 5.25
Allometric power-law relationships. (a and b) Transverse length–longitudinal length for water
bodies and their zones of influence.

(continued)

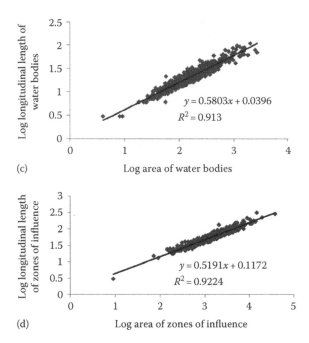

(c)

(d)

FIGURE 5.25 (continued)
Allometric power-law relationships. (c and d) Length–area. (From Sagar, B.S.D., *Water Resour. Res.*, 43(2), W02416, 2007.)

$L_\curvearrowright$, and $L_\perp$ for both water bodies and zones of influence. These relationships yield exponents h and H (Table 5.14). Logarithms of $L_\curvearrowright$ have been plotted as a function of logarithms of areas of water bodies (Figure 5.25c) and zones of influence (Figure 5.25d). Scaling exponents derived include 0.58 and 0.52. Similarly, Hurst exponent (H) derived by plotting $L_\perp$ as a function of $L_\curvearrowright$ for both water bodies and corresponding zones of influence includes 0.87 and 0.96, respectively. h is 0.53, for water bodies, which is relatively larger than 0.52 for zones of influence. H is 0.87, for water bodies, which is higher (0.96) for zones of influence. These comparisons support the notion that the higher the degree of self-affinity, the lower is the exponent in the length–area relationship. From these relationships, it is inferred that the water bodies— which are subsets of their corresponding zones of influence—possess higher values of h and lower values of H than their zones of influence. These exponents are also universal type as they possess similar scaling exponents at all scales, as also exhibited by several environmental phenomena (e.g., Hurst 1951, Hack 1957, Mesa and Gupta 1987, Robert and Roy 1990, Rosso et al. 1991, Ijjasz-Vasquez et al. 1993, Sagar and Rao 1995, Maritan et al. 1996a,b, Rodriguez-Iturbe and Rinaldo 1997, Banavar et al. 1999, Veitzer and Gupta 2000, Maritan et al. 2002, Sagar and Tien 2004). $L_\perp$ and $L_\curvearrowright$ of model and realistic river networks have been considered in their studies. Table 5.14 shows comparative results.

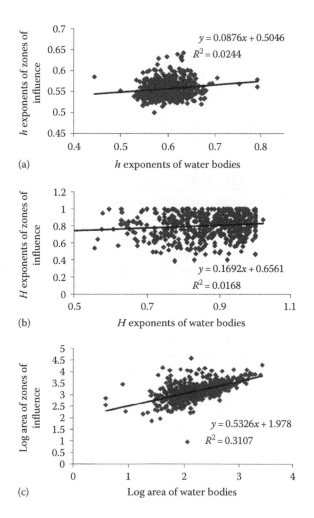

FIGURE 5.26

(a) Hack's exponents, (b) Hurst exponents of water bodies and their zones of influence, and (c) areas of water bodies and ZIs. (From Sagar, B.S.D., *Water Resour. Res.*, 43(2), W02416, 2007.)

From the two popular scaling laws derived for the two interdependent phenomena, viz., water bodies and zones of influence, the following inferences can be drawn:

1. Water bodies have a smaller degree of self-affinity when compared to their zones of influence, which are relatively larger than water bodies.

2. For both water bodies and zones of influence, the scaling exponents, h and H, are in good accord with the results reported elsewhere (Maritan et al. 1996a,b, Rodriguez-Iturbe and Rinaldo 1997, Banavar et al. 1999, Dodds and Rothman 1999, Veitzer and Gupta 2000, Maritan et al. 2002).

TABLE 5.14

Comparative Scaling Relationships

System	h	H	$\alpha = 1 + h$	$h = \dfrac{1}{1+H}$	$1 + \dfrac{2h}{1+H}$	A vs. N
Water bodies	0.60	0.86	1.60	0.54	1.64	1.41
Zones of influence	0.56	0.94	1.56	0.52	1.57	1.70
OCN (rectangular boundary)	0.57	0.84	1.57	0.54	1.57	—
OCN (fractal boundary)	0.56	0.88	1.56	0.53	1.56	—
F-SCN*	0.55	0.95	0.55	0.52	1.54	—

Source: Sagar, B.S.D., *Water Resour. Res.*, 43(2), W02416, 2007; Sagar, B.S.D. et al., *Discrete Dyn. Nat. Soc.*, 2, 77, 1998.

3. As H computed for both water bodies and zones of influence is lesser than 1, these two interdependent topological phenomena are of self-affine, and a well-fitted relationship is observed as $h = 1/1 + H$ (Table 5.14).

4. A value for $1 + 2h/1 + H$ should be higher than $3/2$, which was proposed as a theoretical value, and it is found as 1.64 and 1.57 respectively for water bodies and zones of influence.

5. $\alpha = 1 + h$, a scaling relationship proposed for mass-basin metabolic rate (Maritan et al. 2002), yields 1.60 and 1.56 for water bodies and zones of influence, respectively.

6. Further to the fact that climatically sensitive water bodies possess more elongation ratio, the results indicate that the degree of geometric-similarity preservation is more in the zones of influence than in the corresponding water bodies.

7. No goodness of fit was observed between the h values of water bodies and zones of influence, and also between the H values of water bodies and zones of influence (Figure 5.26a and b).

8. Area of zones of influence plotted as a function of area of water bodies yields a power-law value of 0.5326 implying a relationship as $3/5$ law (Figure 5.26c).

9. There is no correlation between the h, H exponents of water bodies and zones of influence (Figure 5.26a and b) due to the fact that there is no linear relationship observed between the areas of water bodies and zones of influence. It cannot be generalized that the area of water bodies is directly proportional to the area of the corresponding zone of influence.

10. The fact that water bodies and zones of influence belong to different universality classes further enables that certain scaling rules may apply to both exogenically insensitive and exogenically sensitive geomorphologic phenomena.

11. Form and function relationship could be better understood through such scaling laws.

12. Modeling and reconstruction of environmentally dependent geomorphologic system processes could be better handled as water bodies and zones of influence adhere to regular scaling rules.

13. These scaling relationships could be of value also in the understanding of paleo environments and their geomorphologic constitutions.

References

Banavar, J. R., A. Maritan, and A. Rinaldo, 1999, Size and form in efficient transportation networks, *Nature*, 399, 130–134.

Banavar, J. R. et al., 2002, Supply-demand balance and metabolic scaling. *Proceedings of the National Academy of Sciences of the USA*, 99, 10, 506.

Beaucher, S., 1990, Segmentation d-images et morphologic mathematique, These Docteur en Morphologic Mathematique, Ecole des Mines de Paris, Paris, France.

Chockalingam, L. and B. S. D. Sagar, 2005, Morphometry of network and non-network space of basins, *Journal of Geophysical Research*, 110, B08203, doi:10.1029/2005JB003641.

Dodds, P. S. and D. H. Rothman, 1999, Unified view of scaling laws for river networks, *Physical Review E*, 59, 4865–4877.

Feder, J., 1988, *Fractals*, Plenum, New York, 283pp.

Gupta, V. K. and E. Waymire, 1993, A statistical analysis of mesoscale rainfall as a random cascade, *Journal of Applied Meteorology*, 32, 251–267.

Hack, J. T., 1957, Studies of longitudinal profiles in Virginia and Maryland. U.S. Geological Survey Professional Paper 294-B1.

Heijmans, H. J. A. M. and A. V. Tuzikov, 1998, Similarity and symmetry measures for convex shapes using Minkowski addition, *IEEE Transactions on Pattern Analysis and Machine Intelligence*, 20, 980–993.

Horton, R. E., 1945, Erosional development of stream and their drainage basin. Hydrological approach to quantitative morphology, *Bulletin of the Geological Society of America*, 56, 275–370.

Howard, A. D., 1990, Theoretical model of optimal drainage networks, *Water Resources Research*, 26(9), 2107–2117.

Hurst, H. E., 1951, Long-term storage capacity of reservoirs, *Transactions of the American Society of Civil Engineers*, 116, 770–808.

Ijjasz-Vasquez, E. J., R. L. Bras, and I. Rodriguez-Iturbe, 1993, Hack's relation and optimal channel networks. The elongation of river basins as a consequence of energy minimization, *Geophysical Research Letters*, 20, 1583–1586.

Jang, B. K. and R. T. Chin, 1990, Analysis of thinning algorithms using mathematical morphology, *IEEE Transactions on Pattern Analysis and Machine Intelligence*, 12, 541–551.

La Barbera, P. and R. Rosso, 1987, Fractal geometry of river networks, *Eos Transactions AGU*, 68(44), 1276.

La Barbera, P. and R. Rosso, 1989, On the fractal dimension of stream networks, *Water Resources Research*, 25(4), 735–741.

Langbein, W. B., 1947, Topographic characteristics of drainage basins, U.S. Geological Survey Professional Paper 968-C.

Lantuejoul, C., 1978, La sequelettisation et son application aux mesures topologuiques des mosaiques polycristallines, These de Docteur-Ingnieur, School of Mines, Paris, France.

Lantuejoul, C., 1980, Skeletonization in quantitative metallography. In: *Issues of Digital Image Processing*, eds. R. M. Haralick and J. C. Simon, Sijthoff and Neordhoff, Groningen, the Netherlands, pp. 107–135.

Lovejoy, S., 1982, Area-Perimeter relation for rain and cloud areas, *Science*, 216, 185–187.

Mandelbrot, B. B., 1967, How long is the coast of Britain? Statistical self-similarity and fractional dimension, *Science*, 156, 636–638.

Mandelbrot, B. B., 1982, *Fractal Geometry of Nature*, Freeman, New York, p. 468.

Mandelbrot, B. B. D. E. Passoja, and A. Paullay, 1984, Fractal character of fracture surfaces of metals, *Nature*, 308, 721–722.

Marani, A., R. Rigon, and A. Rinaldo, 1991, A note on fractal channel networks, *Water Resources Research*, 27, 3041–3049.

Marani, M., A. Rinaldo, R. Rigon, and I. Rodriguez-Iturbe, 1994, Geomorphological width functions and the random cascade, *Geophysical Research Letters*, 21, 2123–2126.

Maritan, A., F. Coloairi, A. Flammini, M. Cieplak, and J. R. Banavar, 1996a, Universality classes of optimal channel networks, *Science*, 272, 984–988.

Maritan, A., R. Rigon, J. R. Banavar, and A. Rinaldo, 2002, Network allometry, *Geophysical Research Letters*, 29(11), 1508, doi:10.1029/2001GL014533.

Maritan, A., A. Rinaldo, R. Rigon, A. Giacometti, and I. Rodriguez-Iturbe, 1996b, Scaling laws for river networks, *Physical Review E*, 53, 1510–1515.

Masek, J. and D. L. Turcotte, 1993, A diffusion aggregation model for the evolution of drainage networks, *Earth and Planetary Science Letters*, 119, 379–386.

Mesa, O. J. and V. K. Gupta, 1987, On the main channel length: Area relationships for channel networks, *Water Resources Research*, 23, 2119–2122.

Nelson, T. R. and D. K. Manchester, 1988, Modeling of morphogenesis using fractal geometries, *IEEE Transactions on Medical Imaging*, 7(4), 321–327.

Rigon, R., I. Rodriguez-Iturbe, A. Giacometti, A. Maritan, D. Tarbotan, and A. Rinaldo, 1996, On Hack's law, *Water Resources Research*, 32, 3367–3374.

Rinaldo, A., I. Rodriguez-Iturbe, R. L. Bras, and E. Ijjasz-Vasquez, 1993, *Physical Review Letters*, 70, 822–826.

Rinaldo, A., I. Rodriguez-Iturbe, and R. Rigon, 1998, Channel networks, *Annual Review of Earth and Planetary Sciences*, 26, 289–327.

Robert, A. and A. Roy, 1990, On the fractal interpretation of the mainstream length-drainage area relationship, *Water Resources Research*, 26, 839–842.

Rodriguez-Iturbe, I. and A. Rinaldo, 1997, *Fractal River Basins, Chance and Self-Organization*, Cambridge University Press, New York.

Rosso, R., B. Becchi, and P. La Barbera, 1991, Fractal relation of main stream length to catchment area in river networks, *Water Resources Research*, 27(3), 381–387.

Sagar, B. S. D., 1996, Fractal relations of a morphological skeleton, *Chaos, Solitons & Fractals*, 7(11), 1871–1879.

Sagar, B. S. D., 2000, Fractal relation of medial axis length to the water body area, *Discrete Dynamics in Nature and Society*, 4(1), 97.

Sagar, B. S. D., 2007, Universal scaling laws in surface water bodies and their zones of influence, *Water Resources Research*, 43(2), W02416.

Sagar, B. S. D. and L. Chockalingam, 2004, Fractal dimension of nonnetwork space of a catchment basin, *Geophysical Research Letters*, 31, L12502, doi:10.1029/2004GL019749.

Sagar, B. S. D. and K. S. R. Murthy, 2000, Generation of fractal landscape using nonlinear mathematical morphological transformations, *Fractals*, 8, 267.

Sagar, B. S. D., C. Omoregie, and B. S. P. Rao, 1998b, Morphometric relations of Fractal-skeletal based channel network model, *Discrete Dynamics in Nature and Society*, 2, 77–92.

Sagar, B. S. D. and B. S. P. Rao, 1995, Fractal relation on perimeter to the water body area, *Current Science*, 68, 1129–1130.

Sagar, B. S. D. and D. Srinivas, 2002, Estimation of number-area frequency dimensions of surface water bodies, *International Journal of Remote Sensing*, 20, 2491–2496.

Sagar, B. S. D., D. Srinivas, and B. S. P. Rao, 2001, Fractal skeletal based channel networks in a triangular initiator basin, *Fractals*, 9, 429–437.

Sagar, B. S. D. and T. L. Tien, 2004, Allometric power-law relationships of Hortonian fractal digital elevation model, *Geophysical Research Letters*, 31, L06501, doi:10.1029/2003GL019093.

Sagar, B. S. D., M. Venu, and K. S. R. Murthy, 1999, Do skeletal network derived from water bodies follow Horton's laws? *Journal of Mathematical Geology*, 31(2), 143–154.

Sagar, B. S. D. et al., 1998a, Morphological description and interrelationship between force and structure: A scope to geomorphic evolution process modeling, *International Journal of Remote Sensing*, 19, 1341.

Sagar, B. S. D. et al., 2003, Morphological approach to extract ridge valley connectivity networks from digital elevation models (DEMs), *International Journal of Remote Sensing*, 24, 573.

Schroeder, M., 1991, *Fractals, Chaos, Power Laws*, Freeman, San Francisco, CA.

Schumm, S. A., 1956, Evolution of drainage systems and slopes in badlands at Perth Amboy, NJ, *Bulletin of Geological Society of America*, 67, 597–646.

Shlesinger, M. and B. J. West, 1991, Complex fractal dimension of the bronchial tree, *Physical Review Letters*, 67, 2106–2108.

Shreve, R. L., 1967, Infinite topologically random channel networks, *Journal of Geology*, 77, 397–414.

Soille, P., 1998, Grey scale convex hulls: definition, implementation and applications. In: *Proceedings ISMM'98*, Vancouver, British Columbia, Canada, pp. 83–90.

Stark, C. P., 1991, An invasion percolation model of drainage network evolution, *Nature*, 352, 423–425.

Strahler, A. N., 1957, Quantitative analysis of watershed geomorphology, *Eos Transactions AGU*, 38, 913–920.

Strahler, A. H., 1964, Quantitative geomorphology of drainage basin and channel networks. In: *Handbook of Applied Hydrology*, ed. V. T. Chow, Sections 4–11, McGraw-Hill, New York.

Takayasu, H., 1990, *Fractals in Physical Sciences*, Manchester University Press, New York.

Tarbotan, D. G., R. L. Bras, and I. Rodriguez-lturbe, 1990, Comment on "On the fractal dimension of stream networks by Paolo La Barbera and Rosso Renzo," *Water Resources Research*, 26(9), 2243–2244.

Tarboton, D. G. et al., 1988, The fractal nature of river networks, *Water Resources Research*, 24, 1317.

Tay, L. T., B. S. D. Sagar, and H. T. Chuah, 2006, Allometric relationships between travel-time channel networks, convex hulls, and convexity measures, *Water Resources Research*, 42, W06502, doi:10.1029/2005WR004092.

Tay, L. T., B. S. D. Sagar, and H. T. Chuah, 2007, Granulometric analysis of basin-wise DEMs: A comparative study, *International Journal of Remote Sensing*, 28(15), 3363–3378.

Turcotte, D. L., 1997, *Fractals and Chaos in Geology and Geophysics*, Cambridge University Press, New York.

Turcotte, D. L., J. D. Pelletier, and W. I. Newman, 1998, Networks with side branching in biology, *Journal of Theoretical Biology*, 193, 577–592.

Veitzer, S. A. and V. K. Gupta, 2000, Random self-similar river networks and derivations of generalized Horton laws in terms of statistical simple scaling, *Water Resources Research*, 36, 1033–1048.

Veneziano, D., G. E. Moglen, P. Furcolo, and V. Iacobellis, 2000, Stochastic model of the width function, *Water Resources Research*, 36, 1143–1157.

Zunic, J. and P. L. Rosin, 2004, A new convexity measure for polygons, *IEEE Transactions on Pattern Analysis and Machine Intelligence*, 26, 923–934.

6

Size Distributions, Spatial Heterogeneity, and Scaling Laws

Water bodies, unique geomorphologic networks, basins, subbasins, and zones of influence (ZIs) are some of the features of geomorphologic interest. These features that are conspicuous have their spatial organizations that require special emphasis while understanding the complexity involved in terrestrial surfaces and processes. In the previous chapter, morphometric and allometric analyses that led to the derivation of a host of power-laws have been employed. This chapter deals with similar geomorphologic phenomena, whereby granulometric analysis, which has been proved to be a successful approach to quantitatively understand the spatiotemporal organization of varied geomorphologic phenomena.

Granulometries are of two types: binary granulometries and grayscale granulometries. They rely on principles of binary opening and grayscale opening transformations. Opening transformation is idempotent, whereas multiscale opening transformation is a non-idempotent transformation. The basic difference between recursive implementation of opening and multiscale opening is shown in Equations 6.1 and 6.2.

$$(((X \circ B) \circ B) \circ \cdots \circ B) = (X \circ B) \tag{6.1}$$

$$(((X \ominus B) \ominus B) \ominus \cdots \ominus B) \oplus B \oplus B \oplus \cdots \oplus B = (X \circ NB) \tag{6.2}$$

where NB is much larger than B, which is of primitive size.

Performing multiscale morphological opening yields size distribution patterns of considered phenomena. Any phenomenon represented as a thematic information on a raster map can be fragmented, size-distributed, segmented, and vanished under the process of multiscale morphological opening transformation. Distribution pattern of a specific parameter of a phenomenon that is subjected to multiscale opening is an important feature further to compute various quantitative indexes such as granulometric index, pattern spectrum index, morphological entropy, multifractal measures, and power-law values. These indexes could be computed via various approaches, to mention a few fractal, granulometries, $f(\alpha)$-spectra.

In this chapter, the considered phenomena include water bodies, ZIs, basins, and terrestrial surfaces. In different sections of this chapter, multiscale morphological binary opening has been employed to demonstrate on these mentioned phenomena to understand their spatiotemporal complexities.

Size Distributions of Water Bodies and Zones of Influence

Multiscale morphological opening transformation has been employed to distribute a large number of surface water bodies, represented on a raster (Figure 6.1) according to their sizes. One of the important tasks of a geomorphologist is to categorize surface water bodies on the basis of their sizes and shapes. Manual methods to distribute any theme, depicting geomorphologic features, according to their shapes and sizes, are proved to be very tedious and time-consuming. Planimetric analysis–based techniques to compute basic measures of features such as water bodies are laborious and inaccurate (Barker 1975). Size histograms can be constructed by calculating the size of each water body. The three types of size distributions of natural objects— such as grains, water bodies, basins, and subbasins—are size distribution in measure, number, and weight (Delfiner 1972). Earlier, multiscale morphological opening transformation has been successfully applied on sand grains (Delfiner 1972) and cervical cells (Meyer 1980) for categorization based on their sizes and shapes. Multiscale opening is applied, where the size of structuring element increases with increasing cycle of opening. The idea was to distribute a large number of surface water bodies, extracted from IRS-1C LISS III data (Figures 3.16 through 3.18), according to their sizes by employing changing sizes of structuring element in the opening transformation. While performing opening by means of nth size structuring element, denoted by nB, on a section containing water bodies, the water bodies smaller than nB would be vanished by leaving the water bodies larger than nB.

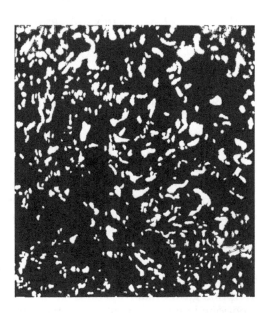

FIGURE 6.1
Data show surface water bodies in digital format. (From Sagar, B.S.D. et al., *Int. J. Remote Sens.*, 16(16), 3059, 1995.)

```
                                        1  1  1
                                     1  1  1  1  1
                 1  1  1         1  1  1  1  1  1  1
              1  1  1  1  1      1  1  1  1  1  1  1  1  1
     1  1  1  1  1  1  1  1  1   1  1  1  1  1  1  1  1  1
  1  1  1  1  1  1  1  1  1  1   1  1  1  1  1  1  1  1  1
  1  1  1  1  1  1  1  1  1  1      1  1  1  1  1  1  1
  1  1  1  1  1     1  1  1  1  1         1  1  1  1  1
     1  1  1           1  1  1            1  1  1
  (a)              (b)                 (c)
```

FIGURE 6.2
Circular structuring elements of different diameters: (a) 5 pixel, (b) 7 pixel, and (c) 9 pixel. (From Sagar, B.S.D. et al., *Int. J. Remote Sens.*, 16(16), 3059, 1995.)

To distribute surface water bodies in number, measure, and weight, circular structuring elements of radii 5, 7, 9, 11, 13, 15, and 17 have been employed in the opening transformation. The first three sizes of circular structuring elements are shown in Figure 6.2. The opening transformation is shown in Equation 6.3, which is another form of representation of Equation 6.2.

$$(X \ominus nB) \oplus nB = X \circ nB \tag{6.3}$$

where X is a section depicting water bodies.

The morphological opened versions of X obtained by structuring elements of diameters 5, 7, 9, and 11 have been shown in Figure 6.3. It is obvious from Figure 6.3a through d that the water bodies smaller than the structuring elements of radii 5, 7, 9, and 11 have been filtered out.

Equations 6.4 through 6.6, which are due to Delfiner (1972), have been used in the computation of size distribution functions in number, measure, and weight of surface water bodies:

$$f(d) = 1 - \frac{\mathbb{N}(X \circ nB)}{\mathbb{N}(X)} \tag{6.4}$$

where $\mathbb{N}(X)$ is the number of water bodies seen in the opened version of X by the structuring element of size n.

$$G(d) = 1 - \frac{V(X \circ nB)}{V(X)} \tag{6.5}$$

where
  $V(X)$ is the total area of water bodies in the section
  $V(X \circ nB)$ is the surface area of the water bodies after opening by structuring element of defined diameters

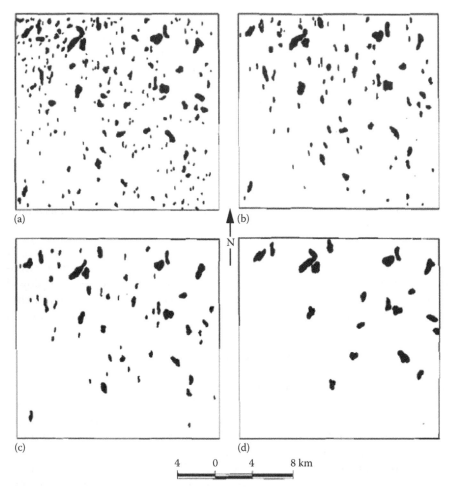

FIGURE 6.3
Evolution of water bodies under opening transformation by circular structuring elements with increasing radii: (a) opening by 3-pixel radii, (b) opening by 5-pixel radii, (c) opening by 7-pixel radii, and (d) opening by 9-pixel radii. (From Sagar, B.S.D. et al., *Int. J. Remote Sens.*, 16(16), 3059, 1995.)

$$\phi(\Delta d) = \frac{\pi (d/2)^2 \{N(X \circ nB) - N(X \circ (n+1)B)\}}{V(X)} \times 100 \qquad (6.6)$$

where $0 \leq d \leq 17$ pixels. In respect to water bodies, the volume is closely related to the parameter weight (mass). On the basis of fractal-perimeter–length–area–volume relations (Mandelbrot 1982), volume can be computed from areas by using the fractal power-law. In Chapter 5, it was shown that the fractal dimension of perimeter of the water bodies considered in the present investigation follows scale-invariant characteristics. Hence, depths of the water bodies possessing similar areas are also assumed to be uniform.

Such an assumption could be validated from the data acquired from the water bodies identified from digital elevation models (DEMs). Depths of the water bodies could be computed more accurately, if water bodies are identified from DEMs. In turn, volumes of water bodies identified from DEMs could be more accurately computed.

In Equation 6.6, d is the diameter of the structuring element and is equivalent to $2n$. This d ranges between 0 and 17 pixels for the water body case considered here. Since the density of water is same in respective categories of water bodies, Equation 6.6 could be used to obtain size distribution in volume. For the considered water bodies (Figure 6.1), the size distribution functions of both number and measure have been recorded (Table 6.1). Histograms are constructed for weights of water bodies (Figure 6.4). These histograms of number, measure, and weight of water bodies explain the extent of similarities and differences. For a diameter of the structuring element $(d) < 11$ pixels, size distribution histograms for number $(F(\Delta d\%))$ is found greater

TABLE 6.1

Size Distribution Functions of a Section of Surface Water Bodies

Diameter of Structuring Element	SD Function F(d)%	Histogram F(Δd)%	SD Function G(d)%	Histogram G(Δd)%	WGTD (ϕd)%
0	0	29.43	0	17.90	7.77
5	29.43	14.71	17.90	17.81	8.04
7	44.14	19.08	35.71	17.77	16.34
9	63.22	16.50	53.48	13.77	21.10
11	79.72	6.36	67.21	9.04	11.36
13	86.08	10.12	76.25	9.02	23.64
15	96.20	3.80	85.27	14.73	12.14
17	100.00	0.00	100.00	0.00	0.00

Source: Sagar, B.S.D. et al., *Int. J. Remote Sens.*, 16(16), 3059, 1995.

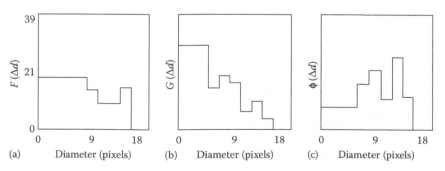

FIGURE 6.4
Histograms show (a) size distribution in number, $F(\Delta d)$, (b) size distribution in measure, $G(\Delta d)$, and (c) size distribution in weight, $\phi(\Delta d)$. (From Sagar, B.S.D. et al., *Int. J. Remote Sens.*, 16(16), 3059, 1995.)

than the size distribution histograms for measure ($G(\Delta d)$), an exception in between the diameters of 5 and 7 pixels. Thereafter, $G(\Delta d)$ is found to be higher because of the fact that a few bigger water bodies contribute with a large area. Except between 5 and 7, and at 9 and 11 pixels of diameter, similar behavior has been observed between the histograms of $G(\Delta d)$ and $\phi(\Delta d)$ (Figure 6.4). To avoid tedious manual methods to distribute water bodies according to their sizes, morphological multiscale opening transformation has been applied on a section containing a large number of surface water bodies. This approach would be of use to geomorphologists, hydrologists, and limnologists in various ways.

Estimation of Number–Area–Frequency Dimension of Surface Water Bodies

Spatially distributed surface water bodies extracted from remotely sensed satellite data (Figures 3.16 through 3.18) are one of the examples of fractals. Such water bodies possess various shapes and sizes. Frequency dimension can be computed through a multiscale morphological opening transformation. It is found that the dimension computed through a correlation plot is well tallied with that of the dimension computed through a power-law relationship: a technique that involves computation of correlation integral, $C(r)$, with increasing cycle of opening transformation. Such a correlation integral would be computed for two parameters, viz., number and area, after respective cycle of opening transformation.

When a fractal phenomenon is spatially distributed in a random fashion, quantifying the degree of randomness in the distribution could be done via computing the frequency dimension. Computation of frequency dimension, which is less than the geometric fractal dimension and information dimension, can be done via a power-law (Equation 6.7).

$$C(r) \sim r^D \tag{6.7}$$

where
  $C(r)$ is the correlational integral
  r is the radius
  D is the correlational dimension

This frequency dimension was proposed by Grassberger and Procaccia (1983), which was demonstrated on time series data of chaotic signals, fractal that come as "dust"—isolated points, thinly sprinkled over some range of space.

Let X be a set that represents randomly situated water bodies of various shapes and sizes. Each water body on this set X is indexed with i, and there are N number of such water bodies on set X. Distribution of water bodies according to their sizes could be done via multiscale opening as shown in Equation 6.8.

$$N(X_i \leq nB) = X \backslash ((X \circ nB) \cap X) \tag{6.8}$$

where

$N(X_i)$ is total number of water bodies on set X

$N(X_i \leq nB)$ is a number of water bodies lesser than the size of nB

$(X \circ nB)$ is an opened version of set X by means of the structuring element B of size n

The value of n is in the range 0, 1, 2, ..., N. $(X \circ nB)$ could be represented morphologically as Equation 6.2, which follows following properties:
(1) $N(X \circ NB) \leq N(X \circ (N-1)B) \leq \cdots \leq N(X \circ 2B) \leq N(X \circ B) \leq N(X \circ 0B)$ and
(2) $A(X \circ NB) \leq A(X \circ (N-1)B) \leq \cdots \leq A(X \circ 2B) \leq A(X \circ B) \leq A(X \circ 0B)$.

$C(r)$ introduced by Grassberger and Procaccia (1983) was redefined by Schroeder (1990) in morphological terms as the ratio between the number (area) of water bodies disappeared after the opening transformation by means of B of size n and the squared total number (area) of water bodies, as shown in Equations 6.9 and 6.10.

$$\text{Number} - C(r \leq nB) = \frac{N(X \backslash X \circ nB)}{(N(X))^2} \tag{6.9}$$

$$\text{Area} - C(r \leq nB) = \frac{A(X \backslash X \circ nB)}{(A(X))^2} \tag{6.10}$$

where $X \backslash X \circ nB$ is the set difference between the original set containing all water bodies of several sizes and shapes, and the filtered water bodies obtained after performing the opening by B of radius (size) n are denoted as nB. D_N and D_A, respectively frequency dimension by number and area, are defined by the initial slope shown as Equation 6.11:

$$D := \frac{\ln C(r \leq nB)}{\ln(r \leq nB)}, \quad nB > (n-1)B > \cdots > 1B > 0B \tag{6.11}$$

For a data set containing 1718 water bodies extracted from remotely sensed satellite data situated between the geographical coordinates of 18°00′–18°30′N and 83°15′–83°45′E (Figure 3.16), recursive multiscale morphological opening is performed by increasing the radius (size) of the structuring element B. Cumulative number and area of water bodies that have been sequentially filtered out with increasing cycle of morphological opening, number, and area-correlational integrals have been computed and tabulated (Table 6.2). Logarithms of these correlational integrals $(\ln C(r))$, for both parameters number and area, are plotted as functions of the logarithms of scale, in other words radius of B, $(\ln C(nB))$, which according to the power-law relationship (6.7) should yield straight lines of positive slopes (Figure 6.5). Figure 6.5 illustrates the experimental determination of number-frequency

TABLE 6.2

Distributed Surface Water Bodies and Correlational Integrals

Diameter of Structuring Element	Number of Vanished Water Bodies	Areas of Vanished Water Bodies in Pixel Units	Number-Correlational Integral	Area-Correlational Integral
5	386	17,019	0.00013078	0.000001718011
7	787	21,299	0.000266642	0.000002150091
9	1059	37,041	0.000358798	0.000003739165
11	1262	67,328	0.000427575	0.000006796537
13	1420	76,583	0.000481107	0.000007730798
17	1718	99,530	0.000582072	0.000010047221

Source: Sagar, B.S.D. and Srinivas, D., *Int. J. Remote Sens.*, 20(13), 2491, 1999.

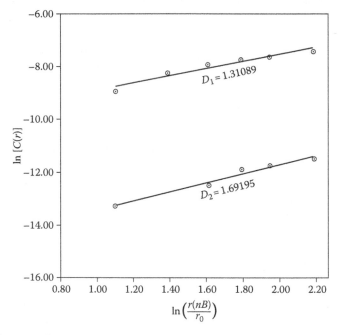

FIGURE 6.5

Determination of frequency dimensions from the number–area–correlation integral. (From Sagar, B.S.D. and Srinivas, D., *Int. J. Remote Sens.*, 20(13), 2491, 1999.)

dimension (D_N) and area-frequency dimension (D_A) for the randomly distributed surface water bodies, which yield straight-line dependence of $\ln N(X \setminus X \circ nB)/(N(X))^2$ and $\ln A(X \setminus X \circ nB)/(A(X))^2$ on $\ln (nB)$ with slopes $(D_N) = 1.3069$ and $(D_A) = 1.7$. In the context of grain size analysis, Delfiner (1972) has explained number and area distribution functions. This analysis has similarity with the technique proposed by Grassberger and Procaccia (1983). This technique to compute frequency dimension can be extended to various phenomena of geomorphologic significance such as river basin, channel networks, islands, and hills. The application of this simple technique can be automatically shown on the information (thematic) retrieved from remotely sensed digital data.

Usually water bodies are observed at topographic depressions of landscape. Such topographic depressions are also treated as local minima. Spatial positions of those local minima, where the presence of water bodies is highly conspicuous, act as sources for floodwater propagation. Spatial organization of those water bodies indicates landscape organization. Hence, quantification of the degree of randomness in the spatial organization of water bodies is important. An approach that provides $f(\alpha)$ spectra construction, which is due to Halsey et al. (1986), has been adopted to compute the degree of randomness in spatial organization of surface water bodies.

Surface water bodies, which are climatically sensitive, are prone to have variations in their geometries with change in time. Hence, understanding not only the spatial organization but also the temporal organization is an important study. Such an understanding in quantitative terms is now possible with recent extension to estimate the local and generalized information dimensions (Halsey et al. 1986). Many models have been available in literature that provide descriptive analysis for qualitative understanding of spatiotemporal organization of water bodies. Multifractal technique is a choice to better quantify and characterize the degree of randomness in the spatiotemporal organization of water bodies. Information dimension that can be computed by using multifractal technique has the capability to capture the characteristic alteration by means of an analytical value.

Self-Similar Size Distributions of Water Bodies by Iterated Bisecting

Let $f(x, y)$ and X be a function depicting a landscape and be a thematic map depicting only water bodies and no-water body regions, decomposed from $f(x, y)$. Total area occupied by water bodies shown in X is denoted by $A(X)$. The set X is bisected in two ways: vertical bisecting and horizontal bisecting. While bisecting X, in horizontal direction, there would be top and bottom sections. How the area of X is distributed in the top and

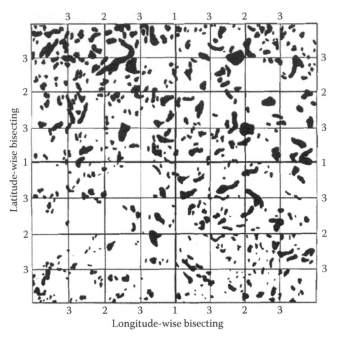

FIGURE 6.6
Section shows water bodies. Different bisecting lines show both latitude and longitude wise. The digits 1, 2, and 3 on latitude and longitude planes represent successive levels of iterated bisecting to construct self-similar distribution of water bodies in Figure 6.1. (From Sagar, B.S.D. et al., *Int. J. Remote Sens.*, 16(16), 3059, 1995.)

bottom pieces after bisecting needs to be computed. One piece of thematic map containing water bodies occupied area of βX, and the other piece occupied $(1 - \beta)X$, where β is a portion of area occupied by water bodies in the normalized scale of 1. Then obviously, β and $(1 - \beta)$ are in the form of fractional values in the range of 0 and 1. Self-similar size distribution process via multiscale morphological opening transformation has been done on another section of water bodies (situated across flood plain region of Gosthani River) (Figure 6.6).

Pietronero and Siebesma (1986) proposed iterated bisecting approach to understand the degree of randomness in the spatially distributed phenomenon like spatial distribution of surface water bodies (Figure 6.1). The iterated bisecting was done in two directions that include horizontal and vertical bisectings. The first three levels of bisecting on both horizontal and vertical planes have been shown in Figure 6.6. It is to be noted that at the zeroth-level bisecting, the area occupied by all the water bodies and the number of water bodies on the section have been treated as the probability with unity (i.e., 1). This is treated as a process that begins with a uniform probability distribution over the unit interval (Figure 6.7a). At the first-level bisecting in horizontal direction, the upper and lower portions are respectively with

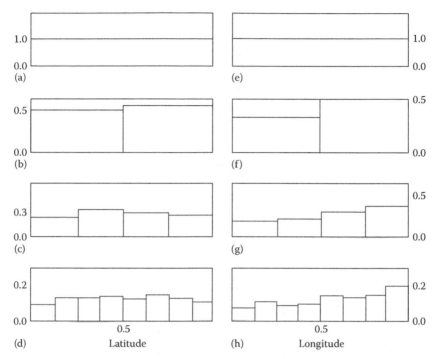

FIGURE 6.7
Self-similar distributions of water bodies by iterated bisecting: (a–d) show latitude-wise bisecting and (e–h) show longitude-wise bisecting. (From Sagar, B.S.D. et al., *Int. J. Remote Sens.*, 16(16), 3059, 1995.)

the probabilities β and (1 − β), with the former greater than the latter are obtained in single-step distribution (Figure 6.7b). Second iterated bisecting resulted in four intervals (viz., upper–upper, upper–lower, lower–upper, and lower–lower). The probability distributions of these four bisected portions have been shown in Table 6.3 and illustrated in Figure 6.7c. The results of these probability distribution values obtained at third-level bisecting have been shown in Figure 6.7d and in Table 6.3. Similar computations have also been done in vertical bisectings. Table 6.3 shows probability distribution values, observed on the data containing a larger number of surface water bodies obtained for both horizontal and vertical bisecting directions. It is obvious that the spatial distribution of water bodies, through an iterated vertical bisecting, is more or less homogeneous, whereas significant heterogeneity is observed from β and (1 − β) computed in horizontal bisecting. The uppermost region is the densest with 19% of the total water body area, and lowermost is found to be sparser with 7% of the total water body area. Interestingly, the distribution of water bodies observed could be simulated via binomial multiplicative process from which the values at successive bisectings could be predicted according to input: β and (1 − β) (at first-level bisecting) yield two portions; at second-level bisecting, β², β(1 − β) and (1 − β)β, (1 − β)² yield four

TABLE 6.3

Self-Similar Distribution of a Section of Water Bodies

	Percentage Area Occupied by the Total Water Body Area					
	Vertical Bisecting			**Horizontal Bisecting**		
First Bisecting	**Second Bisecting**	**Third Bisecting**		**First Bisecting**	**Second Bisecting**	**Third Bisecting**
0.51 (β_1)	0.24 (1 − β_2) 0.26 (β_2)	0.11 (1 − β_3)	0.14 (β_3)	0.63 (β_1)	0.35 (1 − β_2) 0.28 (β_2)	0.19 (1 − β_3) 0.15 (β_3)
0.49 (1 − β_1)	0.27 (β_2) 0.24 (1 − β_2)	0.14 (β_3)	0.13 (1 − β_3)	0.37 (1 − β_1)	0.19 (β_2) 0.17 (1 − β_2)	0.14 (β_3) 0.14 (1 − β_3)
		0.13 (1 − β_3)	0.14 (β_3)			0.10 (1 − β_3) 0.09 (β_3)
		0.13 (β_3)	0.10 (1 − β_3)			0.11 (β_3) 0.07 (1 − β_3)

Sources: Sagar, B.S.D. et al., *Int. J. Remote Sens.*, 16(16), 3059, 1995; Sagar, B.S.D., *Discr. Dyn. Nat. Soc.*, 6(3), 213, 2001.

parts, and the process goes on such that in every bisecting, the distribution of water bodies included was divided in the ratio of $\beta:(1 - \beta)$.

A significant matching was observed between the observed probability distribution values (Table 6.3) and the probability distribution values predicted via computations carried out by binomial multiplicative approach (Sagar et al. 2002; Table 6.4; Figure 6.8). In almost all cases, such matching between the observed and predicted values is possible. By using the probability distribution values β and $(1 - \beta)$, $f(\alpha)$ spectra are computed to further understand the degree of homogeneity quantitatively. From Table 6.3, it is obvious that 51% of water body is covered in the left half, while bisecting vertically, which is considered as β, and then $(1 - \beta)$ is 49%, being the area occupied in the right half of the section. These values have been observed as 63% in the upper half and 37% in the lower half, while horizontal bisectings process was followed. Further bisectings in the vertical and horizontal bisectings have yielded four values from each divided cuts. It is interesting to see the significant similarity in the distributions observed according to the two bisecting directions with that of the values predicted through binomial multiplicative processes (Halsey et al. 1986). The comparisons can be seen from Tables 6.3 and 6.4. Due to this similarity, Equations 6.12 and 61.3, which are due to Halsey et al. (1986), have been employed to compute information and correlation dimensions by taking the values $\beta = 0.51$, $(1 - \beta) = 0.49$ for vertical bisecting, and $\beta = 0.63$, $(1 - \beta) = 0.37$ for horizontal bisecting.

$$D = -\{\beta \log_2 \beta + (1-\beta)\log_2(1-\beta)\} \tag{6.12}$$

$$D_q = -\frac{1}{q-1}\log_2(\beta^q + (1-\beta)^q) \tag{6.13}$$

where D and D_q denote information dimension and global fractal dimensions. These two dimensions have been computed for the sections containing a large number of water bodies as 0.99 and 0.95, respectively for vertical and horizontal bisectings. From Tables 6.3 and 6.4, it was inferred that spatial distribution of water bodies, which considers vertical bisecting, is more homogeneous than that of horizontal bisecting. This qualitative understanding could be quantitatively explained via information dimensions. Higher the information dimension, higher is the homogeneity, and vice versa.

By employing Equations 6.14 and 6.15 to construct $f(\alpha)$ spectra, where the basic inputs are β and $(1 - \beta)$, $f(\alpha)$ spectra could be constructed essentially to derive generalized dimensions (D_0, D_1, D_2, D_3) for a better understanding of spatial distribution of surface water bodies.

$$\alpha_q = -\frac{\beta^q \log_2 \beta + (1-\beta)^q \log_2(1-\beta)}{\beta^q + (1-\beta)^q} \tag{6.14}$$

$$f(\alpha_q) = q\alpha_q + \log_2(\beta^q + (1-\beta)^q) \tag{6.15}$$

where q ranges between any integer values.

TABLE 6.4

Probability Distribution Estimated from Binomial Multiplicative Process

Percentage Area Occupied by the Total Water Body Area

Vertical Bisecting

First Bisecting	Second Bisecting	Third Bisecting
$0.51\ (\beta_1)$	$0.26\ (1-\beta_2)$	$0.13\ (1-\beta_3)$
	$0.25\ (\beta_2)$	$0.13\ (\beta_3)$
		$0.13\ (\beta_3)$
		$0.12\ (1-\beta_3)$
$0.49\ (1-\beta_1)$	$0.25\ (\beta_2)$	$0.13\ (1-\beta_3)$
	$0.24\ (1-\beta_2)$	$0.12\ (\beta_3)$
		$0.13\ (1-\beta_3)$
		$0.11\ (1-\beta_3)$

Horizontal Bisecting

First Bisecting	Second Bisecting	Third Bisecting
$0.63\ (\beta_1)$	$0.40\ (1-\beta_2)$	$0.25\ (1-\beta_3)$
	$0.23\ (\beta_2)$	$0.15\ (\beta_3)$
		$0.09\ (1-\beta_3)$
		$0.09\ (\beta_3)$
$0.37\ (1-\beta_1)$	$0.23\ (\beta_2)$	$0.15\ (\beta_3)$
	$0.14\ (1-\beta_2)$	$0.09\ (1-\beta_3)$
		$0.09\ (\beta_3)$
		$0.05\ (1-\beta_3)$

Source: Sagar, B.S.D., *Discr. Dyn. Nat. Soc.*, 6(3), 213, 2001.

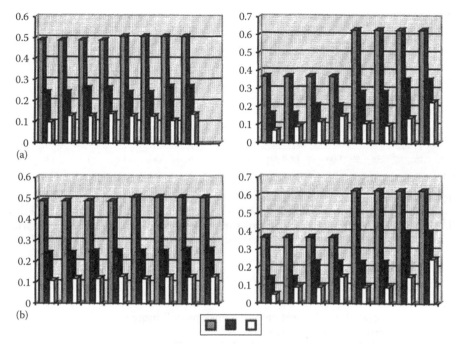

FIGURE 6.8
Comparative self-similar distributions of surface water bodies (a) observed from the data set and (b) estimated through binomial multiplicative process. Gray (long bars), black (medium bars), and white (short bars) indicate the probability distributions at first, second, and third levels of bisectings, respectively. The four right-side long bars (half unit) in the first bisecting indicate the β, and the left side four long bars indicate 1 − β. In the second bisecting, two medium bars are considered as a quarter unit. In the third bisection, each bar is considered as one-eighth of a unit. (From Sagar, B.S.D., *Discr. Dyn. Nat. Soc.*, 6(3), 213, 2001.)

By employing β = 0.51 (for vertical bisecting) and β = 0.63 (for horizontal bisecting) as important parameters in Equations 6.14 and 6.15, $f(\alpha)$ spectra, which are also termed as multifractal spectra, have been constructed (Figure 6.9). It could be seen that $f(\alpha)$ at peak point is equal to capacity dimension (D_0), which is 1 for both bisectings. Information dimension (D_1) could be obtained as the slope of the tangent drawn to the curve of $f(\alpha)$ from the origin of $f(\alpha)$ spectra. D_1 computed are 0.99 and 0.95 for vertical and horizontal bisectings, respectively. D_1 is exactly tallied with D (information dimension) computed according to Equation 6.12. From the multifractal spectra, D_0, D_1, D_2, D_3 for the considered section of spatially distributed surface water bodies could also be seen (Table 6.5).

From the approaches that include self-similar iterated bisecting process, binomial multiplicative process, and construction of multifractal spectra $(f(\alpha))$, it is clear that the degree of spatial randomness of any phenomenon, which could be represented in spatial form, could be well studied to understand that spatial distribution quantitatively. For a phenomenon like surface

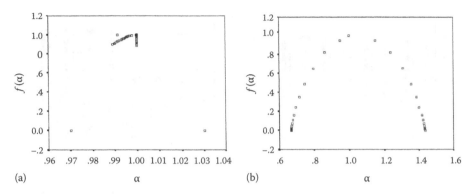

FIGURE 6.9

$f - \alpha$ Spectra for a section of landscape containing a large number of randomly situated surface water bodies. These spectra are constructed for the binomial multiplicative process with (a) $\beta = 0.51$ (vertical bisecting) and (b) $\beta = 0.63$ (horizontal bisecting). (From Sagar, B.S.D., *Discr. Dyn. Nat. Soc.*, 6(3), 213, 2001.)

TABLE 6.5

Generalized Information Dimensions of Randomly Situated Surface Water Bodies

	D_0	D_1	D_2	D_3
Vertical bisection	1	0.9997114	0.000423	0.9991349
Horizontal bisection	1	0.950672	0.9056287	0.8668016

Source: Sagar, B.S.D., *Discr. Dyn. Nat. Soc.*, 6(3), 213, 2001.

water bodies that alter their geometries across time periods, by following these approaches, one can understand not only the spatial complexities but also their spatiotemporal complexities. How this approach could be adapted on size-distributed water bodies has been shown in the following. It is based on a postulate that the larger size water bodies possess distinct degree of spatial randomness than smaller size water bodies.

Is the Spatial Distribution of Smaller Water Bodies More Homogeneous?

The intuitive argument is that the spatial organization of smaller water bodies is more homogeneous than that of larger water bodies. A section containing larger number of water bodies (Figure 6.1) has been considered to first distribute water bodies according to five different size categories to validate the intuitive argument. A multiscale morphological opening transformation has been followed to distribute surface water bodies into following size categories: (1) <7 pixel diameter, (2) 7–11 pixel diameter, (3) 11–15 pixel diameter, and (4) >15 pixel diameter.

The multiscale morphological openings adapted to categorize water bodies as per the earlier ranges have been given in Equation 6.16:

$$\left(X \backslash (X \circ 7B)\right)$$
$$(X \circ 7B) \backslash (X \circ 11B)$$
$$(X \circ 11B) \backslash (X \circ 15B)$$
$$(X \circ 15B)$$

(6.16)

The four sections obtained include water bodies that are smaller than 7 pixel radius, water bodies with sizes that are between the 7 and 11 pixels radii, water bodies with sizes that are in between the 11 and 15 pixels radii, and those water bodies that are larger than the 15 pixel radius. These four sections of water bodies have been shown in Figure 6.3. Figure 6.3a shows water bodies that are smaller than 7 pixel radius, and Figure 6.3b through d follows other three larger categories.

On each of these size-distributed sections, iterated bisecting approach has been implemented and their probability distribution values have been computed (Table 6.6). The probability values computed according to binomial multiplicative process have been found well tallied with that of observed values. These values computed according to binomial multiplicative process have also been given in Table 6.6. By using β and $(1 - \beta)$ for size-distributed water body section (0.8 and 0.2 for >15 pixel diameter; 0.64 and 0.36 for 11–15 pixel diameters, 0.6 and 0.4 for 7–11 pixel diameters, and 0.57 and 0.43 for <7 pixel diameter), $f(\alpha)$ spectra have been constructed (Figure 6.10a through d). It is very interesting to find that the information dimension (D_1) and correlation dimension (D_2) are larger for the smaller-size category water bodies. These two dimensions for the water bodies smaller than 7 pixel diameter include 0.987 and 0.95 (Table 6.6), whereas these dimensions for the water bodies between the sizes of 7 and 11 pixel diameters, 11 and 15 pixel diameters, and >15 pixels diameter, respectively, are 0.97 and 0.89, 0.942 and 0.795, and 0.721 and 0.322. From these quantitative results, it is deduced that the larger the size of water bodies, the higher the heterogeneity and vice versa. This type of analysis would be highly useful to study the spatiotemporal organization of the lakes derived from the multidate remotely sensed data that consist of lakes of various sizes and shapes.

TABLE 6.6

Self-Similar Distribution of Size Distributed Water Bodies, Division Rates, and Estimated Generalized Dimensions

Size Distribution of Surface Water Bodies by a Multiscale Opening Transformation	Number of Water Bodies after Iterated Bisecting with Division Rates in Parentheses			Information (D_1) and Correlation (D_2) Dimensions Derived from $f - \alpha$ Spectra	
	Number of Water Bodies				
Diameter of Structuring Element	Zeroth Bisecting	First Bisection	Second Bisection	(D_1)	(D_2)
>15	20	16 ($\beta = 0.8$) 4 ($1 - \beta = 0.2$)	12 ($\beta^2 = 0.60$) 4 ($\beta(1 - \beta) = 0.20$) 3 ($(1 - \beta)\beta = 0.15$) 1 ($(1 - \beta)^2 = 0.05$)		
11–15	36	23($\beta = 0.64$) 13 ($1 - \beta = 0.36$)	15 ($\beta^2 = 0.42$) 8 ($\beta(1 - \beta) = 0.22$) 8 ($(1 - \beta)\beta = 0.22$) 5 ($(1 - \beta)^2 = 0.14$)	0.942	0.795
7–11	40	24 ($\beta = 0.6$) 16 ($1 - \beta = 0.4$)	14 ($\beta^2 = 0.35$) 10 ($\beta(1 - \beta) = 0.25$) 10 ($(1 - \beta)\beta = 0.25$) 6 ($(1 - \beta)^2 = 0.15$)	0.97	0.89
<7	143	81 ($\beta = 0.57$) 62 ($1 - \beta = 0.43$)	46 ($\beta^2 = 0.32$) 35 ($\beta(1 - \beta) = 0.25$) 25 ($(1 - \beta)\beta = 0.25$) 27 ($(1 - \beta)^2 = 0.18$)	0.987	0.95

Source: Sagar, B.S.D. et al., *Int. J. Remote Sens.*, 23(3), 503, 2002.

Size Distribution–Based Scaling Laws

In Chapter 5, allometric power-law relationships have been shown for a large number of surface water bodies (Figure 5.24a) and their corresponding ZIs (Figure 5.24b). It was also found that these two interdependent phenomena belong to two different universality classes. However, size distributions of these interdependent phenomena yield new scaling relationships.

Multiscale morphological opening transformation has been performed on both water bodies (X) and their ZIs (X_Z) to distribute them according to their sizes. After each cycle of opening on (X) and (X_Z) by means of respective size of structuring element (nB), the areas of water bodies and their ZIs retained have been computed along with the number of retained water bodies and ZIs. N_X, N_{X_Z}, A_X, and A_{X_Z}, respectively, denote number of water bodies, number of ZIs,

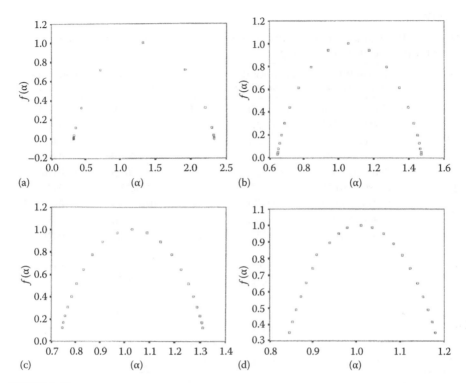

FIGURE 6.10
Multifractal spectra for four different size categories of water body sections. (a) The larger water bodies between 15 and 17 pixel diameter, (b) the water bodies between 11 and 15 pixel diameter, (c) the water bodies between 7 and 11 pixel diameter, and (d) the water bodies less than 7 pixel diameter. (From Sagar, B.S.D. et al., *Int. J. Remote Sens.*, 23(3), 503, 2002.)

area of all water bodies, and areas occupied by all corresponding ZIs. These details have been furnished in Table 6.7. These details satisfy the following properties: (1) $N_X = N_{X_Z}$, (2) $X_i \subseteq X_{Z_i}$, (3) $A(X_i) \le A(X_{Z_i})$, and (4) $A(X) \le A(X_Z)$.

Relationships between N and A of water bodies and their ZIs yield the following power-laws (Equations 6.17 and 6.18).

$$N_X \sim A_X^{1.41} \tag{6.17}$$

$$N_{X_Z} \sim A_{X_Z}^{1.70} \tag{6.18}$$

The earlier power-law relationships have been shown by plotting N_X and N_{X_Z} as functions of their corresponding areas (Figure 6.11a and b). These relationships translate physical phenomenon of drought under the assumptions that (1) the larger water bodies get least affected by drought conditions, and (2) the smaller water bodies get easily affected by drought conditions. Water bodies of larger size category would get vanished with increasing drought conditions. The impact of increasing drought could be mimicked through multiscale opening transformation. However, no significant change in the power-law derived

TABLE 6.7

Areas (in Pixel Units) and the Number of Surface Water Bodies and Their ZIs after Respective Degree of Sifting

SE Size	Area of WBs (and ZIs)	Number of WBs (and ZIs)	SE Size	Area of WBs (and ZIs)	Number of WBs (and ZIs)
0	163,783 (1,234,566)	645 (645)	69	0 (157,351)	0 (24)
5	155,250 (1,213,515)	580 (621)	73	0 (125,092)	0 (17)
9	100,909 (1,171,792)	288 (620)	77	0 (112,670)	0 (15)
13	54,807 (1,127,634)	121 (587)	81	0 (103,365)	0 (14)
17	28,581 (1,061,809)	41 (523)	85	0 (76,682)	0 (10)
21	15,989 (976,057)	15 (445)	89	0 (66,530)	0 (7)
25	12,218 (871,559)	10 (367)	93	0 (65,256)	0 (6)
29	6,801 (752,042)	6 (330)	97	0 (55,234)	0 (5)
33	4,643 (632,944)	3 (225)	101	0 (43,606)	0 (2)
37	1,358 (523,085)	1 (152)	105	0 (40,588)	0 (2)
41	0 (442,187)	0 (109)	109	0 (29,986)	0 (1)
45	0 (369,017)	0 (84)	113	0 (29,710)	0 (1)
49	0 (313,898)	0 (65)	117	0 (29,337)	0 (1)
53	0 (284,865)	0 (62)	121	0 (28,951)	0 (1)
57	0 (239,058)	0 (52)	125	0 (28,306)	0 (1)
61	0 (194,656)	0 (33)	129	0 (27,449)	0 (1)
65	0 (173,284)	0 (28)	133	0 (26,665)	0 (1)

Source: Sagar, B.S.D., *Water Resour. Res.*, 43(2), W02416, 2007.

for N_{X_Z} and A_{X_Z} would be expected as the area of ZI get least affected. From the earlier power-laws (Equations 6.17 and 6.18), it is deduced that power-law exponent for water bodies is smaller than that of their ZIs. This observation may be valid as the ZIs act as super sets for their corresponding water bodies.

To find the extent of deviation between the size categories in terms of power-law derived between the sizes of water body and its corresponding ZIs, size distribution and their density function for X and X_Z have been computed. Equations 6.19 and 6.20 have been followed to compute size distribution functions, normalized size, and number distribution functions for water bodies and their ZIs.

$$S(K) = 1 - \frac{A(X) < nB}{A(X)} \tag{6.19}$$

$$S(K)_Z = 1 - \frac{A(X_Z) < nB}{A(X_Z)} \tag{6.20}$$

where $n \in (0, N)$, numerators and denominators represent the sum of all the water bodies that are smaller than nB and the sum of all the water bodies of

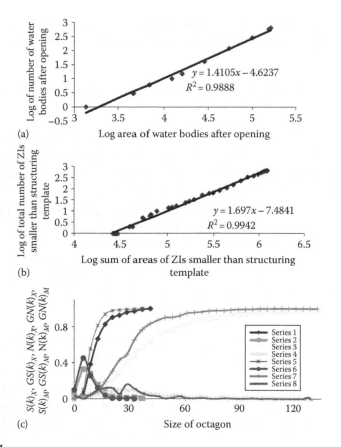

FIGURE 6.11

(a and b) Number–area relationship for water bodies and their ZIs that yield power-laws of 1.41 and 1.70, respectively, and (c) normalized size and number distribution and density functions of water bodies and ZIs plotted against the size of octagon (eight series dots). The size distribution plots give the information about the size distribution of water bodies and their influence zones: approximately 50,000 pixels belong to water bodies of a 5 pixel diameter, 50,000 pixels to water bodies of a 10 pixel diameter, and 25,000 pixels to water bodies of a 11 pixel diameter. Local maxima in the distribution spectra at a given scale indicate the presence of many water bodies and influence zones at that scale. (From Sagar, B.S.D., *Water Resour. Res.*, 43(2), W02416, 2007.)

all sizes respectively. Similarly, $A(X_Z) < nB$ and $A(X_Z)$ respectively represent the sum of all the ZIs that are smaller than nB and the sum of all the ZIs of all size categories.

Probability density functions, to describe the size attributes of individual water bodies and their ZIs, have been computed by using the normalized distribution functions $(S(k))$ and $(S(k)_Z)$ as per Equation 6.21:

$$GS(k) = S(k+1) - S(K) \tag{6.21}$$

The following inferences have been drawn based on normalized size (and number) distributions and their density functions of water bodies and corresponding ZIs:

1. Residues of X and X_Z by removing portions smaller than the threshold structuring element radii.

2. Larger size categories of water bodies and ZIs relatively contribute lesser areas, and significant variations in surface area between two consecutive cycles of openings have been observed (Figure 6.11c), further indicating that the region contains water bodies and ZIs of comparable size to the smaller level of sifting.

3. The first derivative of the surface area at a given scale, which is a local maximum in the size distribution plot (Figure 6.11c), indicating the presence of many water bodies at that scale.

4. In the general trends of various plots (Figure 6.11c) for water bodies and their influence zones, significant similarities have been observed, further indicating spatial relationships between these interdependent topological phenomena (Table 6.7).

Relations between correlation sums ($C(n)$) of area (and number) and radius of structuring element employed as size distribution via multiscale opening transformations have been shown. Radius of structuring element and cycle number of opening are just the same. To avoid confusion, in place of radius (r), cycle number (n) was employed. The ratio between the sum of the areas of water bodies smaller than the structuring element size, nB, and the total area squared of water bodies yields correlation sum of areas of water bodies. This is expressed in Equation 6.22:

$$\text{Area} - C(n)_X = \frac{A\left(\bigcup_{i=1}^{N} X_i < nB\right)}{n^2} \tag{6.22}$$

where X_i denotes water body with index i.

Similar ratio for ZIs could be computed according to Equation 6.23,

$$\text{Area} - C(n)_{X_Z} = \frac{A\left(\bigcup_{i=1}^{N} X_{Z_i} < nB\right)}{n^2} \tag{6.23}$$

where X_{Z_i} is a zone corresponding to X_i.

Correlation sums of the number for water bodies (X) and the ZIs have been computed, by replacing areas with numbers (N) in Equations 6.22 and 6.23, according to Equations 6.24 and 6.25.

$$\text{Number} - C(n)_X = \frac{N(X_i < nB)}{n^2} \tag{6.24}$$

$$\text{Number} - C(n)_{X_Z} = \frac{N(X_{Z_i} < nB)}{n^2} \tag{6.25}$$

where $N[\cdot]$ denotes the number of water bodies (or) ZIs.

Relationships between area–correlation sum of water bodies and the radius of structuring element (nB) yield the following power-law relationships (Equations 6.26 through 6.29).

$$\text{Area} - C(n)_X \sim (n)^{1.26} \tag{6.26}$$

Similar relationship for ZIs yields

$$\text{Area} - C(n)_{X_Z} \sim (n)^{1.16} \tag{6.27}$$

When number $- C(n)_X$ is plotted as a function of the radius of structuring element, the following power-law relationships have been derived:

$$\text{Number} - C(n)_X \sim (n)^{0.94} \tag{6.28}$$

Similar plot for number $- C(n)_{X_Z}$ and the radius of structuring element yielded a power-law relationship like (Equation 6.29)

$$\text{Number} - C(n)_{X_Z} \sim (n)^{1.03} \tag{6.29}$$

The four power-law exponents further indicate that the distribution based on number exhibited more homogeneity than the distribution based on areas (sizes) for both water bodies and ZIs. The graphical plots to derive these four power-law values have been shown in Figure 6.12a.

Ratios of areas of water body and ZIs that are retained after progressive sifting by means of structuring elements with increasing radii and the squared radius $(n)^2$ of structuring element have been computed. Similar ratios of number have also been computed. These ratios are denoted as area $- C(n)_X$, area $- C(n)_{X_Z}$, number $- C(n)_X$, and number $- C(n)_{X_Z}$. These four ratios have been plotted as functions of $1/n$, and other power-law values could be derived (Figure 6.12b). Size distributions of water bodies and their ZIs by performing multiscale opening transformation have provided the number and the area of water bodies and ZIs smaller than the size of the structuring element. These number and areas of the size-distributed water bodies and ZIs have been plotted as functions of size of the structuring element employed in distribution process to derive size distribution–based scaling laws. These scaling laws also reveal different universal scaling characteristics.

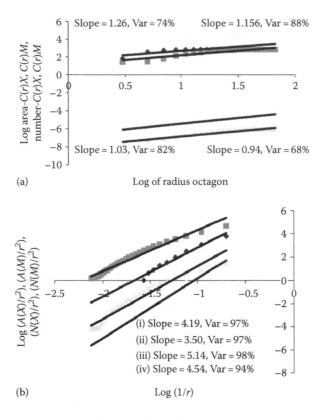

FIGURE 6.12
(a) Power-law relationship between radius and area-correlation sum of water bodies (second plot from bottom) and ZIs (bottommost plot), and number-correlation sum of water bodies (topmost plot) and ZIs (second plot from the top), and (b) power-law relationships between $(1/r)$ and the ratios of (i) $\left[(A_M < nB_r)/r^2\right]$, (ii) $\left[(A_X < nB)/r^2\right]$, (iii) $\left[(N_M < nB)/r^2\right]$, and (iv) $\left[(N_X < nB)/r^2\right]$. (From Sagar, B.S.D., *Water Resour. Res.*, 43(2), W02416, 2007.)

References

Barker, J. L., 1975, Monitoring water quality from Landsat, *Proceedings of NASA Earth Resources Survey Symposium*, Houston, TX, Vol. 1A, pp. 383–418.

Delfiner, P., 1972, A generalisation of the concept of size, *Journal of Microscopy*, 95(2), 203–216.

Grassberger, P. and I. Procaccia, 1983, Estimation of the Kolmogorov entropy from a chaotic signal, *Physical Review*, 28A, 259.

Halsey, T. C., M. H. Jensen, L. P. Kadanoff, I. Procaccia, and B. I. Shraiman, 1986, Fractal measures and their singularities: The characterisation of strange sets, *Physical Review*, A33, 1141–1151.

Mandelbrot, B. B., 1982, *Fractal Geometry of Nature*, W. H. Freeman & Co: San Francisco, CA, p. 468.

Meyer, F., 1980, Feature extraction by mathematical morphology in the field of quantitative cytology, Technical report of Ecole nationale superiere des mines de Paris, Fountainbleau, France.

Pietronero, L. and A. P. Siebesma, 1986, Self similarity of fluctuations in random multiplicative processes, *Physical Review Letters*, 57, 1098–1101.

Sagar, B. S. D., 2001, Quantitative spatial analysis of randomly situated surface water bodies through $f - \alpha$ spectra, *Discrete Dynamics in Nature and Society*, 6(3), 213–217.

Sagar, B. S. D., 2007, Universal scaling laws in surface water bodies and their zones of influence, *Water Resources Research*, 43(2), W02416, 2007.

Sagar, B. S. D., C. B. Rao, and B. Raj, 2002, Is the spatial organization of larger water bodies heterogeneous? *International Journal of Remote Sensing*, 23(3), 503–509.

Sagar, B. S. D. and D. Srinivas, 1999, Estimation of number-area-frequency dimensions of surface water bodies, *International Journal of Remote Sensing*, 20(13), 2491–2496.

Sagar, B. S. D., M. Venu, and B. S. P. Rao, 1995, Distributions of surface water bodies, *International Journal of Remote Sensing*, 16(16), 3059–3067.

Schroeder, M., 1990, *Fractals, Chaos, Power Laws*, Freeman, New York.

7

Morphological Shape Decomposition: Scale-Invariant but Shape-Dependent Measures

Any given foreground space (set) depicting, for instance, Apollonian space, Koch fractal, porous medium, and any other space like nonnetwork space can be fragmented into simple forms of shapes such as squares, rhombus, octagons, triangles, and circles. A popular power-law relationship between the number and radius provides a noninteger dimension that is scale invariant. In this chapter, morphological shape decomposition (MSD) of various sets (foregrounds) mimicking geomorphologic phenomena has been explained to further compute scale-invariant power-laws that are shape dependent.

Introduction on MSD and Its Application in Various Fields

In nature and man-made environment, examples of heterogeneous media to name a few include geomorphologic basins, soils, sandstone, granular media, earth crust, sea ice, and wood (Torquato 2000). The physical phenomenon of interest occurs on various length-scales (e.g., geological media that span from tens of nanometers to meters). Such heterogeneous media can be characterized statistically.

Deformed shapes can be decomposed into regular shapes of several sizes. For instance, a square shape can be decomposed into shapes such as a circle, rhombus, and octagon of various sizes. This type of decomposition facilitates a procedure to estimate the dimension, akin to fractal dimension, through a power-law relationship between size or radius and the number of decomposed and disconnected shapes at a given threshold value. This power-law relationship can be represented as Equation 7.1.

$$N(\leq rB) \sim (r)^D \tag{7.1}$$

where, N, r, and D respectively represent the number of decomposed shapes that is more than the threshold radius, radius, and the fractal dimension. The fractal dimension that could be derived from this power-law relationship can be treated as unique, since only the internal region of the shape of which the dimension is to be estimated is considered. In other words, if the shape of which the dimension is to be estimated is in binary format, for instance shape with 1s, and its background with 0s, only the region with 1s will be considered by leaving the region with 0s. However, in the other procedures, to estimate the fractal dimension, such as box counting method, both shape and its background will be considered.

To estimate the dimension, one can consider a probing rule with which the shape under study is to be decomposed. To estimate the fractal dimension, an alternate method by employing certain mathematical morphological transformations is proposed. Mathematical morphological transformations have been earlier applied to estimate fractal dimensions (Flook 1978, Kanmani et al. 1992a,b, Soille and Rivest 1996). The applications of mathematical morphological transformations that have been employed to deal with the fractal-related studies have been shown (Sagar 1996, 1999, Sagar et al. 1998, 2001, 2002, 2003, Sagar and Murthy 2000).

We employed morphologic transformations in our various earlier studies for decomposing pore structure into nonoverlapping disks (NODs) of various sizes and shapes, and this provided a fair representation and description of pore image morphology. Mathematical morphology-based approaches have shown promising potential for porous medium description due to the fact that the mathematical morphology focuses on the geometric character of pore space. MSD is an elegant approach to decompose binary shapes.

Morphological Shape Decomposition

To estimate the fractal dimension (Mandelbrot 1982) of a shape, a procedure based on morphological decomposition is adapted. This procedure includes systematic use of multiscale *opening* and simple logical operators. A set of equations (7.2) enables on how to decompose a shape by following morphological transformations. The sequential steps involved in decomposing the shape are also depicted in the flowchart (Figure 7.1) as well as in Equation 7.2. Let X be a shape (set) to be decomposed into rhombic shapes X_i of various sizes such that $X_i > X_{i+1} > \dots$ Let B be a symmetrical flat structuring element of primitive size of 3×3, rhombic in shape.

$$X_i = X \setminus (X \circ NB) | (X \circ NB) \neq \varnothing; \quad X \circ (N+1)B = \varnothing$$

$$X_{i+1} = X_i \setminus (X_i \circ NB) | (X_i \circ NB) \neq \varnothing; \quad X_i \circ (N+1)B = \varnothing$$

$$\vdots \tag{7.2}$$

$$X_{i+N} = X_{i+(N-1)} \setminus (X_{i+(N-1)} \circ NB) | (X_{i+(N-1)} \circ NB) \neq \varnothing; \quad X_{i+(N-1)} \circ (N+1)B = \varnothing$$

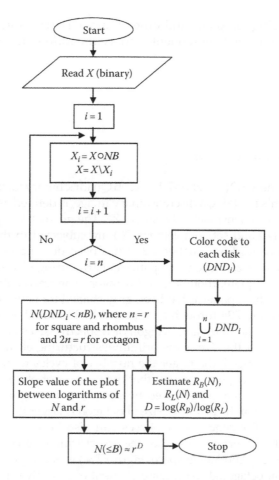

FIGURE 7.1
Flowchart showing the sequential steps. (From Radhakrishnan, P. et al., *Chaos Soliton Fract.*, 21(3), 563, 2004.)

where $X_i, X_{i+1}, \ldots, X_{i+N}$ are connected components decomposed from X. After performing N times of multiscale opening on a shape, which is subjected to for the estimation of fractal dimension, the *opened* shape needs to be subtracted from the original shape. This can be achieved by simple logical operation, which is represented as the symbol (\). If N + 1 times are required to vanish a set (or shape), N times of multiscale openings need to be performed to decompose the shape and successively achieve subtracted portions of the shape. On each subtracted portion, the condition that N + 1 times of multiscale *opening* should vanish the respective shape is taken or the successive subtracted portions are taken into consideration. The number of subtracted portions that may appear while decomposing the shape depends on the size and shape of the primary pattern (or the structuring template) and other characteristic information. These decomposed connected components satisfy the following properties:

1. $X = \bigcup_{i=1}^{N} X_i$

2. $X_i \cap X \neq \varnothing$

3. $X_i \cap X_{i+1} \cap \cdots \cap X_{i+N} = \varnothing$

To demonstrate the MSD (Equation 7.2), a Koch Quadric binary fractal (Figure 3.5a) is considered. This fractal X, a discrete binary image, is defined as a finite subset of Euclidean two-dimensional (2-D) space, Z^2. The geometrical properties of a fractal as a set (X) and set complement (X^c) are subjected to the morphological operations involved in Equation 7.2. It has been decomposed into simpler patterns of various sizes. The three patterns (structuring elements, B) considered here include square, rhombus, and octagon. The fractal after decomposition by means of these patterns has been gray shaded for better understanding and shown in Figure 7.2a through c, respectively. The number of decomposed patterns of square, rhombus, and octagon of respective sizes has been given in Table 7.1. The smaller the size of the primitive structuring element that is used to decompose the fractal, the larger the number of cycles required to decompose the fractal. Hence, it is apparent that the number of phases is more while decomposing with rhombus, followed by square, and octagon. This is due to the fact that the size of the primitive size of octagon is larger than that of square, and of rhombus. The sequence of phases can also be visualized as growth stages of fractal. The primitive size of the structuring elements considered is shown in Figure 2.7. Apollonian space (Figure 7.3a) is decomposed with respect to symmetric flat octagonal structuring element of primitive size 5 × 5. Figure 7.3b shows the gray-shaded superposed connected components decomposed from Apollonian space according to Equation 7.2. This MSD approach has been applied on various phenomena of relevance to geomorphology and petrology, further to estimate power-laws that are scale invariant, but shape dependent.

MSD and Various Power-Laws (Scale Invariant but Shape Dependent)

Estimation of Fractal Dimension

This study enables an alternative procedure to estimate fractal dimensions of planar shapes. It is observed that the estimated fractal dimensions are considerably similar with all the structuring elements. This exercise facilitates to test the relationship between the number of cycles (or radius of the

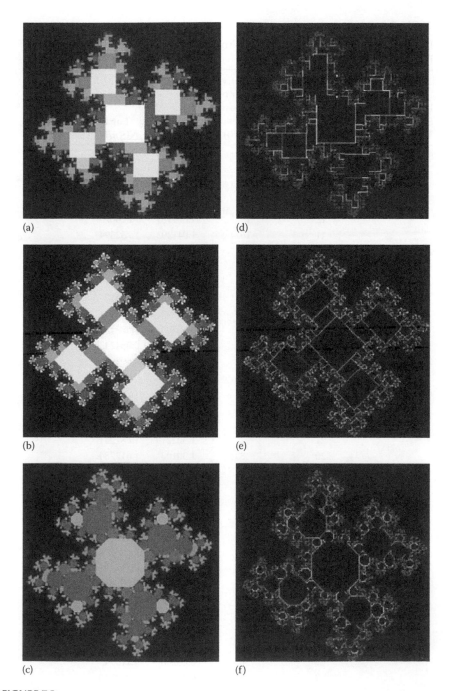

FIGURE 7.2
Fractal decomposition (a–c) by means of square, rhombus, and octagon respectively, and (d–f) the transition lines between the gray-shaded decomposed regions. (From Radhakrishnan, P. et al., *Chaos Soliton. Fract.*, 21(3), 563, 2004.)

TABLE 7.1

Fractal Dimensions Estimated from Number–Radius
Power-Law Relationship

Primitive Structuring Template	Cycle No./ Radius	Cumulative Number of Decomposed Shapes ($N(\leq nB)$)	Log (r)	Log $N(\leq nB)$	Fractal Dimension (D)
Square	42	1	1.623249	2.782473	1.6726
	27	5	1.431364	2.49693	
	15	9	1.176091	2.09691	
	13	12	1.113943	1.929419	
	12	13	1.079181	1.826075	
	11	16	1.041393	1.732394	
	10	17	1	1.716003	
	8	26	0.90309	1.414973	
	7	52	0.845098	1.230449	
	6	54	0.778151	1.20412	
	5	67	0.69897	1.113943	
	4	85	0.60206	1.079181	
	3	125	0.477121	0.954243	
	2	314	0.30103	0.69897	
	1	606	0	0	
Rhombus	63	1	1.799341	1.908485	1.6199
	40	5	1.60206	1.812913	
	22	6	1.342423	1.78533	
	21	7	1.322219	1.681241	
	20	8	1.30103	1.612784	
	15	12	1.176091	1.322219	
	11	21	1.041393	1.079181	
	10	41	1	0.90309	
	9	48	0.954243	0.845098	
	7	61	0.845098	0.778151	
	6	65	0.778151	0.69897	
	5	81	0.69897	0	
Octagon	26	1	1.414973	2.502427	1.6754
	15	5	1.176091	1.892095	
	8	9	0.90309	1.763428	
	6	13	0.778151	1.579784	
	5	21	0.69897	1.322219	
	4	38	0.60206	1.113943	
	3	58	0.477121	0.954243	
	2	78	0.30103	0.69897	
	1	318	0	0	

Source: Radhakrishnan, P. et al., *Chaos Soliton. Fract.*, 21(3), 563–572, 2004.

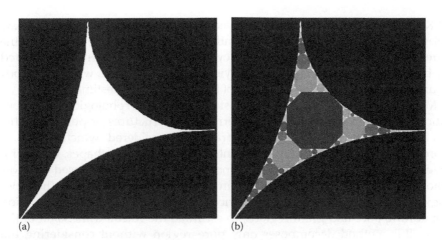

FIGURE 7.3
(a) Apollonian space and (b) Apollonian space after decomposition by means of octagon. (From Teo, L.L. et al., *Chaos Soliton. Fract.*, 19(2), 339–346, 2004.)

structuring element) that could be performed (used) to decompose the fractal at different levels, and the number of shapes that could be fit into the fractal while using the corresponding structuring element. From the number–radius relationship, the fractal dimensions have been estimated, which yield the significantly similar values of 1.67 + 0.05. The number–radius power-law relationship is shown for the fractal that is decomposed with the three considered structuring templates. The power-law relationship is represented as

$$N(\leq rB) \sim (r)^D$$

The power exponent D stands for the fractal dimension. The variable D is estimated from the graphs plotted between the logarithms of radius and the number of decomposed portions of all sizes as 1.67 with all the three structuring elements. The graphical plots are given in Figure 7.4. The fractal dimension

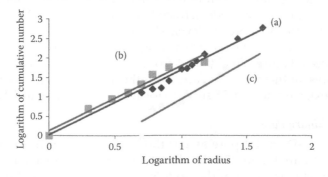

FIGURE 7.4
Fractal plots between the number of decomposed portions and the radius of the structuring elements: (a) square, (b) rhombus, and (c) octagon. (From Radhakrishnan, P. et al., *Chaos Soliton. Fract.*, 21(3), 563, 2004.)

of Koch Quadric binary fractal, estimated from the number–radius power-law relationship, yields the considerably similar values of 1.67 (Table 7.1) with all the three different structuring elements. However, the fractal dimension estimated by box dimension method (Figure 7.4) yields the value of 1.72, whereas the box counting dimension for the boundary of the same fractal is estimated as 1.5.

Method of estimating fractal dimension through morphological decomposition is most appropriate to characterize pore structures or porous media. As a sample study, a Koch Quadric fractal is considered, which is akin to pore structure, and is decomposed into simpler regular shapes, of several sizes, such as square, rhombus, and octagon. This method, based on morphological decomposition, is unique in the sense that it considers the topological region rather than its geometric boundary. For instance, in a section containing pore and grain regions, to estimate the fractal dimension of the pore, this method decomposes only pore region without considering the grain part.

Modeling, Description, and Characterization of Fractal Pore

Decomposition of the Pore Space into Pore Bodies and Pore–Body Network

The aim of this section is to provide a description of fast, simple computational algorithms based up on mathematical morphology technique to extract the description of pore bodies, to represent them in 3-D space, and to produce statistical characterization of their descriptions. Pore bodies are defined as follows: Larger pore space openings in a fluid-bearing rock, where most of the fluid is stored are pore bodies. The pore–body network (PBN) is analogous to the maximal balls. The PBN of pore is obtained via decomposition of 2-D pore space into nonoverlapping bodies of, different and, well-defined sizes. Symbolically, the morphological decomposition procedure is given by Equation 7.2. Information about the scheme followed for pore–body order designations can be seen in the "Ordering Scheme of Morphological Quantities" section. We consider a fractal binary pore of size 256×256 pixels (Figure 7.5a) with 33% porosity level. This pore space took 28 iterative erosions with respect to octagon structuring element of primitive size 5×5. The pore slices (Figure 7.5) obtained from recursive erosions are stacked to further represent them in 3-D space. With the assumption mentioned in the following text, we stack the 55 slices to form the fractal pore in 3-D.

3-D Fractal Binary Pore

We form the 3-D fractal pore as per the scheme shown in the schematic representation in Figure 7.6. In the stack, original fractal binary pore (Figure 7.5a) is embedded at the middle and superposed on both sides of it with the fractal binary pore slice generated by performing increasing cycles of erosion transformation. Each eroded version is superposed on one another as shown in the figure to reconstruct 3-D fractal binary pore.

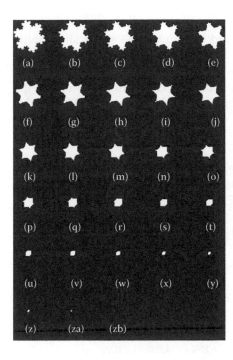

FIGURE 7.5
(a–zb) Fractal pore under increasing cycles of erosion transformation by octagon structuring element. (From Teo, L.L. and Sagar, B.S.D., *Discrete Dyn. Nat. Soc.*, 89280, 2006.)

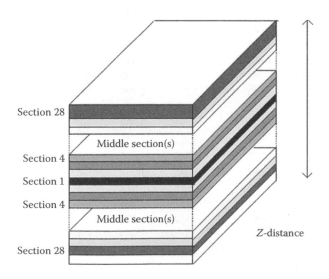

FIGURE 7.6
Schematic of construction of 3-D pore and morphological quantities. For pore sections 1–28, iteratively eroded pore slices shown in Figure 7.5a through zb are considered to form 3-D pore (Figure 7.8a and b). (From Teo, L.L. and Sagar, B.S.D., *Discrete Dyn. Nat. Soc.*, 89280, 2006.)

In turn, in the stack of 3-D pore image, top- and bottom-most pore slices possess less porosity (0.019836%) followed by immediate inner slices with porosity (0.093079%), and so on (Table 7.2). The middle slice in the stack possesses the porosity of 32.49512%. The reason for inserting the pore slices on top and bottom of the middle slices is only to illustrate the model with symmetry. The thickness of each slice is computed as one voxel. Hence, the

TABLE 7.2

Order-Wise Number of Pixels at Each Slice of Fractal Pore Channel, Throat, and Body

Slice (N)	Porosity across Slices	Pore Body	Body		
			1	2	3
1	32.49	21,296	11,083	5,046	3230
2	27.9	18,303	10,309	4,574	2167
3	24.67	16,173	9,563	4,120	1732
4	21.96	14,396	8,845	3,584	1126
5	19.57	12,831	8,155	3,213	726
6	17.52	11,484	7,493	2,669	854
7	15.65	10,262	6,859	1,932	1092
8	13.92	9,128	6,253	1,602	862
9	12.26	8,038	5,675	1,476	520
10	10.65	6,982	5,125	1,114	240
11	9.22	6,048	4,603	985	102
12	7.99	5,241	4,109	780	0
13	6.88	4,510	3,643	558	0
14	5.88	3,855	3,205	494	0
15	5.00	3,279	2,795	388	0
16	4.20	2,753	2,413	252	0
17	3.49	2,293	2,059	102	0
18	2.88	1,890	1,733	80	0
19	2.35	1,545	1,435	60	0
20	1.89	1,243	1,165	42	0
21	1.49	979	923	0	0
22	1.14	751	709	0	0
23	0.84	553	523	0	0
24	0.58	385	365	0	0
25	0.37	247	235	0	0
26	0.21	139	133	0	0
27	0.09	61	59	0	0
28	0.019	13	0	0	0
Total	1,109,016	748,263	219,945	79,459	

Source: Teo, L.L. and Sagar, B.S.D., *Discrete Dyn. Nat. Soc.*, 89280, 2006.

3-D fractal pore data are with the specifications of $256 \times 256 \times 55$. Stack of the pore image slices is represented by the set (Equation 7.3):

$$G = \left\{ X^1, X^2, \ldots, X^N \right\} \tag{7.3}$$

where N is the total number of pore image slices in the stack. For the present case, N is considered as 55. We denote each pore image slice by X^j, where j is the slice index. Entire stack of such pore image slice is depicted by the set G. We followed similar scheme to stack pore object, pore body, further to represent them in 3-D with different views.

MSD on Each Slice

By employing MSD approach, each slice of this pore space is decomposed into pore body (Figure 7.7). To decompose the pore space into pore bodies of various sizes, we employ Equation 7.2. Implementing Equation 7.2 by means of an octagon, we obtain Figure 7.7a through zb. Each eroded fractal binary pore is decomposed into nonoverlapping octagons of different sizes. With the progressive shrinking, it is obvious that the number of octagon categories and its (their) size (sizes) are also reduced. Each order of decomposed category (ies) is gray shaded for better legibility. Change in the size of this template shows impact in terms of scale change. Changes in other characteristic information show implications with changes in the geometric and spatial organization of pore bodies. Since we adopted octagonal element that is symmetric about origin, the spatial organization of fractal pore that is symmetric from left to right and top to bottom yields rather regular spatial organization of these features. The procedure results in size-distributed pore bodies that further facilitate the characterization of pore morphologic complexity. The use of Equation 7.2 to retrieve significant pore bodies is further shown to visualize a 3-D fractal binary pore body (Figure 7.7). For better perception, each level of the decomposed pore bodies are gray shaded (Figure 7.7). We further provide a formulation essentially based on set theory to represent these slice-wise decomposed pore bodies to connect them appropriately across slices. The connected pore bodies are further fragmented to designate each fragmented portion with orders ranging from 1 to N. The basic statistical descriptors of these order-wise fragmented quantities are used then to compute the accurate spatial complexities. The application is illustrated on single 2-D cross section and 3-D fractal binary pores.

- We proposed a way to extend an approach to map similar features from 3-D pore phases. This approach is primarily based on properly connecting the pore-body subsets from respective slices of pore phase.
- We proposed scheme to designate orders for decomposed pore bodies of various sizes. Once these components are designated appropriately with orders, we compute certain complexity measures for porous phase. Fractal characterization of PBN is carried out, and such an analysis facilitates complexity measures.

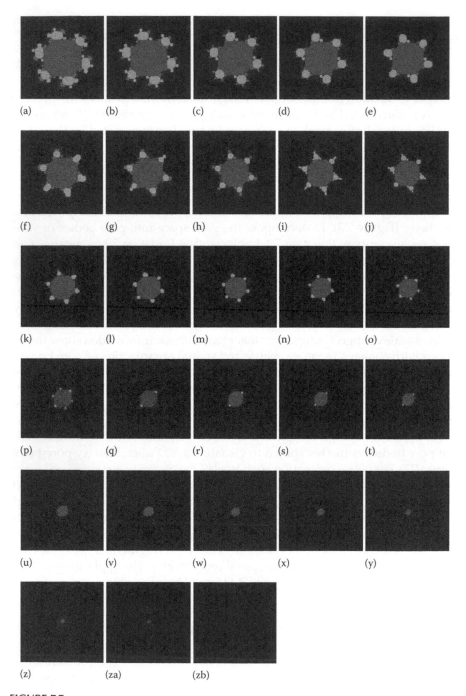

FIGURE 7.7
(a–zb) PBN from eroded versions of fractal pore. (From Teo, L.L. and Sagar, B.S.D., *Discrete Dyn. Nat. Soc.*, 89280, 2006.)

Visualization of PBN in 3-D Space: A Fractal Binary Pore

These recursively eroded pore phases to different degrees are considered as slices, and a 3-D pore fractal pore is constructed systematically. Sequentially eroded versions (Figure 7.5) of fractal pore model are used to represent a 3-D fractal pore model (Figure 7.8). In order to visualize the existing PBN in 3-D space, each eroded slice in 2-D space is parallelly considered to extract slice-wise PBN. Furthermore, such information decomposed from all slices is considered to visualize its 3-D form. And such 2-D pore features are stacked to construct PBN in 3-D space. Nth-level decomposed body subset(s) of each pore slice is (are) used to construct Nth level 3-D decomposed body.

Pore–Body and Its Fragmentation

Stack of the pore bodies of various orders ranging from 1 to N decomposed respectively from each slice of pore image is represented by the set (Equation 7.4):

$$PBN(G) = \left\{ \left[PBN^1(G) \right], \left[PBN^2(G) \right], \ldots, \left[PBN^{N-1}(G) \right], \left[PBN^N(G) \right] \right\} \quad (7.4)$$

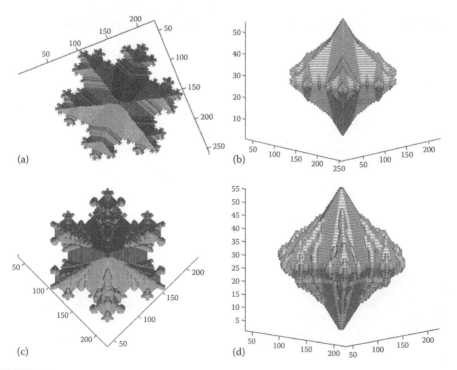

FIGURE 7.8
Top and side views of (a and b) model 3-D fractal binary pore, and (c and d) pore body. (From Teo, L.L. and Sagar, B.S.D., *Discrete Dyn. Nat. Soc.*, 89280, 2006.)

where

$$PBN^1(G) = \left\{(X_1^1 \circ NB), (X_1^2 \circ NB), \ldots, (X_1^{N-1} \circ NB), (X_1^N \circ NB)\right\}$$

$$PBN^2(G) = \left\{(X_2^1 \circ NB), (X_2^2 \circ NX), \ldots, (X_2^{N-1} \circ NB), (X_2^N \circ NB)\right\}$$

$$\vdots$$

$$PBN^N(G) = \left\{(X_N^1 \circ NB), (X_N^2 \circ NB), \ldots, (X_N^{N-1} \circ NB), (X_N^N \circ NB)\right\}$$

where
  superscript and subscript N's respectively denote index of pore image slice
    and order of the decomposed pore body
  NB denotes the size of structuring element

In the current fractal pore model, the PBN is designated with three different orders (Figure 7.9). The order of the decomposed pore body can be determined by a recursive relationship shown in Equation 7.4. The Nth-level decomposed body from all the slices ranging from $j = 1, 2, \ldots, N$ would be put in a separate stack, which is considered as first-order fragmented pore body from G. Similarly, Nth-level decomposed pore bodies that are

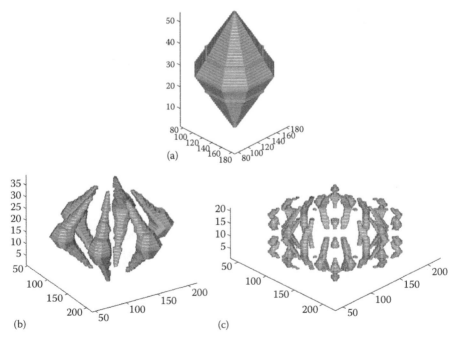

FIGURE 7.9
Order-wise (a–c) pore bodies. (From Teo, L.L. and Sagar, B.S.D., *Discrete Dyn. Nat. Soc.*, 89280, 2006.)

smaller than the Nth-level decomposed disks from the previous order, from the reminder of respective slices, are stacked to visualize the pore bodies of second order in 3-D. This is a recursive process, till all the Nth-level decomposed pore bodies of decreasing sizes retrieved from all levels of pore slices are stacked. The pore bodies are segregated into the first, second, and third orders.

Ordering Scheme of Morphological Quantities

In order to properly designate, 3-D pore bodies of various categories need to be first designated with orders. The pore–body order designation in 3-D is done based on a set of Equation 7.4. Nth-level bodies (Equation 7.4) are considered as order 1, $(N - 1)$th level pore bodies are considered as order 2, and so on. In this fashion, the pore bodies are designated with respective orders.

Estimation of Order-Wise Pore Bodies in 3-D

Volumes of order-wise pore bodies connected across slices as explained in Equation 7.5 and that are fragmented from each 2-D slice of pore image are also estimated to provide graphical relationships.

$$V[PBN^1(G)] = \sum_{j=1}^{55} A\left[X_1^j \circ NB\right], \quad V[PBN^2(G)] = \sum_{j=1}^{55} A\left[X_2^j \circ NB\right], \dots ,$$

$$V[PBN^N(G)] = \sum_{j=1}^{55} A\left[X_N^j \circ NB\right]$$

(7.5)

In other words, the total volume of pore bodies can be derived from $V[PBN(G)] = \sum_{i=1}^{N} V[PBN^i(G)]$. Further, volume fractions are computed for the order-distributed pore bodies (Table 7.2). The order versus number and their corresponding voxel count are plotted as graphs, and complexity measures are estimated for the model fractal pore considered demonstrating the framework.

Relationships between Pore Morphological Quantities: Results and Discussion

Three-dimensional volumes for model pore and corresponding order-wise bodies are computed by stacking the slices that are generated via erosion cycles of pore regions and their slice-wise morphologic quantities extracted at respective phases. Total volumes of pore bodies estimated in voxels with

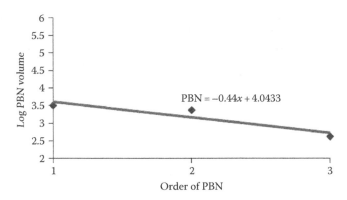

FIGURE 7.10
Distribution of pore body of fractal pore. (From Teo, L.L. and Sagar, B.S.D., *Discrete Dyn. Nat. Soc.*, 89280, 2006.)

slice thickness of one voxel respectively include 5,938, 66,520, and 1,109,016. To characterize this geometrically significant pore bodies, we estimate the volumes occupied by the PBN at respective subset levels. To quantify the spatial complexity of the PBNs, we plot logarithms of volumes of PBN and the corresponding designated orders (Figure 7.10). The volume fractions of order-wise pore bodies are plotted as functions of designated order number. The rates of change in the volume fractions across the orders are relatively more significant in the pore bodies, as revealed from Figure 7.10. The linear plot yields best-fit coefficient 0.44. This value is dependent on the shape of structuring element employed to decompose pore into bodies. This shape dependency provides an insight further to relate empirically between the scale-invariant but shape-dependent power-law and other effective properties of complex porous media.

Use of Scale-Invariant but Shape-Dependent Dimensions for Process Characterization

We presented a morphology-based framework to decompose the pore structure into pore bodies of various orders, which facilitate to compute their accurate size-distribution functions. We documented that an accurate pore morphometry can be carried out, once pore space has been properly decomposed into pore bodies of various orders. This approach is in general useful for analyzing, understanding geometrical properties, and relating them with physical properties. As the choice of template influences their spatial patterns, this study opens a way to understand important shape-dependent topologic properties of porous media. We hypothesize that this morphologically significant decomposed pore bodies at multiple scales can be related with bulk material properties. One can also show new results by employing multiscale morphological transformations that can be treated as a transformation meant for showing systematic variations in porosity. Sparse and intricate PBN would be obvious respectively from simple and complex pore spaces. Variations in decomposed pore morphologic parameters across slices in the tomographic data (e.g., Fontainebleau sandstone) further provide potentially valuable insights. This entire approach can be extended to any 3-D pore image that is constructed by stacking the 2-D slices/tomographic images, to isolate order-wise fragmented pore bodies with appropriate connectivity across slices. The different steps accomplished from this investigation pave a way to relate the statistically derived properties with physical properties.

Morphometry of Nonnetwork Space

Topographically convex regions within a catchment basin represent varied degrees of hillslopes. The nonnetwork space (X), the characterization of which we address in this section, is akin to the space that is achieved by subtracting channelized portions contributed due to concave regions from the watershed space (M). This nonnetwork space is similar to nonchannelized convex region within a catchment basin. We propose an alternative shape-dependent quantity like fractal dimension to characterize this nonnetwork space. Toward this goal, we decompose the nonnetwork space in 2-D discrete space into simple NODs of various sizes by employing mathematical morphological transformations and certain logical operations. Furthermore, we plot the number of NODs of less than threshold radius against the radius and compute the shape-dependent fractal dimension of nonnetwork space.

Why Morphometry of Nonnetwork Space in Place of Morphometry of Networks?

Characterization of branched networks, such as rivers, bronchial trees, vortex dynamic structures, and diffusion-limited aggregation to name a few, is one of the important research areas in geomorphology in recent decades. It is evident, from numerous studies, that various loopless networks ranging from geomorphologic (e.g., Horton 1945, Strahler 1957, Mandelbrot 1982, Turcotte 1997, Rodriguez-Iturbe and Rinaldo 1997), physical (Olson et al. 1998, Mehta et al. 1999), and sociological networks (Arenas et al. 2004) follow Hortonian laws. The Horton–Strahler morphometric statistics of networks that summarize the connectivity and orientation of convex zones of basins offer useful tools for quantitative description of landscapes. From the geophysical context, river networks are characterized via Hortonian laws and fractal-based power-laws. Derivation of these laws based on stream number, mean stream length, and mean areas for river networks facilitates computation of topological quantities, such as bifurcation ratio (R_B), length ratio (R_L), and stream area ratio (R_A) as well as certain scaling laws to further validate and characterize numerous realistic and synthetic network (e.g., Shreve 1967, Mandelbrot 1982, Tokunaga 1984, LaBarbera and Rosso 1987, Tarboton et al. 1988, Takayasu 1990, Howard 1990, Marani et al. 1991, Beer and Borgas 1993, Kirchner 1993, Nikora and Sapozhnikov 1993, Rigon et al. 1993, Rinaldo et al. 1993, Karlinger et al. 1994, Maritan et al. 1996a,b, Sagar 1996, Peckham and Gupta 1999, Rodriguez-Iturbe and Rinaldo 1997, Turcotte 1997, Sagar et al. 1998, 2001, Gupta and Veitzer 2000, Dodds and Rothman 2001, Maritan et al. 2002, Sagar and Tien 2004). Geomorphic processes are explained by relation with the dimension, and certain scaling laws exhibited by networks. The geometric organizations of hillslopes of basins possessing topologically invariant networks may be significantly different. To capture the variations in geometric organizations of basins with topologically invariant networks, alternative method that takes shape into consideration is warranted.

Besides channel network, nonnetwork spaces, the planar forms of hillslopes, are also important features within a basin. If the notion "geometry and topology of the basin have direct relationship with geomorphic processes" has merit, then scaling laws and dimension of the network are of limited use, as they enable less about the geometry and topology. Although the organization of the network is strictly controlled by the spatial organization of concave zones, it is obvious that the Hortonian laws, which provide rich information, and scaling laws have emphasized only little on any shape-dependent quantity. Heuristically, similar networks that exist in an elongated and circular basin provide more or less similar Hortonion quantities. However, the processes involved, respectively, in elongated and circular basins differ significantly due to distinct geometries of nonnetwork space, in other words, planar forms of hillslope morphologies. We argue, as it is intuitively true, that network-based characteristics alone would be insufficient to quantify the sensible variations in the geometric and spatial organization of nonnetworks spaces and to explore links with geomorphic expression and processes. To better explain this argument, we show three synthetic networks (Figure 7.11)

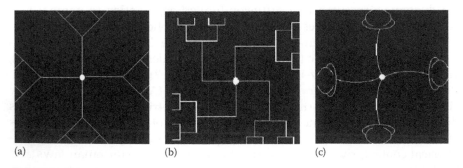

(a) (b) (c)

FIGURE 7.11
(a–c) Schematically represented networks with three different geometric organizations. (From Chockalingam, L. and Sagar, B.S.D., *J. Geophys. Res. Solid Earth (AGU)*, 110, B08203, 2005.)

with distinct topologies and geometric organizations of nonnetwork spaces, possessing similar laws of Horton's number and stream lengths. The typical difference between these three schematic networks (Figure 7.11) is obvious from the diverging angles between the segments and their overall geometry, and also the geometric organization of nonnetwork spaces. As the number of segments and their lengths of these three schematic networks, after designated with Horton–Strahler ordering scheme, are similar, the resultant topological quantities would also be similar. These similarities, irrespective of their dissimilarities in the geometric organization of nonnetwork spaces, mask much of the details.

The quantitative description of concavity of the surface is done through the popularly known slope–area diagram (e.g., Montgomery and Dietrich 1988, 1994, Willgoose et al. 1991, Tarboton and Bras 1992, Moglen and Bras 1995, Whipple and Tucker 1999). Hillslopes, their morphologies, and responses to changes in the tectonic and climatic settings were thoroughly investigated by numerous researchers to explore the characterization of hillslope morphologies via linear transport models (Kirkby 1971, Koons 1989, Fernandes and Dietrich 1997) and nonlinear transport models (Anderson 1994, Howard 1994, Roering et al. 1999). Characterization of the planar form of hillslopes, which we term here as nonnetwork space, and its geometric composition enable rich clues to explore links with geomorphic processes within a basin. The topographically significant regions in the nonnetwork space include regions with varied degrees of slope, narrow regions with steep gradient, and the corner portions adjacent to the stream confluence.

The components of the possible nonnetwork spaces, which may be isolated by subtracting the networks from their corresponding reconstructed basin, can be closely approximated with triangle, square, and circle. We hypothesize that the geomorphic expression and activity depend upon the morphology of the components of nonnetwork spaces. Hence, we propose morphometry of the nonnetwork space. We employ an elegant methodology, proposed by Sagar and Chockalingam (2004), whereby we derive shape-dependent dimensions, which consider the spatial organization of nonnetwork spaces that may be more relevant to relate with geomorphic processes that shape the basin. The decomposition of nonnetwork space throws some light on the classification and characterization of landscape morphology from the point of its surface roughness to further understand about geomorphic activity. The regions of varied degrees of geomorphic activity within a basin can be linked with hillslope processes. Various categories in the nonnetwork space can be better segmented through the various size categories of NODs. The nonnetwork space in between the network segments with lesser diverging angle is the region that we achieve with the decreasing number of multiscale closings. Smaller-category NODs occupy nonnetwork space that is surrounded by dense network segments, the diverging angles between which are relatively less, and the zones adjacent to the channel confluence. The diverging angle of channels determines the topology of confluence. The higher the diverging angle, the larger the disk that can be inscribed, and *vice versa*. This description enables that various categories of NODs can be related to different degrees of topographically convex regions (Dietrich et al. 1992, Montgomery and Dietrich 1992) within a catchment basin. Channel network and nonnetwork spaces are two important features within a catchment basin. Channel and ridge connectivity networks possess scale-invariant properties. Change in lengths of these networks scales as power of resolution that indicates fractality. Despite the fact that there is no significant change in the areal extent of nonnetwork space under the succession of scale changes, its topological organization varies due to change in the network length. The abstract structure, akin to the network connecting the regional *maxima* of topographically convex regions, explains this phenomenon. The topographically convex and concave regions respectively contain hierarchical concavities and convexities with increasing resolution. With increase in magnification, the increase in the observed length is true with both channel network and abstract network of the nonnetwork space (Figure 7.12a and b). This implies that the scale-dependent nonnetwork space that can be represented as abstract structure, from which nonnetwork space can be retrieved, also possesses fractal properties. Verifying Hortonian laws mostly involves the validation of several network models (e.g., Scheidegger 1967, Shreve 1967, Tokunaga 1984, Howard 1990, Rodriguez-Iturbe and Rinaldo 1997, Gupta and Veitzer 2000, Sagar et al. 1998, 2001). In addition to these laws that iron out much of the details of branched networks, in recent past, allometric studies have resulted in several universal power-law relationships (Maritan et al. 1996a,b, Maritan et al. 2002, Sagar and Tien 2004). Network characterization through Hortonian laws, and of late through fractal and multifractal properties, receives notable attention, and various significant results have been accomplished. Several researchers relate fractal dimension of a network within a catchment basin to the bifurcation ratio (R_B) and length ratio (R_L) of idealized Hortonian-network trees as $D = \log R_B / \log R_L$ (Mandelbrot 1982,

(a)   (b)

FIGURE 7.12
Schematic of a catchment basin with channel (dark gray line) and abstract structure (light gray line) of nonnetwork space at varied resolutions. Networks (a) at coarser spatial resolution, and (b) at finer spatial resolution. (From Sagar, B.S.D. and Chockalingam, L., *Geophys. Res. Lett.*, 31(12), 12502, 2004.)

LaBarbera and Rosso 1987, Tarboton et al. 1988, Takayasu 1990, Beer and Borgas 1993, Nikora and Sapozhnikov 1993, Rigon et al. 1993, Sagar 1996, Rodriguez-Iturbe and Rinaldo 1997, Turcotte 1997, Sagar et al. 1998, 2001). This non-shape-dependent dimension based on two morphometric quantities is a space-filling characteristic of the network. Intuitively, it is clear that networks respectively from elongated and radial basins may yield the same fractal dimension D, so it seems that this non-shape-dependent D may be of limited use to relate process with the shape of a watershed. Heuristic argument is that D may be the same for certain elongated and radial basins, as the computed topological quantities do not consider any other properties than network length and number. However, Tarboton et al. (1988) provide a parameter $D = d(\log R_B / \log R_L)$. Karlinger et al. (1994) describe the fractal scaling of river networks in the context of both thin and fat fractals. They characterize the fat-fractal dimension as a scaling exponent derived from the behavior of the river-channel area.

Morphometric analysis of channel network of a basin provides several scale-independent measures. To better characterize basin morphology, one requires, besides channel morphometric properties, scale-independent but shape-dependent measures to record the sensitive differences in the morphological organization of nonnetwork spaces. These spaces are planar forms of hillslopes or the retained portion after subtracting the channel network from the basin space. The principal aim of the "Morphometry of Nonnetwork Space" section is to focus on explaining the importance of alternative scale-independent but shape-dependent measures of nonnetwork spaces of basins. Toward this goal, we explore how mathematical morphology-based decomposition procedures can be used to derive basic measures required to quantify estimates, such as dimensionless power-laws, that are useful to express the importance of characteristics of nonnetwork spaces via decomposition rules.

In this section, we propose a technique to characterize nonnetwork space via a morphological decomposition procedure, which is popular in shape description studies (Serra 1982). This technique provides a shape-dependent power-law (Figure 7.13). We consider this geometric approach to characterize nonnetwork space within catchment basins. Morphological transformations are employed systematically as explained in Equations 2.3 through 2.14 to first achieve the reconstructed basin space (M) from channel network (C) and then to generate nonnetwork space (X). Once X is achieved, we employ these transformations again to convert X into *NODs*, which are simpler convex components. We replace C with X. One can perform these transformations also on nonnetwork space.

Nonnetwork Space of Basins

The channel connectivity networks (Figures 7.14 and 7.15) derived from eight basins are illustrated with Horton–Strahler ordering scheme. The spatial organization of these network patterns determines the basin processes. We employ these channel networks to reconstruct the basins with proper

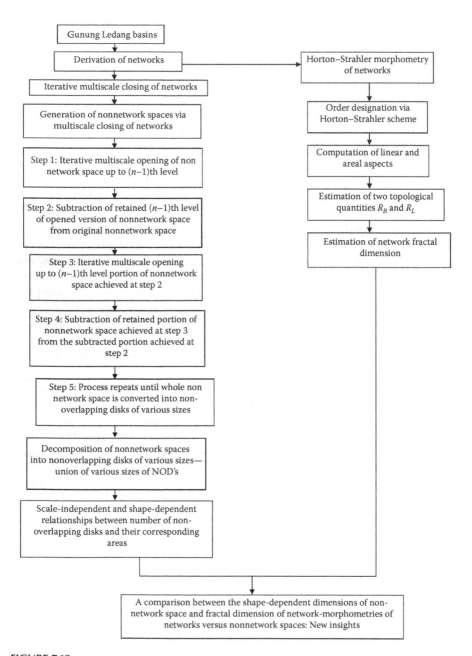

FIGURE 7.13
Flowchart showing sequential steps involved in the derivation of nonnetwork space-based shape-dependent dimensions of eight subbasins of Gunung Ledang region and their comparison with network-based morphometric parameters. (From Chockalingam, L. and Sagar, B.S.D., *J. Geophys. Res. Solid Earth (AGU)*, 110, B08203, 2005.)

FIGURE 7.14
Gunung Ledang DEM after partitioning into eight fourth-order basins. (From Chockalingam, L. and Sagar, B.S.D., *J. Geophys. Res. Solid Earth (AGU)*, 110, B08203, 2005.)

characteristics. A framework (Figure 7.13) based on morphological transformations due to Sagar and Chockalingam (2004) is employed to reconstruct the basins and their internal topological organizations. From such a reconstructed basin, it is also possible to attain a network much similar to the network that is used to reconstruct the basin.

To reconstruct the basin and its topology from channel network, we let C be the channel network (Figure 7.15a) and a structuring element, bounded, convex, symmetric, and containing the origin. Channel networks and their complementary spaces are respectively represented with white and black pixels. To reconstruct the basins from channel networks, we employ multiscale closing as expressed in Equation 7.6.

$$M = \bigcup_{n=0}^{K} C \bullet nB \tag{7.6}$$

where $C \subseteq M$, C, and M are channel network and basin reconstructed from channel network by performing multiscale morphological closing transformation iteratively until M becomes equivalent to $M \bullet B$; in other words, M reaches convergence. Channel networks are subtracted from the reconstructed basins to achieve nonnetwork spaces within basins. We define nonnetwork space (X) within each reconstructed basin (M) as a combination of disconnected, bounded, binary-valued discrete space object as depicted in Equation 7.7.

$$X = M \backslash C \tag{7.7}$$

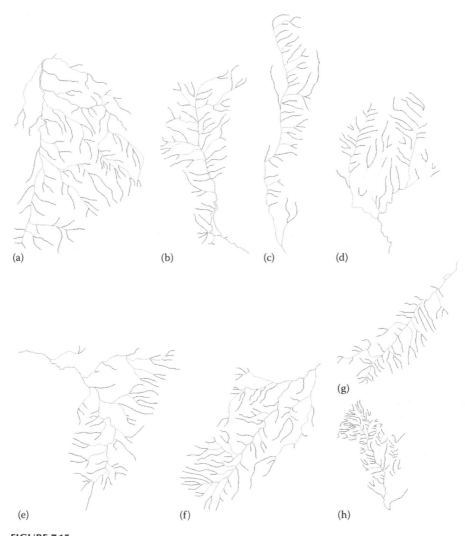

FIGURE 7.15
(a–h) Fourth-order channel networks of eight basins of Gunung Ledang region. (From Chockalingam, L. and Sagar, B.S.D., *J. Geophys. Res. Solid Earth (AGU)*, 110, B08203, 2005.)

where "\" denotes subtraction. By subtracting the channel networks from the bounded reconstructed basins M, we obtain nonnetwork spaces X (Figure 7.16) of the eight basins. For better understanding of basin reconstruction process from the network, we show an evolutionary sequence of network for basin 1 after respective multiscale closings in the inset picture (Figure 7.16). The nonnetwork space (X) is similar to the nonchannelized convex region that consists of varied degrees of topographically convex regions within a basin. As an extension, we emphasize on characterization

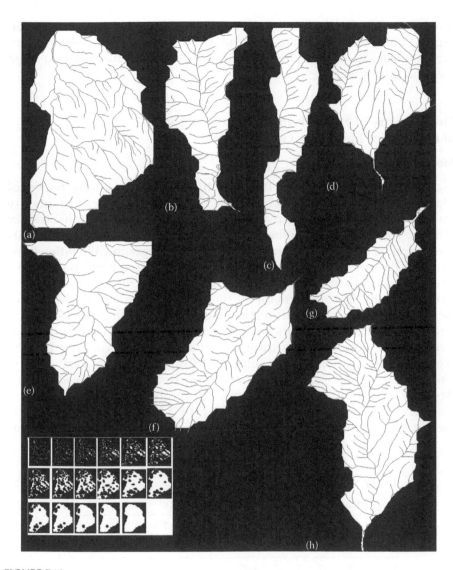

FIGURE 7.16

(a–h) Nonnetwork space white in color and networks black in color within a basin. For basin reconstruction stages, we explain with reference to first basin. Similar approach has been followed to generate topological spaces within the other seven basins. Evolution of networks of first basin after respective multiscale closings is shown in inset. (From Chockalingam, L. and Sagar, B.S.D., *J. Geophys. Res. Solid Earth (AGU)*, 110, B08203, 2005.)

of nonnetwork spaces of the eight basins by involving decomposition rules that are similar to random packing of space, reported elsewhere (Manna and Herrmann 1991, Dodds and Weitz 2002, 2003, Lian et al. 2004, Radhakrishnan et al. 2004). Decomposition of these nonnetwork spaces into NODs of various sizes such that the nonnetwork space within each is filled with NODs of decreasing sizes provides valuable insight for modeling and understanding basins. The characterization of such a scale-dependent topological organization of nonnetwork space has hitherto been received little attention.

Morphometry of Network and Nonnetwork Space of Eight Basins of Gunung Ledang Region

Morphometry of Networks

Eight subbasins are derived from the hilly Gunung Ledang region of Malaysia (Figure 7.14). The channel networks within these basins are traced and designated the stream ordering according to Horton–Strahler scheme. The order-wise number of streams and their lengths in pixel units are computed (Table 7.3). Figure 7.17a and b depicts graphical relationships between the stream order and order-wise stream numbers and lengths. Graphical plots between the stream orders and logarithms of order-wise stream lengths and numbers for all the eight basins (Figure 7.17c and d) facilitate computations of bifurcation and stream-length ratios (Table 7.3). Order-wise stream numbers and lengths are plotted as functions of stream orders for eight fourth-order networks of the Gunung Ledang region. Linear relationships are observed for logarithms of mean stream lengths and number plotted as functions of stream orders. These linearities indicate Hortonity of the networks. From these linear relationships, we derive Hortonian laws of stream lengths and numbers. We compute the antilogarithms of absolute slope values computed from these linear

TABLE 7.3

Basic Measures of Networks of Eight Basins

Basin No.	Order Number				Stream Length (in Pixels)				R_B	R_L
	1	2	3	4	1	2	3	4		
1	85	18	4	2	4891	1611	551	849	3.45	1.90
2	58	15	3	1	2818	775	187	767	3.97	2.33
3	45	11	1	0	2346	594	770	0	6.64	3.87
4	53	11	4	1	2789	748	703	328	3.64	1.90
5	55	17	3	1	2834	961	659	374	3.96	2.07
6	70	18	4	1	3671	1182	518	431	4.16	2.01
7	46	8	1	0	2042	562	479	0	6.78	3.28
8	89	17	3	1	2477	809	194	294	4.57	2.09

Source: Chockalingam, L. and Sagar, B.S.D., *J. Geophys. Res. Solid Earth (AGU)*, 110, B08203, 2005.

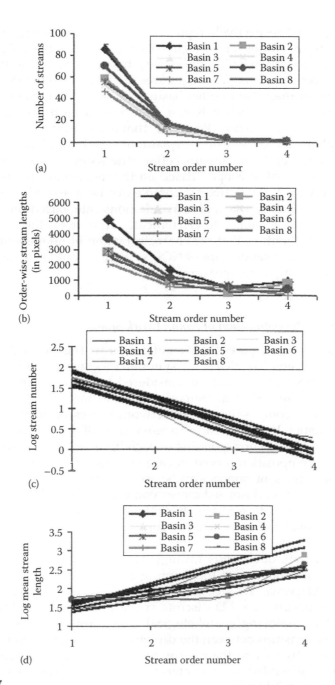

FIGURE 7.17
(a and b) Graphical plots between stream order and order-wise stream number and lengths, and (c and d) stream orders versus logarithms of order-wise numbers, and mean stream lengths of eight basins. (From Chockalingam, L. and Sagar, B.S.D., *J. Geophys. Res. Solid Earth (AGU)*, 110, B08203, 2005.)

relationships that respectively represent basin-wise bifurcation (R_B) and stream length (R_L) ratios for the eight basins (Table 7.3). Basin 7 possesses the highest bifurcation ratio followed by basins 3, 8, and 6, indicating that the underlain geological structures disturb the stream networks relatively lesser than that of other basins 2, 5, 4, and 1. Estimated higher fractal dimensions for basins 8, 6, 4, and 5 indicate higher degrees of space-filling characteristics. We infer that these dimensions derived from morphometry of networks explain space-filling characteristics of networks. However, these measures offer little scope to quantify the geometric complexity of hillslopes. Based on the morphometric statistics of the eight networks, the networks' complexity is in ascending order for the basins 3, 7, 2, 5, 1, 4, 6, and 8. We demonstrate, based on the arguments made with reference to Hortonically similar synthetic networks (Figure 7.11a through c), that the characterization of nonnetwork spaces through statistical relationships of NOD statistics would provide geometric-dependent complexity measures.

Morphological Decomposition of Nonnetwork Space

Complex nonnetwork spaces (X) of eight basins are transformed into "simpler convex polygon-like *NOD*s." A symmetric octagonal structuring element, as a simple probing rule, is considered to convert X into *NOD*s by employing morphological decomposition according to (7.2). Figure 7.18 illustrates the decomposition of nonnetwork spaces of eight basins into NODs. For better legibility, each category of NODs is coded with gray shades. These NODs, corresponding to each basin achieved through morphological decomposition procedure, are considered to quantify the geometric complexities of nonnetwork spaces. Nonnetwork spaces of each basin consist of several isolated *connex* components, which are the planar forms of hillslopes within a basin. It is obvious that the nonconvex connex components consist of more size categories of *NOD*s than that of convex connex components.

We demonstrate our results through characterization of nonnetwork spaces of eight subbasins of the Gunung Ledang region (Figure 7.14) of peninsular Malaysia. We decompose the nonnetwork spaces of eight fourth-order basins in a 2-D discrete space into simple NODs of various sizes by employing morphological transformations. Furthermore, we show relationships between the dimensions estimated via morphometries of the network and their corresponding nonnetwork spaces. This study can be extended to characterize hillslope morphologies, where decomposition of 3-D hillslopes needs to be addressed. In this section, we provide morphometric parameters of both network and nonnetwork spaces of eight basins.

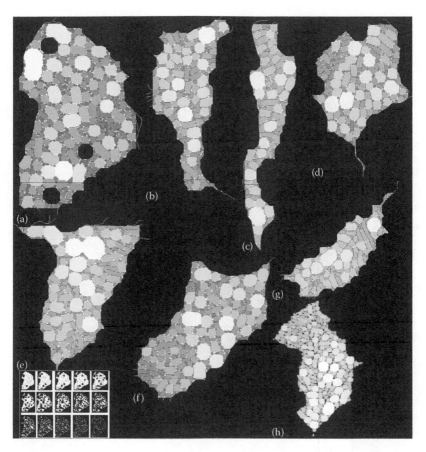

FIGURE 7.18

(a–h) Nonnetwork spaces of eight basins after filling with nonoverlapping octagons of several sizes. Evolution of decay of nonnetwork space of first basin into NODs of decreasing sizes is shown as an inset picture. (From Chockalingam, L. and Sagar, B.S.D., *J. Geophys. Res. Solid Earth (AGU)*, 110, B08203, 2005.)

Morphometry of Nonnetwork Spaces

The geometric complexity of nonnetwork spaces that is computed via fragmentation rules provides four different shape-based measures. We record the number of decomposed NODs, their sizes lesser than the template of specific radius, and their contributing area in pixels (Table 7.4). The statistics of NODs of various sizes that reveal other interesting characteristics for the eight nonnetwork spaces include the number of NODs and their contributing areas. We observe that more number of smaller-size category NODs exist in the eight basins. Decay in the number of NODs in these basins is obvious (Figure 7.19a). Similarly, the distributary patterns in the contributing areas of

TABLE 7.4

Basic Statistics of Distributed Number of Nonoverlapping Disks and Their Contributing Areas of Various Sizes Decomposed from Nonnetwork Space of Eight Basins

Disk Size	N(1)	N(2)	N(3)	N(4)	N(5)	N(6)	N(7)	N(8)	Disk Size	A(1)	A(2)	A(3)	A(4)	A(5)	A(6)	A(7)	A(8)
1	247	139	88	136	148	167	79	197	1	13,201	7,300	4,863	7,642	7,985	9,404	4778	10,011
2	88	50	38	48	50	56	32	81	2	11,358	7,280	5,416	6,228	7,395	8,267	4080	12,807
3	53	23	19	27	41	48	27	56	3	15,444	6,198	4,231	8,609	11,946	13,936	8254	18,218
4	35	18	13	19	19	31	15	36	4	13,888	8,630	5,237	7,858	7,831	13,460	6004	14,990
5	28	19	14	18	14	13	13	24	5	13,785	13,083	10,164	10,304	8,496	7,772	7814	16,167
6	11	9	12	12	12	18	4	12	6	11,648	6,697	9,924	11,033	10,710	14,843	4277	8,743
7	19	7	7	11	8	12	4	12	7	23,316	7,778	7,421	11,195	7,484	12,396	3836	11,741
8	8	7	4	7	7	5	1	11	8	12,216	8,143	4,646	8,350	7,630	5,404	945	16,512
9	11	3	5	3	4	3	0	3	9	18,416	4,802	7,455	4,197	6,020	4,271	0	4,538
10	9	3	0	2	3	4	2	4	10	16,468	5,276	0	3,715	5,815	7,243	3453	6,743
11	5	2	1	3	2	4	1	2	11	13,918	3,702	2,083	6,639	4,199	12,777	3075	4,067
12	2	0	0	1	2	4	1	1	12	4,337	0	0	2,834	4,475	10,031	2360	3,446
13	2	1	2	1	2	2	0	1	13	8,160	2,445	6,711	2,753	7,159	5,978	0	2,632
14	0	1	0	0	2	0	0	0	14	0	3,339	0	0	5,916	0	0	0
15	0	0	0	0	0	0	0	0	15	0	0	0	0	0	0	0	0
16	2	0	0	0	1	1	0	0	16	5,859	0	0	0	3,681	0	0	0
19	0	0	0	0	1	0	0	0	19	0	0	0	0	2,407	0	0	0

Source: Chockalingam, L. and Sagar, B.S.D., *J. Geophys. Res. Solid Earth (AGU)*, 110, B08203, 2005.

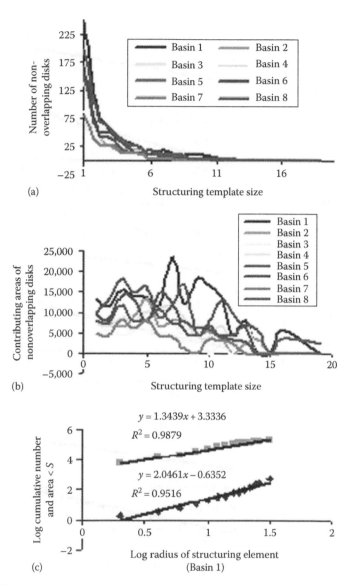

FIGURE 7.19

Morphometric parameter computations achieved through decomposition of nonnetwork space. (a and b) Numbers of NODs of nonnetwork spaces and their corresponding areas as functions of radius of structuring element for considered nonnetwork spaces of eight basins, (c–j) double logarithmic relationships between the radius of template and number of NODs and their contributing areas lesser than the radius of template for eight basins, and (k–r) areas of NODs and number of NODs lesser than the template. The points of these graphs organize themselves into a straight line, the slopes of which for these basins characterize nonnetwork spaces of basins. (From Chockalingam, L. and Sagar, B.S.D., *J. Geophys. Res. Solid Earth (AGU)*, 110, B08203, 2005.)

(continued)

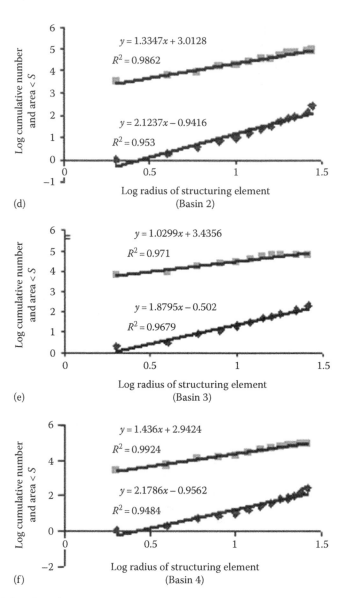

(d)

(Basin 2)

(e)

(Basin 3)

(f)

(Basin 4)

FIGURE 7.19 (continued)
Morphometric parameter computations achieved through decomposition of nonnetwork space.
(a and b) Numbers of NODs of nonnetwork spaces and their corresponding areas as functions
of radius of structuring element for considered nonnetwork spaces of eight basins, (c–j) double
logarithmic relationships between the radius of template and number of NODs and their contributing areas lesser than the radius of template for eight basins, and (k–r) areas of NODs and
number of NODs lesser than the template. The points of these graphs organize themselves into a
straight line, the slopes of which for these basins characterize nonnetwork spaces of basins. (From
Chockalingam, L. and Sagar, B.S.D., *J. Geophys. Res. Solid Earth (AGU)*, 110, B08203, 2005.)

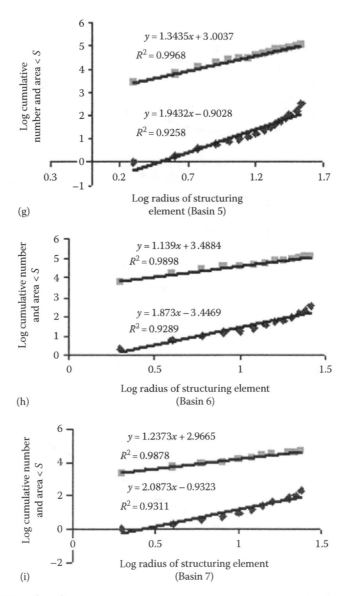

FIGURE 7.19 (continued)
Morphometric parameter computations achieved through decomposition of nonnetwork space.
(a and b) Numbers of NODs of nonnetwork spaces and their corresponding areas as functions
of radius of structuring element for considered nonnetwork spaces of eight basins, (c–j) double
logarithmic relationships between the radius of template and number of NODs and their contributing areas lesser than the radius of template for eight basins, and (k–r) areas of NODs and
number of NODs lesser than the template. The points of these graphs organize themselves into a
straight line, the slopes of which for these basins characterize nonnetwork spaces of basins. (From
Chockalingam, L. and Sagar, B.S.D., *J. Geophys. Res. Solid Earth (AGU)*, 110, B08203, 2005.)

(continued)

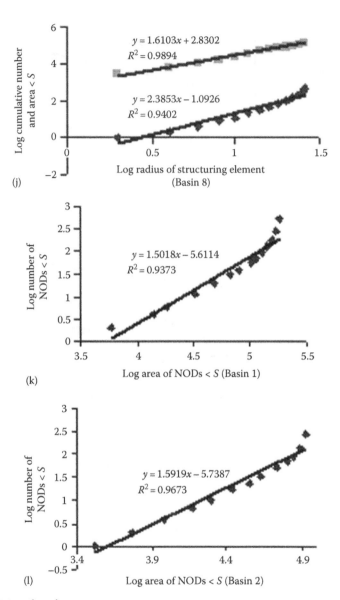

(j)

(k)

(l)

FIGURE 7.19 (continued)
Morphometric parameter computations achieved through decomposition of nonnetwork space. (a and b) Numbers of NODs of nonnetwork spaces and their corresponding areas as functions of radius of structuring element for considered nonnetwork spaces of eight basins, (c–j) double logarithmic relationships between the radius of template and number of NODs and their contributing areas lesser than the radius of template for eight basins, and (k–r) areas of NODs and number of NODs lesser than the template. The points of these graphs organize themselves into a straight line, the slopes of which for these basins characterize nonnetwork spaces of basins. (From Chockalingam, L. and Sagar, B.S.D., *J. Geophys. Res. Solid Earth (AGU)*, 110, B08203, 2005.)

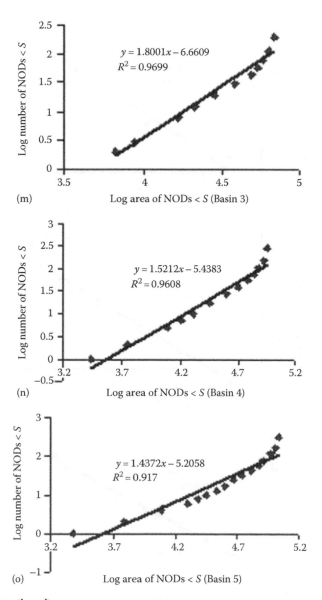

FIGURE 7.19 (continued)

Morphometric parameter computations achieved through decomposition of nonnetwork space. (a and b) Numbers of NODs of nonnetwork spaces and their corresponding areas as functions of radius of structuring element for considered nonnetwork spaces of eight basins, (c–j) double logarithmic relationships between the radius of template and number of NODs and their contributing areas lesser than the radius of template for eight basins, and (k–r) areas of NODs and number of NODs lesser than the template. The points of these graphs organize themselves into a straight line, the slopes of which for these basins characterize nonnetwork spaces of basins. (From Chockalingam, L. and Sagar, B.S.D., *J. Geophys. Res. Solid Earth (AGU)*, 110, B08203, 2005.)

(*continued*)

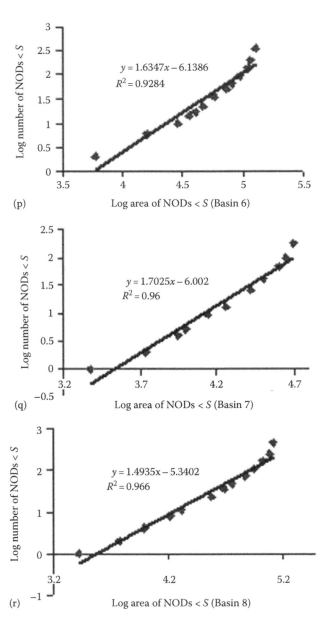

FIGURE 7.19 (continued)
Morphometric parameter computations achieved through decomposition of nonnetwork space. (a and b) Numbers of NODs of nonnetwork spaces and their corresponding areas as functions of radius of structuring element for considered nonnetwork spaces of eight basins, (c–j) double logarithmic relationships between the radius of template and number of NODs and their contributing areas lesser than the radius of template for eight basins, and (k–r) areas of NODs and number of NODs lesser than the template. The points of these graphs organize themselves into a straight line, the slopes of which for these basins characterize nonnetwork spaces of basins. (From Chockalingam, L. and Sagar, B.S.D., *J. Geophys. Res. Solid Earth (AGU)*, 110, B08203, 2005.)

size-wise NODs for these eight basins (Figure 7.19b) show significant oscillations indicating different NOD size categories, which are less in number, occupying larger contributing areas.

The largest templates that could be fit in the eight basins ranging from the first to the eighth basins are respectively of the radii of 32, 28, 26, 26, 34, 26, 24, and 26 pixels (Table 7.4). We estimate the fractal dimension of the nonnetwork space through the following steps. We determine power-law exponents for the NODs' number and size distributions by means of a connection to the decay of nonnetwork space of basin. Based on the assumption that the shape of the nonnetwork space alters the number and size distributions of NODs, these exponents are strongly shape dependent. We compute the number of NODs smaller than the specified threshold radius of the structuring template and their contributing areas (Table 7.5) respectively denoted as $N[\text{NODs}(< nB)]$ and $A[\text{NODs}(<nB)]$. The distribution of number and area of NODs, decomposed from nonnetwork space, depends on the diverging angles of streams. The rate at which the nonnetwork space within a basin gets decayed via morphological decomposition depends on the area, geometric organization, and the outline roughness of connex components of nonnetwork space. We propose that the dimensions derived from the analysis of nonnetwork space provide better reasons to explore links with processes and geomorphic expression of the basin than that of network morphometric characteristics. By employing the numbers of NODs of various sizes, their contributing areas, and the corresponding radius of template, we derive simple power-law relationships for these eight realistic basins. Figure 7.19c through j shows double logarithmic graphs for the cumulative number of NODs (diamond dots) smaller than the threshold radius of the structuring template (disk) and their corresponding contributing areas (square dots) versus the radii of structuring elements n. The slopes of the best-fit lines (α_N and α_A) respectively for number–radius and area–radius relationships (Table 7.6; Figure 7.19c through j) are obtained from the well-fitted relationships as $N[\text{NODs}(<nB)]$ (or) $A[\text{NODs}(<nB)] \sim n^{\alpha_N \,(or)\alpha_A}$, where n is the radius of the template and α is the slope of the best-fit line. These slope values of the best-fit lines provide shape-dependent dimensions as $D_N = \alpha_N - 1$, and $D_A = \alpha_A$ yields D_N and D_A for nonnetwork spaces of eight basins. The slopes are under 1.61 for the number of NODs and are under 2.38 for the contributing areas of NODs. These slope values can be related with erosion laws. These relations can also be linked with slope–area diagram. These statistically derived measures are dependent upon characteristic information of template used to convert the nonnetwork spaces into NODs. The third measure is derived from the plots made by considering the number of NODs as functions of their corresponding areas. The geometric complexities for these eight networks, computed by taking the contributing areas of NODs as functions of radii of templates, are in the ascending order for the basins 3, 6, 7, 2, 1, 5, 4, and 8. It is obvious, from the comparison, that there is no relation between network-based topologic quantities and nonnetwork-based complexity measures. In addition to these statistically derived power-law

TABLE 7.5

Cumulative Number and Corresponding Contributing Areas of Nonoverlapping Disks of Various Sizes Decomposed from Nonnetwork Space of Eight Basins

SE	Basin Number															
	1		2		3		4		5		6		7		8	
	N	A	N	A	N	A	N	A	N	A	N	A	N	A	N	A
34	—	—	—	—	—	—	—	—	316	109,149	—	—	—	—	—	—
32	520	182,014	—	—	—	—	—	—	168	101,164	—	—	—	—	—	—
30	273	168,813	—	—	—	—	—	—	118	93,769	—	—	—	—	—	—
28	273	168,813	282	84,673	—	—	—	—	118	93,769	—	—	—	—	—	—
26	273	168,813	143	77,373	203	68,151	288	91,357	77	81,823	367	125,782	—	—	440	130,615
24	185	157,455	93	70,093	115	63,288	152	83,715	58	73,992	200	116,378	179	48,876	243	120,604
22	132	142,011	93	70,093	115	63,288	104	77,487	44	65,496	144	108,111	100	44,098	162	107,797
20	97	128,123	70	63,895	77	57,872	77	68,878	32	54,786	96	94,175	68	40,018	106	89,579
18	69	114,338	52	55,265	77	57,872	58	61,020	24	47,302	65	80,715	41	31,764	70	74,589
16	58	102,690	33	42,182	58	53,641	40	50,716	17	39,672	52	72,943	41	31,764	46	58,422
14	39	79,374	24	35,485	45	48,404	28	39,683	13	33,652	34	58,100	26	25,760	34	49,679
12	31	67,158	17	27,707	31	38,240	17	28,488	10	27,837	22	45,704	13	17,946	22	37,938
10	20	48,742	10	19,564	19	28,316	10	20,138	8	23,638	17	40,300	9	13,669	11	21,426
8	11	32,274	7	14,762	12	20,895	7	15,941	6	19,163	14	36,029	5	9,833	8	16,888
6	6	18,356	4	9,486	8	16,249	5	12,226	4	12,004	10	28,786	4	8,888	4	10,145
4	4	14,019	2	5,784	3	8,794	2	5,587	2	6,088	6	16,009	2	5,435	2	6,078
2	2	5,859	1	3,339	2	6,711	1	2,753	1	2,407	2	5,978	1	2,360	1	2,632

Source: Chockalingam, L. and Sagar, B.S.D., *J. Geophys. Res. Solid Earth (AGU)*, 110, B08203, 2005.
A, area in pixel units.

relationships for nonnetwork spaces, we also derive shape-based complexity measures by estimating uncertainty index for the number of NODs and their areas. The NODs of various sizes are categorized according to their sizes by performing opening with increasing cycles. For instance, the NODs of non-network space of first basin are segregated into 16 size categories. The distributions of the number of NODs and their contributing areas are computed for these eight basins (Table 7.5). We employ these basic measures of size-distributed NODs to estimate probability distribution functions of number and area (Table 7.5). By employing these normalized plots of number and area, we *estimate* complexity measures (Table 7.5) by following entropy equations

$$H(N)/M = -\sum_{n=1}^{19} p_N(n)\log[p_N(n)] \quad \text{and} \quad H(A)/M = -\sum_{n=1}^{19} p_A(n)\log[p_A(n)],$$

where p_N, p_A, $H(N)/X$, and $H(A)/X$ respectively denote probability distribution functions, and average uncertainty indexes for the number of NODs and their areas. These measures are also scale independent and shape dependent that quantify the degree of randomness in the distributions of the number of NODs and their corresponding areas.

For the considered eight subbasins, we show these shape-dependent and non-shape-dependent dimensions derived respectively from the non-network spaces and network morphometries of eight basins (Figure 7.20). Characterization of network via non-shape-dependent morphometric parameters is not sensitive to sinuosity of stream segments. However, the nonnetwork space characterized via dimensions is sensitive to sinuosity of network (or) curvature and geometric organization of space occupied by varied degrees of convex region within a basin. On the other hand, the dimensions derived from their corresponding nonnetwork spaces are shape dependent.

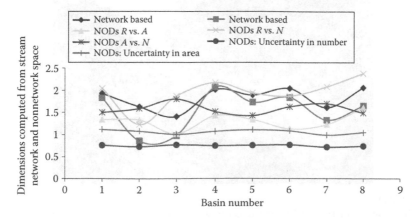

FIGURE 7.20
Basin number versus varied dimensions derived from morphometry of networks and non-network spaces. (From Chockalingam, L. and Sagar, B.S.D., *J. Geophys. Res. Solid Earth (AGU)*, 110, B08203, 2005.)

Additional Case on Durian Tunggal Basin

In addition to these eight basins, an additional basin (Figure 7.21) is also considered, and based on the procedure demonstrated on the eight basins, the nonnetwork space of this basin is also decomposed. The illustrations of sequential steps are given as: (a) network of the Durian Tunggal catchment, (b) nonnetwork space, (c) decomposed-coded nonnetwork space, and (d) transition lines just before X becomes empty. We determine power-law exponents for the NODs' number and size distributions by means of a connection to the decay of nonnetwork space of catchment basin (Table 7.6). Based on the assumption that the shape of the nonnetwork space alters the number and size distributions of NODs, these exponents are strongly shape dependent. We compute the number of NODs smaller than the specified threshold radius of structuring template and their contributing areas respectively denoted as $N[\text{NODs}(<nB)]$ and $A[\text{NODs}(<nB)]$. By employing these numbers, their contributing areas, and the corresponding radius of template, we derive simple power-law relationships for a realistic catchment basin. When we plot double logarithmic graphs, the slopes of the best-fit lines (α_N and α_A) respectively for number–radius and area–radius relationships yield 2.37 and 1.34 (Figure 7.22) from the relationships as $N[\text{NODs}(<nB)]$ (or) $A[\text{NODs}(<nB)] \sim n^{\alpha_N (\text{or}) \alpha_A}$, where n is the radius of template and α is slope of the best-fit line. These slope values of the best-fit lines provide shape-dependent dimensions as $D_N = \alpha_N - 1$ and $D_A = \alpha_A$.

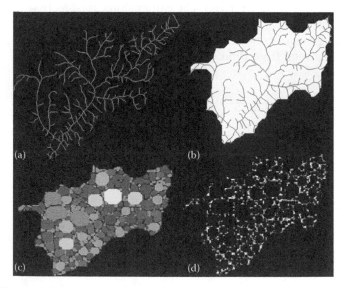

FIGURE 7.21
(a) Fifth-order channel network (C) of Durian Tunggal catchment basin: basin M is reconstructed from this channel network via multiscale morphological closing transformation, (b) $X = M \backslash C$ is nonnetwork space within a catchment basin, (c) decomposition of nonnetwork space (X) into NODs of octagon shape of several sizes, and (d) transition lines between the packed objects. (From Sagar, B.S.D. and Chockalingam, L., *Geophys. Res. Lett.*, 31(12), 12502, 2004.)

TABLE 7.6

Dimensions Derived from Morphometry of Network and Power-Laws Derived from Nonoverlapping Disks of Nonnetwork Space, and Shape Complexity Measures Estimated for NODs' Number and Their Corresponding Areas

Basin Number	Network		Nonnetwork Space				
	Network FD ($\text{Log } R_B/\text{Log } R_L$)		R vs. A	R vs. N	A vs. N	$H(N)/X$	$H(A)/X$
1	1.83	193	1.34	2.04	1.50	0.76	1.116
2	0.86	1.63	1.33	1.23	1.59	0.73	1.078
3	0.98	1.41	1.02	1.87	1.80	0.77	1.009
4	2.07	2.01	1.43	2.17	1.52	0.75	1.075
5	1.73	1.90	1.34	1.94	1.43	0.76	1.108
6	1.84	2.04	1.13	1.87	1.63	0.77	1.086
7	1.33	1.61	1.23	2.08	1.70	0.72	0.991
8	1.65	2.06	1.61	2.38	1.49	0.74	1.050

Source: Chockalingam, L. and Sagar, B.S.D., *J. Geophys. Res. Solid Earth (AGU)*, 110, B08203, 2005.

We compute D_N and D_A for nonnetwork space (Figure 7.21b), which yield 1.38 and 1.34 (Figure 7.22a). We also show a power-law relationship, with an exponent value 1.79, between the area and the number of NODs observed with increasing radius of structuring template (Figure 7.22b). However, the ratio of logarithms of bifurcation and mean length ratios of the network yield fractal dimension of 1.77. This shape-dependent dimension provides an insight, if it can be related to other dimensions estimated via linear aspects of the branched networks. We also find that the dimensions computed by means of two topological quantities and area–number relationship are significantly similar. We infer that these dimensions of 1.77 and 1.79 for this case explain space-filling characteristics of networks.

Morphometry of Networks versus Morphometry of Nonnetwork Spaces

This section addresses four aspects: (1) reconstruction of the basin from channel networks, (2) generation of nonnetwork spaces (X) from the basins (M) reconstructed from channel network such that the channel networks are contained in M, (3) decomposition of X into NODs to compute morphometry of nonnetwork spaces, and (4) derivation of relationships among several parameters of morphometries of networks and their nonnetwork spaces. To achieve these goals, we use set theory and topology-based mathematical transformations that have hitherto been relatively less employed in geophysics. This framework and the results derived from realistic cases allow systematic characterization and validation of the topological properties of the nonnetwork space of various realistic and simulated networks

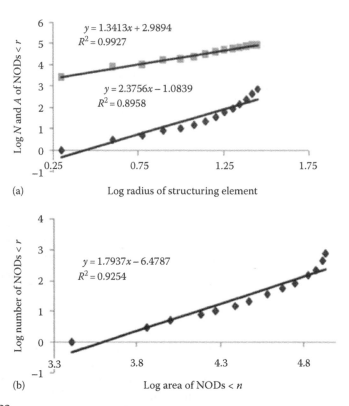

(a)

(b)

FIGURE 7.22
(a) Double-logarithmic plot between the radii of structuring templates and corresponding number and area of NODs, and (b) double logarithmic plot between area and the number of NODs with increasing radius of structuring element. (From Sagar, B.S.D. and Chockalingam, L., *Geophys. Res. Lett.*, 31(12), 12502, 2004.)

via shape-dependent measures. This systematic framework to quantify the organization of hillslope morphologies would be useful in modeling the landscape evolution. We conclude that morphological decomposition of non-network space into *NODs* facilitates new measures based on the general statistical relationships and probability distribution functions of the number of NODs and their corresponding areas. We argue that these shape-dependent measures, which are useful to capture the basic dissimilarities between Hortonically similar basins and to adequately characterize the Hortonian and non-Hortonian basin (e.g., Scheidegger 1967) morphologies, are better indicators than Hortonian-based measures. It would be interesting to compute a spectrum of similar quantities in both two and three dimensions, by employing a family of various symmetric and asymmetric probing rules, for the basins possessing varied degrees of self-affinity (decreasing circularity ratio) to establish a relationship between the shape-dependent dimension, and the geometric and morphometric characteristics. This framework allows systematically characterizing and validating the topological properties of the nonnetwork space of various realistic and simulated networks *via* shape-dependent dimension.

Intuitively, the hypotheses are that (1) the involved morphologic process in a circular nonnetwork space is different from that of an irregular nonnetwork space, and (2) the rate of erosion would be relatively lesser in the connex components with higher degree of convexity. In turn, the number distribution functions of NODs would provide insights to explore links with morphologic organization of hillslopes and erosion laws. In order to quantify the basic differences in terms of geomorphic process and landscape response to perturbation due to tectonic and/or climatic settings, shape-dependent measures are particularly useful. This provides an additional important procedure for shape-based classification of landscape. A broader implication is that the nonnetwork spaces within basins with lesser relief ratio (e.g., tidal basins and braided channels) can be better quantified through these shape-based measures. This approach has important yet unexplored implications for how hillslopes can be classified based on geometric organization in a 3-D space. Further implications of such a classification would provide insightful ideas toward exploring links between quantitative results and the morphological processes of basins.

References

Anderson, R. S., 1994, Evolution of the Santa Cruz Mountains, California, through tectonic growth and geomorphic decay, *Journal of Geophysical Research*, 99, 20161–20174.

Arenas, A. L., L. Danon, A. Diaz-Guilera, P. M. Gleiser, and R. Guimera, 2004, Community analysis in social networks, *European Physical Journal B*, 38(2) 373–380.

Beer, T. and M. Borgas, 1993, Horton's laws and the fractal nature of streams, *Water Resources Research*, 29, 1457–1487.

Chockalingam, L. and B. S. D. Sagar, 2005, Morphometry of networks and non-network spaces, *Journal of Geophysical Research-Solid Earth (American Geophysical Union)*, 110, B08203, doi:10.1029/2005JB003641.

Dietrich, W. E., C. J. Wilson, D. R. Montgomery, and J. McKean, 1992, Erosion thresholds and land surface morphology, *Journal of Geology*, 3, 161–173.

Dodds, P. S. and D. H. Rothman, 2001. Geometry of river networks II: Distributions of component size and number, *Physical Review E* 63: 016116.

Dodds, P. S. and J. S. Weitz, 2002, Packing of limited growth, *Physical Review E*, 65, 056108.

Dodds, P. S. and J. S. Weitz, 2003, Packing-limited growth of irregular objects, *Physical Review E*, 67 (1), article number: 016117, Part 2.

Fernandes, N. F. and W. E. Dietrich, 1997, Hillslope evolution by diffusive processes: The timescale for equilibrium adjustments, *Water Resources Research*, 33, 1307–1318.

Flook, A. G., 1978, The use of dilation logic on the quantimet to achieve fractal dimension characterization of textured and structured profiles, *Powder Technology*, 21, 295–298.

Gupta, V. K. and S. Veitzer, 2000, Random self-similar networks and derivations of Horton-type relations exhibiting statistical simple scaling, *Water Resources Research*, 36, 1033–1048.

Horton, R. E., 1945, Erosional development of stream and their drainage basin: hydrological approach to quantitative morphology, *Bulletin Geophysical Society America*, 56, 275–370.

Howard, A. D., 1990, Theoretical model of optimal drainage networks, *Water Resources Research*, 26(9), 2107–2117.

Howard, A. D., 1994, A detachment-limited model of drainage basin evolution, *Water Resources Research*, 30, 2261–2285.

Kanmani, S., Rao, C. B., and B. Raj, 1992a, On the computation of the Minkowski dimension using morphological operations, *Journal of Microscopy*, 170, 81–85.

Kanmani, S., Rao, C. B., Bhattacharya, D. K., and B. Raj, 1992b, Multifractal analysis of stress corrosion cracks, *Acta Stereologica*, 11, 349–354.

Karlinger, M. R., T. M. Over, and B. M. Troutman, 1994, Relating thin and fat-fractal scaling of river-network models, *Fractals*, 2(4), 557–565.

Kirchner, J. W., 1993, Statistical inevitability of Horton's laws and the apparent randomness of stream channel networks, *Geology*, 21, 591–594.

Kirkby, M. J., 1971, Hillslope process-response models based on the continuity equation, *Institute of British Geographers Special Publication*, 3, 15–30.

Koons, P. O., 1989, The topographic evolution of collisional mountain belts: A numerical look at the Southern Alps, New Zealand, *American Journal of Science*, 289, 1041–1069.

LaBarbera, P. and R. Rosso, 1987, The fractal geometry of river networks, *Eos Transactions American Geophysical Union*, 68 (44), 1276.

Lian, T. L., Radhakrishnan, P., and B. S. D. Sagar, 2004, Morphological decomposition of sandstone pore-space: Fractal power-laws, *Chaos Solitons & Fractals*, 19(2), 339–346.

Mandelbrot, B. B., 1982, *Fractal Geometry of Nature*, W.H. Freeman, San Francisco, CA, p. 468.

Manna, S. S. and H. J. Herrmann, 1991, Precise determination of the dimension of Appollonian packing and space filling bearings, *Journal of Physics A: Mathematical and General*, 24, L481–L490.

Marani, A., R. Rigon, and A. Rinaldo, 1991, A note on fractal channel network, *Water Resources Research*, 27, 3041–3049.

Maritan, A., F. Coloairi, A. Flammini, M. Cieplak, and J. R. Banavar, 1996b, Universality classes of optimal channel networks, *Science*, 272, 984–986.

Maritan, A., R. Rigon, J. R. Banavar, and A. Rinaldo, 2002, Network allometry, *Geophysical Research Letters*, 29(11), 1508, doi:10.1029/2001GL014533.

Maritan, A., A. Rinaldo, R. Rigon, A. Giacomatti, and I. Rodriguez-Iturbe, 1996a, Scaling laws for river networks, *Physical Review*, E 53, 1510–1515.

Mehta, A. P., C. Reichhardt, C. J. Olson, and F. Nori, 1999, Topological invariants in microscopic transport on rough landscapes: Morphology, hierarchical structure, and Horton analysis of river like networks of vortices, *Physical Review Letters* 82 (18), 3641–3644.

Moglen, G. E. and R. L. Bras, 1995, The effect of spatial heterogeneities on geomorphic expression in a model of basin evolution, *Water Resources Research*, 31, 2613–2623.

Montgomery, D. R. and W. E. Dietrich, 1988, Where do channels begin? *Nature*, 336, 232–234.

Montgomery, D. R. and W. E. Dietrich, 1994, Landscape dissection and drainage area-slope thresholds. In: *Processes Models and Theoretical Geomorphology*, ed. M. J. Kirkby, John Wiley, Hoboken, NJ, pp. 224–246.

Montgomery, D. R. and W. E. Dietrich, 1992, Channel initiation and the problem of landscape scale, *Science*, 255, 826–832.

Nikora, V. I. and V. B. Sapozhnikov, 1993, River network fractal geometry and its computer simulation, *Water Resources Research*, 29(10), 3569–3575.

Olson, C. J., C. Reichhardt, and F. Nori, 1998, Fractal networks, braiding channels, and voltage noise in intermittently flowing rivers of quantized magnetic flux, *Physical Review Letters*, 80(10), 2197–2200.

Peckham, S. and V. Gupta, 1999, A reformulation of Horton's laws for large river networks in terms of statistical self-similarity, *Water Resource Research*, 35(9), 2763–2777.

Radhakrishnan, P., B. S. D. Sagar, and L. L. Teo, 2004, Estimation of fractal dimension through morphological decomposition, *Chaos Solitons & Fractals*, 21(3), 563–572.

Rigon, R., A. Rinaldo, I. Rodriguez-Iturbe, R. L. Bras, and E. Ijjasz-Vasquez, 1993, Optimal channel networks: A framework for the study of river basin morphology, *Water Resources Research*, 29, 1635–1646.

Rinaldo, A., I. Rodriguez-Iturbe, R. L. Bras, and E. Ijjasz-Vasquez, 1993, Self-organized fractal river networks, *Physical Review Letters*, 70, 822–826.

Rodriguez-Iturbe, I. and A. Rinaldo, 1997, *Fractal River Basins: Chance and Self-Organization*, Cambridge University Press, New York.

Roering, J. J., J. W. Kirchner, and W. E. Dietrich, 1999, Evidence for nonlinear, diffusive sediment transport on hillslopes and implications for landscape morphology, *Water Resources Research*, 35, 853–870.

Sagar, B. S. D., 1996, Fractal relations of a morphological skeleton, *Chaos Solitons Fractals*, 7(11), 1871–1879.

Sagar, B. S. D., 1999, Estimation of number-area-frequency dimensions of surface water bodies, *International Journal of Remote Sensing*, 20, 2491–2496.

Sagar, B. S. D. and L. Chockalingam, 2004, Fractal dimension of non-network space of a basin, *Geophysical Research Letters*, 31(12), 12502, doi:10.1029/2004GL019749.

Sagar, B. S. D. and K. S. R. Murthy, 2000, Generation of a fractal landscape using nonlinear mathematical morphological transformations, *Fractals*, 8(3), 267–272.

Sagar, B. S. D., M. B. R. Murthy, C. B. Rao, and B. Raj, 2003, Morphological approach to extract ridge-valley connectivity networks from digital elevation models (DEMs), *International Journal of Remote Sensing*, 24(3), 573–581.

Sagar, B. S. D., C. Omoregie, and B. S. P. Rao, 1998, Morphometric relations of fractal-skeletal based channel network model, *Discrete Dynamics in Nature and Society*, 2, 77–92.

Sagar, B. S. D., C. B. Rao, and B. Raj, 2002, Is the spatial organization of larger water bodies heterogeneous? *International Journal of Remote Sensing*, 23(3), 503–509.

Sagar, B. S. D., D. Srinivas, and B. S. P. Rao, 2001, Fractal skeletal based channel networks in a triangular initiator basin, *Fractals*, 9(4), 429–437.

Sagar, B. S. D. and T. L. Tien, 2004, Allometric power-law relationships of Hortonian fractal digital elevation model, *Geophysical Research Letters*, 31(6), L06501, doi:10.1029/2003GL019093.

Scheidegger, A. A., 1967, A stochastic model for drainage patterns into an intramontane trench, *Bulletin Association of Scientific Hydrology*, 12, 15–60.

Serra, J., 1982, *Image Analysis and Mathematical Morphology*, Academic Press, London, U.K., p. 610.

Shreve, R. L., 1967, Infinite topologically random channel networks, *Journal of Geology*, 75, 178–186.

Soille, P. and J. F. Rivest, 1996, On the validity of fractal dimension measurements in image analysis, *Journal of Visual Communication and Image Representation*, 7(3), 217–229.

Strahler, A. N., 1957, Quantitative analysis of watershed geomorphology, *EOS Transactions, American Geophysical Union*, 38(6), 913–920.

Takayasu, H., 1990, *Fractals in Physical Sciences*, Manchester Univ. Press, Manchester, U.K.

Tarboton, D. G. and R. L. Bras, 1992, A physical basis for drainage density, *Geomorphology*, 5, 59–76.

Tarboton, D. G., R. L. Bras, and I. Rodriguez-Iturbe, I., 1988, The fractal nature of river networks, *Water Resources Research*, 24, 1317–1322.

Teo, L. L. P. Radhakrishnan and B. S. D. Sagar, 2004, Morphological decomposition of sandstone pore-space: Fractal power-laws, *Chaos Solitons & Fractals*, 19(2), 339–346.

Teo, L. L. and B. S. D. Sagar, 2006, Modeling, description and characterization of fractal pore via mathematical morphology, *Discrete Dynamics in Nature and Society*, Article ID, 89280, DOI 10.1155/DDNS/2006/89280.

Tokunaga, E., 1984, Ordering of divide segments and law of divide segment numbers, *Transactions Japanese Geomorphological Union*, 5, 71–77.

Torquato, S., 2000, Modeling of physical properties of composite materials, *International Journal of Solids and Structures*, 37, 411–422.

Turcotte, D. L., 1997, *Fractals in Geology and Geophysics*, Cambridge University Press, New York.

Whipple, K. X. and G. Tucker, 1999, Dynamics of the stream power river incision model: Implications for height limit of mountain ranges, landscape response time scales, and research needs, *Journal of Geophysical Research*, 104, 17661–17674.

Willgoose, G. R., R. L. Bras, and I. Rodriguez-Iturbe, 1991, The relationship between catchment and hillslope properties: Implications of a catchment evolution model, *Geomorphology*, 5(1/2), 21–38.

8

Granulometries, Convexity Measures, and Geodesic Spectrum for DEM Analyses

Approaches to compute quantitative characteristics of features derived from terrestrial data that have been provided in Chapters 5 through 7 rely on thematic information that is essentially in binary form. This chapter provides three different approaches to characterize terrestrial surface data that are in the grayscale form. These three approaches include (1) grayscale granulometries to characterize foreground and background roughness of terrestrial surfaces, (2) computation of convexity measures that are akin to channel density, and (3) computation of geodesic spectrum that provides one-dimensional (1-D) geometric support of terrestrial basins.

Grayscale Granulometric Analysis

In Chapters 4 through 7, multiscale morphological opening transformation has been employed to distribute surface water bodies, extracted from remotely sensed satellite data, according to their sizes. Besides, zones of influence of water bodies have also been size distributed using opening transformation. The parameters taken from size distributions have been considered as basic inputs to derive scaling relationships and certain quantitative indexes that provide insights into understanding the degree of heterogeneity in the spatial distribution. All through these studies, the opening transformation employed was binary opening transformation. Entire analysis carried out in Chapters 4 through 7 comes under a topic on "Applications of binary granulometries."

As the idea is to characterize the terrestrial complexity, instead of considering the phenomenon extracted from terrestrial data (e.g., remotely sensed data, digital elevation models [DEMs]), DEMs, which are in grayscale format, are considered to demonstrate on how terrestrial surfaces could be characterized. Then, for the transformation involved to characterize terrestrial surfaces that are available as DEMs, grayscale granulometries were employed. Grayscale granulometric analysis involves the grayscale morphological opening transformation.

Several terrestrial characteristics and processes have relationships with terrain roughness. Terrain roughness was earlier quantified based on several methods (e.g., Horton 1945, Stone and Dugundji 1965, Daniels et al. 1970, Franklin 1987, Ackeret 1990). Out of these methods, morphometry, fractals, and allometric scaling analysis (Horton 1945, Langbein 1947, Rodriguez-Iturbe and Rinaldo 1997, Turcotte 1997, Sagar et al. 1998a, Maritan et al. 2002, Sagar and Chockalingam 2004, Sagar and Tien 2004) have been receiving wide attention as they can provide quantitative characterization tools. Most of these characterization techniques are mainly feature (theme) based and emphasize the spatial organization of certain features that are decomposed from topographic maps and/or channel and ridge connectivity networks (Sagar et al. 2003), watersheds, basins, mountain objects, etc. Earlier, terrain roughness has been quantified via different indexes (e.g., Goodchild 1980, Dubuc et al. 1989, Gilbert 1989, Cherbit 1991, Fatale et al. 1994, Nikora 2005).

With the availability of DEMs, significant breakthroughs in terrain characterization studies have emerged. The importance of DEMs in understanding the geophysical and geomorphologic processes has been highlighted by various researchers (e.g., Montgomery and Foufoula-Georgiou 1993, Rodriguez-Iturbe and Rinaldo 1997, Whipple et al. 1999, Whipple and Tucker 1999, Snyder et al. 2000, Dall et al. 2001, Baratoux et al. 2002, Rodriguez et al. 2002, Tay et al. 2005, 2007). In quantifying shape and size content of terrestrial surfaces possessing geometrical structures, spectral and fractal analyses offer very little. Hence, quantitative characterization to derive roughness parameters has proven difficult because of the morphological complexity of terrestrial surfaces.

One of the best approaches to quantify the shape and size content of terrestrial surfaces (terrestrial basins) is unquestionably the grayscale granulometry. Grayscale granulometries by opening and closing respectively quantify the roughness and mean size of foreground and background of DEMs. Granulometries via opening and closing would explore (1) the composition of bright (higher elevation regions) and dark (lower elevation regions) regions in a DEM, (2) how bright and dark regions of a DEM get transformed with multiscale opening and closing, (3) roughness characterization of surface, and (4) shape–size complexity measures of foreground and background that provide quantitative characteristics of terrestrial surfaces.

DEMs of a part of Cameron Highlands (Figure 3.12a) and Petaling region of Malaysia (Figure 3.12b) have been considered to apply grayscale granulometries to derive shape–size content of DEMs. Shape–size content provides quantitative characteristics of terrestrial surfaces. These shape–size complexity measures, in other words, morphological entropy, are scale invariant by shape dependency. To demonstrate this further, rhombus-, square-, and octagonal-shaped structuring elements have been employed in granulometric analysis.

Granulometries via Multiscale Opening

Multiscale opening has been performed on a DEM (Figure 3.12a and b) by increasing the size (scale) of the structuring element nB, where $n = 0, 1, 2, ..., N$. This multiscale opening on $f(x, y)$, in other words, DEM, is represented as Equation 8.1:

$$(((f \ominus B) \ominus B) \ominus \cdots \ominus B) \oplus B \oplus B \oplus \cdots \oplus B = (f \ominus nB) \oplus nB = f \circ nB \quad (8.1)$$

where
　f is a DEM, a function usually denoted as $f(x, y)$
　B is a structuring element
　$\ominus$, $\oplus$, and $\circ$, respectively, denote symbols for morphological erosion, dilation, and opening

Opening of f by increasing sizes of B transforms f in such a way that the brighter regions (higher elevation regions of a DEM) would get merged into the darker regions. Multiscale grayscale opening of f by nB satisfy the following properties:

1. $f(x, y) = Z; A(f) = \sum\limits_{x,y} f(x, y)$
2. $A(f \circ 0B) = A(f)$
3. $A(f \circ nB) \geq A(f \circ (n+1)B)$
4. $A((f \circ nB) - (f \circ (n+1)B)) = A(f \circ nB) - A(f \circ (n+1)B)$
5. $A(f \circ NB) = A(f \circ (N+1)B) =$ Morphological convergence

Multiscale grayscale opening smoothens the protrusions in a DEM. The larger the size of the structuring element employed in the opening transformation, the larger the protrusions that filtered out. This multiscale opening has been applied on 14 subbasins partitioned from 2 DEMs of 2 different regions (Figure 3.12a and b). Basic measures of such as basin size, height, and maximum number of iterative openings to transform the basin to reach morphological convergence have been given in Table 8.1. Flowchart depicting sequential steps to compute shape–size content is given in Figure 8.1.

　For demonstration, grayscale opened versions of 1 of the 14 basins have been generated. These opened versions include basin opened with respect to B, 40 B, 80 B, 120 B, 160 B, and 200 B, where B is of rhombus shape, with the primitive size of 5×5 (Figure 8.2). For these computations, the primitive sizes of square, octagon, and rhombus structuring elements are taken as 5×5 (Figure 8.2). The elements in primitive square, octagon, and rhombus structuring elements include 25, 21, and 13, respectively. It is worth mentioning that the iterative openings required to make each subbasin darker are

TABLE 8.1

Basic Measures of Basin Size, Height, and Maximum Number of Iteration for All 14 Basins

Basin Number N	Basin Size (No of Pixel)	Dem Height				Maximum Number of Iteration		
		Max (m)	Min (m)	Max–Min (m)	Relief Ratio	Square	Octagon	Rhombus
1	105,400	1280.1	540.6	739.5	0.422	170	227	340
2	137,463	1596.5	591.8	1004.7	0.371	208	277	415
3	131,517	1695.9	587.4	1108.5	0.346	214	285	427
4	107,625	1594.5	570.5	1024.0	0.358	188	250	375
5	89,300	1745.2	503.0	1242.2	0.288	190	254	380
6	60,520	1667.7	483.4	1184.3	0.290	178	238	356
7	36,814	929.6	475.9	453.7	0.512	117	156	233
8	134,400	208.0	54.3	153.7	0.261	210	280	420
9	57,950	155.7	50.7	105.0	0.325	153	204	305
10	48,000	193.5	48.7	144.8	0.251	160	214	320
11	68,800	192.7	40.7	152.0	0.211	160	214	320
12	72,000	215.7	32.9	182.8	0.152	180	240	360
13	72,500	153.2	31.7	121.5	0.207	145	194	290
14	42,000	169.1	27.6	141.5	0.163	150	200	300

Source: Tay, L.T. et al., *Int. J. Remote Sens.*, 28(15), 3363, 2007.

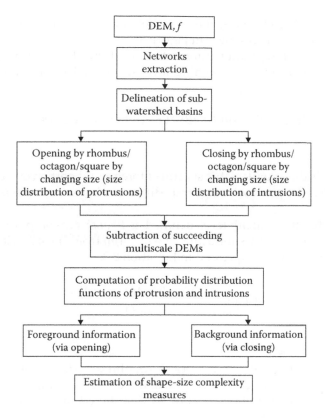

FIGURE 8.1
Flowchart depicting the sequential steps adapted in this investigation. (From Tay, L.T. et al., *Int. J. Remote Sens.*, 28(15), 3363, 2007.)

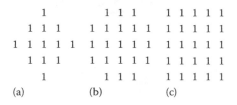

FIGURE 8.2
Structuring elements of primitive size 5 × 5. (a) Rhombus, (b) octagon, and (c) square.

dependent not only on the size, shape, origin, and orientation of considered primitive structuring elements but also on the size of the basin. More number of opening cycles are required when structuring element rhombus (with 13 elements) is considered, and it is followed by octagon (with 21 elements) and square (with 25 elements).

It is conspicuous that the grayscale values represent elevations have been progressively becoming dark with increasing cycle of opening. The areas of these opened versions are in decreasing trend. By following octagon and square structuring elements, the snapshots of the opened versions of this subbasin have also been shown in Figure 8.3b and c.

The changing foreground information is derived by subtracting each opened version from the preceding level of opened version according to Equation 8.2:

$$PS_f(+n, B) = A((f \circ nB) - (f \circ (n+1)B)), \quad 0 \le n \le N \tag{8.2}$$

where
$PS_f(+n, B)$ is the granulometric spectra of foreground portion of f relative to B
$a(x) - b(x)$ is the point-wise algebraic difference between the two functions

Probability functions at nth level, denoted as $PS(n,f)$, are computed by dividing the area obtained by subtracting $(f \circ (n+1)B)$ from $(f \circ nB)$ with the total area of f, $A(f)$, which is expressed in Equation 8.3:

$$ps(n, f) = \frac{A(f \circ nB) - A(f \circ (n+1)B)}{A(f)} \tag{8.3}$$

where $0 \le ps(n, f) \le 1$.

Average size ($AS(f/B)$) and average roughness ($H(f/B)$) of foreground, where protrusions are highly conspicuous, could be computed by taking probability size distribution function of distributed protrusions according to Equations 8.4 and 8.5:

$$AS(f/B) = \sum_{n=0}^{N} nps(n, f) \tag{8.4}$$

$$H(f/B) = -\sum_{n=0}^{N} ps(n, f) \log ps(n, f) \tag{8.5}$$

According to Equations 8.3 through 8.5, probability distribution functions, average size, and average roughness parameters have been computed for all the 14 subbasins with respect to square, octagon, and rhombus structuring elements. Average sizes and average roughness for these 14 basins have been plotted as functions of basin numbers (Figure 8.4). With rhombus, the average sizes of foregrounds of 14 subbasins have been found larger than that of octagon and square. These average size values rely on the total size of protrusions filtered via granulometric analysis. Larger average size values of protrusions have been obtained for basins 1, 2, 3, and 4 by means of rhombus structuring element. These average size values are scale invariant but shape dependent.

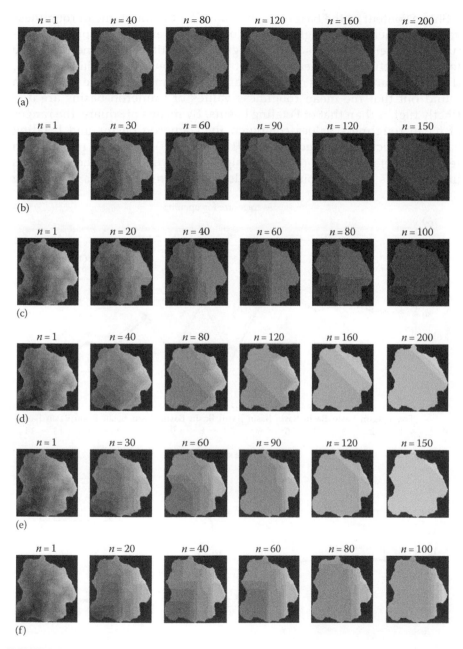

FIGURE 8.3

One subbasin example at multiple scales generated via closing and opening. Basin boundaries are superimposed on the DEM to depict the basin-wise multiscale characteristics. (a–c) DEM at multiple scales generated via opening by means of rhombus, octagon, and square, (d–f) multi-scale DEMs generated via closing by means of rhombus, octagon, and square. (From Tay, L.T. et al., *Int. J. Remote Sens.*, 28(15), 3363, 2007.)

Shape content of the basins can be quantified through mean roughness values. Higher mean roughness for a basin with respect to a specific structuring element indicates higher degree of surficial roughness relative to that structuring element. Ratio of average roughness value and N_{max} (Figure 8.4c and d) yields mean roughness in normalized scale. It has been found out that the mean roughness values of Cameron basins are distinctly higher than that of Petaling basins. By means of square, the ranges of mean roughness values computed for Cameron basins and Petaling basins respectively include 0.88–0.91 and 0.74–0.85. These ranges with respect to octagon and rhombus are significantly different. The variations in the ranges of mean roughness values between Cameron and Petaling

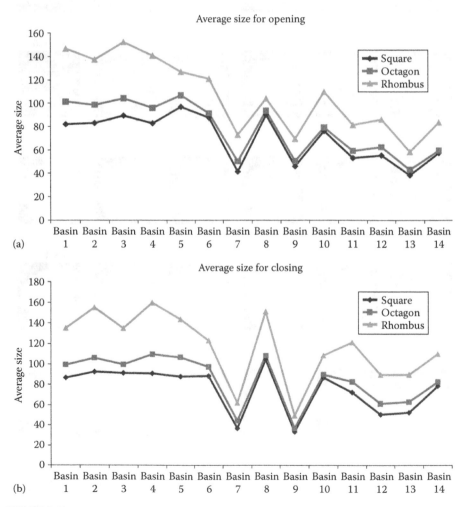

FIGURE 8.4

Mean size and roughness values vs. basin number. (a and b) Average size values computed for foregrounds and backgrounds of 14 basins by means of square, octagon, and rhombus.

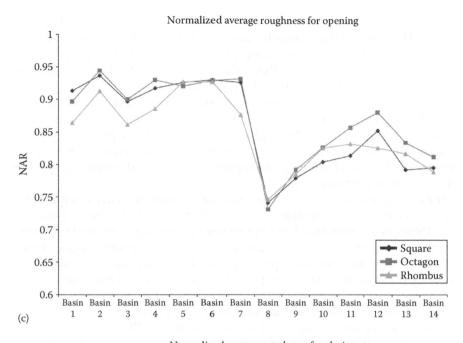

(c)

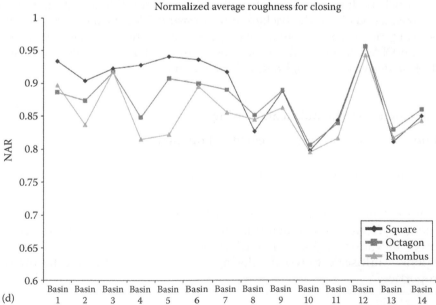

(d)

FIGURE 8.4 (continued)
Mean size and roughness values vs. basin number. (c and d) Normalized mean roughness values computed for foregrounds and backgrounds of 14 basins by means of square, octagon, and rhombus. (From Tay, L.T. et al., *Int. J. Remote Sens.*, 28(15), 3363, 2007.)

basins indicate that Cameron basins are more complex than Petaling basins. This observation is true because of the fact that all Cameron basins are high-altitude basins possessing greater relief difference, when compared to Petaling basins. All these shape–size complexity measures are structuring element dependent, further supporting that shape matters more than the scale. If basin region and structuring element have significant characteristic similarities, then the average roughness yields lower value. On the contrary, high roughness values would be produced, which indicate that the basin is rough with respect to the structuring element. In general, all the 14 subbasins yield higher roughness values with respect to square structuring element.

The average roughness and average size values have been derived quantifying the shape–size content of foregrounds of the basin-DEMs. The basis to compute these measures stems from the recursive multiscale grayscale opening that filters the protrusions that are conspicuous from foreground regions of the basin. These protrusions have geomorphologic relationship with valley connectivity network. In order to compute mean size and mean roughness values of backgrounds of the basin-DEMs, anti-granulometric analysis is required. Precisely, anti-granulometric analysis could be performed by performing multiscale grayscale closing transformation on basin-DEMs. Multiscale closing, which is dual operation of multiscale opening, filters intrusions that are conspicuous from the background of basin-DEMs. Geomorphologically speaking, ridge connectivity network possesses relationship with intrusions of basin-DEMs. In what follows includes anti-granulometric analysis of basin-DEMs to derive mean roughness of background.

Granulometries via Multiscale Closing

Multiscale closing of f is represented as Equation 8.6:

$$(((f \oplus B) \oplus B) \oplus \cdots \oplus B) \ominus B \ominus B \ominus \cdots \ominus B = (f \oplus nB) \ominus nB = f \bullet nB \quad (8.6)$$

where $\bullet$ denotes symbol for closing. Closing of f by B of increasing sizes n ranging from 1 to N transforms f in such a way that the darker regions (lower elevation regions of a DEM) would get merged into brighter regions. Multiscale grayscale closing of f by nB satisfies the following properties:

1. $f = (f \bullet 0B)$
2. $A(f \bullet 0B) = A(f)$
3. $A(f \bullet (n + 1)B) \geq A(f \circ nB)$
4. $A((f \bullet (n + 1)B) - (f \bullet nB)) = A(f \bullet (n + 1)B) - A(f \bullet nB)$
5. $A(f \bullet KB) = A(f \bullet (K + 1)B) \Rightarrow$ Close hull

Multiscale grayscale closing transformation smoothens the intrusions in DEMs. Larger-size intrusions get filtered when larger-size structuring element is employed in the closing transformation. Multiscale closing has been performed on 14 subbasin-DEMs (Figure 8.3d).

Granulometries by closing (anti-granulometries): Anti-granulometric spectra of f in relation to B for different size n could be computed according to Equation 8.7:

$$PS_f(-n, B) = A(f \bullet nB) - A(f \bullet (n-1)B), \quad 1 \le n \le K \tag{8.7}$$

where
$PS_f(-n, B)$ denotes pattern spectra of background portion of f in relation to B
$(f \bullet nB) - (f \bullet (n-1)B)$ is point-wise algebraic difference

The difference between the area of nth-level closed basin and the area of $(n-1)$th-level closed basin (where n ranges from 1 to K) is divided by $A(f \bullet KB) - A(f)$ to compute probability function at nth level, $ps(-n, f)$. This computation of probability function is expressed as Equation 8.8:

$$ps(-n, f) = \frac{A(f \bullet nB) - A(f \bullet (n-1)B)}{A(f \bullet KB) - A(f)}, \quad 1 \le n \le K \tag{8.8}$$

where $0 \le ps(-n, f) \le 1$. The Kth level of closing is decided according to Equation 8.9:

$$K = \min\{K : A(f \bullet KB) = A(f \bullet (K+1)B)\} \tag{8.9}$$

Based on the probability size distribution function of distributed intrusions, average roughness of background is estimated for all 14 basin-DEMs by incorporating probability function relative to B as shown in Equation 8.10:

$$H(f / B) = -\sum_{n=-K}^{n} ps(-n, f) \log ps(-n, f) \tag{8.10}$$

The multiscale grayscale closing transformation has been applied on 14 subbasin-DEMs, and the snapshots of certain closed versions of 1 of the 14 subbasin-DEMs have been shown in Figure 8.3d through f. It is obvious that the basin-DEM becomes brighter with increasing cycle of closing. Figure 8.3d through f, respectively, shows snapshots of closed versions of basin-DEMs obtained with respect to rhombus, octagon, and square (Figure 8.2).

Average sizes and average roughness parameters have been computed for backgrounds of all 14 basins with respect to square, octagon and rhombus structuring elements (Figure 8.2). These parameters have been plotted as functions of basin numbers (Figure 8.4b and d). Larger average size values of intrusions are obtained for basins 2, 4, 5, and 8 by means of rhombus structuring elements. The ranges of normalized roughness values for background by means of square structuring element are 0.84–0.89 and 0.79–0.83 for Cameron and Petaling basins, respectively. These ranges yielded by octagon and rhombus structuring elements differ significantly. Further, these ranges for Petaling basins and Cameron basins include 0.73–0.83 and 0.73–0.89, which indicate that the backgrounds of Petaling basins are smoother than Cameron basins. Protrusions and intrusions that are conspicuous from DEM basins testify the presence of valleys and ridges. The average roughness values describing the complexities of foreground (protrusions) and background (intrusions) could be related with valley and ridge connectivity networks.

In order to compare the foreground and background roughness values, estimated, respectively, via granulometric and anti-granulometric analyses, of 14 subbasins with fractal dimensions of networks of 14 subbasins, box counting approach (Feder 1988) has been followed. The networks extracted from 14 subbasin-DEMs have been shown in Figure 5.16a and b. Fractal dimensions of the networks belonging to 14 subbasins have been tabulated (Table 8.2). Fractal dimensions of networks and average

TABLE 8.2

Fractal Dimensions Calculated via Different Approaches for the 14 Subbasins

Basin	Fractal Dimension (via Box Counting)
1	1.5141
2	1.5506
3	1.5814
4	1.4692
5	1.4519
6	1.4776
7	1.3192
8	1.3140
9	1.2398
10	1.2445
11	1.1817
12	1.2946
13	1.1706
14	1.1721

Source: Tay, L.T. et al., *Int. J. Remote Sens.*, 28(15), 3363, 2007.

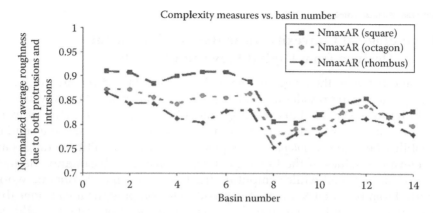

FIGURE 8.5
Fractal dimensions and complexity measures vs. basin number. (From Tay, L.T. et al., *Int. J. Remote Sens.*, 28(15), 3363, 2007.)

roughness values of both foreground and background were plotted as functions of basin numbers (Figure 8.5).

Shape–size complexity measures of foreground and background have been computed by filtering protrusions (from foreground) and intrusions (from background) via granulometric analyses by opening and closing, respectively. These measures that exhibit scale-invariant characters are shape dependent. Relating terrestrial processes with these complexity measures is a potentially valuable study that needs further investigations.

Quantitative characterizations of terrestrial surfaces, basins, and their associated features via approaches like morphometric analysis, allometric scaling analysis, binary, and grayscale granulometric analyses have been shown in Chapters 5 through 8. A basin-DEM with a clear biogeographic boundary consists of various geomorphologic features such as valley and ridge connectivity networks, mountain objects, hill slopes, and outlets. Shapes of such basins provide important clues about the type of processes involved within. Earlier, basin width function, which is a 1-D geometric support, used to be computed to understand the basin processes. Computation of basin width function involved basin in planar form. However, the elevation regions within the basin have many controls that impact the basin processes. In conventional width function, the elevation regions within a basin have been ignored. If one considers all threshold elevation regions (TERs) within a basin, then there would be a possibility to derive a geodesic spectrum based on geodesic flow fields that could be simulated across TERs of a basin by using geodesic dilations and certain logical operations.

Morphological Convexity Measures for Terrestrial Basins Derived from Digital Elevation Models

The ratio between the length of channel network (L) and the area of basin (A) in planar form provides a quantitative index, the channel density, which has hitherto been related to various geomorphologic processes. This index is one kind of convexity measure. Such a measure fails to capture the spatial variability between *homotopic* basins possessing different altitude ranges as the elevation values of the topological region within a basin and channel network are ignored while computing the basin area and channel network length. From basin-DEMs, now it is possible to compute the area under the basin, network, and convex hull functions. The elevation values would be taken into consideration while estimating these three basic measures, which further provide an approach to compute two types of convexity measures that have potential to capture the terrain elevation variability. The two types of convexity measures—which are altitude dependent and could capture the spatial variability across the *homotopic* basins of different altitudes—are the ratios of (1) length of channel network function and area of basin function and (2) areas of basin and its convex hull functions. Estimation of these two convexity measures are demonstrated on (1) synthetic basin functions, (2) fractal basin functions, and (3) realistic DEMs of two regions of peninsular Malaysia. The relationships between these convexity measures and other quantitative indexes such as fractal dimensions and complexity measures (roughness indexes) have been shown.

Channel Density, Convexity Measure, and Importance of Elevation Values

A popular quantitative index termed as channel density (Horton 1945), which is termed here as convexity measure of basin in two dimensions, is the ratio between the planar length of the channel network and the planar area of the basin. In the context of hydrogeomorphology, channel density is related to climate, geology, rainfall, erosion rate, and relief (e.g., Kirkby 1980, 1993, Schumm et al. 1987, Montgomery and Dietrich 1989, 1994, Howard 1997). Importance of DEM analysis in understanding the landscape state and process interactions is realized, and Tucker and Bras (1998) have shown how drainage density is related to topographic relief by the sign of the predicted relationship between drainage density and relief. Most of the approaches available to estimate drainage density were meant for fluvial basins (e.g., Tucker et al. 2001), and of late, an approach to estimate drainage density of tidal basins was addressed by Marani et al. (2003).

Due to the presence of valleys and ridges, no basin could be treated as a fully convex basin. A parameter that computes the degree of convexity of a basin is the convexity measure, which is related to channel (drainage) density.

The basic parameters such as the length of the network and the area of the basin required to estimate this measurement are drawn from plan view of the basin. Hence, this definition, from the point of convexity measure, has a limitation as it cannot capture the elevation variability among different drainage basins. Due to this limitation, the maps of drainage density do not carry much information about terrain morphology as high drainage density values may be seen in both flat, low-relief basins and mountainous, high-relief basins. The convexity measures of the seemingly alike (*homotopic*) basins with different altitude ranges should be different in such a way that it reflects the changes in the altitudes involved. Such a distinction could be shown through alternate measures (Lim et al. 2011)—where the inputs are represented as 3-D functions and not as planar sets—that are shown in this section. The three inputs include areas of basin $f(x, y)$, its convex hull $CH(f)$ functions, and the length of the network function $g(x, y)$, which are respectively denoted as $A(f)$, $A(CH(f))$, and $A(g)$. These three basic measures follow the property: $A(g) < A(f) < A(CH(f))$. These two convexity measures that could be computed are, respectively, the ratio (1) between the length of channel network function $A(g)$ and the area of basin function $A(f)$ and (2) between the area of basin function $A(f)$ and the area of its corresponding convex hull $A(CH(f))$.

Data Used and Their Specifications

To demonstrate the estimations of the two convexity measures, two types of data, namely, synthetic DEMs (simple synthetic functions and fractal basin functions) and real-world DEMs, were considered. These synthetic DEMs are denoted as basin functions f_1 and f_2, respectively, as shown in Figure 8.6a and b. The synthetic basin functions (f_1 and f_2) that have similar geometrical arrangement depict different topographic elevation ranges. Basin f_1 has higher elevation range than basin f_2: 15–20 versus 10–15. Fractal basin functions indicated as basin functions f_3 and f_4 that are simulated by transforming a binary fractal shape into fractal basin functions that mimic DEMs are depicted in Figure 8.7a and b. Iterative morphologic erosions by means of structuring element of octagonal shape (Sagar and Tien 2004, Chockalingam and Sagar 2005) have been performed on the binary fractal shape, the network of which follows Hortonian laws of morphometry (Sagar et al. 2001). Eleven iterative erosions are performed to transform binary fractal shape of size 256 × 256 into 11 eroded versions. Each eroded version is gray shaded separately to generate two fractal basin functions (Figure 8.7a and b). The gray-shade numbers employed to respectively denote these two functions are in the ranges of 1–11 and 5–15. These ranges are used to show that these two *homotopically* similar synthetic fractal basin functions with similar geometric organizations possess different altitude ranges. These two functions are also shown in 3-D representation (Figure 8.7c and d). In these basin functions, each discrete element with specific numerical value represents elevation at spatial

20	20	20	20	20	20	20	20	20	20	20
20	19	19	19	19	19	19	19	19	19	20
20	19	18	18	18	18	18	18	18	19	20
20	19	18	17	17	17	17	17	18	19	20
20	19	18	17	16	16	16	17	18	19	20
20	19	18	17	16	15	16	17	18	19	20
20	19	18	17	16	16	16	17	18	19	20
20	19	18	17	17	17	17	17	18	19	20
20	19	18	18	18	18	18	18	18	19	20
20	19	19	19	19	19	19	19	19	19	20
20	20	20	20	20	20	20	20	20	20	20

(a)

15	15	15	15	15	15	15	15	15	15	15
15	14	14	14	14	14	14	14	14	14	15
15	14	13	13	13	13	13	13	13	14	15
15	14	13	12	12	12	12	12	13	14	15
15	14	13	12	11	11	11	12	13	14	15
15	14	13	12	11	10	11	12	13	14	15
15	14	13	12	11	11	11	12	13	14	15
15	14	13	12	12	12	12	12	13	14	15
15	14	13	13	13	13	13	13	13	14	15
15	14	14	14	14	14	14	14	14	14	15
15	15	15	15	15	15	15	15	15	15	15

(b)

1	0	0	0	0	0	0	0	0	0	1
0	1	0	0	0	0	0	0	0	1	0
0	0	1	0	0	0	0	0	1	0	0
0	0	0	1	0	0	0	1	0	0	0
0	0	0	0	1	0	1	0	0	0	0
0	0	0	0	0	1	0	0	0	0	0
0	0	0	0	1	0	1	0	0	0	0
0	0	0	1	0	0	0	1	0	0	0
0	0	1	0	0	0	0	0	1	0	0
0	1	0	0	0	0	0	0	0	1	0
1	0	0	0	0	0	0	0	0	0	1

(c)

1	1	1	1	1	1	1	1	1	1	1
1	1	1	1	1	1	1	1	1	1	1
1	1	1	1	1	1	1	1	1	1	1
1	1	1	1	1	1	1	1	1	1	1
1	1	1	1	1	1	1	1	1	1	1
1	1	1	1	1	1	1	1	1	1	1
1	1	1	1	1	1	1	1	1	1	1
1	1	1	1	1	1	1	1	1	1	1
1	1	1	1	1	1	1	1	1	1	1
1	1	1	1	1	1	1	1	1	1	1
1	1	1	1	1	1	1	1	1	1	1

(d)

FIGURE 8.6

(a and b) Synthetic basins depicted as discrete functions, in which the higher the value, the higher is the elevation. In turn, these functions are treated as two different basins with two different altitude setups, (c) typical planar form of drainage network that summarizes the connectivity and shape of these two functions. It is extracted by following morphology-based transformations (e.g., Sagar et al. 2000). In Figure 8.6c, 1s are channel subsets and 0s represent non-channel regions, (d) planar form of the basin areas of the two synthetic basin functions, threshold value employed is <20 and <15 (respectively for two functions shown in (a and b)) and converted into 1s and 0s for other values.

(e)

20	0	0	0	0	0	0	0	0	0	20
0	19	0	0	0	0	0	0	0	19	0
0	0	18	0	0	0	0	0	18	0	0
0	0	0	17	0	0	0	17	0	0	0
0	0	0	0	16	0	16	0	0	0	0
0	0	0	0	0	15	0	0	0	0	0
0	0	0	0	16	0	16	0	0	0	0
0	0	0	17	0	0	0	17	0	0	0
0	0	18	0	0	0	0	0	18	0	0
0	19	0	0	0	0	0	0	0	19	0
20	0	0	0	0	0	0	0	0	0	20

(f)

15	0	0	0	0	0	0	0	0	0	15
0	14	0	0	0	0	0	0	0	14	0
0	0	13	0	0	0	0	0	13	0	0
0	0	0	12	0	0	0	12	0	0	0
0	0	0	0	11	0	11	0	0	0	0
0	0	0	0	0	10	0	0	0	0	0
0	0	0	0	11	0	11	0	0	0	0
0	0	0	12	0	0	0	11	0	0	0
0	0	13	0	0	0	0	0	13	0	0
0	14	0	0	0	0	0	0	0	14	0
15	0	0	0	0	0	0	0	0	0	15

(g)

20	20	20	20	20	20	20	20	20	20	20
20	20	20	20	20	20	20	20	20	20	20
20	20	20	20	20	20	20	20	20	20	20
20	20	20	20	20	20	20	20	20	20	20
20	20	20	20	20	20	20	20	20	20	20
20	20	20	20	20	20	20	20	20	20	20
20	20	20	20	20	20	20	20	20	20	20
20	20	20	20	20	20	20	20	20	20	20
20	20	20	20	20	20	20	20	20	20	20
20	20	20	20	20	20	20	20	20	20	20
20	20	20	20	20	20	20	20	20	20	20

(h)

15	15	15	15	15	15	15	15	15	15	15
15	15	15	15	15	15	15	15	15	15	15
15	15	15	15	15	15	15	15	15	15	15
15	15	15	15	15	15	15	15	15	15	15
15	15	15	15	15	15	15	15	15	15	15
15	15	15	15	15	15	15	15	15	15	15
15	15	15	15	15	15	15	15	15	15	15
15	15	15	15	15	15	15	15	15	15	15
15	15	15	15	15	15	15	15	15	15	15
15	15	15	15	15	15	15	15	15	15	15
15	15	15	15	15	15	15	15	15	15	15

FIGURE 8.6 (continued)
(e and f) The elevation values from basin functions shown in Figure 8.6a and b corresponding to the channel subsets shown in Figure 8.6c, and (g and h) convex hulls of two synthetic basin functions constructed according to a procedure due to Soille (1998). (From Lim, S.L. et al., *Comput. Geosci.*, 37, 1285, 2011.)

coordinates (x, y). These two pairs of synthetic basins possess similar spatial organization of networks, but it is obvious that they belong to two different altitudes. It is assumed that these two pairs of functions possess different spatial organizations of hillslopes as they belong to different elevation ranges. The planar views of basin and its corresponding channel

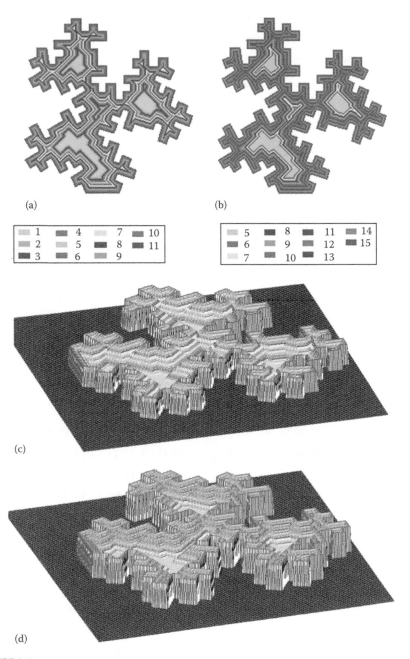

FIGURE 8.7
(a and b) Fractal basin functions with elevation ranges of 1–11 and 5–15 and (c and d) 3-D representation of fractal basin functions shown in (a and b). (From Lim, S.L. et al., *Comput. Geosci.*, 37, 1285, 2011.)

network are like sets that are decomposed from a function (f) where its landscape is represented as DEM (e.g., Figure 8.7c through d).

Methodology

Derivation of a Channel Network from a Basin Function

By implementing an approach explained in Chapter 4 in the synthetic basin, the channel networks could be extracted (e.g., Figure 8.6c). By performing suprema operation ($\vee$) between the network (subsets derived in the form of a planar set) and their corresponding points from the basin function, channel network function is obtained (e.g., Figure 8.6a and b). Such maxima form the network function (e.g., Figure 8.6e through f). Similar approach is followed to obtain network functions from synthetic DEMs, fractal basins, and real-world DEMs of both fluvial and tidal basins. The planar forms of networks extracted from fractal functions (Figure 8.7a and b) and real-world DEMs (Figure 8.8a and b) by the following approach from Sagar et al. (2000) are shown in Figures 8.7a and 8.8c and d.

Derivation of a Convex Hull of a Basin Function

Grayscale convex hulls of the basin functions are obtained according to the approach (Soille 1998, Lim and Sagar 2008a) demonstrated on the cloud surfaces in Chapter 4. The computed convex hulls for discrete basin functions f_1 and f_2 (Figure 8.6a and b) are shown, respectively, in Figure 8.6g and h. The reason for obtaining convex hulls with all 20s and 15s, respectively, in Figure 8.6g and h is that the highest elevation values of 20s and 15s surround the outlets, which are located at the centers of the discrete basin functions and of lower elevation values. Thus, the convex hull of these basin functions would look like "closed lids" with higher-altitude values than the centers of the basin functions. Here, rectangles with highest elevation of the basins form the convex hulls of basins f_1 and f_2. The convex hulls generated by similar approach for fractal basin functions (f_3 and f_4) shown in Figure 8.7a and b are shown in Figure 8.8e and f. It is obvious that the convex hulls (Figures 8.6g and h and 8.8e and f) obtained for synthetic basin functions (f_1 and f_2) and fractal basin functions (f_3 and f_4) are convex.

Area Estimations for Functions and Convexity Measure Computation

The areal extent of functions that are evidently elevation dependent are estimated as $A(f) = \sum_{(x,y)} f(x,y)$, $A(g) = \sum_{(x,y)} g(x,y)$, and $A(CH) = \sum_{(x,y)} CH[f(x,y)]$. These elevation-dependent measures are more

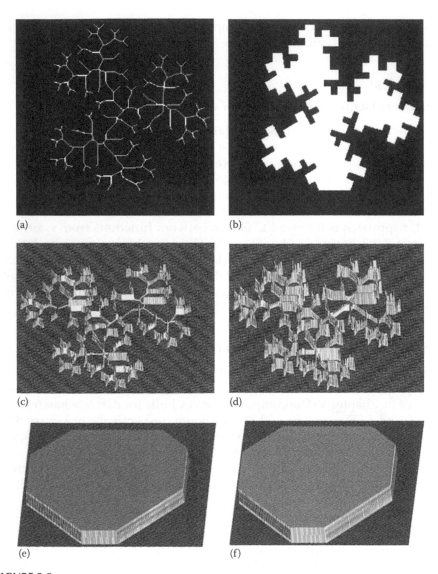

FIGURE 8.8
(a) Planar view of the network that represents channel network from both fractal basin functions, (b) planar view of the threshold basin region of both fractal basin functions, (c and d) 3-D representation of channel network functions of the two fractal basin functions, and (e and f) 3-D representation of convex hull functions of the two fractal basin functions. (From Lim, S.L. et al., *Comput. Geosci.*, 37, 1285, 2011.)

appropriate to estimate convexity measures that capture the basic spatial variability between the basins of different altitudes. This is unlike the Hortonian drainage density computation that does not consider the altitudes of the DEMs and thus shows similar result for *homotopic* DEMs with different heights, as shown in simple synthetic DEMs in Figure 8.6a through d, and the results are shown in Table 8.3. Equations 8.11 and 8.12 are employed to compute convexity measures (CM_f):

$$CM_f = \frac{A(g)}{A(f)} \qquad (8.11)$$

$$CM_f = \frac{A(f)}{A(CH)} \qquad (8.12)$$

Hereafter, we denote Equation 8.11 as method-1, and Equation 8.12 method-2.

Demonstrations and Comparisons

Although both basins possess different elevation ranges, the conventional Hortonian drainage density estimated for both basins f_1 and f_2—with similar planar area, which is 121, and with similar length of the planar networks, which is 21—is $(21/121) = 0.1736$. This clearly shows that Horton drainage density, where $A(g_1) = A(g_2)$ and $A(f_1) = A(f_2)$ on planar view respectively denote the length of networks and area of basins on planar view, is the same for both functions f_1 and f_2. Here, we show synthetic DEMs (Figure 8.6a and b) in which the spatially distributed numerical values represent topographic elevations: the higher the numerical values, the higher the elevation, and *vice versa*. The DEM is represented as a grayscale function (e.g., Figure 8.7a and b), where each gray value (intensity I) at respective spatial coordinates (x, y) denotes elevation value. Area is nothing but the sum of those elevation values across all the spatial positions. Such areas for the synthetic DEMs shown in Figure 8.6a and b are $A(f_1) = 2255$ and $A(f_2) = 1650$ in pixel units, respectively. Similarly, the lengths of the channel network functions $A(g_1)$ and $A(g_2)$ are 375 and 270, respectively. The areas of convex hulls of these two functions, $A[CH(f_1)]$ and $A[CH(f_2)]$, are obtained, respectively, as 2420 and 1815. The convexity measures computed via two proposed methods (Equations 8.11 and 8.12) clearly capture the spatial variability (see Table 8.3). Consider the two synthetic basins f_1 and f_2, the drainage densities of which according to Equations 8.11 and 8.12 include 0.9318 and 0.9091. The ranges of elevation values for functions f_3 and f_4 include 1–11 and 5–15. These ranges are used to show that these two *homotopically* similar synthetic fractal basin functions with similar geometric organizations possess different altitude ranges. These two functions are also shown in 3-D representation (Figure 8.7c and d). The lengths of planar network (Figure 8.7a), and also areas of planar view of these two functions (Figure 8.8b), are found to be the same. As a result, the Hortonian drainage densities computed for f_3 and f_4 are the same (0.0904) although they exhibit different altitude ranges. In contrast, the lengths of network functions (Figure 8.8c and d) and areas of basin functions and their corresponding convex hull functions (Figure 8.8e and f) show distinction in

TABLE 8.3

Comparison between Drainage Density and Convexity Measure of Synthetic and Fractal DEMs

Basin	Areas of Planar Forms		Areas of Functions			Convexity Measure		
	Basin	Network	Basin	Network	Convex Hull	Horton-DD	Method-1	Method-2
f_1	121	21	2,255	375	2420	0.1736	0.1663	0.9318
f_2	121	21	1,650	270	1815	0.1736	0.1636	0.9091
f_3	20,334	1838	152,844	12,132	396,814	0.0904	0.0794	0.3852
f_4	20,334	1838	234,180	19,484	541,110	0.0904	0.0832	0.4328

Source: Lim, S.L. et al., *Comput. Geosci.*, 37, 1285, 2011.

the convexity measures of these two fractal basin functions. As shown in Table 8.3, the convexity measures, respectively, for fractal basin functions f_3 and f_4 are 0.0794 and 0.0832 according to method-1 and are 0.3852 and 0.4328 according to method-2. As fractal basin function f_3 has lower altitude range than f_4, its convexity measures computed through methods-1 and 2 are lower than that of f_4. In general, it is seen from Tables 8.3 and 8.4 that values from method-1 are always smaller or equal to traditional drainage density, the equal case will occur only for a basin that is flat with slightly incised valleys; the smaller the values from method-1 in comparison with the traditional drainage density, the more incised the valleys. Hence, the convexity measures estimated according to the two proposed methods clearly exhibit spatial variability of the basins, especially those *homotopically* similar basins with different altitude ranges. The Hortonian drainage densities computed for 14 Cameron and Petaling basins, respectively, range from 0.0539 to 0.0673 and from 0.0226 to 0.0245 (Table 8.4). Similar trend is also observed from the convexity measures obtained from methods-1 and -2. These convexity measures yield the ranges of 0.051–0.065 and 0.0160–0.0207, and 0.5748–0.7047 and 0.4773–0.6191, for Cameron basins and Petaling basins, respectively. These results match the trend observed from convexity measures computed via methods-1 and -2 for the cases of fractal basin functions, that is, convexity measure varies with the altitude ranges of the basins. The higher the altitude range of the basin, the greater the convexity measure, and *vice versa*. A comparison is shown (Table 8.4) between these convexity measures and complexity measures (roughness values), generated by following the method explained in the "Grayscale Granulometric Analysis" section. Significantly, a clear distinction exists in the complexity measures between the Cameron basins and Petaling basins as Cameron basins are highland and mountainous region while Petaling basins are relatively low and flat terrains. As such, the roughness values of Cameron basins are generally greater than that of Petaling basins. This statement is justified by the result shown in Table 8.4, where the ranges of normalized complexity measures for Cameron basins are 0.8963–0.9362, and 0.7413–0.8516 for Petaling basins. Besides, the fractal dimensions of the basin-wise channel networks (Figure 8.8c and d) of Cameron Highlands and Petaling regions also indicate clear distinction between these two regions (see last column of Table 8.4). Figure 8.9a and b shows a better view on the relationships among these various parameters. From these graphs, we can infer that Cameron basins, which have higher-altitude basins than low-lying Petaling basins, show higher drainage densities and convexity measures (whether Horton, method-1, or method-2), higher normalized complexity measures, and also higher fractal dimension values than that of Petaling basins. Besides, unlike the case of synthetic basin and fractal basin functions, the convexity measures obtained from methods-1 and -2 for Cameron and Petaling basins correspond well with Horton drainage density. Furthermore, it is interesting to note from Figure 8.9b that the convexity measure from method-1 follows closely with the Horton drainage density.

TABLE 8.4

Comparisons among Drainage Density, Convexity Measure, Complexity Measure, and Fractal Dimension of Realistic DEMs

Basin	Areas of Planar Forms		Areas of Functions			Convexity Measures			Normalized Complexity Measure	Fractal Dimensions
	Basin	Network	Basin	Network	Convex Hull	Horton-DD	Method-1	Method-2		
1	71,045	3826	60,291,000	3,072,600	85,558,000	0.0539	0.0510	0.7047	0.9130	1.5141
2	77,780	4612	73,903,000	4,204,400	125,490,000	0.0593	0.0569	0.5889	0.9362	1.5506
3	84,699	4775	83,499,000	4,452,000	122,740,000	0.0564	0.0533	0.6803	0.8963	1.5814
4	55,912	3227	50,863,000	2,774,300	80,163,000	0.0577	0.0545	0.6345	0.9165	1.4692
5	41,253	2583	43,913,000	2,662,800	76,397,000	0.0626	0.0606	0.5748	0.9255	1.4519
6	31,226	2101	30,471,000	1,981,400	45,184,000	0.0673	0.0650	0.6744	0.9291	1.4776
7	19,780	1156	14,265,000	772,550	20,828,000	0.0584	0.0542	0.6849	0.9255	1.3192
8	66,824	1629	8,124,200	167,870	14,854,000	0.0244	0.0207	0.5469	0.7413	1.3140
9	25,164	588	2,605,000	46,830	5,458,100	0.0234	0.0180	0.4773	0.7788	1.2398
10	31,779	767	3,769,600	75,553	6,088,900	0.0241	0.0200	0.6191	0.8038	1.2445
11	35,805	808	3,703,100	65,298	7,216,900	0.0226	0.0176	0.5131	0.8134	1.1817
12	36,953	884	3,798,300	62,811	7,609,700	0.0239	0.0165	0.4991	0.8516	1.2946
13	40,845	933	3,189,600	50,907	6,578,400	0.0228	0.0160	0.4849	0.7921	1.1706
14	23,497	576	1,786,700	31,969	3,268,300	0.0245	0.0179	0.5467	0.7951	1.1721

Source: Lim, S.L. et al., *Comput. Geosci.*, 37, 1285, 2011.

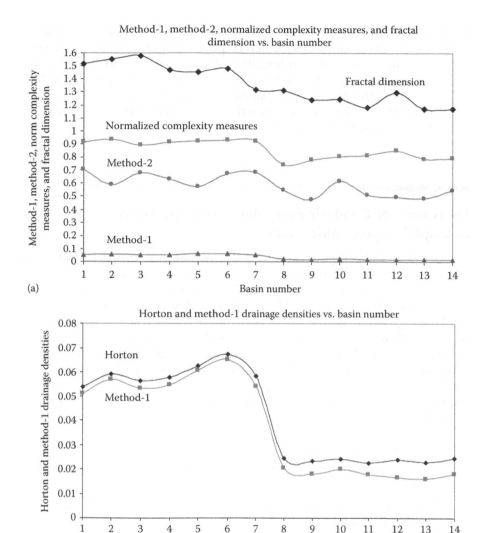

FIGURE 8.9
(a) Convexity measures computed from method-1, method-2, and normalized complexity measures and fractal dimension (via box counting method) for all 14 basins, and (b) channel densities from Horton–Strahler method-1. (From Lim, S.L. et al., *Comput. Geosci.*, 37, 1285, 2011.)

Conclusion

In this section, we show the changes in convexity measures due to different elevation ranges of basins with similar geometrical arrangement. Function-based convexity measures that capture the spatial variability between basins with different altitude are appropriate quantitative geomorphometric parameters. The convexity measures that are computed through these function-based approaches are clearly elevation dependent. An interesting open problem is to validate the relationship between these convexity measures of realistic basins that possess different physiographic setups. These convexity measures—related with fractals and granulometric analysis that are derived via geometry-based techniques—provide new insights to explore further links with various other established morphometric measures. The study to draw scale-invariant characteristics from these convexity measures is a potential scope for this investigation.

Derivation of Geodesic Flow Fields and Spectrum in Digital Topographic Basin

DEMs and digital bathymetric models (DBMs) are two terrestrial functions that are represented as raster images respectively representing elevations above mean sea level and depth below mean sea level. Basins within the biogeographic boundaries could be demarcated, and the flow fields within the basins provide geometrically significant details. Areas between the successive flow fields within a basin consisting of various TERs would be of use to construct geodesic spectrum, a 1-D geometric support. Such a geodesic spectrum that could be compared with as in width function that provides rich clues about the complexity of the processes involved in shaping the basins. Computation of geodesic spectrum involves the following sequential steps:

1. $f(x, y)$ depicting a basin
2. Decomposition of basin into channelized and non-channelized regions
3. Indexing of successively decomposed regions and sorting them into markers and masks
4. Extraction of geodesic flow fields
5. Construction of geodesic spectrum

Morphodynamics of topographies could be efficiently studied with the advent of remote sensing technologies that pave approaches to provide DEMs and DBMs at multiple spatial and temporal scales. The features that could be mapped with precision include topographies of basins, bays, lakes, estuaries that are conspicuous at land–sea confluences. Quantitative characterizations of such mapped topographies in spatial and temporal scales are important from the point of understanding the involved topographical processes. This aspect has been addressed in several different ways. However, a new descriptor, a geodesic spectrum, with geometric significance has been proposed in this section.

Various techniques have been proposed to characterize basin-wise digital topographies that include both surficial and bathymetric (Stone and Dugundji 1965, Fatale et al. 1994, Turcotte 1997, Tay et al. 2005, 2007). Other important descriptors that could be derived from DEMs are mainly based on morphometry, hypsometry, width function, and cumulative-area distribution analyses (Perera 1997). In the context of understanding the mathematical properties of random topographic functions such as DEMs and DBMs, various other techniques that have been proposed recently include geomorphologic width function, convexity measures of travel time networks, granulometries, and morphometry of nonnetwork spaces (Marani et al. 1994, Rodriguez-Iturbe and Rinaldo 1997, Veneziano et al. 2000, Chockalingam and Sagar 2005, Tay et al. 2006). The dynamical processes and type of agents controlling dynamical processes in fluvial systems and tidal systems are distinctly different. The basins within tidal environment are with less topographic undulations compared to fluvial systems. There are numerous techniques to characterize the basins belonging to fluvial systems, but many of these techniques could not be used to characterize the basins of tidal systems. Because of the fact that the topographic undulations are relatively low in floodplains and tidal regions, evidently new basin descriptors are required. If the basins are relatively with flat surface, basic flow fields could be simulated by employing shallow water equations (Fagherazzi et al. 2003, Blondeaux and Vittori 2005a,b). It is important to generate flow fields in shallow water regimes (e.g., floodplains, tidal environments) and coastal environments (e.g., bays, estuaries, lagoons) to further quantify the morphologic organization of topographic zones. In view of the fact that the existing techniques cannot be generalized for both regions that possess flat and undulated topography either above the mean sea level or below the mean sea level, a geodesic spectrum is proposed to characterize the topography within a basin. The main objective of this section includes the construction of geodesic spectrum that acts as a basin descriptor for which the geodesic flow fields need to be simulated.

Through geodesic morphological transformations, elevation-dependent flow fields have been simulated. Further, to quantitatively characterize the basin function, probabilities of areas embedded between the successive flow fields have been estimated.

Model and motivation: Flood propagation that can be simulated within a basin through dilation process is the principle involved in simulating geodesic flow fields. Floodwaters flow from lower-elevation regions into higher-elevation regions. Geodesic morphological transformations could be employed to simulate such flood propagation by considering the digital topographic data (e.g., DEMs, DBMs, Figure 3.15). Basin topographies can be characterized via geodesic flow fields. Within a basin-DEM, the velocity of floodwater front propagating from lower- to higher-elevation regions would vary with the TERs. In any geomorphologic basin, there would be broadly two regions: (1) channelized region and (2) non-channelized regions. Channelized regions are the first-level unique topographic depressions, for which the floodwater propagation speed is more than that of non-channelized regions. This is due to the fact that the non-channelized regions are with higher gradient. In view of these variations in the directions of floodwater, and also in the velocity of floodwater front between the channelized and non-channelized regions of a basin, it is essential to decompose the basin regions into TERs to simulate geodesic flow fields. The geometric organization of geodesic flow fields would be determined by the geometry of TERs and the rules followed to simulate floodwater fronts. To model the waterfront propagation within a basin consisting of regions with varied viscous characteristics, geodesic balls (structuring elements) of different radii synchronizing the changes in surficial characteristics are required. Sequential steps involved in the process of computing geodesic spectra via probabilities of geodesic flow fields are shown in Figure 8.10. These steps have been explained in the following.

Decomposition of Basin into TERs

A function f is treated as a basin (e.g., Figure 3.14a through c) represented by a nonnegative 2-D sequence $f(x, y)$, which assumed $I + 1$ possible intensity values: $i = 0, 1, 2, \ldots, I$. $I = 255$ for 8 bit/pixel digital topographic data. The f (basin) is discrete, defined on a (rectangular) subset of the discrete plane Z^2. The lower the intensity value, the lower is the topographic elevation, and *vice versa*. This f, in other words digital topographic data, consists of various bounded sets that include inlet point, channelized zones, and non-channelized zones. Threshold decomposed elevation regions could be obtained from f at all possible intensity levels $0 \leq t \leq T$, by thresholding: $f^t = \begin{cases} 1, f(x,y) \geq t \\ 0, f(x,y) < t \end{cases}$. Such decomposed elevation regions, threshold sets, include both channelized and non-channelized regions that take values 0 and 1 (the pixels with 1 and 0 represented with white and black shades denote, respectively, sets and their compliments). The sets (f_i) form a sequence of sets that characterize f entirely and are such that for any threshold elevations i and $i + 1$, $(i + 1) > (i) \Rightarrow (f_{i+1}) \subseteq (f_i)$, for i ranging between 1 and I—as illustrated in Figure 2.20. From the point of basin's physiography, the elevations of inlet point(s) are lesser than that of the mean elevation of successively decomposed channels and basin's inland. Marker and mask set(s) can be obtained by simple logical difference between the successive threshold decomposed sets, which is according to $[f_i - f_{i+1} = X_i]$, where $i = 1, 2, \ldots, I$ and $i = 1, 2, \ldots, I$. Each set obtained via logical difference is denoted by X_i. This process of subtraction of successive TERs is followed to decompose a synthetic function (basin) (Figure 3.14c) consisting of nine zones into sets with designated set orders ranging from

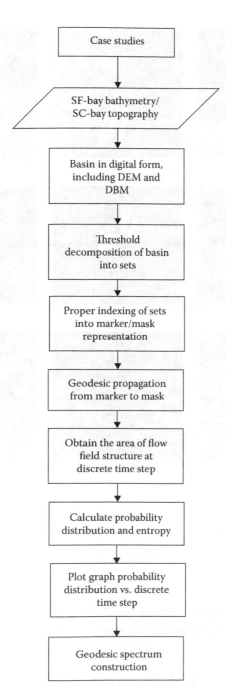

FIGURE 8.10
Flowchart depicting various steps involved in modeling and simulation. (From Lim, S.L. and Sagar, B.S.D., *Discrete Dyn. Nat. Soc.*, 2008, 26, 2008b.)

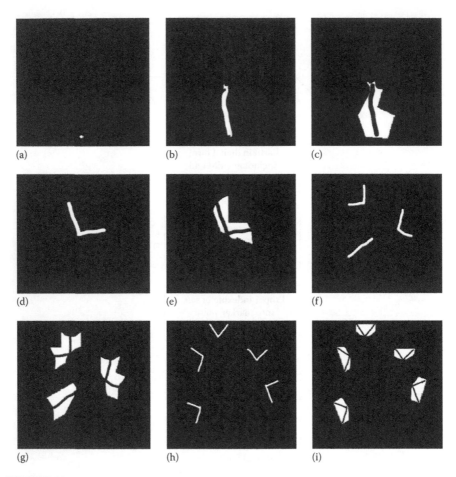

FIGURE 8.11
Synthetic tidal basin shown in Figure 3.14c—which consists of channelized and non-channelized regions—after decomposing into sets. (a–i) Sets representing channelized and non-channelized regions of which the mean elevations increase from X_1 to X_9. The sets with even- and odd-numbered indexes respectively represent the zones occupied by channelized and non-channelized regions. (From Lim, S.L. and Sagar, B.S.D., *Discrete Dyn. Nat. Soc.*, 2008, 26, 2008b.)

1 to 9 (Figure 8.11). The union of these sets (f_i) and (X_i) satisfies the inclusion relationship (Maragos and Ziff 1990) as shown in Equation 8.13:

$$f = \bigcup_{i=1}^{I} f_i \quad \text{and} \quad f = \bigcup_{i=1}^{I} X_i \tag{8.13}$$

For clarity, each set obtained by $[f_i - f_{i+1}]$ is denoted with X_i, X_{i+1}, X_{i+2}, ..., X_I, with i ranging from 1 to I (see Figure 8.11). Set X_1 denotes the inlet point.

The set with immediate higher index acts as mask set to the marker set with preceding index. For better understanding of threshold decomposition and isolation of sets, see Figure 2.20.

Geodesic Propagation: Methods

Geodesic morphological dilation (Serra 1982) is adopted. By considering basin as a mask, and inlet point—through which water flows into basin during the high tide/flood—as a marker from which the flow propagates as the tide-level increases, geodesic morphological dilation is implemented to simulate geodesic propagation of marker in a mask. To implement geodesic propagation with uniform velocity within a mask (set X_{i+1}) with certain boundary conditions from the source point (set X_i), (1) a superficially simple morphological dilation has been performed iteratively on X_i (marker) with a structuring element (B) of primitive size 5×5, which is symmetric about the origin, (2) a logical intersection is employed between the dilated marker X_i and mask X_{i+1} sets, and (3) a logical union of flow fields is applied at respective discrete times.

Logical intersection between $(X_i \oplus nB)$ and (X_{i+1}) as shown in Equation 8.14 is employed to retrieve the flow fields at respective discrete times:

$$(X_i \oplus nB) \cap (X_{i+1}) \tag{8.14}$$

where
  nB is a symmetric probing rule with size n
  X_i, X_{i+1}, respectively, denote marker set and mask set (e.g., Figure 8.12a)

The characteristic information of the structuring element B can be related with the discharges per unit width in x- and y-directions. Size of primitive structuring element and number of dilations performed on X_i to simulate

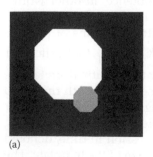 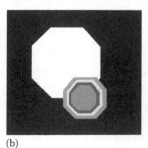 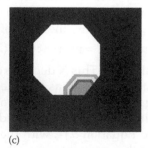

(a) (b) (c)

FIGURE 8.12
(a) Marker set X_i (in gray) and mask set X_{i+1} (in white), (b) after iterative dilations up to fourth level superposed on the mask set X_{i+1}, and (c) the dilated marker set for four iterations intersected with mask set X_{i+1}. (From Lim, S.L. and Sagar, B.S.D., *Discrete Dyn. Nat. Soc.*, 2008, 26, 2008b.)

flow fields between the successive order-designated sets is related to the velocity of geodesic flow field. If the intersection of mask set with which the flow field that should propagate from the marker set to mask set produces an empty set, we should not consider such a mask set. Then we have to proceed further to consider mask set that is decomposed with the next higher threshold value. X_i and X_{i+1} are taken as marker and mask sets, respectively, if $(X_i \oplus nB) \cap (X_{i+1}) \neq \emptyset$. If $(X_i \oplus nB) \cap (X_{i+1}) = \emptyset$, then X_i and X_{i+1} should not be treated as marker and mask sets. Instead one needs to check whether the intersection, $(X_i \oplus nB) \cap (X_{i+2})$, yields nonempty set to consider X_i and X_{i+2}, respectively, as marker and mask sets. If there are two marker sets with similar threshold value, then one of the marker sets that is fully surrounded by the mask set needs to be ignored.

There will be no distinction between channelized and non-channelized regions in terms of elevations/depths in a flat basin (e.g., Figure 3.14a and d). In such a flat basin, the total flow field can be defined in morphological terms as the intersecting portion of gradient of successively dilated marker set with the mask set. In Equation 8.15, nth time-step gradient $\delta^n(X_i)$ is defined as the boundary between the successively dilated marker sets:

$$\partial^n(X_i) = [((X_i \oplus (n+1)B) \setminus (X_i \oplus nB))] \tag{8.15}$$

where
  $\partial^n(X_i)$ is the nth time-step gradient
  nB, $(n+1)B$, respectively, denote forcing in terms of structuring element with different sizes

The primitive structuring template of size 5×5 is considered as a unit to simulate flood/tidal forcing. The tidal/flood flows into channel from the inlet point as time progresses. This progression in the flow is simulated by B by iterative dilations by means of B until idempotence, in other words, until the intersection of gradient with mask set becomes an empty set. Alternatively, the state of convergence can be written in terms of positive integer $N = \min\left\{ n : \left[\partial^{n-1}(X_i) \cap X_{i+1} \right] \neq \emptyset; \left[\partial^n(X_i) \cap X_{i+1} \right] = \emptyset \right\}$, for the flow for all $n > N$. This N depends on X_i, X_{i+1}, and B. The flow field at nth discrete time step is defined as the line that is obtained by intersecting the nth time-step gradient $\partial^n(X_i)$ with the mask set (X_{i+1}). The propagation of flow field after nth time step is simulated by intersecting the gradient computed according to Equation 8.14, with the mask set (X_{i+1}). The progression in time, denoted with increments of $n = 0, 1, 2, \ldots, N$, is related to the size of the template. The larger the cumulative effect of tidal/flood forcing at successive discrete time steps, the larger the size of the template. Then the total flow field, which refers to any water frontline propagating toward immediate spatially distributed elevation regions, in the simplest case 1 (Figure 3.14a and d) could

be achieved by Equation 8.16. To visualize the flow fields within the channelized and non-channelized zones (or sets), a logical union operation is
considered in Equation 8.16:

$$TB_{flow} = \bigcup_{\substack{n \geq 0 \\ i \geq 1}}^{\substack{I \\ N}} \left\{ \left[\delta^n(X_i) \right] \cap X_{i+1} \right\} \tag{8.16}$$

where $i = 1, 2, \ldots, I$; n = discrete time (with time effect of cumulative tidal/
flood forcing increases), and also n denotes the size of the structuring element and the discrete time, and the limit of N is the iteration step at which
the convergence is reached. The increment in n defines the increase in
the size of structuring element, in other words the cumulative tidal/flood
forcing. The gradients between the successively dilated sets are intersected with mask set X_{i+1}. Once this process reaches convergence, the flow
propagation simulation proceeds further from the marker set X_{i+1} into the
mask set X_{i+2}.

Simulations of Geodesic Flow Fields

In the floodplains and coastal and tidal environments, the flow fields'
structure will be greatly influenced by fluctuating tidal forcing and river
water inflows. In such environments, it is important to describe the spatiotemporal structure of the flow fields to further study the morphodynamic
problems. By applying a proposed geodesic dilation-based algorithm on
an input of digital topographic function available in raster format, the flow
field simulations are carried out. The flow fields propagate from inlet point
(initial marker) into channels along the medial axis direction with greater
velocity along the medial axis line than along the channel walls. In order
to justify this hydrodynamically viable assumption, a template of octagon
in shape, symmetric about origin and of primitive size 5 × 5 (Figure 8.13),
has been considered to simulate flow field propagation. Tide with exceed-
ing velocity (forcing) inundates the tidal basin's inland. By tuning characteristics of structuring element, the velocity variations can be imposed
while changing the (1) medium from channelized set to non-channelized
set, (2) tidal forcing, (3) elevation, (4) spatial positions of source(s) of inlet(s),
and (5) direction of flow. For instance, due to these factors, the geodesic
ball that would be used to model propagation within the non-channelized
regions would be with relatively larger radius compared to that of required
radius of geodesic ball to model the water propagation in the channelized region. The propagation is necessarily isotropic as B (e.g., Figure 8.13)
employed is disk like. However, a directional propagation can be obtained
when B is a unit segment in the direction of propagation. One can alternate

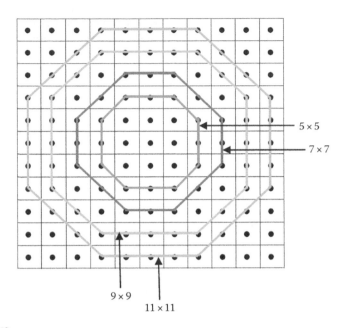

FIGURE 8.13
Octagonal symmetric structuring elements of various primitive sizes ranging from 5 × 5 to 11 × 11. These primitive sizes can be considered as *B* in the employed equations to simulate flow fields with various velocities.

unit segment and unit disk if a mixture is required, but at the idempotent limit, the result will be that of the disk propagation.

Case 1: Unidirectional propagation of flow fields can be formulated in tidal basins in which the tidal channels and inlands are of same elevation. The flow of propagating water synchronizing the tidal forcing is like a sheet of water flowing in unidirectional way on a flat surface from the inlet source. Equations 8.14 and 8.15 can be adapted to simulate unidirectional flow fields within a tidal basin in which there is one inlet point and no channelized sets. The gradients of such flow fields at discrete intervals are shown in Figure 8.14a, where the circular path and inlet set act like respectively the boundary conditions and flow propagation source. Flow field complexity depends not only on basin shape and general bathymetry, but also on the spatial organization of tidal channels. Tidal flows in tidal channels and non-channelized regions of tidal basins are simulated by following geodesic propagation methods. The latter two cases are modeled based on the assumptions that the tidal channels are the first-level zones that get affected by fluctuating tides and followed by nontidal channelized regions that are relatively with less depth. Flow field propagation would be in the channelized regions of lower threshold decomposed region (X_i) first followed by that in the non-channelized region of X_{i+1}. Eventually, in the channelized region of say X_1, and then in the non-channelized region

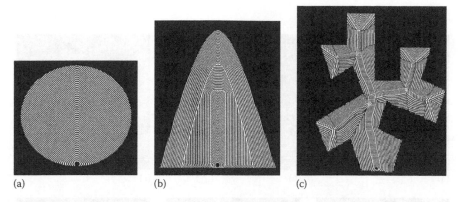

(a)   (b)   (c)

FIGURE 8.14
(a) Flow fields with isotropic propagation, (b) isotropic flow fields, and orthogonality between the flow fields of channelized and non-channelized zones is obvious, and (c) flow fields within the tidal basin. (From Lim, S.L. and Sagar, B.S.D., *Discrete Dyn. Nat. Soc.*, 2008, 26, 2008b.)

of X_2, and so on. The flow propagation pattern in tidal basin is categorized as (1) propagation in channelized region and (2) propagation in non-channelized zones.

Case 2: In this case, the propagation flow fields in channelized sets are orthogonal to that of corresponding non-channelized sets. The geometric and spatial organizations of flow fields within channelized regions are different from that of their corresponding non-channelized regions. Hence, the equations governing the flow fields are indexed-set dependent. For flow field simulations, Equations 8.14 through 8.16 are considered for case 2 in which marker and mask sets are recursively changed in the fashion of ith and $i + 1$th sets acting respectively as marker and mask sets (Figure 8.14b).

Case 3: The channels, the first-level zones that are affected by fluctuating stream flow discharges and/or tides, and non-channelized regions surrounding the channels are relatively with different depths/heights. Flow fields' directions and spatial complexity depend not only on basin shape and general bathymetry, but also on the spatial organizations of tidal channels and inlands. The flow fields in the basin's inland propagate in the direction perpendicular to that of channels. For the third case of flow field simulations, we consider set with index $i = 1$ as a marker and is allowed to geodesically propagate (e.g., Figure 8.12) within the mask set indexed with $2i$. For simplified representation, the threshold decomposed sets thus obtained (Figure 8.11) are denoted as $X_i = X_1, X_2, X_3, ..., X_I$ with i ranging from 1 to I. This notation is done to explicitly write the equations in such a way that the tidal channels and their corresponding tidal inlands can be respectively represented with even and odd ith values. In case 3 (Figures 3.14c and f and 8.11), certain sets are order-designated as sets with indexes $2i$ (for i ranging from 1 to I), denoting those sets that occupy tidal channels. The diameter

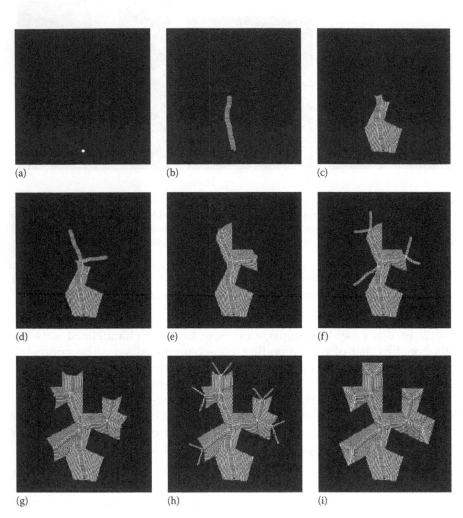

FIGURE 8.15
Result of simulation at different time instances. (a) Inlet point or set X_1 from which the water flows into tidal basin, (b) water propagation from X_1 (marker set) into X_2, (c) flow fields propagating from marker set X_2 into set X_3—non-channelized set or the influence zone of set X_2—that acts as a mask set, (d) set X_4 the mask set that gets flooded due to the water flowing from the marker set X_1 after sets X_2 and X_3 are completely flooded, (e) set X_5—non-channelized (influence) zone of set X_4 gets flooded from the marker set X_4, (f) set X_6 that acts as mask set to allow the water flows from the extreme tips of set X_4, (g) set X_7 the influence zone of the channelized set X_6—here the mask set X_7 would be progressively flooded from the water flowing from the marker set X_6, (h) channelized mask set X_8 in which the water flows from the extremities of set X_6, and (i) mask set X_9—influence zone of set X_8—gets progressively flooded due to water flowing from set X_8 that acts as a marker set to fill the water in its corresponding mask set X_9. (From Lim, S.L. and Sagar, B.S.D., *Discrete Dyn. Nat. Soc.*, 2008, 26, 2008b.)

of tidal channel with index $2i$ is larger than that of $2(i + 1)$, and so on. Other sets that occupy non-channelized zones are order-designated as sets with indexes $(2i + 1)$. In turn, the relationship between order-designated channelized and non-channelized sets is in such a way that set with index $2i$ is surrounded by set with index $(2i + 1)$. Sets indexed with even and odd numbers respectively represent channelized and non-channelized regions of tidal subbasins. This distinction in indexing sets denoting channelized and non-channelized regions represented with even and odd numbers is shown to simulate flow propagation in channelized and non-channelized sets subsequently. This separation is physically acceptable as the directions of flow propagation are perpendicular/orthogonal to each other. With this reordering of simple indexing, a set of equations (8.17) to simulate flow fields, alternatively in channelized and non-channelized regions by incrementing the set indexes, are proposed:

When $i = 1$:

$$C_{\text{flow}} = \bigcup_{n=0}^{K}\left\{\left[\partial^n(X_1)\right] \cap X_{2i}\right\} \quad \text{and} \quad NC_{\text{flow}} = \bigcup_{n=0}^{N}\left\{\left[\partial^n(X_{2i})\right] \cap X_{2i+1}\right\}$$

When $i = 2$:

$$C_{\text{flow}} = \bigcup_{n=K+1}^{P}\left\{\left[\partial^n(X_1)\right] \cap X_{2i}\right\} \quad \text{and} \quad NC_{\text{flow}} = \bigcup_{n=0}^{N}\left\{\left[\partial^n(X_{2i})\right] \cap X_{2i+1}\right\}$$

When $i = 3$:

$$C_{\text{flow}} = \bigcup_{n=P+1}^{Q}\left\{\left[\partial^n(X_1)\right] \cap X_{2i}\right\} \quad \text{and} \quad NC_{\text{flow}} = \bigcup_{n=0}^{N}\left\{\left[\partial^n(X_{2i})\right] \cap X_{2i+1}\right\} \quad (8.17)$$

When $i = 4$:

$$C_{\text{flow}} = \bigcup_{n=Q+1}^{N}\left\{\left[\partial^n(X_1)\right] \cap X_{2i}\right\} \quad \text{and} \quad NC_{\text{flow}} = \bigcup_{n=0}^{N}\left\{\left[\partial^n(X_{2i})\right] \cap X_{2i+1}\right\}$$

where in non-channelized flow $0 \leq n \leq N$ and in channelized flow, $n = 0 \leq k \leq p \leq q \leq \cdots \leq N$. The positive integers N, K, P, and Q are dependent upon mask and marker sets' size and shape characteristics. N varies from one cycle to another cycle with changing i, and $0 \ll K \ll (K + n) \ll (P + n) \ll (Q + n) \ll \cdots \ll N$.

By following Equation 8.17, the basic flow fields are simulated for this synthetic tidal basin (Figures 3.14c and f, and Figure 8.15). The following eight sequential steps have been followed to systematically generate time sequential waterfront propagation:

Step 1	Inlet point or set X_1 from which the water flows into basin
Step 2	Mask set (X_2) that would be flooded from the water flowing from X_1 (marker set)
Step 3	Set X_3—non-channelized set or the influence zone of set X_2—that acts as a mask set that gets flooded due to the water propagates from marker set X_2
Step 4	Set X_4—the mask set that gets flooded due to the water flows from the marker set X_1 that already fills the set X_2 and X_3 completely
Step 5	Set X_5—non-channelized (influence) zone of set X_4 gets flooded from the marker set X_4
Step 6	Set X_6 that acts as mask set to allow the water flows from the extreme tips of set X_4
Step 7	Set X_7—the influence zone of the channelized set X_6—here the mask set X_7 would be progressively flooded from the water flowing from the marker set X_6
Step 8	Channelized mask set X_8 in which the water flows from the extremities of set X_6, and mask set X_9—Influence zone of set X_8—gets progressively flooded due to water flowing from set X_8 that acts as a marker set to fill the water in its corresponding mask set X_9

This is a recursive process—until the process reaches convergence—in which the sets with odd- and even-numbered indexes respectively represent the zones occupied by channelized and non-channelized regions. The direction-specific flow fields are obvious, which could be seen from Figure 8.14b and c. The characterization of these direction-specific flow fields separately in time-sequential mode would offer potentially innovative insights further to understand (1) the relationship between the flow fields that are orthogonal to each other and (2) the relationship between the induced tidal forcing and spatial organization of flow fields. With increasing degree of tidal forcing, it is intuitively true that the evolution of flow fields and their spatiotemporal organization can be better linked with time-dependent morphological processes that occur due to time-dependent endogenic processes. In general, the velocity of flow fields in non-channelized zones is usually lesser than that of channelized zones, as these two zones act like two different media with variations in (1) surficial roughness characteristics, (2) topographic effects, and (3) depths. Hence, defining the size and the other characteristics of the structuring element synchronizing the velocity characteristics of flow fields is an important task that needs to be addressed.

Central San Francisco Bay: To generate flow fields using geodesic (marker-mask) propagation approach, a part of Central San Francisco (SF) Bay bathymetry (Figure 3.15a and b) has been considered. Permissions to use the images (Figure 3.15a and b) have been obtained from USGS team. A part essentially at the mouth of the bay from which the tidal flow fields enter into bay is considered (blocked region in Figure 3.15b). This part has various depth zones ranging from the depths of −115 to −14 m. This bathymetric map is available in gray-scale form, with darker zones representing more depth than brighter zones that are shallower, and is converted broadly into seven regrouped zones by the following thresholding technique. The gray-level ranges with the depth ranges include 0–33 = (−115) to (−106 m); 34–59 = (−105) to (−91 m); 60–100 = (−90) to (−68 m); 101–150 = (−67) to (−46 m); 151–201 = (−45) to (−27 m); 202–233 = (−26) to (−15 m); and 234–255 = (−14) to (0 m). By choosing threshold values from the upper limits of these ranges, the considered bathymetric image is decomposed into threshold bathymetric zones as X_1, X_2, ..., X_7 as seven threshold grayscale values have been chosen. Considering X_i as marker set, and X_{i+1} as mask set, flow fields are simulated in each of the threshold bathymetric zones according to the algorithm detailed in the "Geodesic Propagation: Methods" and "Simulations of Geodesic Flow Fields" sections (Figure 8.16b).

Coastal Santa Cruz region: A minor basin (Figure 3.15d and e) of which the discharges are flowing into sea and consists of elevation ranges between 1 and 263 m is considered. By choosing certain threshold ranges, this basin (Figure 3.15e) is decomposed into sets (Table 8.5). Flow fields generated by following the framework implemented on previous cases are shown in Figure 8.16c. In the dilation process, an octagonal structuring template is opted to simulate flow fields in both San Francisco Bay and Santa Cruz DEM cases (Figures 8.16b and c). Table 8.5 provides basic details, such as the types of basins, the elevation ranges with

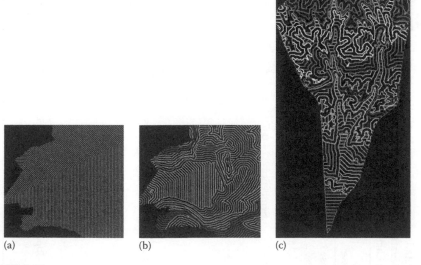

FIGURE 8.16
Flow fields simulated by considering only water surface and also the bathymetry. (a) Flow field simulated on SF Bay without considering bathymetry, (b) flow field simulated on SF Bay bathymetry by using octagon, and (c) flow field simulated on Santa Cruz DEM by using octagon. (From Lim, S.L. and Sagar, B.S.D., *Discrete Dyn. Nat. Soc.*, 2008, 26, 2008b.)

TABLE 8.5

Details of Synthetic and Realistic Digital Topographies Considered with Their Gray Levels and Corresponding Elevation/or Depth Ranges and Entropy Values Estimated for Each Threshold Elevation/Depth Decomposed Set of Each Digital Topographic Basin

Case	Type	Dyn Range	No. Dec	Gray Value Range	Elevation Range (m)	Used SE	No. Flow Field	Entropy
Case 1	Synthetic	0–1	1	0–1	0–1	Rhombus	113	2.014109
Case 2	Synthetic	0–3	3	0–1	0–1	Rhombus	97	0.335195
				1–2	2		39	0.666177
				2–3	3		46	0.987891
Case 3	Synthetic	0–7	8	0–1	0–1	Rhombus	108	0.174197
				1–2	2		39	0.421361
				2–3	3		67	0.136298
				3–4	4		32	0.272975
				4–5	5		90	0.164091
				5–6	6		29	0.562372
				6–7	7		14	0.122462
				7–8	8		17	0.332124
SF Bay	Bathymetry	0–255	7	0–33	−115 to −106	Octagon	34	0.048562
				34–59	−105 to −91		146	0.593921
				60–100	−90 to −68		57	0.365169
				101–150	−67 to −46		57	0.604285
				151–201	−45 to −27		23	0.304051
				202–233	−26 to −15		56	0.321996
				234–255	−14 to 0		22	0.120496
SC-Topo	Topography	0–255	14	0–1	0–1	Octagon	60	0.084891
				2–14	2–14		65	0.150806
				15–34	15–35		36	0.163969

Source: Lim, S.L. and Sagar, B.S.D., *Discrete Dyn. Nat. Soc.*, 2008, 26, 2008b.

corresponding gray values, and ranges of threshold values employed to decompose the basins into sets, type of structuring element used to generate geodesic flow fields, and the number of flow fields generated within each decomposed set.

Geodesic Flow Function Analysis

Properties of Geodesic Flow Fields in Geophysical Basin

For basins like simple cases 1, 2, the case of central SF Bay and Santa Cruz region for all $n \geq 1$ and $i \geq 1$:

$$(X_i) \subseteq \left[(X_i \oplus nB) \cap X_{i+1}\right] \subseteq \left[(X_i \oplus (n+1)B) \cap X_{i+1}\right] \subseteq \cdots \subseteq \left[(X_i \oplus NB) \cap X_{i+1}\right]$$

$$\subseteq [X_{i+1}] \subseteq \left[(X_{i+1} \oplus nB) \cap X_{i+2}\right] \subseteq \left[(X_{i+1} \oplus (n+1)B) \cap X_{i+2}\right] \subseteq \cdots$$

$$\subseteq \left[(X_{i+1} \oplus NB) \cap X_{i+2}\right] \subseteq [X_{i+2}] \subseteq \left[(X_{i+2} \oplus nB) \cap X_{i+3}\right]$$

$$\subseteq \left[(X_{i+2} \oplus (n+1)B) \cap X_{i+3}\right] \subseteq \cdots \subseteq \left[(X_{i+2} \oplus NB) \cap X_{i+3}\right] \subseteq \cdots$$

For case 3, when $i = 1$ and $0 \leq k \leq p \leq q \leq N$:

$$C_{flow} = (X_1) \subseteq \left[(X_1 \oplus kB) \cap X_{2i}\right] \subseteq \left[(X_1 \oplus (k+1)B) \cap X_{2i}\right] \subseteq \cdots \subseteq$$

$$\left[(X_1 \oplus KB) \cap X_{2i}\right] \subseteq [X_{2i}]$$

$$NC_{flow} = \left[(X_{2i} \oplus nB) \cap X_{2i+1}\right] \subseteq \left[(X_{2i} \oplus (n+1)B) \cap X_{2i+1}\right] \subseteq \cdots$$

$$\subseteq \left[(X_{2i+1} \oplus NB) \cap X_{2i+1}\right] \subseteq [X_{2i+1}]$$

when $i = 2$ and $0 \leq k \leq p \leq q \leq N$:

$$C_{flow} = \left[(X_1 \oplus KB) \cap X_{2i}\right] \subseteq \left[(X_1 \oplus (K+1)B) \cap X_{2i}\right] \subseteq \cdots$$

$$\subseteq \left[(X_1 \oplus pB) \cap X_{2i}\right] \subseteq [X_{2i}]$$

$$NC_{flow} = \left[(X_{2i} \oplus nB) \cap X_{2i+1}\right] \subseteq \left[(X_{2i} \oplus (n+1)B) \cap X_{2i+1}\right] \subseteq \cdots$$

$$\subseteq \left[(X_{2i+1} \oplus NB) \cap X_{2i+1}\right] \subseteq [X_{2i+1}]$$

when $i = 3$ and $0 \leq k \leq p \leq q \leq N$:

$$C_{flow} = \left[(X_1 \oplus pB) \cap X_{2i}\right] \subseteq \left[(X_1 \oplus (p+1)B) \cap X_{2i}\right] \subseteq \cdots$$

$$\subseteq \left[(X_1 \oplus QB) \cap X_{2i}\right] \subseteq [X_{2i}]$$

$$NC_{flow} = \left[(X_{2i} \oplus nB) \cap X_{2i+1}\right] \subseteq \left[(X_{2i} \oplus (n+1)B) \cap X_{2i+1}\right] \subseteq \cdots$$

$$\subseteq \left[(X_{2i+1} \oplus NB) \cap X_{2i+1}\right] \subseteq [X_{2i+1}]$$

when $i = 4$ and $0 \leq k \leq p \leq q \leq N$:

$$C_{\text{flow}} = \left[(X_1 \oplus QB) \cap X_{2i}\right] \subseteq \left[(X_1 \oplus (Q+1)B) \cap X_{2i}\right] \subseteq \cdots \subseteq \left[(X_1 \oplus NB) \cap X_{2i}\right] \subseteq \left[X_{2i}\right]$$

$$NC_{\text{flow}} = \left[(X_{2i} \oplus nB) \cap X_{2i+1}\right] \subseteq \left[(X_{2i} \oplus (n+1)B) \cap X_{2i+1}\right] \subseteq \cdots$$

$$\subseteq \left[(X_{2i+1} \oplus NB) \cap X_{2i+1}\right] \subseteq \left[X_{2i+1}\right]$$

Geodesic Flow Spectrum

Area of each TER $A[X_i]$ and area of all the TERs are estimated respectively according to $\sum_{x,y} X_i(x,y)$ and $\left(A\left[\sum_{i=1}^{I} \left(\sum_{x,y} X_i \right) \right] \right)$. For simplicity, we write these areas respectively as $A[X_i]$ and $A\left[\sum_{i=1}^{I} (A(X_i)) \right]$. The total time (iterations) taken to have complete flow fields is computed as $\sum_{i=1}^{K} N$. Thus, $A\left[(X_i \oplus nB) \cap X_{i+1} \right]$ increases as n(cumulative effect of flood forcing after nth-time step) increases, where $A(\cdot)$ denotes finite set of cardinality. These areas are normalized by the area $A\left[\sum_{i=1}^{I} (A(X_i)) \right]$ of basin (f). For a flat basin with no distinction in the mean elevations of channelized and non-channelized regions (e.g., Figure 3.14a and d), the cumulative area flooded after nth time is estimated as $A\left[(X_i \oplus nB) \cap X_{i+1} \right]$, where nB is a symmetric probing rule with certain characteristic information, and X_i, X_{i+1} respectively denote marker set and mask set (e.g., Figures 8.11 and 8.12a). Whereas when the elevation distinction between the channelized and non-channelized regions (e.g., Figure 3.14c and f) is realized, the cumulative area flooded in channelized region after nth time (iteration) is estimated as $A\left[(X_1 \oplus nB) \cap X_{2i} \right]$. Similarly, in the non-channelized region, the area is estimated as $A\left[(X_{2i} \oplus nB) \cap X_{2i+1} \right]$, where $i = 1, 2, \ldots, N$, and the marker sets for flow field propagation simulations are the sets indexed with $2i$, and these sets are geodesically dilated with reference to the mask sets indexed with $(2i + 1)$. The area of f, $A(f) = \sum_{(x,y)} f(x,y)$. These calculations for all the cases are plotted as functions of time (Figure 8.17f).

The areas embedded between the successive flow fields are considered to construct geodesic flow spectrum. This spectrum of decomposed elevation set (X_i) with structuring element B of radius n denoted as $GS_{X_i(n,B)}$ is defined as follows: $GS_{X_i(n,B)} = A\left[(X_i \oplus (n+1)B) \cap (X_{i+1}) \right] - A\left[(X_i \oplus (n)B) \cap (X_{i+1}) \right]$. Then the probability is derived as follows: $P_{X_i(B)} = GS_{X_i(n,B)} \Big/ A\left[\sum_{i=1}^{I} X_i \right]$, where $i = 1, 2, 3, \ldots, I$. The decomposed set-wise entropy with respect to total area of all the sets—decomposed from the function—is defined as

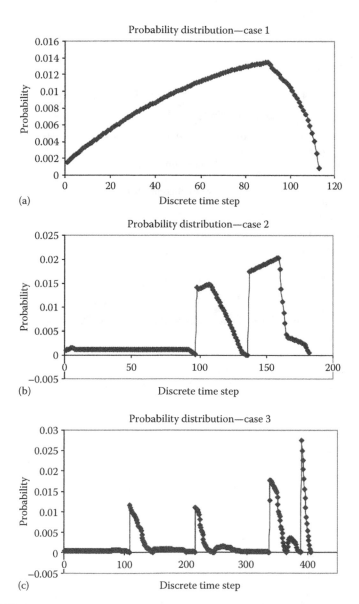

FIGURE 8.17
Probability of estimated area flooded at each discrete time step. The flow propagation for the three cases are simulated by using rhombus as structuring element, while flow fields for SF Bay and Santa Cruz are simulated with the use of octagon as structuring element. (a) Case 1, (b) case 2, (c) case 3.

(continued)

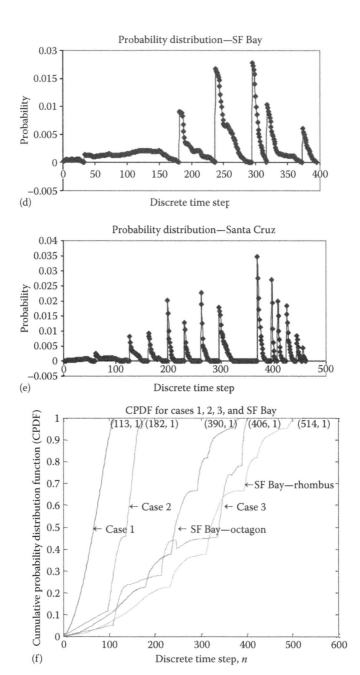

FIGURE 8.17 (continued)
Probability of estimated area flooded at each discrete time step. The flow propagation for the three cases are simulated by using rhombus as structuring element, while flow fields for SF Bay and Santa Cruz are simulated with the use of octagon as structuring element. (d) SF Bay, (e) Santa Cruz, and (f) cumulative probability for total area flooded. (From Lim, S.L. and Sagar, B.S.D., *Discrete Dyn. Nat. Soc.*, 2008, 26, 2008b.)

$H/(X_i, B) = -\sum_{n=0}^{N} P_{X_i(n,B)} \log P_{X_i(n,B)}$. Entropy values estimated by considering the probabilities that are computed with respect to the whole basin are given in the last column of Table 8.5. The morphological organizations of TERs can be better understood through the interpretation of this geodesic spectrum, a 1-D path support of different TERs and adjacent TERs. A potentially valuable insights and links with instantaneous unit hydrography can be explored. These geodesic spectra provide general geodesic distribution pattern between the TERs, as each geodesic spectrum exhibits distinct pattern that further explains that geodesic spectrum of each TER is someway similar to geomorphic width function.

Results and Discussion on Geodesic Spectrum

Probability distribution values to further compute the entropy values are estimated by dividing the areas embedded between the successive flow fields with the total area of corresponding threshold bathymetry zone. These probability distribution values and hence entropy values are marker–mask sets dependent. The geometric relationship between the marker and mask sets as well as the structuring elements' characteristic information influence the general flow fields' spatial organization, which further affects the probabilities and entropy values. Total area flooded after each cycle of geodesic propagation is estimated and plotted as a function of discrete time for all the five considered cases (Figure 8.17). It is obvious that the rates of change in the flow fields' pattern in the considered cases are different. Such variations are attributed to the spatial and topographic complexities of basins. To understand the rates of change in the areas between the flow fields of corresponding threshold bathymetry region (TBR), the probability distribution values of each TBR are plotted as functions of discrete time steps (Figure 8.17a through e). From these plots, it is obvious that the larger the peak, the wider is the area embedded between the successive flow fields. This analysis facilitates new insights to explore links between general statistical measures (e.g., probabilities, entropy values), and dynamics of sediment inflow patterns within each TBR and the morphological constitution of tidal and floodplain basins across times, since the surficial process involved therein is highly time dependent. It is hypothesized that the zones with abrupt changes in the probability patterns attribute to the fact that these zones support the occurrence of unusual suspended sediment patterns, due to high degree of spatial complexity of the flow fields. These zones as demarcated in the graph(s) further facilitate proper categorization of either surficial or bottom topographic zones—in terms of zones that are prone to have varied degrees of sensitivities to perturbation from dominating inflows such as tidal flow, river flow, and flow due to flooding.

In cases where the topography or bathymetry of basins is not available, and instead remote sensing data are used, a simplified method of the estimation of the tidal flow fields within the basin, neglecting the bathymetry, becomes helpful. However, the flow fields simulated by merely considering basin as flat surface would be directly determined by the boundary of the basin alone. If one cross-checks the flow fields simulated with an assumption that the bottom topography is completely flat with that of the flow fields estimated from the bathymetric data, one realizes how the former is entirely dependent on the boundary of the basin. In fact, the two flow fields are highly contrasting in nature as shown in Figure 8.16a and b. If topographic data of basins (or inlets, estuaries, and bays) are available at multitemporal mode, the flow field can be simulated via geodesic method as proposed here for the study of the spatiotemporal dynamics of the bottom topography for understanding the coastal dynamics (dynamics of tidal environments). This framework can be tested on any basin by using very high-resolution DEM/DBM data at different time periods (perhaps during pre- or post-flood times and low- or high-tide time periods) that generally influence the surficial morphology of the floodplain or tidal basin as the process involved there are time dependent in contrast to that of basins in fluvial environment. Very high-resolution DEM (e.g., retrieved from Shuttle Radar Topography Mission) provides subtle changes in topographic elevation. Usually, the elevation differences within floodplain environment and in tidal environment are minor. However, the morphological variations within such environments are highly time dependent. This time-dependent morphological changes may synchronize the fluctuating hydrological flows that are usually influenced due to flooding/tide patterns of the system. Furthermore, this framework has been applied to generate flow fields on three simulated basins and on two digital topographies of SF Bay and Santa Cruz region. Space–time structures of flow fields in basins that occur due to changes in inflow patterns can be treated as a coupled dynamical system. Spatial organization of flow fields is sensitive to such spatiotemporal changes.

Why Geodesic Spectrum?

Based on geodesic morphologic transformations, a framework has been proposed in the "Geodesic Flow Spectrum" section to characterize discrete geophysical basins of surficial and bathymetric types. The three phases of the framework include (1) the decomposition of digital topographic basin into sets through thresholding technique, (2) the generation of geodesic flow fields within each set successively, and (3) the estimation of probabilities of areas being embedded between flow fields of each set and the successive sets. This three-phase framework has been demonstrated on several synthetic and realistic digital topographic (bathymetric) basins. A new basin descriptor, a construction of geodesic spectrum for basin functions, that can be further linked with geomorphic width function has been derived. Geodesic spectra of basin functions depend on the (1) general structure of basin function, (2) the geometric organization, and their internal spatial relationships of TERs, and (3) the structure of geodesic propagation (frontlines). These geodesic spectra provide insights into studies related to (1) modeling the sediment transport and deposition processes, (2) morphologic processes that control the morphologic development of basin function, and (3) understanding of morphodynamical processes in a quantitative fashion when topographies of basins are available at higher spatial resolutions.

References

Ackeret, J. R., 1990, Digital terrain elevation data resolution and requirements study, Interim Report ETL-SR-6, U.S. Army Corps of Engineers, Washington, DC.

Baratoux, D., N. Mangold, C. Delacourt, and P. Allemand, 2002, Evidence of liquid water in recent debris avalanche on Mars, *Geophysical Research Letters*, 29, 1156.

Blondeaux, P. and G. Vittori, 2005a, Flow and sediment transport induced by tide propagation—1: The flat bottom case, *Journal of Geophysical Research*, 110(C7), Article ID C07020, 13.

Blondeaux, P. and G. Vittori, 2005b, Flow and sediment transport induced by tide propagation—2: The wavy bottom case, *Journal of Geophysical Research*, 110(C8), Article ID C08003, 11.

Cherbit, G., 1991, *Fractals Non-integral Dimensions and Applications*, John Willey, Chichester, U.K.

Chockalingam, L. and B. S. D. Sagar, 2005, Morphometry of network and nonnetwork space of basins, *Journal of Geophysical Research*, 110(B8), Article ID B08203, 15.

Dall, J., S. N. Madsen, K. Keller, and R. Forsberg, 2001, Topography and penetration of the Greenland ice sheet measured with airborne SAR interferometry, *Geophysical Research Letters*, 28(9), 1703–1706.

Daniels, R. B., L. A. Nelson, and E. E. Gamble, 1970, A method of characterizing nearly level surfaces, *Zeitscheift fur Geomorphologies*, 14, 175–185.

Dubuc, B., S. W. Zucker, C. Trikot, J. F. Quiniou, and D. Wehbi, 1989, Evaluating the fractal dimension of surface, *Proceedings of the Royal Society of London*, A425, 113–127.

Fagherazzi, S., P. L. Wiberg, and A. L. Howard, 2003, Tidal flow field in a small basin, *Journal of Geophysical Research*, 108(C3), 3071, 10.

Fatale, L., J. R. Ackeret, and J. Messmore, 1994, Impact of digital terrain elevation data-DTED-resolution on army applications: Simulation vs. reality, in *Proceedings of the American Congress on Surveying and Mapping (ACSM'94)*, Reno, NV, pp. 89–104.

Feder, J. 1988. *Fractals*. Plenum Press, New York.

Franklin, S., 1987, Geomorphometric processing of digital elevation models, *Computers & Geosciences*, 13, 603–609.

Gilbert, L. E., 1989, Are topographic data sets fractal? *Pure and Applied Geophysics*, 131, 241–254.

Goodchild, M. F., 1980, Fractals and accuracy of geographical measures, *Mathematical Geology*, 12, 85–98.

Horton, R. E., 1945, Erosional development of streams and their drainage basins: Hydrological approach to quantitative morphology, *Bulletin of the Geophysical Society of America*, 56, 275–370.

Howard, A. D., 1997, Badland morphology and evolution: Interpretation using a simulation model, *Earth Surface Processes Landforms*, 22, 211–227.

Kirkby, M. J., 1980, The stream head as a significant geomorphic threshold. In: Coates, D. R. and J. D. Vitek, *Thresholds in Geomorphology*, Allen and Unwin, London, U.K., pp. 53–73.

Kirkby, M. J., 1993, Long term interactions between networks and hillslopes. In: Beven, K. J. and M. J. Kirkby, *Channel Network Hydrology*, John Wiley, New York, pp. 255–293.

Langbein, W. B., 1947, Topographic characteristics of drainage basins, U.S. Geological Survey Professional Paper, 968-C, 125–157.

Lim, S. L. and B. S. D. Sagar, 2008a, Cloud field segmentation via multiscale convexity analysis, *Journal of Geophysical Research*, 113(D13208), doi:10.1029/2007JD009369.

Lim, S. L. and B. S. D. Sagar, 2008b, Derivation of geodesic flow fields and spectrum in digital topographic basins, *Discrete Dynamics in Nature and Society*, 2008(2008), Article ID 312870, 26, doi:10.1155/2008/312870.

Lim, S. L., B. S. D. Sagar, V. C. Koo, and L. T. Tay, 2011, Morphological convexity measures for terrestrial basins derived from Digital Elevation Models, *Computers & Geosciences*, 37, 1285–1294.

Maragos, P. A. and R. D. Ziff, 1990, Threshold superposition in morphological image analysis systems, *IEEE Transactions on Pattern Analysis and Machine Intelligence*, 12(5), 498–504.

Marani, M., E. Belluco, A. D'Alpaos, A. Defina, S. Lanzoni, and A. Rinaldo, 2003, On the drainage density of tidal network, *Water Resources Research*, 39(2), 4.1–4.11, doi: 10.1029/2001WR001051.

Marani, M., A. Rinaldo, R. Rigon, and I. Rodriguez-Iturbe, 1994, Geomorphological width functions and the random cascade, *Geophysical Research Letters*, 21(19), 2123–2126.

Maritan, A., R. Rigon, J. R. Banavar, and A. Rinaldo 2002, Network allometry, *Geophysical Research Letters*, 29(11), 1508, doi:10.1029/2001GL014533.

Montgomery, D. R. and W. E. Dietrich, 1989, Source areas, drainage density, and channel initiation, *Water Resources Research*, 25(8), 1907–1918.

Montgomery, D. R. and W. E. Dietrich, 1994, Landscape dissection and drainage area-slope thresholds. In: *Process Models and Theoretical Geomorphology*, John Wiley, New York, pp. 221–246.

Montgomery, D. R. and E. Foufoula-Georgiou, 1993, Channel network source representation using digital elevation models, *Water Resources Research*, 29, 3925–3934.

Nikora, V. I., 2005, High-order structure functions for planet surfaces: A turbulence metaphor, *IEEE Geoscience and Remote Sensing Letters*, 2, 362–365.

Perera, J. H., 1997, The hydrogeomorphic modeling of sub surface saturation excess run-off generation, PhD thesis, University of Newcastle, Newcastle, NSW, Australia.

Rodriguez, Z. F., E. Maire, P. Courjault-Rade, and J. Darrozes, 2002, The black top hat function applied to a DEM: A tool to estimate recent incision in a mountainous watershed (Estibere Watershed, Central Pyrenees), *Geophysical Research Letters*, 29, Art. No. 1085.

Rodriguez-Iturbe, I. and A. Rinaldo, 1997, *Fractal River Basins: Chance and Self-Organization*, Cambridge University Press, Cambridge, U.K.

Sagar, B. S. D. and L. Chockalingam, 2004, Fractal dimension of non-network space of a catchment basin, *Geophysical Research Letters*, 31, L12502.

Sagar, B. S. D., M. B. R. Murthy, C. B. Rao, and B. Raj, 2003, Morphological approach to extract ridgevalley connectivity networks from digital elevation models (DEMs), *International Journal of Remote Sensing*, 24, 573.

Sagar, B. S. D., C. Omoregie, and B. S. P. Rao, 1998a, Morphometric relations of fractal-skeletal based channel network model, *Discrete Dynamics in Nature and Society*, 2, 77–92.

Sagar, B. S. D., D. Srivinas, and B. S. P. Rao, 2001, Fractal skeletal based channel networks in a triangular initiator basin, *Fractals*, 9(4), 429–437.

Sagar, B. S. D. and T. L. Tien, 2004, Allometric power-law relationships in a Hortonian Fractal DEM, *Geophysical Research Letters*, 31, L06501.

Sagar, B. S. D., M. Venu, and D. Srivinas, 2000, Morphological operators to extract channel networks from Digital Elevation Models, *International Journal of Remote Sensing*, 21(1), 21–30.

Schumm, S. A., M. P. Mosley, and W. E. Weaver, 1987, *Experimental Fluvial Geomorphology*, John Wiley, New York.

Serra, J., 1982, *Image Analysis and Mathematical Morphology*, Academic Press, London, U.K.

Snyder, N. P., K. X. Whipple, G. E. Tucker, and D. J. Merritts, 2000, Landscape response to tectonic forcing: Digital elevation model analysis of stream profiles in the Mendocino triple junction region, northern California, *Geological Society of America Bulletin*, 112, 1250–1263.

Soille, P., 1998, Gray scale convex hulls: Definition, implementation and application, *Proceedings of ISMM'98*, Kluwer Academic Publishers, Amsterdam, the Netherlands.

Stone, R. and J. Dugundji, 1965, A study of microrelief: Its mapping, classification, and quantification by means of a Fourier analysis, *Engineering Geology*, 1, 89–187.

Tay, L. T., B. S. D. Sagar, and H. T. Chuah, 2005, Derivation of terrain roughness indicators via granulometries, *International Journal of Remote Sensing*, 26(18), 3901–3910.

Tay, L. T., B. S. D. Sagar, and H. T. Chuah, 2006, Allometric relationships between traveltime channel networks, convex hulls, and convexity measures, *Water Resources Research*, 42(6), W06502, 8.

Tay, L. T., B. S. D. Sagar, and H. T. Chuah, 2007, Granulometric analyses of basin-wise DEMs: A comparative study, *International Journal of Remote Sensing*, 28(15), 3363–3378.

Tucker, G. E. and R. L. Bras, 1998, Hillslope processes, drainage density, and landscape morphology, *Water Resources Research*, 34(10), 2751–2764, doi:10.1029/98WR01474.

Tucker, G. E., F. Catani, A. Rinaldo, and R. L. Bras, 2001, Statistical analysis of drainage density from digital terrain data, *Geomorphology*, 36, 187–202.

Turcotte, D. L., 1997, *Fractals and Chaos in Geology and Geophysics*, 2nd edn., Cambridge University Press, Cambridge, U.K.

Veneziano, D., G. E. Moglen, P. Furcolo, and V. Iacobellis, 2000, Stochastic model of the width function, *Water Resources Research*, 36(4), 1143–1157.

Whipple, K. X., E. Kirby, and S. H. Brocklehurst, 1999, Geomorphic limits to climate-induced increases in topographic relief, *Nature*, 401, 39–43.

Whipple, K. X. and G. E. Tucker, 1999, Dynamics of the stream-power river incision model: Implications for height limits of mountain ranges, landscape response timescale, and research needs, *Journal of Geophysical Research*, 104, 17661–17674.

9

Synthetic Examples to Understand Spatiotemporal Dynamics of Certain Geo(morpho)logical Processes

Several systems of geomorphological interest undergo morphological changes with time. In the process of changing morphological organization, systems traverse various phases. It is understood through numerous studies that the geomorphological systems undergo nonlinear processes. One equation that explains several phases that a system could undergo is logistic equation, which is also termed as first-order nonlinear difference equation. In this equation, important parameters include the strength of the nonlinearity parameter that controls the dynamics of a system and the state of the systems (e.g., initial condition). Logistic equation is based on a wonderful recipe to simulate the processes in such a way that when the system attains a higher value (e.g., population of a specie, fractal dimension, area of a lake), this equation reduces this to a lower value in the next time step, and vice versa. Using the values obtained via the iteration of the logistic equation, by changing the strength of control parameters, several spatiotemporal dynamics that we studied under strong theoretical assumptions include behaviors of (1) water bodies under controlled stream flow discharges, (2) highly ductile fold dynamics, and (3) pyramidal sand dune dynamics and avalanche size distributions.

Logistic Map: A Toy Model

Logistic Map as a Viable Model to Simulate Dynamical Behaviors of Certain Geomorphological Processes

Geomorphological processes that were aimed at include geometric changes in the lake morphologies, folds, and sand dunes. Such geomorphological systems can be reduced to simple systems that still capture the salient features of the original systems. In various other fields, the logistic maps have been taken as the basis to simulate the dynamical behaviors (May 1976, Feigenbaum 1980, Abraham and Shaw 1982, Devaney 1986). The application of logistic map–based simulation in abstract understanding of the dynamical behaviors of various geomorphological processes has been addressed (Sagar and Rao 1995a,b, Sagar 1998, 1999b, 2001b, 2005, Sagar et al. 1998b, 2003a, Sagar and Venu 2001). Application of logistic maps to understand and simulate the dynamical behaviors of certain geomorphological phenomena (e.g., lakes, sand dunes, folds, geomorphological structures) is included in the following sections of this chapter.

First-Order Nonlinear Difference Equation: Logistic Map

The difference equation that can be used to study a simple geomorphological dynamical system is written as Equation 9.1:

$$X_{t+1} = F(X_t) \tag{9.1}$$

where X_{t+1}, X_t denote parameters of a specific variable that are changing with time (e.g., areal extent of water bodies, steepness of sand dunes and folds). Equation 9.1 enables the iterative process, where the output $F(X_t)$ becomes input. The roles of the first-order nonlinear difference equations (e.g., Equation 9.2) and the bifurcation theory have been lucidly explained in seminal paper by Robert May (1976).

$$X_{t+1} = \lambda X_t (1 - X_t) \tag{9.2}$$

where
   X_t and X_{t+1} are the parameters (e.g., population, areal extents of water bodies) at time periods, t and $t + 1$, respectively
   λ is the nonlinearity parameter or threshold control parameter, which determines the magnitude of variation

By tuning λ parameter, different types of dynamical behaviors of dynamical systems could be modeled. The relation between the magnitude of the parameter at a definite time and the magnitude of that parameter at preceding time is shown in Equation 9.2. In Equation 9.2, the first and the second terms respectively are linear and nonlinear, and the λ that ranges between 1 and 4 denotes the strength of the nonlinearity parameter that explains the magnitude of variation in the parameter. X_{t+1} and X_t are in the range between 0 and 1 (normalized scale). Strength of nonlinearity can be estimated by plotting X_{t+1} versus X_t. This determines the future areal extent of the lake, say, for example, X_{t+1}, and X_{t+2}, ..., and so on, at time steps $t + 1$ and $t + 2$, ..., respectively, from the previous value. Equation 9.2 defines an inverted parabola with intercepts at $X_t = 0$ and 1, and a maximum value at $X_t = 0.5$, which would be $X_{t+1} = \lambda/4$. The heights of the humps of inverted parabolas defined based on Equation 9.2 for λ values of 4, 3, 2, and 2.5, respectively, include 1, 0.75, 0.5, and 0.625. The parameter λ gives complete description of the system.

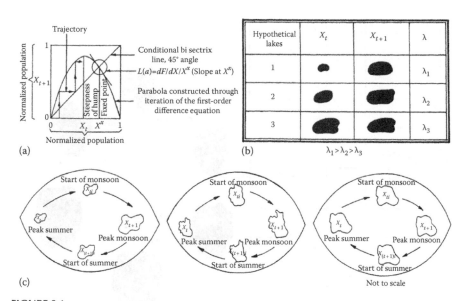

FIGURE 9.1
(a) Logistic map and its essential parameters, (b) involvement of the population of water body pixels at different time periods and its strength of nonlinearity, and (c) conceptual cycle of water body behavior of different regions. (From Sagar, B.S.D. and Rao, B.S.P., *Curr. Sci.*, 68, 950, 1995b.)

In Equation 9.2, the strength of nonlinearity, λ, gives the entire description of the changing parameter (e.g., population, areal extents of water bodies) of a dynamical system. For a higher value of X_t, the expression $(1 - X_t)$ reduces the output value, and vice versa. The essential parameters to construct a logistic map are the initial value, X_t, represented in a normalized scale, and the strength of nonlinearity, λ. Essential parameters involved in the construction of logistic maps are identified in Figure 9.1a. As an example to show how areal extents of a lake change with seasons, and to show the possibility of using logistic maps to understand the dynamical behavior of lakes, Figure 9.1b and c has been illustrated. Figure 9.2a through c shows hypothetical water bodies with magnitudes of variations (λ) of >3.8, <1, and 2. Figure 9.2 shows a qualitative representation of a water body undergoing changes in areal extents by synchronizing peak monsoon—peak summer—peak monsoon with cascades of contraction–expansion. Figure 9.2e through g shows typical logistic maps constructed for different λ values. For small values of λ, the system is stable (i.e., $X_{t+1} = X_t$ and so on). As λ increases, the system moves away from its stable state. An equation to compute the strength of nonlinearity in structures may be derived by considering the structure at times t and $t + 1$. Equation 9.2 has been studied extensively and is considered to be a simple model to explain the dynamics in one-dimensional (1-D) maps, where increasing λ induces period-doubling bifurcation leading to chaos. This equation possesses one equilibrium point and the stability of the fixed point, and the consequent dynamics exhibited by the systems are dependent

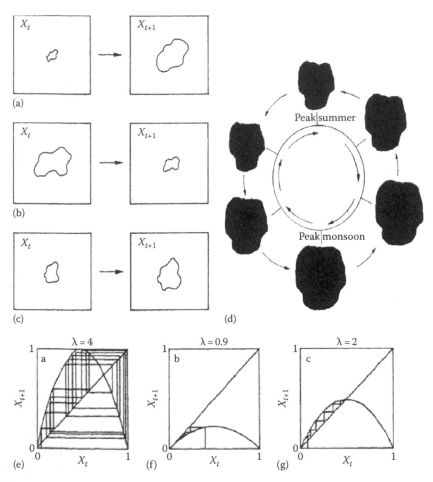

FIGURE 9.2

Hypothetical water bodies at times t and $t + 1$ of two different years. (a) The magnitude of variation $\lambda > 3.8$, (b) magnitude of variation $\lambda < 1$, (c) the amount of nonlinearity is exactly 2, (d) a qualitative representation of morphological evolution of a lake from peak summer–peak monsoon–peak summer (clockwise direction, C-EC process), peak monsoon–peak summer (anticlockwise direction, C-CE process), and (e–g) the return maps constructed by taking the areal extents from the possibilities given in Figure 9.2a through c and the computed strength of nonlinearity into account where X_t and X_{t+1} are populations of the water bodies at different times (e.g., peak summer and peak rainy seasons, respectively). (From Sagar, B.S.D. and Rao, B.S.P., *Int. J. Remote Sens.*, 16, 365, 1995a; Sagar, B.S.D. and Rao, B.S.P., *Curr. Sci.*, 68, 950, 1995b; Sagar, B.S.D. et al., *Int. J. Remote Sens.*, 19(7), 1341, 1998b.)

on λ alone. Equation 9.2 was used in many studies to quantify several natural processes (May 1976, Jenson 1987, Sagar and Rao 1995a,b). We apply Equation 9.2 as a basis to model the fluctuations in the (1) areal extents in lakes, (2) ductile symmetrical folds, and (3) sand dunes. The idea of considering this equation is that it can simulate several possible behaviors, ranging from periodic, quasi-periodic to chaotic, of various physical systems.

Logistic Equation in Modeling the Geomorphological Phenomena (Lakes)

Ranking of Lakes: Logistic Models

One of the simplest systems a limnologist can study is the periodical variations in the areal extent of lakes. As lakes exhibit fluctuations, it is interesting to categorize them according to their dynamical behavior from stable to
unstable to apparently random fluctuations. Such fluctuations in the areal
extents of lakes vary with time and physiographic conditions of the terrain.
Therefore, it is important to develop a model for a better understanding
of the role of certain parameters that exhibit significant variations across
time periods. Thus, certain parameters of some lakes may behave chaotic
or periodic. This system may be represented in a single difference equation
$X_{t+1} = \lambda X_t(1 - X_t)$. Based on this equation, logistic maps have been constructed
to predict the size of changing areal extent of water and marsh in the Chilka
lake. The logistic maps provide an approximate model, at least, to study the
evolution of the lake at periodical intervals. They may also help segregate
lakes according to their magnitude of variation in areal extents.

Investigations on the behavior of lakes constitute one of the important
aspects of limnology. Nonlinear fluctuations in areal extents of lakes, an
important aspect of limnology, are quantified by the use of simplified mathematical models like the first-order difference equation (9.2). In (9.2), X_t and
X_{t+1} represent areal extents of a lake, with subscripts t and $t + 1$ indicating
successive discrete periods. For instance, if the lake in period, t would reach
X_{t+1} area, then the areal extent, X_{t+1}, in the next period is the product λX_t.
Equation 9.2, modeling in the simplest way the decline in growth factor of
the Chilka lake, serves to keep the areal extent of lakes below the limit [1],
as one of the parameters is $(1 - X_t)$. When $\lambda < 1$, the areal extent decreases.
Without any calculations, the successive areal extent of the lakes may be
determined, through return/logistic maps—described by the single difference equation (9.2), provided that the strength of nonlinearity in lakes is
properly predicted. The periodical fluctuations in the areal extent of lakes,
a natural phenomenon, may be due to meteoro-geo-physiographical conditions, the magnitude of variation being dependent upon the intensity of the
factors. Hence, both variations in areal extent and intensity of factors should
be quantified in order to provide a better understanding of the lake behavior. When the original areal extent of lake X_0 is small (much less than 1 on
a normalized scale, where 1 stands for any number, such as one million
square kilometers), the nonlinear term can initially be neglected. Then the
areal extent at time step $X_0 = 1$ will be approximately equal to λX_0. If $\lambda > 1$,
the areal extent increases. If $\lambda < 1$, the areal extent decreases. Therefore, the
linear term in Equation 9.2 can be interpreted as a linear growth rate that by
itself would lead to exponential growth in areal extents. If $\lambda > 1$, the areal

extent eventually increases to a value large enough for the nonlinear term $-\lambda X_t^2$. Since this term is negative, it represents a nonlinear decrement rate, which dominates when the areal extent in the lake becomes too large. Limnologically, this decrement in lakes could be due to cultural eutrophication or environmental effect.

Sample Study

The changes in areal extents of Lake Chilka, in southern Orissa, India, have been computed using a multi-date Landsat multispectral scanning system and thematic mapper (MSS TM) data. Table 9.1 shows the data and amount of nonlinearity for both water and marsh of Lake Chilka from 1978 to 1987. As the main factor affecting the areal extent of water in Lake Chilka is the rapid growth of marsh (Murthy et al. 1988), an attempt was made to relate marsh to the changes in areal extent. Figure 9.3a and b shows logistic maps of changes in the areal extent and marsh, respectively, in Lake Chilka. The areal extent at successive periods was determined by tracing the lines on the return maps. The trajectory of the logistic map for the areal extent is found attracted to the initial conditions (Figure 9.3a), while that of the marsh is attracted to a fixed point (Figure 9.3b). Table 9.1 shows that the areal extent of both water and marsh, predicted from logistic maps by tracing lines, does not differ considerably from that of the remotely sensed data. However, more data, over longer time periods and sampled at more regular intervals, would give more precise results for the modeling of lake areas on the basis of logistic equations.

Morphological Description: A Scope to Geomorphic Evolution Process Modeling

The reaction of a geomorphic feature to a perturbation caused due to endogenic and/or exogenic nature of forces (the collective effect of which created as a morphological force) is discussed and could be seen in terms of changing geometries. The morphological dynamics of a system subjected to undergo morphological processes due to such perturbations can be qualitatively modeled through graphic analysis.

Introduction of Morphological Behavior

The morphological behavior of certain objects depends upon the original morphological constitution, the type of force they are subjected to, and the type of process undergone. During the evolution process, some systems may disintegrate and then disappear. Some morphologically stable systems may become unstable, and vice versa. Many possibilities from stable to unstable, and/or chaotic, behavior in the morphology may be encountered during the evolution process of certain geomorphic features. Such geomorphic features can be

TABLE 9.1

Remote Sensing and Logistic Map Data—Areas of Water and Marsh in Chilka Lake at Different Times and Their Strength of Nonlinearities

Time	Water (km²)	Marsh (km²)	Strength of Nonlinearity in Water Spread	Strength of Nonlinearity in Marsh	Time	Area of Water Spread Predicted through Logistic Map (km²)	Area of Marsh Spread Predicted through Logistic Map (km²)
1978 (March)	$0.02 \times (4 \times 10^4)$	0.109×10^3	0.9	1.6	1978 (March)	802	0.109×10^3
1981 (May)	$0.019 \times (4 \times 10^4)$	0.171×10^3			1981 (May)	745	0.191×10^3
1984 (March)	$0.018 \times (4 \times 10^4)$	0.203×10^3			1984 (March)	680	0.240×10^3
1987 (April)	$0.018 \times (4 \times 10^4)$	0.221×10^3			1987 (April)	640	0.250×10^3

Source: Sagar, B.S.D. and Rao, B.S.P., *Int. J. Remote Sens.*, 16, 365, 1995a.

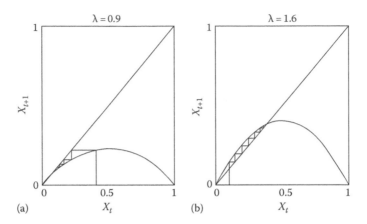

FIGURE 9.3
(a) Logistic map shows fluctuations in areal extents at tri-annual intervals (March 1978–May 1981). The computed strength of nonlinearity is 0.9. The trajectory is found attracted toward initial condition. (b) Logistic map shows the rate of increase in marsh at tri-annual intervals. The computed strength of nonlinearity is 1.6. Trajectory is found attracted to a fixed point. (From Sagar, B.S.D. and Rao, B.S.P., *Int. J. Remote Sens.*, 16, 365, 1995a.)

broadly categorized as the features where the changes can be noticed at short time intervals (e.g., lake), and those in which changes can be noticed at long time intervals. The difference between short and long time periods is yet to be defined. To model the dynamically changing morphology of a geomorphic feature by incorporating the concepts that follow in this section, various types of remotely sensed data available in temporal sequence to model the morphological changes are essential. What follows in this section includes the use of multitemporal satellite data to make an attempt to model the morphological dynamics. To model the morphological changes that have been collated from multitemporal satellite data, the application of mathematical morphological concepts, fractal geometry, and chaos theory is foreseen in the modeling and simulation studies of the geomorphic evolution process. The study of a specific geomorphic feature undergoing transformation across times enables the evolution of the type of force and the process undergone. The force responsible for the transformation can be defined and designed in morphological terms. The sequential steps to model the morphological dynamics include the following:

It is quite obvious that a feature generally undergoes either one or any combination of the four possible morphological processes (details of these processes are described in the "Numerical Simulations Through First-Order Nonlinear Difference Equation to Study Highly Ductile Symmetric Fold Dynamics: A Conceptual Study" section). Depending upon the complexity in the morphological dynamics of the feature, certain conditions have to be imposed on the force to which the feature is subjected.

Both homogeneous and nonhomogeneous effects are likely to operate simultaneously in the evolution of a geomorphic system. The recognition of such concurrent characters in geomorphic evolution is significant due to endogenic forces, i.e., tectonic character (systematic) and exogenic forces (nonsystematic). It is assumed that the geomorphic evolution process, in general, is based on such forces acting concurrently. The impact of such concurrent forces in a geomorphic system can be studied by taking into account the morphological changes that have occurred in temporal sequence. It is also assumed that the degree of deformation depends upon the intensity of collective or concurrent forces. Hence, the deformed portion of the geomorphic feature is taken as the basis to study the geomorphic dynamics systematically.

Laws of Structures

Structures will undergo contraction, expansion, expansion followed by contraction (C-EC), or contraction followed by expansion (C-CE) or any combination of these processes. Based on the properties of structures when subjected to perturbation by any force, five laws of structures have been proposed as follows:

1. A structure reaches the state of convergence during the process of continuous expansion by a force of size more than that of the structure.

2. During the process of continuous contraction by a force of size of more than that of structure, the structure either disappears or disintegrates and then disappears.

3. A Euclidean type of structure will not undergo any change under the process of C-EC. However, there will be a variation in the transformed structure if any of the characteristics of force to expand is different from that to contract.

4. Under the process of C-CE, if the cumulative force acting upon the structure is less than the size of the structure, the transformed structure is geometrically similar to the force. As long as the structure does not disintegrate during this process, if the force to contract is different from that to expand, the morphology of the resultant structure depends upon the succeeding force and the structure remaining just before it gets vanished during the subprocess of contraction. The structure disappears if the cumulative force is more than the structure.

5. Under any process, when both structure and force are geometrically similar, and also the cumulative force does not dominate the structure, the transformed structure will be geometrically similar to both original structure and force.

Based on these laws, a critical point can be defined for the expansion and the cascade processes. The critical point is the iteration number (time) at which the structure reaches the state of convergence. This point depends upon the process, characteristics of force, and the original structure.

Geomorphic Evolution Modeling: A Scope

Evolution of Lake Morphology

In the natural phenomena, the endogenic and exogenic forces show an impact on lake morphology. For instance, a hypothetical representation of the lake morphology at peak summer is shown in Figure 9.2d. This figure also shows the intermediary evolution phase of the lake as time progresses toward the peak monsoon season. It is also apparent that the morphological process that the lake has undergone is a continuous expansion. In order to design the morphological force that is responsible for the continuous expansion process (peak summer to peak monsoon season), the difference portion between the lake at time t (X_t) and that of the lake at time $t + 1$ (X_{t+1}) is considered. For the cycle of hypothetical lake evolution process from peak summer–peak monsoon–peak summer (Figure 9.2d), the ensuing process is C-EC. This process can be represented mathematically as $((X \oplus B) \oplus \cdots \oplus B) \ominus B \ominus \cdots \ominus B$.

The reverse process can also be visualized as two successive phases (peak monsoon–peak summer–peak monsoon), i.e., cascade of contraction–expansion (C-CE). The study of nonhomogeneous nature of these basic morphological processes sheds light on the study of the dynamical process in the natural lake evolution.

Modeling of Morphological Dynamics of a Lake: A Qualitative Study

The morphological conditions of a geomorphic feature play a significant role in predicting its morphological behavior, the descriptive analysis of which is of limited use. In this section, an attempt is made to analyze the geomorphic evolution process systematically through mathematical morphological transformations. From a topological point of view, a circular type of system is more stable than the system with a nonstandard morphological form. A geomorphic system at equilibrium state is more stable than that at disequilibrium state. The morphological stability of a geomorphic system at equilibrium state may be defined by the morphological behavior of the system when it is subjected to a small perturbation. A system is said to be stable if it returns to its original state and unstable if it continues to move away from equilibrium state as a result of a perturbation caused by a homogeneous morphological force. Thus, the qualitative analysis has great significance to understand the geomorphic structural dynamics.

Logistic map analysis: The following is a maiden attempt to model morphological dynamics that follows an ideal condition. The variation in the structure under specific transformation can be quantified by constructing a 1-D map (logistic map). As the theory of 1-D maps constructed by iterating the first-order difference equation is well established (May 1976), it will be useful if an appropriate 1-D map can be constructed from the structure under study. In this study, instead of the population X_t and X_{t+1}, the fractal dimensions (Mandelbrot 1982) in normalized scale α_t and α_{t+1}, useful to quantify the degree of irregularity of the generated shapes before and after every process and iteration, are considered. To represent the fractal dimensions in a normalized scale, the topological dimension (D_T) is subtracted from the fractal dimension (D) as $\alpha = D - D_T$.

If a structure transforms itself by following a specific rule, the resultant transformation at time $t + 1$ can be determined from the structure at initial time t, according to Equation 9.2, which is $\alpha_{t+1} = f(\alpha_t)$ or $\alpha_{t+1} = f'(\alpha_t)$. This functional iteration could be shown as $\alpha_{t+n} = f^n(\alpha_t)$, where function f is defined in 1-D space and λ is the magnitude of variation from time t to $t + 1$. This function depends upon λ, which can be incorporated as a single description to predict the behavior of a structure. Considering a specific process, the future structure may be predicted through intensive studies of the first-order difference equation. This study mainly depends upon the computation of the strength of nonlinearity (λ), which varies with the structure, and type of force (disturbance) acting on the structure. To construct a logistic map to a set of observations $\alpha_t, \alpha_{t+1}, \ldots$, a plot between α_{t+1} and α_t is necessary to estimate the value needed to plot the curve $y = \alpha_t(1 - \alpha_t)$.

Attracting to a fixed point: In this section, a deterministic approach has been followed to model the dynamics of a transcendentally generated fractal lake under specified morphological transformations. To explain and quantify the entire morphological evolution process of a hypothetical lake going toward extinction, a fractal lake (Figure 9.4a) is allowed to undergo the C-CE process iteratively by means of an octagonal morphological force up to four cycles. Iterations beyond the fourth cycle are not considered to construct the model as the fractal lake is getting vanished. Figure 9.4b through d shows the representations after respective cycles of the C-CE process (e.g., contraction phase of lake evolution from peak monsoon to peak summer, and expansion phase from peak summer to peak monsoon) by means of an octagonal structuring element. The textural and structural variations in the sequence of the transformed fractal lakes are due to the increase in the size of the force during this cascade process. As the force is smaller in the initial phase of transformation, a textural variation is observed, while in the latter phase, a structural variation is observed due to an increase in the force. The deformation in the transformed

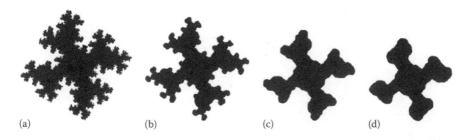

(a) (b) (c) (d)

FIGURE 9.4
(a) Fractal lake, (b) lake after one cycle of C-CE, (c) lake after two cycles of C-CE, and (d) lake after three cycles of C-CE. (From Sagar, B.S.D. et al., *Int. J. Remote Sens.*, 19(7), 1341, 1998b.)

TABLE 9.2

Morphological Dynamics: Computed and Predicted
Fractal Dimension Values

Morphological Process	Computed Fractal Dimension and α Values		Predicted Fractal Dimension and α Values	
	Fractal Dimension	α	Fractal Dimension	α
$(X \ominus B) \oplus B$	1.53	0.53	1.53	0.53
$(X \ominus 2B) \oplus 2B$	1.38	0.38	1.386	0.386
$(X \ominus 3B) \oplus 3B$	1.36	0.36	1.36	0.36
$(X \ominus 4B) \oplus 4B$	1.34	0.34	1.35	0.35

Source: Sagar, B.S.D. et al., *Int. J. Remote Sens.*, 19(7), 1341, 1998b.

fractal lakes from iteration to iteration is quantified in terms of fractal dimension (Table 9.2) computed through box counting method proposed elsewhere (Feder 1988). The computed fractal dimensions for the four transformed structures obtained from the respective cycles are 1.53, 1.38, 1.36, and 1.34 (Table 9.2). From the normalized fractal dimensions (NFDs), at discrete time periods, the strength of nonlinearity, λ, is estimated as 1.53. A logistic map (Figure 9.5) is constructed for this process using the strength of nonlinearity and the initial NFD. The trajectory of the logistic map is traced to predict the successive values of α. The predicted fractal dimensions ($\alpha + D_T$) of the fractal lake inferred from the logistic map are close to computed fractal dimensions for the respective cycles (Table 9.2). Though this model is qualitative, this approach, where

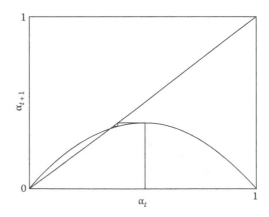

FIGURE 9.5
Representation of dynamical changes in the morphology of evolving fractal lake (shown in Figure 9.4a through d) through logistic map, $\lambda = 1.53$. (From Sagar, B.S.D. et al., *Int. J. Remote Sens.*, 19(7), 1341, 1998b.)

mathematical morphology, fractals, and chaos theory are integrated, helps to provide cogent models for certain geomorphic evolution processes.

To model the morphological dynamics of a real-world geomorphic structure, the best source of information is the various types of data acquired by remote sensing satellites in multitemporal domain. The other way of modeling the morphological dynamics is perhaps by recording the morphological changes episodically or continuously using GPSs. Further, the scope of the work in modeling the dynamics of real-world geomorphic data using multitemporal satellite data by following the concepts of mathematical morphology is foreseen positively.

Discrete Simulations of Spatiotemporal Dynamics of Small Water Bodies Under Varied Stream Flow Discharges

In this section, we provide a simple scheme to relate the time series of stream flow discharge with templates to simulate the various possible morphological dynamics of small water bodies (SWBs) in 2-D space. We employ mathematical morphological transformations to show such a relationship under certain theoretical assumptions. These assumptions are primarily based on a postulate that enlargements and contractions of SWBs are due to fluctuations in the stream flow discharge pattern. In the present investigation, the studies on spatiotemporal organization of randomly situated SWBs are carried out in discrete space under the influence of various stream flow discharge behaviors.

Large floods and intense droughts are capable of inducing spectacular changes in the morphological configuration of water body and its surroundings. Multi-date earth-observing remotely sensed satellite data of various resolutions are of use to monitor the climatically sensitive SWBs (e.g., Harris 1994). Hitherto, many studies emphasize characterizing the time series of 1-D stream flow discharge data to understand its behavioral pattern. The impacts of varied types of such patterns on the spatial phenomena (e.g., water bodies, streams) can be observed via 2-D maps retrieved from various remotely sensing sources at temporal intervals. By considering the SWBs that could be precisely retrieved from multi-date remotely sensed data, one can understand the spatiotemporal organization of the climatically sensitive SWBs to further validate the theoretical discrete models. The changes in planar shapes and sizes of climatically sensitive lakes can be better mapped from multi-date remotely sensed satellite data. These changes can be spatially correlated with the changes in stream flow discharge pattern. Toward this direction, this chapter gives a new insight into investigations, by stressing the importance of the geometry and topology of the randomly distributed SWBs. Spatiotemporal patterns of SWBs under the influence of temporally varied stream flow discharge are simulated in discrete space by employing geomorphologically realistic expansion and contraction transformations. Cascades of expansion–contraction are systematically performed by synchronizing them with stream flow discharge simulated via the logistic map. Templates with definite characteristic information are defined from stream flow discharge pattern as the basis to model the spatio-temporal organization of randomly situated surface water bodies of various sizes and shapes. These spatiotemporal patterns under varied parameters (λs) controlling stream flow discharge patterns are characterized by estimating their fractal dimensions. At various λs, nonlinear control parameters, we show the union of boundaries of water bodies that traverse the water body and non-water-body spaces as geomorphic attractors. The computed fractal dimensions of these attractors are 1.58, 1.53, 1.78, 1.76, 1.84, and 1.90, respectively, at λs of 1, 2, 3, 3.46, 3.57, and 3.99. These values are in line with general visual observations.

Introduction of Spatiotemporal Dynamics

Flood plain is found in an area of ecological transition from wet to dry and is characterized by flood and low water regimes. Flood and drought are two extreme events that show impact on climatically sensitive SWBs. Flood and drought are the effects, respectively, due to stream flow discharge that is more or less than mean stream flow discharge (MSD). Stream flow discharge dynamics controls the morphological dynamics of ephemeral SWBs that exist within a basin with less relief ratio, as there would not be much difference between the highest and lowest observed elevations in the floodplain basins. Due to this low relief ratio, the expansions and contractions of SWBs under the influence of variations in the stream flow discharge pattern are assumed isotropic. They tend to merge with each other when peak stream flow discharge is much larger than MSD. In contrast, due to intense drought during which the stream flow discharge is much lesser than MSD, the spatiotemporal organization of water bodies will be disturbed.

The homogeneous and heterogeneous progressive and retrogressive growths of lakes depend on various physical, meteorological, and physiographic factors. In a single dynamical system, these two phases may occur successively under the influence of peak stream flow discharge followed by low stream flow discharge that, respectively, lead to flood and drought. The homogeneous the stream flow discharge behavioral pattern, the more is the predictability of morphological dynamics of SWBs. The heterogeneous the stream flow discharge over a time period, the more complex is the morphological evolution.

In general, varied degrees of two types of morphological changes that we visualize include isotropic expansion and contraction. This investigation is based on the following postulates: (1) variations in stream flow discharge pattern cause modifications in geomorphic organization; (2) expansion and contraction depend on original spatial organization, as sparser phenomenon is worst affected due to low stream flow discharge compared to denser phenomena; and (3) morphological evolutionary pattern in these phenomena follows the stream flow discharge behavioral pattern. The heterogeneities in these morphological processes may be attributed to topographic effects.

It is reported in several studies that the behavior of such stream flow discharges may produce low-dimensional attractor that depicts chaoticity (e.g., Savard 1990, 1992, Jayawardena and Lai 1994, Tsonis et al. 1994, Beauvais and Dubois 1995, Pasternack 1999, Sivakumar 2004). In these studies, correlation dimension of attractors governing the trajectories of stream flow discharge dynamics describes the degree of chaoticity in the stream flow discharge time series. In the present investigation, time series of such mean discharges are simulated through first-order nonlinear difference equation, the logistic map, further to relate the impact of changing stream flow discharges on spatiotemporal organization of randomly situated SWBs.

Although this study, on simulation of morphological dynamics of SWBs by employing mathematical morphological transformations, is first of its kind, the applications of morphological transformations in the context of geomorphology and geophysics are common in the extraction of significant geomorphological features from digital elevation models (DEMs) (Sagar 2001, Sagar et al. 2001, 2003, Chockalingam and Sagar 2003), estimation of basic measures of water bodies (Sagar et al. 1995a,b) and roughness indexes of terrain (Tay et al. 2007), modeling and simulation of geomorphic processes (Sagar et al. 1998, Sagar 2001), generation of fractal landscapes (Sagar and Murthy 2000), and fractal relationships among various parameters of geomorphological interest (Sagar and Rao 1995, Sagar 1996, 1999, 2000, Sagar et al. 1998, 1999, 2001, Chockalingam and Sagar 2004, Sagar and Chockalingam 2004, Sagar and Tien 2004, Tay et al. 2007).

Expansion–Contraction due to Flood–Drought

During progressive and retrogressive growths, SWBs respectively flood and vanish or disintegrate and then vanish. These two processes are simulated under the influence of various stream flow discharge behavioral patterns in discrete space by employing geomorphologically realistic expansion and shrinking transformations. These transformations of varied degrees are termed as the two succeeding phases of a geomorphic system. These transformations are popularly known as *dilation* and *erosion* (Matheron 1975, Serra 1982), hereafter referred to as flood and drought transformations.

The neighboring water bodies are connected under continuous flood process, and the clustered water bodies are disconnected during continuous drought process. Water bodies merge together during the process of continuous expansion by incessant stream flow discharge. During the process of continuous contraction by B of size more than that of water body, the water body either disappears or first disintegrates and then disappears. To generate higher degrees of drought or flood, these transformation processes are iterated. Instead of using a larger B (for peak stream flow discharge), with the use of smaller B repeatedly, one will get the same effect. Consecutive drought and flood transformations for n times are, respectively, represented as $(X \ominus nB)$ and $(X \oplus nB)$. The role of B that functions as an interface between water body and stream flow discharge is to simulate the effects of flood and drought.

Unique Connectivity Networks

Two unique connectivity networks include flow direction network (*FDN*) and self-organized critical connectivity network map (*SOCCNM*), which are one pixel wide caricatures that summarize the overall shape, size, orientation, and association of regions respectively occupied by water bodies (X) and their complimentary spaces (X^c). The *FDN* is mathematically defined as $FDN(X) = \bigcup_{n=0}^{N} FDN_n(X)$, where $FDN_n(X) = (X \ominus nB) \setminus (X \ominus nB) \circ B$. $FDN(X)$ is exactly similar to that of skeleton network extraction explained in Equations 2.15 and 2.16.

The sequential steps to extract *FDN* of (X) are similar to that illustrated in Figure 2.14. Similar steps are needed to perform on non-water-body space (X^c) to extract *SOCCNM* of (X). Flooding process, during which randomly situated surface water bodies of various sizes and shapes self-organize (Sagar 2001), is simulated mathematically as high degree of flood intensity makes the distant water bodies contact together to achieve *SOCCNM* (Figure 9.6). In a way, *SOCCNM* depicts the extinguishing points of self-organized water bodies at a critical state. These two topographically significant networks (Figure 9.6) enable the structural composition of water bodies and their complementary space. *SOCCNM* of (X) is mathematically expressed as $SOCCNM(X) = \bigcup_{n=0}^{N} SOCCNM_n(X)$, where $SOCCNM_n(X) = (X^c \ominus nB) \setminus (X^c \ominus nB) \circ B$.

Erosion and dilation mechanisms are employed to simulate the flood and drought impacts in discrete space. The importance of unique networks lies in the aspects of synchronizing the stream flow discharges with the travel time required for reaching the varied flood frontlines propagating from *FDN* to *SOCCNM*.

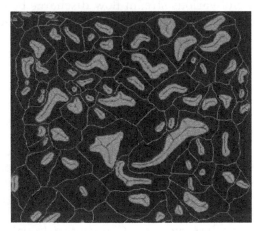

FIGURE 9.6
Randomly distributed surface water bodies (in gray-shaded objects) at their full capacity under the presence of MSD. Topological quantities *FDN* and *SOCCNM* are also shown. (From Sagar, B.S.D., *Nonlinear Process. Geophys., Am. Geophys. Union*, 12, 31, 2005.)

The travel time is nothing but the size of B. To simulate the effect of travel time in terms of the size of the template, we follow the postulate: A larger size of the template is required to simulate a flood propagating at higher speed. However, we follow in this investigation that the propagation is uniform. These unique networks (e.g., Figure 9.6) such as *FDN* and *SOCCNM* are employed to derive a template with certain characteristic information to synchronize the stream flow discharges required to simulate complete flood and complete drought.

Diameter of B that is required to construct a large cell of *SOCCNM* from *FDN* is considered as $B_{max} = NB$ that makes all the SWBs merge. *NB* is defined as the template large enough to fill the largest cell achieved. The flood and drought transformations, and the transformations to extract *FDN* and *SOCCNM*, are of use in visualizing all possible dynamical behaviors of SWBs under various stream flow discharge behavioral patterns.

Impact of Stream Flow Discharge on Spatial Organization of SWBs: Numeric versus Graphic

Patterns of orderly, periodically, and chaotically changing stream flow discharges at discrete time intervals are simulated through a first-order nonlinear difference equation. By employing these patterns, simulations and computations are performed on a large number of randomly situated and climatically sensitive SWBs (Figure 9.6) of various sizes and shapes. These water bodies are from a flood plain region of the Gosthani River, one of east-flowing rivers of India, within the geographical coordinates 18°07′ and 18°12′ north latitudes and 83°17′ and 83°22′ east longitudes. These SWBs under respective cycles of contraction and expansion are shown in Figure 9.7a and b, respectively, which depicts all possible frontlines from the origin to the *SOCCNM*.

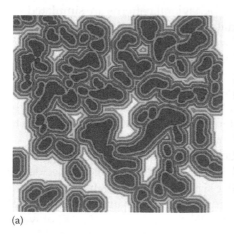
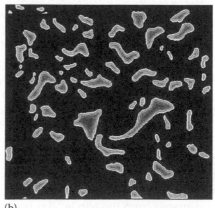

(a)   (b)

FIGURE 9.7
(a) Water bodies under continuous flooding due to continuous peak stream flow discharge, and (b) drought due to low stream flow discharge input. (From Sagar, B.S.D., *Nonlinear Process. Geophys.*, Am. Geophys. Union, 12, 31, 2005.)

We study the morphological evolution of these SWBs under the influence of various stream flow discharge inputs that are simulated according to Equation 9.2 (May, 1976): $A_{t+1} = \lambda A_t(1 - A_t)$, where λ is an environmental parameter $0 \leq \lambda \leq 4$, $0 \leq A_t \leq 1$, and $A_t(t \to \infty) \to 0$. Stream flow discharge in the normalized scale ranges from 0 to 1. It is shown that for $1 \leq \lambda \leq 3$ as $A_t(t \to \infty) \to$ constant value, the discharge value reaches a stable state and remains there. The environment provides enough stream inflow to sustain SWBs. This facilitates visualization of various possible spatiotemporal organizations of SWBs (Figure 9.6). For example, the interplay between numerically simulated stream flow discharge and its impact on the spatially distributed SWBs is assumed for a case when $\lambda = 3.99$ as the amount of stream flow discharge in succeeding times is oscillating chaotically between high and low to dissimilar degrees. It means that SWBs are undergoing cascade of flood–drought (C-FD) transformation, in which the flood followed by drought is not of the same degree. If we see the whole process in a reverse way, then the SWBs undergo cascade of drought–flood (C-DF) to varied degrees. For further changes in λ, one can visualize the other discrete spatiotemporal patterns of SWBs. In a sense, the SWBs' morphological dynamics is a coupled system that depends on the dynamics of stream flow discharge. While considering the MSD as the basis, a heuristically true argument is that the reduction in stream flow discharge that is capable of vanishing the water bodies of all sizes may be equivalent to the amount of stream flow discharge that is capable of making multiple water bodies merge together. In support of this argument, the number of drought cycles due to B required to vanish the SWBs in a floodplain basin is equivalent to the number of flood cycles due to B required to merge SWBs. The amount of stream flow discharge much lesser than or much greater than MSD respectively indicates the presence of drought and floods. By presuming A_{t+1} (areal extent) much lesser than MSD, i.e., 0, and A_{t+1} much greater than MSD, i.e., 1, various stream flow discharge behavioral patterns are simulated.

To link the stream flow discharge data simulated under varied λs, with the degree of either flood or drought, we adapt the procedure by considering a relationship between *SOCCNM* and *FDN*. This relationship explains the time required by two neighboring SWBs to merge. For the present case, *FDN* and *SOCCNM* are shown in loopless and looplike networks for SWBs (Figure 9.6). It is presumed that *NB* to attain $FDN_N(X)$ is equivalent to *NB* to attain $SOCCNM_N(X)$. This *NB* is considered as the template in matrix form to simulate either complete flooding or drought.

The *NB* is related to the largest stream flow discharge value that is able to merge the neighboring water bodies. Further, this *NB* is taken as the basis to decide the other possible templates of various smaller sizes, correspondingly to relate with other stream flow discharge values. For instance, the maximum distance that is estimated from the *N*th level subsets of $FDN_N(X)$ and $SOCCNM_N(X)$ of water bodies is $NB \oplus NB = 1$ (stream flow discharge in normalized scale) and $NB = 0.5$ (stream flow discharge in normalized scale) that makes water bodies attain their full capacity. Similarly, when there is absolutely no stream flow discharge, such an aspect is linked to the minimum value in the time series of simulated stream

TABLE 9.3

Hypothetically Represented Flood and Drought Transformations by means of B of Specified Diameter and Their Relation with Normalized Stream Flow Discharge Values

Stream Flow in Normalized Scale	Diameter of *B* in Pixels	Process	Stream Flow in Normalized Scale	Diameter of *B* in Pixels	Process
1	10	Flood	0.4	6	Drought
0.9	9	Flood	0.3	7	Drought
0.8	8	Flood	0.2	8	Drought
0.7	7	Flood	0.1	9	Drought
0.6	6	Flood	0.0	10	Drought
0.5	0	No process	—	—	—

Source: Sagar, B.S.D., *Nonlinear Process. Geophys. Am. Geophys. Union*, 12, 31, 2005.

flow discharge in the normalized scale, i.e., 0. The water bodies at stability state attain their full capacity filled due to the presence of consistent MSD (Table 9.3). With an environmental parameter (λ) value of 2, all the SWBs attain stability as there would be no change in the simulated stream flow discharges across discrete time intervals. When the pattern of stream flow discharge is unusual, the climatically sensitive SWBs behave differently. For the present case, the MSD is assumed as 0.5, which makes all water bodies attain their full level. This explains the impact of variations in the stream flow discharge pattern on the spatial organization of the water bodies that are assumed to be at stable state under the availability of stream flow discharge of 0.5.

A stream flow discharge less than MSD makes the SWBs contract, while a stream flow discharge greater than MSD makes the SWBs expand. By means of this template, MSD (i.e., 0.5 in normalized scale), and stream flow discharge value simulated from the logistic equation, we impose an appropriate morphological transformation. To determine the involved morphological transformation, we check the stream flow discharge values at discrete time intervals with reference to 0.5 (i.e., MSD). These relationships are depicted as

$$\text{If } A_{t+1} > 0.5, \quad \text{then } X \oplus NB$$

$$\text{If } A_{t+1} < 0.5, \quad \text{then } X \ominus NB \tag{9.3}$$

$$\text{If } A_{t+1} = 0.5, \quad \text{then } X \oplus 0B$$

In other words, maximum level of flood that merges all the water bodies under the availability of stream flow discharge that is much higher than the MSD is expressed with a morphological relationship as follows:

$$(FDN(X) \oplus NB \oplus NB) = (X \oplus NB) \tag{9.4}$$

where NB and $0B$ are the sizes of templates that are equated with normalized stream flow discharge values, respectively, at 1 and 0. Similarly, the stream flow discharges that keep the water bodies at stable levels and vanish, respectively, are morphologically related as

$$[FDN(X) \oplus NB] = [X \oplus 0B] \tag{9.5}$$

$$(X \ominus NB) = (FDN(X) \oplus 0B) \tag{9.6}$$

Geomorphological Attractors

For SWBs, NB is derived as the template with a radius of 15 pixels. The template with a radius of 15 pixels is required to vanish the water bodies of various sizes in the section under the drought transformation. Hence, to relate the template with a radius of 15 pixels to the normalized stream flow discharge values, 0.5 is divided by 15, which yields each cycle of either drought or flood transformation with the interval of 0.03333. Table 9.4 depicts these

TABLE 9.4

Morphological Transformations due to Stream Flow Discharge Template Derived from Varied Normalized Stream Flow Discharge Values

N	Stream Flow Discharge	Notation	Environmental Phase	N	Stream Flow Discharge	Notation	Environmental Phase
15	1.0000	$X \oplus 15B$		1	0.4666	$X \ominus 1B$	
14	0.9666	$X \oplus 14B$		2	0.4333	$X \ominus 2B$	
13	0.9333	$X \oplus 13B$		3	0.4000	$X \ominus 3B$	
12	0.9000	$X \oplus 12B$		4	0.3666	$X \ominus 4B$	
11	0.8666	$X \oplus 11B$		5	0.2222	$X \ominus 5B$	D
10	0.8333	$X \oplus 10B$	F	6	0.3000	$X \ominus 6B$	r
9	0.8000	$X \oplus 9B$	l	7	0.2666	$X \ominus 7B$	o
8	0.7666	$X \oplus 8B$	o	8	0.2333	$X \ominus 8B$	u
7	0.7333	$X \oplus 7B$	o	9	0.2000	$X \ominus 9B$	g
6	0.7000	$X \oplus 6B$	d	10	0.1666	$X \ominus 10B$	h
5	0.6666	$X \oplus 5B$		11	0.1333	$X \ominus 11B$	t
4	0.6333	$X \oplus 4B$		12	0.1000	$X \ominus 12B$	
3	0.6000	$X \oplus 3B$		13	0.0666	$X \ominus 13B$	
2	0.5666	$X \oplus 2B$		14	0.0333	$X \ominus 14B$	
1	0.5333	$X \oplus 1B$		15	0.0000	$X \ominus 15B$	
0	0.5000	$X \oplus 0B$	Stable				

Source: Sagar, B.S.D., *Nonlinear Process. Geophys., Am. Geophys. Union*, 12, 31, 2005.

details with the involved morphological processes at respective stream flow discharge values. The higher the number of cycles that a section containing the water bodies requires to establish either *FDN* or *SOCCNM*, the closer the comparison with the values in normalized scale.

Variations in stream flow discharges are due to several factors that include rainfall pattern and landscape topological organization. Time series of such fluctuating stream flow discharges is simulated according to the first-order nonlinear difference equation as observed stream flow records are insufficient. These simulated data are considered to study how the boundaries of the SWBs are modified. The morphological behaviors of SWBs under varied simulated stream flow discharge behavioral pattern, by considering initial stream flow discharge $A_0 = 0.5$ for all the cases and $\lambda \in (1,4)$, are simulated. This phenomenon is better explained through the logistic map (Figure 9.8). Various types of morphological behaviors of SWBs that include attracting to initial conditions, stable, periodically changing, and chaotically changing are simulated (Figure 9.9a through f). The water bodies' boundaries (δ_X) at the next time period is defined as a function of that of the preceding time period and given as $\delta_{X_{t+1}} = f(\delta_{X_t})$, where δ_{X_t} is $(X_t - (X_t \ominus B))$. The union of boundaries of the dynamically changing SWBs, which are superimposed patterns, is termed as attractor describing the morphological dynamics of SWBs under varied stream flow discharge dynamics. The attractor of SWBs' space–time morphological dynamics is defined as

$$\bigcup_{t=0}^{n} (\delta_{X_t}) \tag{9.7}$$

where X_0 is a section consisting of water bodies during the presence of MSD. The impact of varied stream flow discharge dynamics, simulated numerically via first-order nonlinear difference equation, on the SWBs is visualized by synchronizing with appropriate degrees of contraction and expansion (Figure 9.9).

The computed fractal dimensions of the spatiotemporal patterns of SWBs that are simulated by considering the time series of stream flow discharge simulated at $\lambda = 1, 2, 3, 3.46, 3.57$, and 3.99, respectively, are $1.58, 1.52, 1.78, 1.72, 1.84$, and 1.90 (Figure 9.10; Table 9.5). It is apparent from the fractal dimensions of these attractors that the higher the fractal dimension, the higher is the randomness in the morphological behavior. When λ is 1 and 3, the spatiotemporal patterns of SWBs exist or completely occupy the region within the SWBs that is attained under the availability of MSD. Hence, the fractal dimensions are higher than that of the succeeding threshold control parameters, e.g., $\lambda = 2$ and 3.46. The spatiotemporal patterns, which have aroused under the influence of λ values of 2 and 3.46 are, respectively, one or two patterns. Hence, the fractal dimensions are lesser than their preceding λ values. The higher the fractal dimension, the greater is the difficulty in predicting the behavior. The rises and falls of the levels of water bodies lead to a dynamic sequence of adjustment throughout the year.

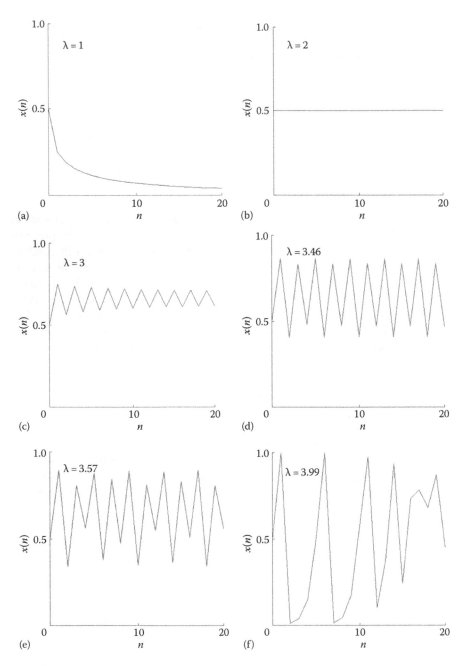

FIGURE 9.8
Stream flow discharge behavioral pattern at different environmental parameters. (a–f) $\lambda = 1$, 2, 3, 3.46, 3.57, and 3.99. (From Sagar, B.S.D., *Nonlinear Process. Geophys., Am. Geophys. Union*, 12, 31, 2005.)

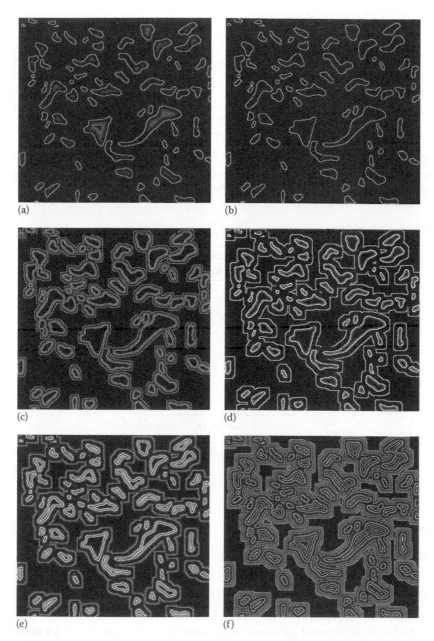

FIGURE 9.9
Spatiotemporal organization of the surface water bodies under the influence of various stream flow discharge behavioral patterns at the environmental parameters at (a–f) $\lambda = 1, 2, 3, 3.46, 3.57$, and 3.99 are shown up to 20 time steps. In all the cases, the considered initial MSD, $A_0 = 0.5$ (in normalized scale), is considered under the assumption that the water bodies attain their full capacity. It is illustrated only for the overlaid outlines of water bodies at respective time steps with various λs. (From Sagar, B.S.D., *Nonlinear Process. Geophys.*, *Am. Geophys. Union*, 12, 31, 2005.)

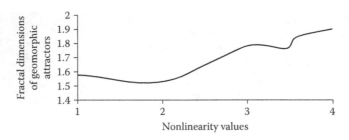

FIGURE 9.10

Relationship between λ and fractal dimension of geomorphological attractors. (From Sagar, B.S.D., *Nonlinear Process. Geophys.*, Am. Geophys. Union, 12, 31, 2005.)

TABLE 9.5

Fractal Dimensions
of SWB Attractors

Environmental Parameter (λ)	SWB
1	1.58
2	1.53
3	1.78
3.46	1.76
3.57	1.84
3.99	1.90

Source: Sagar, B.S.D., *Nonlinear Process. Geophys.*, Am. Geophys. Union, 12, 31, 2005.

Numerical Simulations Through First-Order Nonlinear Difference Equation to Study Highly Ductile Symmetric Fold Dynamics: A Conceptual Study

The study of deformation in geological materials is one of the important tasks in structural geology. Fold one of such geological formations may be transformed due to mechanical properties. These transformations may be according to a rule through which one can predict the dynamical changes in folds. Several papers have emerged during the last decade, which cast the application of fractal concepts to study the fold mechanism. Several models are developed to study the folding processes and mechanisms (Chapple 1968, Dieterich and Carter 1969, Dieterich 1970, Parrish 1973, Means 1976, 1990, Ramsay and Huber 1987, Price and Crosgrove 1990). Behavior of various systems of geoscientific interest such as electrical conductivity and fractures of rocks to the microcrack population (Maden 1983), coalescence of fractures (Allegre et al. 1982, Newman and Knopof 1982) and stick–slip behavior (Smalley et al. 1985) through renormalization group approach, and the fault models using fractals and homogenization concepts (Davy et al. 1990) were studied. The rate of deformation depends not only on the rock mechanical properties of the geological formations and the energy acting on it but also on the antecedent morphological state of the fold. The shortening and amplification in the symmetric folds can be seen due to variations in the stress and ductility of the fold. Ductile folds are precarious to stresses.

Moreover, fluctuations in the stress dynamics result in variations in the dynamical behavior of a symmetric fold ranging from steady state to periodicity and chaotic state. The random behavior of fold, from its inception of the formation, is due to stress dynamics and the internally exerting forces (IEFs) that randomly influence the fold. The ductile folds of vertical axial type are subjected in the present qualitative investigation. The significant point is that this study is based on the assumption that the deformation in the ductile fold is not permanent and also that it will not ensue the state of brittleness during the influence of stress dynamics. In particular, this section deals with a continuous phase transition in a symmetric fold under dynamical conditions by considering Equation 9.2. The logic behind using Equation 9.2 in regard to understanding the fold morphological dynamics is as given in the following paragraph.

The intensity of the cause can be derived from the effect. Such a derived cause might be in terms of various physical forces (stress and internally exerting force). The collectively acting coexisting physical forces are the cause to see the effect. This effect is in terms of deformation. Such a deformation can be quantified by means of an analytical value (e.g., fractal dimension [Mandelbrot 1982]). By considering this quantified parameter at discrete time intervals, the term called stress regulatory force can be derived. These fractal dimensions at discrete time intervals enable that the dynamics of fold is of nonlinear type. However, based on the instinctive argument, it is apprehended that the fold dynamics follows nonlinear rules. This intuitive argument may be endured by the fact that due to the heterogeneous nature of external and internal stress influences, folds may undergo compression, amplification, cascade of compression–amplification, and shear over a time interval. This argument is also supported by a postulate that the successive phases of a fold undergoing dynamics may be nonoverlapping; moreover, the output in terms of fractal dimension of the fold undergoing dynamics may not be directly proportional to its input. This phenomenon is due to the fact that the stresses and internally exerting forces are divergently balanced at discrete time intervals. These unequally balanced forces act against each other. Therefore, it is also visualized that the morphological dynamics of a fold is nonlinear. To carry out computer simulations to visualize distinct possible behaviors concerning a change in control parameter, a first-order nonlinear difference equation (see May 1976), which has physical relevance as the simplest possible model of a highly ductile symmetric fold (HDSF) undergoing morphological changes, is considered as the basis to further derive Equations 9.9 through 9.12 and 9.16 through 9.19. Hence, qualitative studies have been carried out for understanding the fold morphological dynamics and the acting stress dynamics of the fold by considering the first-order nonlinear difference equation (9.2). The definition of symmetric folds and the basic equations that are considered to study these symmetric folds are described in the "Logistic Equations to Study Fold Dynamics" and "Computation of IA (θ) of Corresponding NFD (α) of a Symmetrical Fold Under Dynamics" sections, respectively.

Symmetric Folds with Three (Fold Type I) and Two (Fold Type II) Limbs

The description of the morphology of a fold pattern is mainly concerned with the outcrop of its profile. Generally, the nose of the fold is described as round or angular. If the limbs of a fold are of equal lengths, the fold is said to be symmetric (e.g., chevron or concordian fold) (Hobbs et al. 1976). A typical asymmetric fold pattern is shown in Figure 9.11b, where one limb length differs from that of two other limbs. In this section, two types of upright symmetric folds of vertical axial type (e.g., zigzig, chevron, or accordion folds) with rigid limbs (Figure 9.11a and c) are considered. An upright

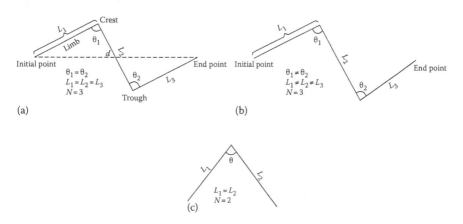

(a)   (b)

(c)

FIGURE 9.11
(a) Symmetric fold with three limbs, (b) an asymmetrical fold pattern, and (c) symmetric fold with two limbs. (From Sagar, B.S.D., *Discrete Dyn. Nat. Soc.*, 2, 181, 1998.)

symmetric fold (i.e., dip of the axial surface) with three limbs (Figure 9.11a) with the following specifications is studied:

- Fold pattern should have three limbs ($N = 3$) with equal lengths (L), ($L_1 = L_2 = L_3$), forming an anticline and a syncline.
- The angles (θ_1, θ_2) between the two successive limbs should be equal ($\theta_1 = \theta_2$).
- The distance of vertical projection, d, should be greater than the length of a rigid limb ($d > L_1 = L_2 = L_3$).

An upright symmetric fold with two limbs (Figure 9.11c) with the following specifications is also studied:

- Fold pattern should have two limbs ($N = 2$) with equal lengths (L), ($L_1 = L_2$) forming an anticline or a syncline.
- The distance of vertical projection, d, should be greater than the length of a rigid limb ($d > L_1 = L_2$).

The length of the fold limb (L) is considered as rigid when stress is acting on it. The stress concerned here is referred to horizontal stress only. Barring this, d varies with the difference in the stress. The four possibilities of fold transformation that may arise in nature are presented (Table 9.6). If stress at discrete time intervals $\lambda_t > \lambda_{t+1}$ or $\lambda_{t+1} > \lambda_t$ play successively, the morphology of the HDSF changes, which is obvious in geological context.

TABLE 9.6

Four Possible Dynamics of Symmetrical Fold

Probable Circumstances	Probable Dynamical Process	Trajectory Behavior
A fold with high sinuosity index may become straight	Due to dominating internally acting exerting force	Attracting to an initial condition
A fold with medium sinuosity index may increase as time progresses and then converge to a point from which any two patterns will overlap	Due to unequal stress and the internally acting force	Attracting to a fixed point
A fold oscillating between two sinuosity indexes	Fold shape oscillating between two points periodically—shortening and amplification, and vice versa	Oscillating between two points
A fold with either low or high tortuosity may behave chaotically such that no two patterns overlap	Cascade of aperiodic stress and internally exerting forces	Chaotically behaving

Source: Sagar, B.S.D., *Discrete Dyn. Nat. Soc.,* 2, 181, 1998.

Logistic Equations to Study Fold Dynamics

The dynamical rule is visualized in the present investigation in two ways. They are according to the first-order difference Equation 9.2 and a modulated logistic Equation 9.8. In the former case, the stress regulatory parameter λ is a constant stress control parameter, which acts against the internally exerting forces, whereas in the latter rule, λ_t is controlled by the strength of the stress modulated parameter (SSMP) μ to understand the time-dependent stress control parameter, λ_t, which describes the time-dependent evolution of the fold morphological dynamics. Two types of fold dynamical systems are studied here:

1. One that undergoes constant stress dynamics (CSD)
2. One that undergoes time-dependent stress dynamics (TDSD)

First-Order Difference Equation as a Dynamical Rule

The fold morphological dynamics is controlled by a time-dependent stress regulatory parameter.

The general form of the difference equation is taken as the dynamical behavior of symmetric folds under different total effective stresses, which is studied by following a function shown as the nonlinear first-order difference equation (9.2). From the knowledge of the strain states of the fold at specific time intervals, the condition of the stress can be calculated. Force per unit area is stress. This is used to study the agents responsible for the deformation in the rock as it progressively changes shape. Such a study needs to investigate the nonlinear equations in which the stress that controls the fold dynamical system is constant during the evolution. To carry out such a study, Equation 9.2 may be considered as a dynamical rule: $\alpha_{t+1} = \lambda \alpha_t (1 - \alpha_t)$. The limits of λ are 1 and 4, and the strains at respective states are quantified by α as 0 and 1. The numerical representation 4 for λ, and α stand for any number, say 1000 kbar and the upper limit of fractal dimension in normalized scale respectively.

Computation of SSM

The SSM can be considered either as a constant or as a time-dependent parameter that controls the fold morphological dynamics. Rather than computing the physical forces that alter the fold dynamics, from the strain, the dynamics of the stress regulatory parameter can be computed. The collective impact of such stresses (cause) that alter fold morphology can be defined by studying the (degree of deformation) effect due to the cause at discrete time intervals. As the fractal dimension enables the characteristic of the fold that is shortened as well as amplified, the parameter representing the strength of the regulatory force can be defined as a numerical value. From the degree of deformation states at discrete time intervals, one can tell whether the stress influence is constant or not by fitting α_{t+1} versus α_t to fit the curve $\alpha_{t+1} = \lambda(1 - \alpha_t)$. This derived stress is the slope value of the fitted curve. Such a value, $1 < \lambda < 4$, is considered as a constant stress. This constant stress can also be computed from the fractal dimensions of a fold at discrete time intervals. The fluctuations in the fold morphology depend on the changes both in the stress intensity and in the original constitution of the fold. If one knows the stress states at different time intervals, say, λ_t, λ_{t+1}, ..., λ_{t+N}, the SSM (μ) can be derived to compute the time-dependent stress states by plotting λ_{t+1} versus λ_t to fit the curve $\lambda_{t+1} = \mu(1 - \lambda_t)$: It is hypothesized that as the time-dependent stress regulatory parameter attains higher value, at subsequent times, it is controlled by the factor $(1 - \lambda_t)$. It is visualized that if the stress regulatory force is high, make it small, and vice versa. This is a wonderful recipe to carry out simulation numerically. The time-dependent stress that, in turn, controls the fold morphology can be computed from the stress states in a time series form. This aspect is to study the coupled systems. In this coupled system, which is detailed in the sequel, the stress and fold morphological dynamics are interdependent.

Symmetric Fold Dynamics Under the Influence of Constant Stress

A fold with high sinuosity will have an interlimb angle (IA) of $\theta = 60°$ (for three-limb fold) and $\theta = 90°$ (for two-limb fold), and for a linear fold, $\theta = 180°$. A fold with high sinuosity will have a value of α approaching 1, and for a straight line, $\alpha = 0$. The upper and lower limits of α, viz., 0 and 1, arise at lowest and greatest stress states, viz., $\lambda = 1$ and 4, respectively. The parameter λ gives total description of the dynamics of fold. The impact of the unequal compressive forces on a symmetric fold in terms of its dynamical behavior is investigated through the first-order difference equation of the form $\alpha_{t+1} = f(\alpha_t)$; the fractal dimension in normalized scale at $t + 1$, α_{t+1}, is given as some function, f, of the fractal dimension at time t, α_t. If this equation were linear ($f = \lambda\alpha$), the fractal dimension would simply increase or decrease exponentially if $\lambda < 1$. Moreover, the fractal dimension tends to increase when at low α and to crash at high α value, corresponding to some nonlinear function, with a hump, of which the quadratic is $f = \alpha_{t+1} = \lambda\alpha(1 - \alpha)$. It does mean that there is a tendency for the variable α to increase from time "t" to the next when it is small and for it to decrease when it is large. When the symmetric fold possesses less fractal dimension, there may be a possibility for it to get compressed due to stresses that dominate internal force. When it possesses high fractal dimension, due to internal forces that dominate the stress acting against, this may lead to a decline of the fractal dimension. This tendency is due to the fact that the internally exerting forces dominate the impact of stresses. The impacts of internal forces fluctuate. These fluctuating impacts depend on the α values. The reason behind this possibility may be the fact that during the fold dynamics, unequal internal forces influence the fold at discrete time intervals and also the variations in the strength of the fold itself. This tendency is preserved due to $(1 - \alpha_t)$ in Equation 9.2. Equation 9.2, to compute $\alpha_t + 1$, $\lambda\alpha_t(1 - \alpha_t)$, explains that the normalized status of a symmetric fold dynamics, if α starts at larger than 1, immediately goes negative at one time step. Moreover, if $\lambda > 4$, the hump of the parabola exceeds 4, thus enabling the initial α value near 0.5 to shear in two time steps. Therefore, the analysis is restricted to value of λ, α between 0 and 1. It is also interesting to study the critical states from which the internal forces dominate the external stresses (CSD). The impacts of such internal forces acting alternatively are predominant at larger threshold regulatory stresses. This idea can be seen from the depicted bifurcation diagrams in the "Bifurcation Diagrams" section.

In qualitative understanding of the dynamical behavior, the value α_{t+1} is obtained from the previous value α_t by multiplying it by $\lambda(1 - \alpha_t)$; it is clear that for $\lambda(1 - \alpha_t)$ greater than 1, the successive values, viz., α_{t+2}, α_{t+3}, α_{t+4}, ..., α_{t+N}, will grow bigger, i.e., a change in α_t will get amplified. This is the fold shortening due to relatively high stress. However, α_t cannot increase indefinitely because of the mechanical properties of the geological material makeup of the stratum. $\lambda(1 - \alpha_t)$ becomes smaller than 1, and the subsequent values must diminish. In the context of fold dynamics, this is fold stretching (amplification) due to high impact of exerting forces that dominate the stress. To determine the stability concerning incessantly acting stress with different magnitudes, a linearized analysis may be conducted through the studies of the dynamical behaviors of a model that is described by the first-order difference equation, which consists in finding constant equilibrium solutions.

Fold Morphological Dynamics Under the Influence of Time-Dependent Stress

In contrast to the fold dynamics, under the influence of constant stress, the behavior variations may be observed when stress is made time dependent. This idea is induced from the following statement of Ruelle (1987). It states that the behavior of a dynamical system can be studied with adiabatically fluctuating parameters where the control parameter has a very slow variation in time and this time dependence itself might be determined by a dynamics. This is the origin to consider stress as a time-dependent parameter that controls the fold morphological dynamics.

Besides this, the logic behind using the TDSD is that the complexity of fold morphological dynamics depends on the complexity of stress dynamics. Hence, in understanding the fold dynamics, the dynamics of the stress should also be understood.

The dynamics of the time-dependent stress is a possibility for stress being a time-dependent parameter, which may be confirmed from the fact that the stress influence is not homogeneous in the time domain. In such a case, understanding the dynamics of stress is an important event. However, we assumed that the stress at time $t + 1$ is not directly proportional to the stress at time t. This engendered to consider the first-order nonlinear difference equation as a rule to understand the stress dynamics also (Equation 9.8). In Equation 9.8, λ_t is a time-dependent stress and μ is the SSMP that controls the TDSD. By considering this time-dependent stress (λ_t), the degree of deformation at discrete time intervals may be studied by the modulated logistic equation (9.8). To show the effect of time-dependent stress regulatory parameter on the fold dynamical system, Equation 9.8 is considered to carry out numerical simulation.

In Equation 9.8, the behavior of α is controlled by the behavior of λ. This is explored as the fold and the stress, which is represented in numerical form $1 < \lambda < 4$, dynamical systems, in which the behavior of the fold morphology depends on the behavior of λ. It means that this coupled system contains two dynamical systems, in which the dynamical parameters are α and λ_t. The equation to describe this coupled system is written from Equation 9.2 as Equation 9.8:

$$\alpha_{t+1} = \lambda_t \alpha_t (1 - \alpha_t), \quad \lambda_{t+1} = \mu \lambda_t (1 - \lambda_t) \tag{9.8}$$

Various phases that fold dynamics can undergo, under the influence of constant and time-dependent stresses, can be studied by following Equations 9.2 and 9.8, respectively. In Equations 9.2 and 9.8, a detailed form of forces and fluxes will be indirectly represented by λ (CSD) or μ (strength of stress modulation to model the time-dependent stress).

Computation of IA(θ) of Corresponding NFD(α) of a Symmetrical Fold Under Dynamics

By considering the parameters such as fractal dimension (Mandelbrot 1982), in normalized scale α to describe the change in morphology of the fold, and the constant (λ) and the time-dependent stress regulatory parameter (λ_t) to describe the detailed form of forces and fluxes in the proposed equations (9.9 through 9.12 and 9.16 through 9.19), the dynamical behavior of symmetric fold types I and II that may behave from stable to chaotic can be quantified. Fold type I Equations 9.9 and 9.10 are proposed, which include certain specifications of a symmetric fold type I under evolution according to Equation 9.2 to record the changing IAs (θ) for both constant (Equation 9.9) and time-dependent (Equation 9.10) stress regulatory parameters:

$$\theta_{t+1} = \cos^{-1}\left(\frac{5 - 10^{2\log N/[\lambda \alpha_t (1-\alpha_t) + D_T]}}{4} \right) \tag{9.9}$$

$$\theta_{t+1} = \cos^{-1}\left(\frac{5 - 10^{2\log N/[\lambda_t \alpha_t (1-\alpha_t) + D_T]}}{4} \right) \tag{9.10}$$

Fold type II Equations 9.11 and 9.12 are proposed to compute the IA for the symmetric fold type II, which is under evolution according to a rule of Equation 9.2. Equations 9.11 and 9.12 are proposed respectively for both constant (λ) and time-dependent (λ_t) stress regulatory parameters:

$$\theta_{t+1} = \sin^{-1}\left(\frac{10^{\text{Log}\,N/[\lambda\alpha_t(1-\alpha_t)+D_T]}}{2} \right) \tag{9.11}$$

$$\theta_{t+1} = \sin^{-1}\left(\frac{10^{\text{Log}\,N/[\lambda_t\alpha_t(1-\alpha_t)+D_T]}}{2} \right) \tag{9.12}$$

Relation between α and θ

The variables α and θ are, respectively, denoted for the fractal dimension in normalized scale and the IA of the symmetric fold. As the fold is contracted horizontally in such a way that the limbs (L) will not change and by having the change in d, the IAs (θ) will be changed. A symmetric fold with high degree of linearity (straight) approximately possesses 180° IA. A fold with high sinuosity such that it is self-avoiding at any higher magnifications possesses 60° IA. A symmetric fold with 60° and 180° of IAs possesses fractal dimensions 2 and 1, respectively. However, these two limits of IAs for the type II fold are, respectively, 90° and 180°. A symmetric fold under dynamics will reach to criticality where the ratio between $\text{Log}(N)$ and $\text{Log}(d/L)$ becomes 2. At this critical state, the IA becomes 60, which is called critical angle, θ_{crit}. This critical angle for the symmetric fold type II is 90°. A symmetric fold under study is self-avoiding if and only if $\theta > \theta_{\text{crit}}$. With $\theta < \theta_{\text{crit}}$, fold pattern gets sheared. At the critical angle, θ_{crit}, the parameter α attains its peak value, $\alpha = 1$. The corresponding fractal dimension is at its criticality, i.e., $\alpha + D_T = 2$, for intersecting. With α_t and λ as 0.5 and 4, respectively, the α value of the fold under evolution at time $t + 1$ enables at one single time step, and the θ will be found at its criticality. Once the IA reaches its criticality, the symmetric fold may become stable, get stretched, or break as the influence of the stress continues. Equations 9.9 through 9.12 and 9.16 through 9.19 help to observe how the IAs are restricted between 180° and 60° and 180° and 90° for the fold types I and II, respectively, under the influence of CSD and TDSD. The latter values, 60° and 90°, are critical angles beyond which the folds self-intersect. The magnitude of variation in the θs from time t to $t + 1$ depends on the intensity of the stress and the internally exerting forces that the fold is subjected to. As shown in Equation 9.2, $\alpha \in [0,1]$, representing the fold with linearity and with the greatest possible contortion, respectively. The corresponding θs at $\alpha = 0$ and 1 are computed as 180° (lower limit) and 60° (upper limit), and 180° (lower limit) and 90° (upper limit) for symmetric folds with three and two limbs, respectively. It is worth mentioning that the fold, possessing parasitic

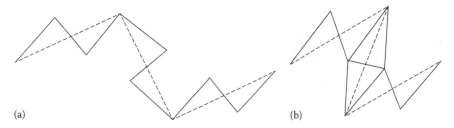

FIGURE 9.12
Symmetric folds with several folds of different IAs are shown schematically. (a) Schematic of self-avoiding symmetric fold profile with second-order folds. The IA of first-order fold (shown in dotted line) is greater than 60°, and (b) a schematic of self-intersecting symmetric fold profile with second-order folds. The IAs of first-order fold shown (as dotted line) is lesser than critical angle, i.e., 60°. Hence, it is self-intersecting. The intersecting second-order folds may be seen. (From Sagar, B.S.D., *Discrete Dyn. Nat. Soc.*, 2, 181, 1998.)

folds, will self-intersect at less than the upper limits, viz., 60° and 90°, for the two types of folds. The lower and upper limits represent the most probable contorted fold at which the parasitic folds will self-intersect and the linear structure before getting folded, respectively. It is essential to mention that the first-order fold at various magnifications contains parasitic folds that contain still minor folds and so on. Up to 60° of IA of a symmetric fold at any higher magnification, minor folds that possess exact self-similarity will not self-intersect. With the IA of a first-order symmetric fold with lesser than the critical angle, minor folds will self-intersect. For better comprehension, this phenomenon is represented diagrammatically in Figure 9.12. From θ, the IA, the corresponding NFD can be calculated for the symmetric folds with three and two limbs, respectively, from Equations 9.13 and 9.14:

$$\alpha = \frac{2\mathrm{Log}N}{\left[\mathrm{Log}(5 - 4\cos\theta)\right]} - D_T \quad (\text{for } N = 3) \tag{9.13}$$

$$\alpha = \frac{\mathrm{Log}N}{\left[\mathrm{Log}\left(2\sin(\theta/2)\right)\right]} - D_T \quad (\text{for } N = 2) \tag{9.14}$$

These expressions give the NFD of the symmetric folds with three limbs and two limbs.

The corresponding NFDs for these folds with θ > 60°, 90° < 180° are 0 < α < 1.

Iteration by Considering θs at Discrete Time Intervals

Instead of considering the αs, one can consider the θ values to carry out simulations for fold modeling. Equations 9.16 through 9.19 are proposed in which the IAs are considered instead of the NFDs to compute the IAs of the fold undergoing dynamics according to the first-order difference equation as a dynamical rule. These equations are similar to Equations 9.9 through 9.12. It is intended to compute the IAs at time $t + 1$ by considering θ at time t as some function from the relation between α and θ described in the "Results of Simulations" section. The following generalized equation (9.15), which is akin to that of Equation 9.1, is considered to perform functional iteration:

$$\theta_{t+1} = f(\theta_t) \tag{9.15}$$

The function in Equation 9.15 is expanded as Equation 9.16 by substituting Equations 9.9 and 9.13 for the fold with three limbs that is undergoing dynamics as

$$\theta_{t+1} = \cos^{-1}\left\{\frac{5 - 10^{\frac{2\log N}{\{\lambda\{\log N/[\log[2\sin\theta_t/2]]-D_T\}\{1-\{\log N/[\log[2\sin\theta_t/2]]-D_T\}\}+D_T}}}{4}\right\} \tag{9.16}$$

The expression as an exponent is based on the first-order nonlinear difference equation. In the earlier equation, the strength of stress regulatory force is a constant stress regulatory parameter.

However, the emphasis is also given in the present investigation to carry out the iterations to understand the possible dynamics by understanding the dynamics of the time-dependent stress regulatory parameter. This function for the time-dependent stress regulatory parameter is defined as Equation 9.17 in which Equations 9.15 and 9.13 are considered:

$$\theta_{t+1} = \cos^{-1}\left\{\frac{5 - 10^{\frac{2\log N}{\{\lambda_t\{\log N/[\log[2\sin\theta_t/2]]-D_T\}\{1-\{\log N/[\log[2\sin\theta_t/2]]-D_T\}\}+D_T}}}{4}\right\} \tag{9.17}$$

The function expressed in Equation 9.15 is expanded as Equation 9.18 by considering Equations 9.11 and 9.14 as follows for the symmetric fold with two limbs:

$$\theta_{t+1} = 2\sin^{-1}\left\{\frac{10^{\frac{\log N}{\{\lambda\{\log N/[\log[2\sin\theta_t/2]]-D_T\}\{1-\{\log N/[\log[2\sin\theta_t/2]]-D_T\}\}+D_T}}}{2}\right\} \tag{9.18}$$

By substituting the time-dependent stress regulatory parameter (λ_t), Equation 9.18 is rewritten as Equation 9.19:

$$\theta_{t+1} = 2\sin^{-1}\left\{\frac{10^{\frac{\log N}{\{\lambda_t\{\log N/[\log[2\sin\theta_t/2]]-D_T\}\{1-\{\log N/[\log[2\sin\theta_t/2]]-D_T\}\}+D_T}}}{2}\right\} \quad (9.19)$$

Symmetric fold dynamical behaviors can be studied by these equations.

Computation of Metric Universality by Considering the AIAs of Symmetric Folds Under Dynamics

The critical states are broadly categorized as attracting to initial state, attracting to a fixed point state, oscillating between two points period 2 and period 3, and chaotic state of fold dynamics. The threshold stress regulatory parameter is the value at which the symmetric fold under dynamics produces critical state(s) or attractor(s). These are threshold stress regulatory parameters, for CSD (λ) and TDSD (μ): λ_1, $\mu_1 = 3.00$; λ_2, $\mu_2 = 3.46$; λ_3, $\mu_3 = 3.569$; λ_4, $\mu_4 = 3.57$. The parameters, λ and μ, respectively, represent the constant and SSM to simulate time-dependent stress regulatory parameters considered for fold dynamical systems respectively. Feigenbaum (1980) proposed the universality constant, i.e., 4.669... for the celebrated nonlinear first-order difference equation (9.2). Similarly, the distance between the openings of attractors at respective threshold stress regulatory parameters is considered to compute metric universality (δ), which converges to 2.5069 (Feigenbaum 1980). The attractor interlimb angles (AIAs) are computed (Tables 9.8 and 9.9) for both the types of fold systems that are controlled by both constant and time-dependent stress regulatory parameters. By considering these AIAs of coupled and non-coupled fold dynamical systems, Equations 9.20 and 9.21 to compute Feigenbaum's metric universality constant for both the types of fold morphological dynamics are proposed.

Fold type I The parameter (δ) that converges to 2.5069 can be computed for the symmetric fold under dynamics by considering the AIAs by Equation 9.20:

$$\delta \sim \frac{\left\{\mathrm{Log}\left(5-4\cos\theta^*_{N+1}\right)-\mathrm{Log}\left(5-4\cos\theta^*_N\right)\right\}\left\{\mathrm{Log}\left(5-4\cos\theta^*_{2N+2}\right)\right\}\left\{\mathrm{Log}\left(5-4\cos\theta^*_{2N+3}\right)\right\}}{\left\{\mathrm{Log}\left(5-4\cos\theta^*_N\right)\right\}\left\{\mathrm{Log}\left(5-4\cos\theta^*_{N+1}\right)\right\}\left\{\mathrm{Log}\left(5-4\cos\theta^*_{2N+3}\right)-\mathrm{Log}\left(5-4\cos\theta^*_{2N+2}\right)\right\}}$$

$$(9.20)$$

where
$N = 2, 4, 6, 8, 16, \ldots$
$\theta^* = \text{AIA}$

Fold type II AIAs are liable to vary with the type of fold. The parameter δ can be computed for the symmetric fold type I under dynamics by considering the AIAs by Equation 9.21. For this type of fold, the AIAs for the two dynamical rules will be computed according to Equations 9.18 and 9.19:

$$\delta \sim \frac{\left\{ Log\left(2\sin\left(\theta^*_{N+1}\right)/2\right) - Log\left(2\sin\left(\theta^*_N\right)/2\right) \right\} Log\left(2\sin\left(\theta^*_{2N+2}\right)/2\right) Log\left(2\sin\left(\theta^*_{2N+3}\right)/2\right)}{Log\left(2\sin\left(\theta^*_N\right)/2\right) Log\left(2\sin\left(\theta^*_{N+1}\right)/2\right) \left\{ Log\left(2\sin\left(\theta^*_{2N+3}\right)/2\right) - Log\left(2\sin\left(\theta^*_{2N+2}\right)/2\right) \right\}}$$

$$(9.21)$$

where
$N = 2, 4, 6, 8, 16, \ldots$
$\theta^* = $ AIA

Results of Simulations

The recent advancement is that the nonlinear differential equations are used to represent the motion of the actual processes in the form of "maps." Several natural phenomena of geoscientific interest are modeled. The cogency of the model can be justified provided the large amount of time series data are procurable. Such time series data enable one to find whether the attractor that describes the evolutionary pattern of the folds possesses low dimensionality. However, in this section, the time series data that reveal the possible dynamics of the stress and the fold morphology are simulated to show the qualitative characteristics. Two cases have been considered, of which the first one is by following the CSD, and in the second one, the TDSD is followed.

Fold Dynamical System Under the Influence of Constant Stress

A case study is shown by considering the symmetric fold type I for better understanding. By changing λ, the constant stress, with a fixed initial value, two possible states of dynamical behaviors are simulated qualitatively and illustrated in Figure 9.13a and b. Based on Equation 9.9, two sets of conditions are considered to transform a symmetric fold qualitatively with α as 0.0636314 and control parameter λ as 3.9 (chaotic attractor) and 2.8 (fixed point attractor). The IAs (θ) of dynamically changing symmetric fold are computed by Equation 9.9, and the parameters of the symmetric fold under study are presented in Table 9.8. Figure 9.13a and b shows a simulated fold at successive stages of evolution under different constant stress control parameters represented as λ. To illustrate the chaotic fluctuations in the symmetric fold evolution, with $\lambda = 3.9$, the evolution process is simulated on a computer (Figure 9.13a). During this evolution, progressive compressions are followed by amplification randomly. In Figure 9.13a, the fold was progressively compressed, which is due to the horizontal stress up to discrete time, $t = 6$. The fold at

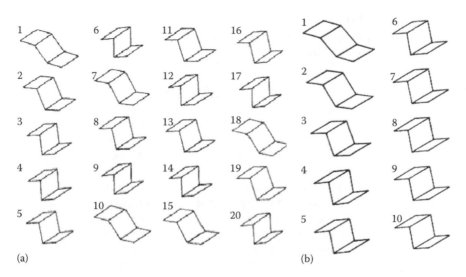

FIGURE 9.13
Evolution of a fold type with the strength of nonlinearities: (a) $\lambda = 3.9$ and (b) $\lambda = 2.8$. The numbers represent the discrete times. (From Sagar, B.S.D., *Discrete Dyn. Nat. Soc.*, 2, 181, 1998.)

discrete time $t = 6$ (approaching critical angle, $\theta = 61°$) gets amplified due to dominating internal forces at $t = 7$. At discrete times 7, 10, 15, and 18, the fold amplification in the fold profile can be seen due to higher internal forces than the CSD parameter. These observations can be seen from the numerically represented parameters depicted in Table 9.7. This fold evolution process is represented qualitatively through graphic analysis. It represents the qualified dynamical behavior of the evolving fold in a quantitative manner. Figure 9.14a shows the return map, in which chaotic behavior of the trajectory can be seen. In Figure 9.13b, symmetric fold was compressed progressively. The compression is due to horizontal stress. It may be observed that after discrete time $t = 5$, the fold has reached equilibrium state. This evolution is also qualitatively represented through graphic analysis in Figure 9.14b, in which the trajectory is attracting to a fixed point. Instead of the fractal dimensions in normalized scale, their corresponding IAs are represented on return maps. It is observed that when the α values lie between 0 and 1, their corresponding IAs will be between 180° and 60°, respectively. AIAs at respective threshold stress regulatory parameters are computed (Table 9.8) for the fold type I under dynamics by considering the initial fold specification with $\alpha = 0.00001$ ($\theta = 179.43028°$). The number of iterations performed is 3×10^4 time steps.

Fold Dynamical System Under the Influence of Time-Dependent Stress

It is assumed that the fold dynamical system is controlled by the TDSD. Hence, the study of the fold dynamical system is treated as a coupled system. The stress dynamics is simulated by considering the first-order

TABLE 9.7

Certain Essential Parameters of the Fold
Behavior Model

t	α	D	% Shortening	θ (°)
$L = 2.1364248; \lambda = 3.8$				
1	0.0636314	6	0	136.20495
2	0.2323667	5.2101002	13.17	103.69863
3	0.6956524	4.0838214	31.94	70.334921
4	0.8257085	3.8996213	35.01	65.350395
5	0.5612645	4.3180416	28.04	76.777388
6	0.960362	3.7417248	37.64	61.108436
7	0.1484606	5.5607296	7.3212	116.33816
8	0.4930383	4.4591445	25.681	80.740878
9	0.974811	3.7264139	37.893	60.697772
10	0.0957626	5.8225257	2.958	127.36545
11	0.3377092	4.8565565	19.06	92.41023
12	0.8722808	3.8416856	35.99	63.791861
13	0.4344873	4.5951013	23.42	84.636437
14	0.9582616	3.7439746	37.6	61.168786
15	0.1559856	5.5262096	7.9	115.00552
16	0.511379	4.419505	26.34159	79.587900
17	0.974495	3.726746	37.88757	60.682251
18	0.096933	5.816304	3.0616	127.028912
19	0.341393	4.845945	19.23425	92.039776
20	0.876891	3.836153	36.06412	63.6174582
$L = 2.1364248; \lambda = 2.8$				
1	0.0636314	6	0	136.20495
2	0.1668308	5.4776101	8.706498	113.1671
3	0.3891952	4.7112823	21.478628	88.037037
4	0.6656225	4.1318048	31.136587	71.643445
5	0.6262268	4.198355	30.027417	73.46716
6	0.655387	4.1456897	30.85517	72.105131
7	0.6322937	4.1876539	30.205768	73.173163
8	0.6509213	4.1561442	30.73093	72.309177
9	0.635064	4.183054	30.282433	73.017503
10	0.648922	4.159500	30.675	72.371934

Source: Sagar, B.S.D., *Discrete Dyn. Nat. Soc.*, 2, 181, 1998.

nonlinear difference equation as the basis to further generate the time-
dependent stress regulatory parameter. With this simulated time-dependent
stress regulatory parameter, the fold dynamics is controlled. With different
possible TDSD, the symmetric fold dynamics is studied, and sets of equations are proposed in which the dynamically changing parameters are IAs

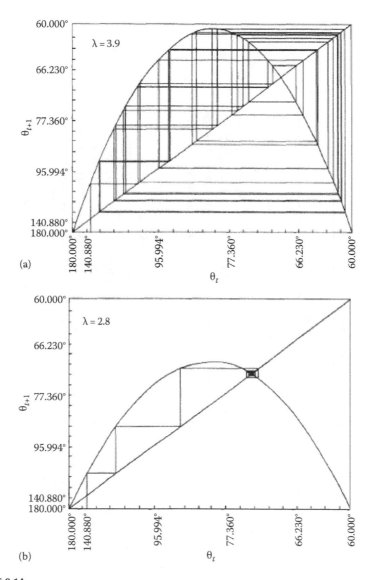

FIGURE 9.14
Logistic maps for the qualitative dynamical behavior of symmetric folds under evolution shown in Figure 9.13a and b. It may be seen that the values mentioned on the abscissa are IAs in degrees for the symmetric fold with three limbs: (a) $\lambda = 3.9$ and (b) $\lambda = 2.8$ (From Sagar, B.S.D., *Discrete Dyn. Nat. Soc.*, 2, 181, 1998.)

and the AIAs. Some interesting results have been arrived at when the stress regulatory parameter is made time dependent. At the threshold stress regulatory parameter in the coupled system, i.e., μ, the time-dependent stress regulatory parameter, λ_t, the attractor NFDs, and the corresponding AIAs are computed and compared with the results for the autonomous fold

TABLE 9.8

AIAs at the Threshold Regulatory Forces after 3×10^4 Time Steps

Threshold Control Parameter	AIAs (°)							

Symmetric Fold Type I ($\alpha = 0.000001$, $\theta_{init} = 179.43028$)

λ_1	3	$\theta_1 = 71.663691$			$\theta_2 = 71.53066$			
λ_2	3.46	$\theta_3 = 64.896443$	$\theta_4 = 86.185576$		$\theta_5 = 64.14859$	$\theta_6 = 82.378284$		
λ_3	3.569	$\theta_7 = 63.154877$	$\theta_8 = 80.241149$		$\theta_9 = 65.166519$	$\theta_{10} = 89.678203$		
		$\theta_{11} = 63.451428$	$\theta_{12} = 77.326159$		$\theta_{13} = 65.932067$	$\theta_{14} = 91.542719$		
λ_4	3.57	63.15	80.533198	65.095265	89.514049	63.473466	77.1892	65.969598	91.646881
		63.188252	81.513418	63.101017	88.917082	63.558405	76.831445	66.076159	91.887804

Symmetric Fold Type II ($\alpha = 0.000001$, $\theta_{init} = 179.43028$)

λ_1	3	98.601081			98.502297			
λ_2	3.46	93.588105	109.48		93.037603	106.61775		
λ_3	3.569	92.307411	105.01353		93.787106	112.10937		
		92.525161	102.82903		94.351735	13.5141		
λ_4	3.57	92.302608	105.23	97.734594	111.98573	92.541349	102.72656	94.379436	113.5926
		92.33191	105.96831	92.267878	111.53614	92.603746	102.45876	94.458098	113.77415

Source: Sagar, B.S.D., *Discrete Dyn. Nat. Soc.*, 2, 181, 1998.
Note: The dynamical rule is simple first-order nonlinear difference equation with constant stress control parameter.

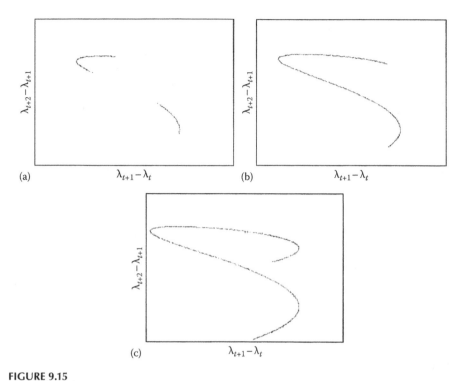

FIGURE 9.15
Return map of the dynamics of time-dependent stress regulatory parameter $(\lambda_{t+1} - \lambda_t)$ versus $(\lambda_{t+2} - \lambda_{t+1})$. The stress modulation control parameter (μ) that controls the time-dependent stress regulatory parameter: (a) $\mu = 3.6$, (b) 3.8, and (c) 3.57. The initial normalized stress parameter $\lambda_t = 0.00001$. (From Sagar, B.S.D., *Discrete Dyn. Nat. Soc.*, 2, 181, 1998.)

dynamical system, which is controlled by the non-time-dependent stress regulatory parameter. Return maps are plotted for the low-dimensional deterministic randomness of the dynamical system of time-dependent stress regulatory parameter $(\lambda_{t+1} - \lambda_t)$ versus $(\lambda_{t+2} - \lambda_{t+1})$ (Figure 9.15a through c) and the fold morphological dynamical system that is controlled by the time-dependent stress regulatory parameter $(\theta_{t+1} - \theta_t)$ versus $(\theta_{t+2} - \theta_{t+1})$ (Figure 9.16a through c). Plots are constructed by considering the differences of successive θ values in the time domain t in θ_t and θ_{t+1} phase space (Figure 9.16a through c). These return maps indicate the characteristic behavior of the simulated time-dependent stress and fold dynamical systems. This demonstrates that one can analyze the temporal aspects of a system in the same manner as used to analyze time series data of a system variable. These return maps are plotted by considering the variables $\lambda_t = 0.00001$; $\mu = 3.6, 3.80, 3.97$; $\alpha_t = 0.00001$, or $\theta_t = 179.43028$; number of iterations is 10×10^6 time steps. The AIAs are also computed by iterating Equations 9.16 through 9.19, respectively, for the two symmetric fold dynamical systems under the influence of constant and time-dependent stresses (Tables 9.8 and 9.9). The difference in the AIAs from the type I to type II symmetric folds is apparent. The variation is also observed in

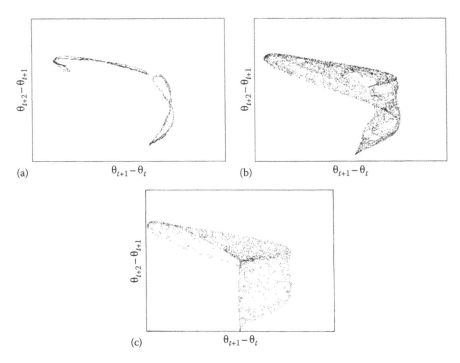

FIGURE 9.16
Return map of the modulated fold morphological dynamics by time-dependent stress regulatory parameter $(\theta_{t+1} - \theta_t)$ versus $(\theta_{t+2} - \theta_{t+1})$. The control parameter is (a) $\mu = 3.6$, (b) 3.8, and (c) 3.57. The specifications for the simulation are $\lambda_t = 0.00001$, $\theta_t = 179.43028$, or $\alpha_t = 0.00001$ and iteration number is 10×10^6 time steps. This map is plotted in θ-parameter space. (From Sagar, B.S.D., *Discrete Dyn. Nat. Soc.*, 2, 181, 1998.)

the AIAs in these two types of folds when they are subjected to the dynamical rules that include TDSD and CSD (Table 9.9). These AIAs are liable to vary with the variations in the fold specifications and dynamical rules involved in the fold morphological dynamics and in the stress dynamics. For instance, if the rule that controls the stress dynamics is a linear equation, contrary to the dynamical rule considered in this study, the AIAs are liable to vary. This important point can be further justified by considering the natural data in relation to stress and the changes in the fold morphologies in a temporal domain. Such a justification explains whether the HDSFs will change their phases. Periodic locking is observed at the μ values between 3.392 and 3.64 (Table 9.10).

This analysis is shown to have a better understanding that these data are following deterministic randomness; that is, each successive value depends on the value of its predecessor. The time-dependent stress dynamical system is also represented as return maps $(\lambda_{t+1} - \lambda_t)$ versus $(\lambda_{t+2} - \lambda_{t+1})$ for the μ values of 3.6, 3.8, and 3.57. Figure 9.15a through c illustrates these return maps. These illustrations allow for qualitative understanding of the stress dynamics that follow the deterministic randomness.

TABLE 9.9

AIAs at the Threshold Regulatory Forces after 3×10^4 Time Steps

Threshold Control Parameter	AIAs (°)							
Symmetric Fold Type I ($\alpha = 0.000001$, $\theta_{init} = 179.43028$, $\lambda = 0.00001$)								
μ_1 3		73.327164				73.728954		
μ_2 3.46	107.68174		64.294614		82.599485		77.825268	
μ_3 3.569	80.778449		84.660144		115.27091		63.503065	
	79.025155		88.477219		118.38197		63.266815	
μ_4 3.57	80.991449	84.305693	114.9721	63.528255	78.985088	88.734899	118.60113	63.24672
	81.746702	83.107759	113.92723	63.613464	78.985146	89.642031	119.39317	63.158243
Symmetric Fold Type II ($\alpha = 0.000001$, $\theta_{init} = 179.43028$, $\lambda = 0.00001$)								
μ_1 3		99.840737				100.14045		
μ_2 3.46	125.68524		93.145029		106.7839		103.20273	
μ_3 3.569	105.41666		108.33262		131.40715		92.563091	
	104.10171		111.20491		133.75145		92.389588	
μ_4 3.57	105.57651	108.06612	131.18194	92.581572	104.07168	111.39894	133.91655	92.374946
	106.14345	107.16576	130.39438	92.644198	104.07172	112.08213	134.51323	92.309882

Source: Sagar, B.S.D., *Discrete Dyn. Nat. Soc.*, 2, 181, 1998.

Note: The dynamical rule is simple first-order nonlinear difference equation with time-dependent stress control parameter.

TABLE 9.10

AIAs of the Fold Dynamical System Following Time-Dependent Stress Control Parameters ($\lambda_0 = 0.00001$; $\alpha_0 = 0.00001$ or $\theta_0 = 179.43028°$; Number of Iterations 3×10^4)

Stress Modulation Parameter (μ) to Control TDSD	$\lambda_{t+1} = \mu\lambda_t(1 - \lambda_t)$ Attractor Time-Dependent Stress Control Parameters (λ^*)	AIAs (θ^*) Fold Type I	Fold Type II
0.848	0.43991	92.389892	114.15255
		41.528587	95.894165
	0.84082	92.3629	114.13221
		41.563547	95.903425
0.860	0.442194	96.82515	117.49648
		66.579149	94.829506
	0.848505	89.600981	112.05121
		70.063627	97.411428

Source: Sagar, B.S.D., *Discrete Dyn. Nat. Soc.*, 2, 181, 1998.

Period Locking

Period locking is identified between the dynamics of the stress regulatory parameter and the dynamics of the fold system. From the fold dynamics that is being controlled by time-dependent stress regulatory parameter, one can see that the dynamics of the time-dependent stress regulatory parameter is enslaved to the dynamics of the fold system. The dynamics of the stress regulatory parameter is following period 2; however, the dynamics of the fold system that is being controlled by this controlled stress dynamics follows period 4. This "period locking" is observed between the μ, the stress regulatory parameter in the modulated logistic system, values 3.392 and 3.44. This possibility of the periodic locking in the modulated fold dynamical system needs to be described by analyzing the physical forces of specific range. This needs to be compared in a meaningful way with the stress regulatory control parameter represented as a numerical value (i.e., $\mu < 4 > 1$). It is interesting to see how the dynamics of the stress regulatory parameter is enslaved to the dynamics of fold morphological behavior between the values 3.392 and 3.44 (Table 9.10).

Bifurcation Diagrams

Fold Dynamics Under the Influence of Constant Stress

In Figure 9.17a, a bifurcation diagram is shown for various possible dynamical behaviors of the symmetric folds under dynamics, viz., stable, unstable, and chaotic. The evolution types of fold transformations can be segregated

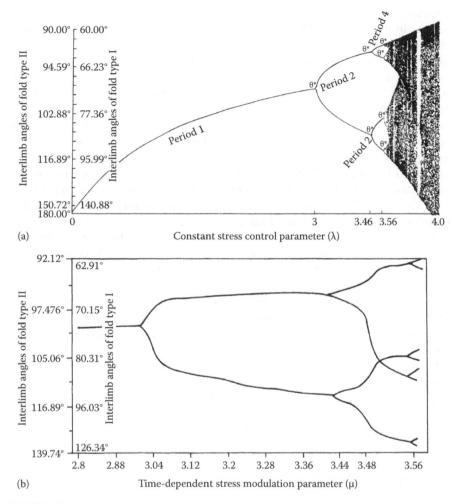

FIGURE 9.17
(a) Bifurcation diagram showing various possibilities of fold transformations: stable, unstable, and chaotic. The value of λ measures the constant strength of stress regulatory parameter that controls the fold. The evolution of the fold system can be segregated as period 0, period 1, period 2, and chaotic. Period 0: A contorted fold with $\alpha_t = 0.06363$ becomes straight when α_{t+1} approaches zero, λ is between 0 and 1. This is possible under the process of continuous fold amplification. Period 1: When λ is between 1 and 3, the fold pattern shortens, and the pattern reaches a fixed point attractor. It means that the fold reaches the equilibrium state. Period 2: The fold pattern oscillates between two points when λ is between 3 and 3.569. The fold amplification and compression will occur periodically. Chaotic: The behavior of fold is such that the fold shapes at different time periods do not overlap. Here, the fold amplification and compressions may occur, as time progresses, randomly. The values on both sides of the Y-axis represent the IAs of the symmetric folds with three limbs and two limbs, respectively. (b) Bifurcation diagram of Equation 9.9 that describes the fold dynamics under the influence of time-dependent stress regulatory parameter. The branches crossing over each other in the four-cycle region result in a complete modification of the structure. (From Sagar, B.S.D., *Discrete Dyn. Nat. Soc.*, 2, 181, 1998.)

as period 0, period 1, period 2, and chaotic. As the parameter λ is varied, changes in the qualitative behavior of the system can occur. Such qualitative behavior can be seen in the bifurcation diagram (Figure 9.17a) in which the attractor set against the control parameter is plotted. In this bifurcation diagram, as $\lambda \in [1,4]$, the dynamical behavior possesses one stable fixed point. As λ is increased past 3, the behavior becomes unstable, and two new stable periodic points appear. Fold behavior follows periodicity where both amplification and shortening of folds are subsequently involved. The dynamics become unstable, each originating two new stable periodic points of period 4 as λ is further increased from 3.569. Through this bifurcation diagram, the fold dynamical behavior path can be found with respect to the control parameters. This diagram (Figure 9.17a) portrays not only the type of dynamical behaviors of the fold with respect to the control parameter but also the critical states in terms of IAs of the fold under dynamics. The number of critical states that a fold reaches under the dynamics depends on the initial fold state and the control parameter (λ). For every value of λ, there will be an attracting point. These attracting points are represented in two ways: (1) the fractal dimension in normalized scale and (2) critical states shown as IA of a symmetric fold under dynamics. Instead of the fractal dimensions in normalized scale, their corresponding IAs are represented. If the fold under dynamics is according to the rule of the first-order difference equation (9.2), Figure 9.17a shows various behavior paths and their stability with respect to the initial fold state, and λ values were shown by means of respective critical states represented by θs. The important contribution of this diagram (Figure 9.17a) is the information regarding the history of folding that can be studied, provided the initial state of the symmetric fold and the control parameter that controls the fold dynamics are precisely computed.

Dynamical System Under the Influence of TDSD

The influence of TDSD on fold dynamical behavior is depicted through bifurcation diagram (Figure 9.17b). This is controlled by considering Equations 9.17 and 9.19. In these equations, the value of the parameter λ at any instance is a single nonlinear function of its value in the previous instances. In this, μ plays the role of the control parameter, which is thought of as the SSMP. For $1 < \mu < 3$, stress dynamics follows attracting fixed point for λ_t and here for $\theta_t, \theta_{t+1}(180°, 60°, 90°)$ and $\lambda \in [1,4]$. The bifurcation diagram in Figure 9.17b is generated by starting from a parameter value $\mu = 3$ and increasing it in steps of 0.001, by initial values of θ_t and λ_t, say 179.43028 and 0.00001, respectively. Due to modulation by TDSD, changes between the bifurcation diagrams (Figure 9.17a and b) are observed. The fundamental difference is that the bifurcation occurs earlier than in the case of the fold dynamics under the influence of CSD from the observed bifurcation orders. The normal feature in the modulated system is the crossing-over of the inner bifurcation branches in the four-cycle region. It lacks the symmetry of the bifurcation structure of the fold dynamical systems that is influenced by the CSD.

Results and Discussion

Changes in the morphology of a geological fold are due to stress and IEFs. Such morphological changes can be quantified in terms of fractal dimensions. Stress and the fractal dimension are depicted in normalized scale as dimensionless parameters. Incorporating these parameters in a first-order nonlinear difference equation that has physical relevance as the simplest viable model of a symmetric fold sustaining morphological changes, numerical simulations are carried out that are analogous to creep experiments. In the first experiment, the constant stress (λ) is employed to model the morphological dynamical behavior of HDSFs that are postulated as they are precarious to stress and IEF and will not supervene the state of brittleness during the evolution. In the second experiment, the time-dependent stress that is changed according to a dynamical rule is used to model distinct dynamical behaviors of these HDSFs. The results arrived through computer simulations are the AIAs. Bifurcation diagrams are also depicted to show the dynamical behaviors concerning the change in the stress dynamics.

We have studied the highly ductile nature of symmetric fold dynamical behaviors that are controlled by the constant and the time-dependent stress modulated parameters respectively through numerical simulations. In particular, we discuss the computations of the changing AIAs at respective stress modulated parameters that are used to control the behavior of fold dynamical systems. Equations are proposed to compute IA of these symmetric folds undergoing dynamical changes, which encompass the rule that is ensued to transform the folds and certain specifications of the folds. Bifurcation diagrams are described to show how these symmetric folds under dynamics behave under the change of constant stress control parameter, λ, and the SSMP, μ, to control TDSD, λ_t. The AIAs (θ^*) are shown on the bifurcation diagrams. By considering these AIAs, equations are also proposed to compute metric universality. The periodic nature of the phase changes in the fold morphological dynamics is studied using the time-dependent and constant stresses that follow a dynamical rule. Interesting conclusions are arrived at in terms of variations in the AIAs of the fold following these two dynamical rules. These theoretical conclusions have an important bearing when considering strategies for the understanding of geological fold dynamics, and more generally, when considering the behaviors of natural time series data in a range of geological situations where folding is taking place. This type of time series data indicates that the possibility of predicting predictability depends on the degree of randomness in the behavior of the dynamical system. From the time series data, attractor can be constructed in phase space. The dimension of the attractor provides the possibility of *predicting predictability*.

Low-dimensional attractors of dynamical systems allow the behavior to be predicted through some nonlinear equations. However, as the dimensionality of the attractor that describes the behavior of dynamical system is high, the predictability becomes difficult. These two types of systems are termed as the dynamical systems that follow deterministic randomness and the natural randomness in their behaviors. Generally, the system that follows deterministic randomness will possess the strange attractor of which the dimensionality is low. The assumption considered as the basis is that the dynamics of both fold morphology and the acting stress possesses the low-dimensional attractors. To infer whether the attractor of fold morphological dynamics possesses the low dimensionality, long time series data are required. This deterministic approach emphasizes to give certain possible behaviors of fold dynamics with the respective critical states represented by IAs. It is concluded that the critical states of symmetric folds under dynamics depend on the stress that influences the fold and the initial state of the fold. With the aid of the SSM parameter and the specifications of initial state of symmetric fold, graphic analysis may be carried out to investigate the history of folding. Such an investigation, to find out the critical states of several possible behaviors, will shed light on predicting the fold dynamical behaviors. The dynamically transforming symmetric fold with different time-dependent and constant stress controlling parameters was shown for a better qualitative understanding. This qualitative study is an attempt as an example for academic interest to furnish the interplay between numerical experiments and analytical theory. This maiden attempt is considered as a preliminary effort to introduce bifurcation theory for the understanding of the dynamical behavior of symmetric folds. In brief, this chapter presents a maiden attempt to show how a symmetric fold can modify its shape, in particular the IA, through a nonlinear first-order difference equation. This approach could be valid as a potential application of these equations to a geological problem to resolve real fold cases. However, with historical data available, the phase that the fold has undergone can be studied by investigating the fold at different time intervals to fit the equations. From such derived equations, assumed to be first-order nonlinear difference equations, as the underlying dynamical rule in the present qualitative investigation, our understanding of the fold dynamics will certainly be enhanced.

Logistic Equation in Sand Dunes

Certain possible morphological behaviors with respective critical states represented by inter-slipface angles of a sand dune under the influence of nonsystematic processes are qualitatively illustrated by considering the first-order difference equation that has the physical relevance to model the morphological dynamics of the sand dune evolution as the basis. It is deduced that the critical state of a sand dune under dynamics depends on the regulatory parameter that encompasses exodynamic processes of random nature and the morphological configuration of the sand dune.

With the aid of the regulatory parameter, and the specifications of initial state of sand dune, morphological history of the sand dune evolution can be investigated. As an attempt to furnish the interplay between numerical experiments and theory of morphological evolution, the process of dynamical changes in the sand dune with a change in the threshold regulatory parameter is modeled qualitatively for a better understanding. Avalanche size distribution in such a numerically simulated sand dune dynamics has also been studied in this section.

Sand falls on the supply area in the form of particles of various shapes. The description of the morphology of a sand dune is mainly concerned with its profile that may be described as angular. Such a sand dune is assumed as pyramidal if its slipfaces are of equal lengths. Pyramidal sand dunes form due to convection and interferential types of wind. Such sand dunes are common in the Central Asia and Africa. There is less scope for the movement of sand dune due to convection and interferential type of wind conditions. Such sand dunes cover limited areal extent and owe their composition of the interference of air waves caused by wind reflecting from mountain barriers (Alonso and Hermann 1996). The transitions in the sand dune profile may be observed under different types of conditions. It is heuristically justifiable that the degree of unsteady state to fall over is more in the steep sand dunes.

The accumulation of thick strata of sand and its transformation into a sand dune are a lengthy and complicated process proceeding under the effect of various exodynamic processes. These processes are the direct causes for sanddrift, sand withdrawal, and sand assemblage, and eventually the effect of these processes is the oscillations in the morphology of a sand dune profile. Due to these effects, sand dunes undergo flattening and protrusion. The spatiotemporal organization of such a sand dune can show many different morphological dynamics because of different morphological constitutions that the sand dune traverses and also due to the type of wind actions. These morphological changes may be according to a rule through which one can explore the morphological dynamics. Moreover, these oscillations are dependent on the regulatory parameter that plays a vital role in the present investigation. This parameter can be derived by studying the morphology of a sand dune at specific time intervals. Several papers have appeared during the last decade that address the application of fractal concepts in the studies of geoscientific interest. Behavior of various systems of geoscientific interest such as electrical conductivity and fractures of rocks to the microcrack population (Maden 1983), coalescence of fractures (Allegre et al. 1982, Newman and Knopof 1982), and stick–slip behavior (Smalley et al. 1985) through renormalization group approach was studied. In particular, several models have been proposed to comprehend the dynamical processes of sand dunes in two dimensions (Manna 1991, Alonso and Hermann 1996). This section aims to provide a qualitative model for morphological dynamics of a pyramidal sand dune through bifurcation theory.

Morphological Evolution of a Pyramidal Sand Dune through Bifurcation Theory: A Qualitative Model

Equation 9.2 that has the physical basis also to model several possible morphological dynamical behaviors of dunes is considered to carry out numerical simulations further to understand the dune dynamical behaviors. Equation 9.2 that has physical relevance also to model the morphological evolution of a pyramidal sand dune is used to simulate distinct possible behaviors. As an attempt to furnish the interplay between numerical experiments and theory of morphological evolution, numerical simulations are performed by iterating Equation 9.2 2093 time steps to illustrate several possible morphological dynamical behaviors of a sand dune by changing the regulatory parameter (λ) that explains the detailed form of exodynamic process. Bifurcation diagram is described as a model to illustrate how the sand dune under dynamics behaves concerning the change of regulatory parameters. Computed attractor inter-slipface angles (AISFAs) at respective threshold regulatory parameters are depicted on the bifurcation diagram. By considering these (θ^*s), an equation is also proposed to compute metric universality.

Definition of a Profile of a Sand Dune

The description of the morphology of a sand dune is concerned with its profile that is described as angular, the slipfaces of which are of equal lengths. A typical linear dune profile is shown in Figure 9.18.

- Profile of a dune should have a heap with two slipfaces each of the same length ($L_1 = L_2$). The profile is symmetric with respect to the origin at the center of the base of the dune.
- Width (d) of the base of the dune must be greater than the length of the slipface. This assumption is valid due to the fact that the length of the slipface is not greater than the width (d) in the case of real-world dunes.

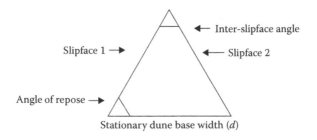

FIGURE 9.18
Pyramidal sand dune profile. (From Sagar, B.S.D., *Chaos Soliton. Fract.*, 10(9), 1559, 1999b; Sagar, B.S.D. et al., *Fractals*, 11(2), 183, 2003.)

- Width of the dune is considered as rigid during the progressive dune evolution. However, the length of the slipface (*L*) varies with the continuous accretion of sand. Dune base length is stationary since the characteristic of supply area does not change. However, due to continuous sand supply, the slipface length tends to change, in turn the sand dune morphological dynamics. In other words, the slipface length is dynamic, whereas the base width is static. Inter-slipface angle is the diverging angle of a sand dune profile with two slipfaces. Characteristics of the simulated sand dune include that the profile of the dune has two slipfaces and hence an inter-slipface angle (θ), the base length (*d*), which is the same for all the profiles of a dune under dynamics. The lesser the inter-slipface angle, the more is the height of the dune from the base, and *vice versa*.

The degree of sand dune steepness can be quantified by fractal dimension (Mandelbrot 1982). The shape of a generator (Mandelbrot 1982) incited us to use fractal dimension as a main parameter to simulate dune dynamics numerically, as the profile of which is compared with the generator morphologically. The fractal dimension is used as the main property of the sand dune undergoing dynamical changes. From a profile of a sand dune undergoing dynamics, the characteristics that substantiate the morphological constitution of sand dune at specific time interval include angle of repose, inter-slipface angle, dune height from the middle point of the sand dune base, and slipface lengths. The morphological dynamics of an ideal sand dune, of the type considered in the present study, can be modeled by considering any one, or the combination, of the characteristics. It is understood that by considering any two characteristics mentioned one can derive the other characteristics. However, the fractal dimension of the profile of an ideal sand dune is a unified property from which one can define the other characteristics. For the profile of a sand dune, the NFD determines the steepness.

Rule to Perform Numerical Simulation of Dune Morphological Dynamics by Incorporating Normalized Fractal Dimensions

If the slope of the dune is initially very small, only a few slides may occur, and so the dune will steepen. If the slope is very large, huge avalanches will sweep over the edges of the dunes, and the slope will then become less steep. It is intuitively justifiable that the morphological change in the sand dune is a nonlinear phenomenon, since the fractal dimensions of the successive profiles of a sand dune undergoing dynamics are not directly proportional to each other at successive time intervals. The intuitive argument may be endured by the fact that the sand dunes steepen and flatten over a time interval due to distinct nature of sand dune structures. This argument may be supported by a postulate that the fractal dimension of successive profiles of a sand dune undergoing dynamics may be nonoverlapping and hence may be nonlinear.

This phenomenon is due to the relatively divergent behavior of the sand that is accumulated and also due to the change in morphological constitution at discrete time intervals. It is intuitively apparent that the degree of unsteady state to fall over is more in the steep sand dune that possesses high fractal dimension. Hence, as the steepness of sand dune increases, the degree of fall over of sand becomes more when compared to the sand dune of lesser steepness. This phenomenon can be compared with the *overcrowding* parameter in the context of population dynamics described in the logistic equations. This statement supports the argument that (α), the NFD, tends to increase when it is small and to decrease when it is large.

Several assumptions of the morphological dynamics seem to be cogent by the fact that the exodynamic processes are always nonsystematic, which alter the morphological behavior of a sand dune. As the accretion process continues, several possible sand dune dynamical behaviors can be observed. To quantify these dynamical behaviors, of interest to certain geodynamicists, Equation (9.2) could be taken as a basis. Based on Equation 9.2, certain other equations have been derived to estimate the attracting inter-slipface angles. The morphological dynamics of a sand dune profile, with two slipfaces and a fixed base length (d), has been modeled (Sagar 1999) through bifurcation theory (May 1976). To carry out computer (numerical) simulation to visualize distinct possible behaviors concerning a change in the strength of nonlinearity, Equation 9.2, proposed elsewhere, and several possible phase changes of a sand dune, undergoing dynamics, are considered as the basis—$\alpha_{t+1} = \lambda \alpha_t (1 - \alpha_t)$, where α is the NFD of a sand dune profile, $0 \leq \alpha \leq 1$, and λ is the strength of regulatory parameter, $1 \leq \lambda \leq 4$. The NFD α of the sand dune can be obtained by subtracting the topological dimension (D_T) from the fractal dimension as shown in Equation 9.22:

$$\alpha = \left[\frac{\log(N)}{\log(d/L)} \right] - D_T \qquad (9.22)$$

where
  N is the number of slipfaces (2 for the present case)
  d is the width of the stationary base of the sand dune
  L is the length of the slipface, $L \leq d$
  D_T is the topological dimension
  α is the NFD of a sand dune profile
  $\alpha + D_T$ is the fractal dimension (D)

A sand dune with high degree of steepness will have a value of $\alpha = 1$, and with no steepness, it will have a value of $\alpha = 0$. Exodynamic processes that determine changes of a sand dune undergoing dynamics can be quantified by means of fractal dimension. To examine the long-term behavior of the sand dune morphology, or of the fractal dimension of the dune profile,

Equation 9.2, which has physical viability to understand the various phases, is considered. In particular, we are interested in how this behavior depends upon the strength of nonlinearity parameter, λ. To keep the fractal dimensions of the profiles of a sand dune undergoing dynamics, and their corresponding inter-slipface angles between 180° and 90°, we limit our examination to values of λ between 1 and 4.

To study morphological dynamical behavior of a sand dune, it is necessary to know how much of the total morphological change is accommodated across time intervals. The rates of change in the fractal dimension of dynamically changing sand dune at discrete time intervals depend upon the exodynamic processes. The collective impact of exodynamic processes (cause) that alter sand dune morphology can be defined as the strength of regulatory parameters by studying the (degree of deformation) effect due to the cause at discrete time intervals. As the fractal dimension enables the characteristic of the sand dune profile that is steepened as well as flattened, the parameter λ can be defined as a numerical value. From the theoretical standpoint, λ may be computed by considering α_t and α_{t+1} to fit the curve $\lambda\alpha_t(1 - \alpha_t)$. The parameter λ gives the total description of the dynamics of the sand dune. The impact of nonsystematic exodynamic processes on a sand dune in terms of its dynamical behavior is investigated through the first-order difference equation (9.2) of the form, $\alpha_{t+1} = f(\alpha_t)$; the NFD at $t + 1$, α_{t+1}, is given as some function f of the α_t at time t. If this equation were linear (e.g., $f = \lambda\alpha$), α would just increase or decrease exponentially if $\lambda < 1$. Moreover, the fractal dimension tends to increase at low α and to crash at high α value, corresponding to some nonlinear function with a hump of which the quadratic $f = \alpha_{t+1} = \lambda\alpha(1 - \alpha)$. It does mean that there is a tendency for the variable α to increase from time t to the next when it is small and for it to decrease when it is large. This tendency is preserved due to the term $(1 - \alpha_t)$ in Equation 9.2. In Equation 9.2, to compute α_{t+1}, $\lambda\alpha_t(1 - \alpha_t)$ explains that the normalized status of a sand dune dynamics, in the case of α starting at larger than 1, immediately goes negative at one time step. If λ is less than 1, the sand dune is in an inhospitable environment that its fractal dimension diminishes at every discrete time interval. For values of λ below 1, the eventual fractal dimensions in normalized scale are zero of which the inter-slipface angle is zero (or it does not exist). Moreover, if $\lambda > 4$, the hump of the parabola exceeds 1, thus enabling the initial α value near 0.5 to exceed criticality in two time steps. Therefore, there is a need to restrict the analysis to values of λ between 1 and 4, and values of α between 0 and 1. In the qualitative understanding of dynamical behavior, value α_{t+1} is obtained from the previous value of α_t by multiplying it by $\lambda(1 - \alpha_t)$. It is clear that for $\lambda(1 - \alpha_t)$ to be greater than 1, the successive values, viz., α_{t+2}, α_{t+3}, α_{t+4} ..., α_{t+N}, will grow bigger—that is, a change in α_t will get amplified. This is the sand dune steepness due to sand assemblage. If $\lambda(1 - \alpha_t)$ becomes smaller than 1, then subsequent values must diminish. This is sand dune flattening due to fall over of sand.

Relationship between Normalized Fractal Dimension and Inter-Slipface Angle

As the sand dune crest reaches to critical inter-slipface angle, i.e., 90° (steepest of the sand dune) at which the NFD $\alpha = 1$, there is a tendency for α to decrease due to the fact that the degree of unsteady state of sand to fall over is more in steeper sand dunes. On the contrary, when the sand dune profile possesses less fractal dimension, there may be a possibility for it to get steepened due to sand assemblage and due to more *sand holding* capacity in the supply area. When it possesses a high fractal dimension, the degree of unsteady state to fall over is more, and this may lead to a decline of the fractal dimension. However, it may also lead to oscillations, or even chaotic fluctuations, depending on the nature of exodynamic processes and the sand dune characteristics.

Equation 9.22, a part of which is due to Mandelbrot (1982) to compute the fractal dimension of Koch generator that is similar to the sand dune profile, can be written as Equation 9.14. This equation computes the NFD of the sand dune profile by considering the inter-slipface angle (θ). From the θ, the corresponding NFD can be calculated for the profile of a sand dune that has been considered by using Equation 9.14, which is $\alpha = \{[\log(N)/\log[2\sin(\theta/2)]] - D_T\}$. From Equation 9.14, it can be understood that the profiles of the sand dunes with $\theta = 90°$ (steepest) and $\theta = 180°$ (zero steepness) of inter-slipface angles possess NFDs 1 and 0, respectively. The simplest profile of a simple sand dune, one could imagine, is with two slipfaces ($N = 2$) making an angle θ that satisfies $90° \leq \theta \leq 180°$. The limit case $\theta = 180°$ generates a dune at the initial state ($t = 0$); the case $\theta = 90°$ generates a dune at the unstable state, the fractal dimension of which has been estimated as 1 (when $\theta = 180°$) and 2 (when $\theta = 90°$). For a better understanding, the profiles of a sand dune are illustrated with NFDs and their corresponding inter-slipface angles (Figure 9.19).

Computation of Inter-Slipface Angle of a Sand Dune Under Dynamics

For the profile of a sand dune under dynamics with two slipfaces, the variable θ_{t+1} is a function of θ_t. Instead of α, one can consider θ values to carry out simulations for modeling. The *ISFA* at time t is considered instead of the *NFD* to compute the *ISFAs* at time $t + 1, \ldots, t + n$ of the sand dune undergoing dynamics according to first-order difference equation as a dynamical rule. The *ISFAs* at time $t + 1$ can be computed by considering θ at time t as some function defined as $\theta_{t+1} = f(\theta_t)$. The function is defined as

$$\theta_{t+1} = 2\sin^{-1}\left\{\frac{10^{\{\lambda\{\log N/[\log[2\sin\theta_t/2]]-D_T\}\{1-\{\log N/[\log[2\sin\theta_t/2]]-D_T\}\}+D_T}}{2}\right\}$$

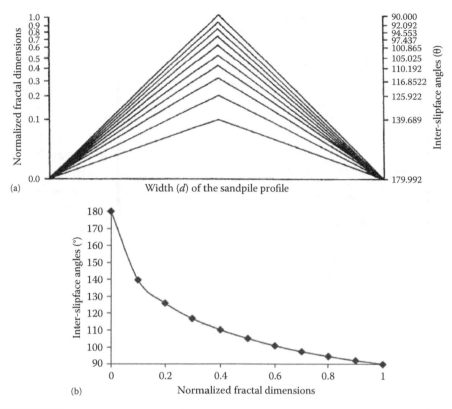

FIGURE 9.19
(a) Sand dune profiles with different NFDs and their corresponding inter-slipface angles, and (b) graphical plot between the two parameters. (From Sagar, B.S.D. et al., *Fractals*, 11(2), 183, 2003.)

where θ_t and θ_{t+1} are inter-slipface angles at discrete times t and $t+1$, respectively. The limits of various parameters are $0 < \alpha < 1$, $180° > \theta > 90°$, $1 < D < 2$. Iterating the function—that is similar to that of fold type I explained through Equation 9.18—produces time series of inter-slipface angles (θ) of a simulated sand dune undergoing morphological changes dynamically. A 1-D map (Figure 9.20a) has been plotted by considering the time series of inter-slipface angles computed at $\lambda = 4$. Figure 9.20b illustrates the return map plotted for $\theta_{t+1} - \theta_t$ versus $\theta_t - \theta_{t-1}$ for sand dune with $\lambda = 4$. This type of map enables the region of avalanche occurrence. This time series data can be used to compute the sizes of avalanches occurred. The inter-slipface angles computed by iterating the function, similar to Equation 9.18, can be used to represent them in θ-space to visualize them in the form of a sand dune phase map (Sagar and Venu 2001). More details on avalanche size distribution can be seen in the "Avalanches in a Numerically Simulated Sand Dune Dynamics" section.

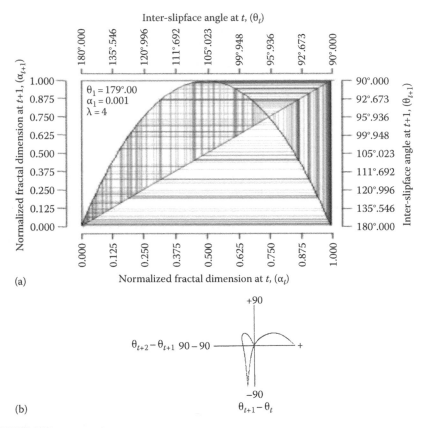

FIGURE 9.20
(a) A 1-D map plotted between θ_{t+1} versus θ_t for sand dune case $\lambda = 4$ and (b) return map plotted between $\theta_{t+1} - \theta_t$ versus $\theta_{t+2} - \theta_{t+1}$ for sand dune case with $\lambda = 4$. (From Sagar, B.S.D. et al., *Fractals*, 11(2), 183, 2003; Sagar, B.S.D. and Venu, M., *Discrete Dyn. Nat. Soc.*, 6(1), 64, 2001.)

Attracting Inter-Slipface Angles

The threshold regulatory parameter is the value at which the sand dune under dynamics produces critical state (s) or attractor (s). The threshold regulatory parameters are—$\lambda_1 = 3$; $\lambda_2 = 3.46$; $\lambda_3 = 3.569$; and $\lambda_4 = 3.57$. The AISFAs (θ^*s) at these regulatory parameters are computed (Table 9.11) by considering the initial sand dune specification with $\alpha = 0.00001$ or $\theta = 179.57334°$. The corresponding attractor sand dune profiles at respective regulatory parameters are arrived through numerical simulations that are performed by iterating 3×10^4 time steps. Figure 9.21a shows the initial profile of the sand dune. This profile is allowed to undergo morphological changes with different threshold regulatory parameters. Figure 9.21b through e shows attractor sand dune profiles under the threshold regulatory parameters. During the evolution, variations in the length of slipfaces may be seen. In Figure 9.21b, the sand dune crest was progressively raised.

TABLE 9.11

AISFAs of Sandpile at Respective Threshold Regulatory Parameters after 3×10^4 Iterations

λ^*	$\alpha_0 = 0.00001; \theta_0 = 179.57334°$							
3	98.601081°							
3.46	93.588105°		109.48°					
3.569	92.307411°	105.01353°	93.787106°	112.10937°				
3.57	92.3°	105.23263°	93.735048°	111.98573°	92.541161°	102.72656°	94.379436°	113.5926°

Source: Sagar, B.S.D., *Chaos Soliton. Fract.*, 10(9), 1559, 1999b.

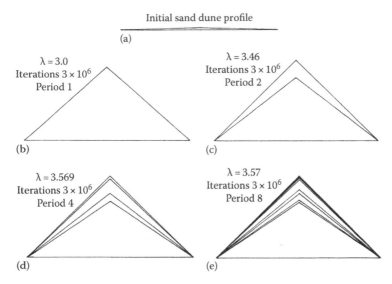

FIGURE 9.21

(a) Initial sand dune profile with α = 0.00001 or θ = 179.57334. The attractor sand dune pro-files at various threshold regulatory parameters: (b) λ = 3, fixed point attractor sand dune; (c) λ = 3.46, period 2 attractor sand dunes; (d) λ = 3.569, period 4 attractor sand dunes; and (e) λ = 3.57, period 8 attractor sand dunes. The attractor sand dune profiles shown in (b–e) are arrived after simulations performed by iterating 3×10^4 time steps. The AISFAs of these attractor sand dune profiles were shown in Table 9.11. (From Sagar, B.S.D., *Chaos Soliton. Fract.*, 10(9), 1559, 1999b.)

Periodically changing morphologies of sand dune under dynamics with regulatory parameter 3.46 is shown in Figure 9.21c. Similarly, periods 3 and 7 are also shown in Figure 9.21d and e, respectively.

Bifurcation Phenomenon

Changes in the regulatory parameter result in variations in the behavior of the morphological dynamics of a sand dune ranging from steady state to periodicity and chaotic state. The random behavior of sand dune, from its inception of the formation, is not only due to randomness of the exody-namic processes that influence the sand dune incessantly but also due to the morphological constitution of the sand dune. This study opens a way to reconstruct the initial sand dune by considering the equilibrium status and the critical angle of the sand dune. The historical track along which the morphology of sand dune evolves as the regulatory parameter grows can be characterized by a sequence of stable regions where deterministic laws dominate and of unstable ones near the bifurcation points where the system can choose between or among them one possible future. This mixture com-poses the morphological account of the sand dune formation. In Figure 9.22, bifurcation diagram is shown for various possible dynamical behaviors,

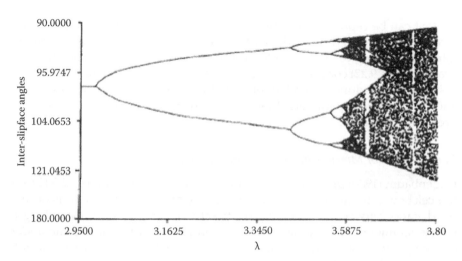

FIGURE 9.22

Bifurcation diagram showing various possibilities of sand dune morphological dynamics. The value λ measures the regulatory parameter that controls the morphological dynamics of sand dune. The evolution of the sand dune system can be segregated as period 1, period 2, and chaotic. Period 1: When λ is between 0 and 2, the sand dune profile protrudes, and the dynamics attracting to a fixed point. Period 2: The sand dune profile oscillates between two points when λ is between 2 and 3. The sand dune protrusion and flattening will occur periodically. Chaotic: The behavior of sand dune is such that the sand dune profiles at different time periods do not overlap exactly. Here, the sand dune protrusion and flattening may occur randomly as time progresses. (From Sagar, B.S.D., *Chaos Soliton. Fract.*, 10(9), 1559, 1999b.)

viz., stable, unstable, chaotic, of a sand dune under dynamics. The evolution types of sand dune transformations can be segregated as period 1, period 2, and chaotic. As λ is varied, changes in the qualitative behavior of the system can occur. Such qualitative behavior can be seen in the bifurcation diagram, Figure 9.22, in which the AISFAs are plotted against regulatory parameter λ. In this bifurcation diagram, as λ ∈ [1,3], dynamical behavior possesses a stable fixed point. As λ is increased past 3, the behavior becomes unstable, and two new stable periodic points appear. The morphological behavior of sand dune under dynamics follows periodicity where both protrusion and flattening are involved subsequently. The dynamics become unstable each spawning two new stable periodic points of period 4 as λ is further increased from 3.46. Through this bifurcation diagram, the dynamics of morphological behavior path can be found out with respect to the regulatory parameters. This diagram (Figure 9.22) portrays not only the type of dynamical behavior of the sand dune with respect to the regulatory parameter but also the critical states in terms of inter-slipface angles of the sand dune under dynamics. The number of critical states that a sand dune reaches under the dynamics during the influence of exodynamic process depends on the starting initial sand dune configuration and the regulatory parameter (λ). For every value of λ, there will be an attracting point. This attracting point in the present context can be represented in two ways: (1) the NFD and (2) the inter-slip-face angle of a sand dune under dynamics. The values on abscissa represent inter-slipface angles. These angles are computed by Equation 9.9. This diagram (Figure 9.22) contributes the information regarding the morphological history of sand dune formation, provided the initial state of the sand dune and the regulatory parameter that controls the morphological dynamics are precisely computed.

Computation of Metric Universality Considering $\theta*s$

Feigenbaum (1980) proposed the metric universality constant, i.e., 2.5029, for the celebrated nonlinear first-order difference equation (9.2) by considering the distance between the openings of attractors at respective threshold regulatory parameters. The variable δ can be computed for the sand dune under dynamics by considering the AISFAs by Equation 9.21, which is given as follows to avoid confusion:

$$\delta \sim \frac{\left\{\text{Log}\left(2\sin\left(\theta^*_{N+1}\right)/2\right) - \text{Log}\left(2\sin\left(\theta^*_N\right)/2\right)\right\}\text{Log}\left(2\sin\left(\theta^*_{2N+2}\right)/2\right)\text{Log}\left(2\sin\left(\theta^*_{2N+3}\right)/2\right)}{\text{Log}\left(2\sin\left(\theta^*_N\right)/2\right)\text{Log}\left(2\sin\left(\theta^*_{N+1}\right)/2\right)\left\{\text{Log}\left(2\sin\left(\theta^*_{2N+3}\right)/2\right) - \text{Log}\left(2\sin\left(\theta^*_{2N+2}\right)/2\right)\right\}}$$

where
  $N = 2,4,6,16, \ldots$
  $\theta^* = $ AISFA

This equation is derived by computing the differences at respective attractor NFDs. Instead of attractor NFDs, their corresponding AISFAs are considered to compute δ.

Avalanches in a Numerically Simulated Sand Dune Dynamics

Windblown sand forms the dunes of various types. In a seminal work, Bagnold (1941) discussed several aspects of dynamics of sand dunes. Since the introduction of the Bak–Tang–Wiesenfeld's concept (Bak et al. 1987) of self-organized criticality (SOC), a number of automaton models have been developed to study avalanche dynamics (Vandewalle and Ausloos 1996, Kadanoff et al. 1989, Takayasu 1989, Zhang 1989, Dhar 1990, Priezzhenev et al. 1996). Lattice models of granular materials have previously been examined in many contexts, particularly in terms of self-organization and generalized in sand dunes (Kadanoff et al. 1989, Takayasu 1989, Zhang 1989, Dhar 1990, Priezzhenev et al. 1996, Vandewalle and Ausloos 1996). Coulomb noticed that a granular system with a slope of angle larger than an angle of repose would be unstable (Nedermann 1992). A detailed scenario for the avalanches is discussed by deGennes (1998). Power-law distribution of avalanche size–number has been studied by Herrmann (1999) and Vandewalle (1999). It is, however, interesting to model the distinct morphological dynamical behaviors of dunes. This study (Sagar 1999) has been further extended here to compute the number of avalanches of varied sizes by tuning the strength of nonlinearity parameter. Avalanches of various sizes occur due to instability during sand dune dynamics. To visualize the size distribution of avalanches, sand dune dynamics has been numerically simulated by changing the strength of nonlinearity parameter (λ) that shows different impacts on dune dynamics. An equation to compute avalanche diameters, by considering the inter-slipface angles of a simulated sand dune under dynamics, is proposed. Using this equation, avalanche diameters have been computed from a dynamically changing simulated sand dune.

Is There Any Sand Dune That Possesses the Angle of Repose of More than 45°?

Sand dunes may have the range of angles of repose from 10° to 45° depending upon the sand particle properties such as shape, size, and interlocking, which are subjected to change with exogenic nature of forces. This upper limit of the angle of repose, i.e., 45°, incited us to consider the lower and upper limits of the NFDs and their corresponding inter-slipface angles within the range of $0 \leq \alpha \leq 1$ and $180° \leq \theta \leq 90°$, respectively. The upper limit of fractal dimension of the sand dune is 2, and the corresponding inter-slipface angle and the angle of repose are, respectively, 90° and 45°. Since the upper limit of angle of repose is 45°, the validity of the range of fractal dimensions between $90° \leq \theta \leq 180°$ of sand dune with two slipfaces undergoing dynamics is reasonable and logical. In simple terms, if there exists a sand dune with angle of repose or inter-slipface angle, at any time as the sand dune undergoing dynamical changes exceeds the limit of 45° (angle of repose) or 90° (inter-slipface angle), this simple model, where the NFD of an ideal sand dune that has been considered as a unified quantity, fails.

Avalanches in a Simulated Sand Dune

Up to some state of sand accretion process, no avalanches in the sand dune will occur. From the critical state, which can be determined by an angle of repose, the avalanches of several sizes will occur. As the steepness of sand dune increases, the possibility of avalanche occurrence will also increase. A recipe to understand the sand dune dynamical behavior, while the sand supply is continuous, is that a higher value for the NFD (α) of the sand dune profile at discrete time t than at time $t + 1$ is an indication of occurrence of an avalanche of a specific size. The avalanche size is computed as the distance between the two peaks, with a condition that the $\theta_{t+1} > \theta_t$, of successive profiles. This rolling starts from the state that the angle of repose of the dune reaches critical state. The avalanche size is defined as the diameter of

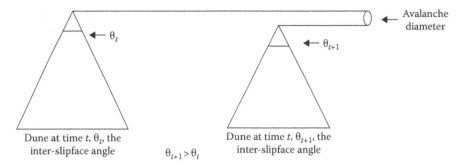

FIGURE 9.23
Diameter of an avalanche. (The inter-slipface angles at discrete time intervals are shown with a possible avalanche since $\theta_{t+1} > \theta_t$.) (From Sagar, B.S.D. et al., *Fractals*, 11(2), 183, 2003.)

the out-scribed circle of the avalanche (Figure 9.23). This schematic diagram shows the dune at two discrete time intervals with a possible avalanche. The avalanche size or diameter can be computed by Equation 9.23:

$$\text{Avalanche size} = \frac{d}{2}\left\{\left[\cot\left(\frac{\theta_t}{2}\right)\right] - \left[\cot\left(\frac{\theta_{t+1}}{2}\right)\right]\right\} \qquad (9.23)$$

It will be considered that there is an avalanche of particular size only if $\theta_{t+1} > \theta_t$. The size of the avalanche depends on the difference between the θ values at successive discrete time intervals. In contrast, if the θ value is lesser than its preceding value in the time series data, then it will not be considered as an avalanche. However, it can be said that the sand dune steepens further. The trajectory parts between the region below the conditional bisectrix line and that above the inverted parabola of 1-D map shown in Figure 9.20a indicate the occurrence of avalanches. The larger the length of the trajectory part in this region, the larger is the avalanche size.

Sample Study and Results

The dune profiles and corresponding inter-slipface angles are generated in discrete time intervals. From these time series of inter-slipface angles, the distance between the two peaks of the successive dune profiles at discrete time intervals can be computed. From the simulated time series of embedded θ values with a condition that $\theta_{t+1} > \theta_t$, the changing out-scribed diameter of an avalanche is computed by using Equation 9.23. It is interesting to observe the number of avalanches of varied sizes by changing various parameters, in Equation 9.2, such as λ, α, and d. In Table 9.12, the size distribution of avalanches has been shown by changing d with $\lambda = 4$. These results are discussed.

TABLE 9.12

Total Avalanche Count and Avalanche Distribution (with $\alpha = 0.1$; Dune Base Length = 9 m; Number of Iterations = 1500)

λ	Total Avalanche Count	Distribution of Avalanches according to Diameters (m)								
		<0.5	0.5–1.0	1.0–1.5	1.5–2.0	2.0–2.5	2.5–3.0	3–3.5	3.5–4	>4
2.1	2	2	0	0	0	0	0	0	0	0
2.2	4	4	0	0	0	0	0	0	0	0
2.3	5	5	0	0	0	0	0	0	0	0
2.4	6	6	0	0	0	0	0	0	0	0
2.5	9	9	0	0	0	0	0	0	0	0
2.6	13	13	0	0	0	0	0	0	0	0
2.7	19	19	0	0	0	0	0	0	0	0
2.8	30	30	0	0	0	0	0	0	0	0
2.9	63	63	0	0	0	0	0	0	0	0
3.0	748	748	0	0	0	0	0	0	0	0
3.1	748	748	0	0	0	0	0	0	0	0
3.2	748	4	744	0	0	0	0	0	0	0
3.3	748	5	743	0	0	0	0	0	0	0
3.4	749	3	746	0	0	0	0	0	0	0
3.5	749	1	374	374	0	0	0	0	0	0
3.6	749	76	220	453	0	0	0	0	0	0
3.7	715	271	156	128	160	0	0	0	0	0
3.8	655	150	131	101	171	102	0	0	0	0
3.9	606	76	61	134	86	122	127	0	0	0
4.0	492	74	65	62	62	43	47	52	50	36

Source: Sagar, B.S.D. et al., *Fractals*, 11(2), 183, 2003.

Strength of Nonlinearity versus the Avalanche Size Distribution

It is worthwhile to study the relation between the strength of nonlinearity and the avalanche size distributions. To deal with this exercise, a unified diagram may be shown to understand the avalanche dynamics in this simulated sand dune dynamics. This simulated sand dune dynamics enables all possible behaviors of a sand dune that undergoes morphological changes with a given strength of nonlinearity. In the present model, the avalanches started being observed at the angle of repose 37.4°. The sand dune under dynamics with a strength of nonlinearity 2.1 will attain critical state from which the avalanches are being observed; the angle of repose of such sand dune under dynamics is 37.4°.

- The avalanche count is found to increase and then decrease with an increase in strength of nonlinearity. As the strength of nonlinearity is increased, it is observed that the number of avalanche size categories has increased. It is also observed during the investigations that when the strength of nonlinearity is less than 2, no avalanches were observed in this numerically simulated sand dune dynamics. Avalanche size distribution has been carried out by changing the λ value, and the results are given in Table 9.12.

• All slopes below some critical value seem to be stable. After some time, the shape does not change anymore, and all additional grains just flow along the surface to the rim of the base where they fall off. While for spherical particles it is reported that the angle of repose is typically 10°–20°, dry sand exhibits ~30°–40°, and the humidity can make it rise much more. However, the computed angle of repose, from the model thus simulated, is 37.4°. This is in conformity with the specified range, i.e., 30°–40° of angle of repose for the dry sand, proposed by Herrmann (1999). From the study, it is inferred that the critical angle of repose is 37.4°. This angle of repose will be attained when the strength of nonlinearity (λ) that has been used in the model is >2.

Classification of Dunes Based on Occurrence of Avalanches

Certain characteristics of the dune dynamics at threshold strength of nonlinearities have been given in Table 9.13.

TABLE 9.13

Dune Classification Based on the Occurrence of Avalanches
($\alpha = 0.1$; $d = 9$ m; Iterations = 1500)

Threshold Strength of Nonlinearity (λ)	Occurrence of Avalanches during Active State of Dune	No. of Avalanches during Active State of Dune	Avalanche Diameter(s) (m)	Stability Type	Active/ Inactive over a Period of Time	No. of Angle(s) of Repose
2	No	0	0	Stable	Inactive	Nil
3	Yes	748	0.027514	Initially period 2	Inactive	One
3.46	Yes	749	1.086073 and 0.87355	Periodically changing	Active	Two
3.57	Yes	749	Many avalanches of various diameters ranging from 0.54 to 1.4	Chaotically changing	Hyperactive	Many

Source: Sagar, B.S.D. et al., *Fractals*, 11(2), 183, 2003.

- Stability of the sand dune is defined in terms of occurrence of avalanches. Continuous accretion of sand keeps the dune active. Such dunes are called *active dunes*. However, a dune is said to be *inactive*, if there are no avalanches after certain discrete time intervals. Due to absence of sand supply or winds capable of transporting sand, an *active dune* may turn into *inactive dune*. In real case, such a phenomenon might arise after a long period. On the Mars, such *inactive dunes* that were once active can be seen. It is observed that for the strength of nonlinearity parameter $\lambda > 3$, dunes are active.

- From this numerically simulated sand dune dynamics, the dune dynamics are categorized, based on the avalanche occurrences with discrete time, as super-stable, semi-stable, and chaotically behaving dunes.

- Sand dune behaviors can be visualized as phase changes. Conventionally, it has been defined that a sand dune will have one angle of repose. Once the dune reaches to this critical state, avalanches will be observed. However, it is also true that there may be numerous angles of repose in a sand dune undergoing dynamics. This can be schematically represented through a bifurcation diagram. We can argue this phenomenon of having different angles of repose in a sand dune undergoing dynamics as changing properties of sand. With changing sand properties, the interlocking parameters will be changed, hence the angle (s) of repose when the interlocking properties of sand particles (of a dune) change due to the reason that the sand characteristics are primarily subjected to exogenic nature of processes. In turn, the angle of repose is not just one; there will be numerous angles of repose as the sand dune undergoes dynamical changes. It is reported by several researchers that the angle of repose varies with the change in characteristics of sand particles, and also with the fluctuations in the wind strength. It can be said that these characteristic changes may be due to exogenic nature of processes in general. There will be an angle of repose variation during the process of sand dune dynamics. Dune changes its phases with dynamically varying sand particle interlocking properties, strength of wind, etc. While traversing several phase changes, a dynamically changing dune possesses one or more angles of repose. It is observed that the avalanche diameter gets reduced with discrete time for the strength of nonlinearity $2 < \lambda \leq 3$. As the iterative process is progressing, the dune becomes stable for the strength of nonlinearity between 2.1 and 3. Avalanches of fixed sizes are observed with a periodic interval for the strength of nonlinearity $3 < \lambda \leq 3.46$. It is interesting to note that the avalanches of two different diameters are observed periodically when the strength of nonlinearity is $3.46 \leq \lambda < 3.57$. Such a case can be visualized when the strength of nonlinearity is in the range $3.46 \leq \lambda < 3.57$. For the strength of nonlinearity >3.57, the avalanche diameter and the number of distributary patterns are chaotic. The number of angle(s) of repose varies with the strength of nonlinearity, of a dynamically changing simulated sand dune, and has (have) been given in Table 9.13.

Avalanche Distribution in Different Sizes of Dunes

Results with stationary base lengths of 3, 6, and 9 m with initial NFD of 0.1 and the number of iterations of 12,000 have been given in Table 9.14 to understand the distributary pattern of avalanche diameter–number of a chaotically behaving sand dune.

- It is observed that the total avalanche count remains the same in spite of a change in the base length. Graphs have been plotted between the avalanche size and number for the sand dunes with base width of 6 and 9 m (Figure 9.24a and b). No significant variation has been found in these graphs. However, the distributed avalanche count varies.
- For 3 m base width, avalanches up to 1.5 m, while for 6 m base width, the avalanche size up to 3 m, and it is more than 4 m for 9 m base width were observed.
- The number of avalanches of a specific size reduces as the base width is increased.
- With base widths 6 and 9 m, it is observed that the number of avalanches of different diameters initially reduces, reaches a minimum, and then increases again before the process extincts.

This study of theoretical interest can be validated by incorporating the interslipface angles of corresponding profiles of a real-world sand dune undergoing dynamics, the retrieval of which is possible with the advent of the

TABLE 9.14

Avalanche Distribution

	$\alpha = 0.1$; $\lambda = 4$; Iterations = 12,000								
Total Count	<0.5 m	0.5–1.0 m	1.0–1.5 m	1.5–2.0 m	2.0–2.5 m	2.5–3.0 m	3–3.5 m	3.5–4 m	>4 m
$d = 3$ m									
3905	1611	1154	1140	0	0	0	0	0	0
$d = 6$ m									
3905	885	722	613	554	557	574	0	0	0
$d = 9$ m									
3905	592	569	446	417	385	365	416	400	288

Source: Sagar, B.S.D. et al., *Fractals*, 11(2), 183, 2003.

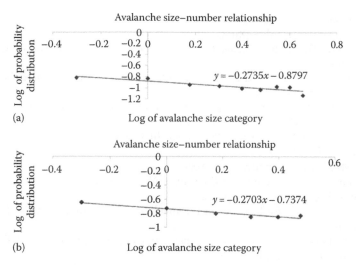

FIGURE 9.24
Graphical plots between the logarithms of avalanche size and avalanche number for (a) a dune width of 9 m and (b) a dune width of 6 m. (From Sagar, B.S.D. et al., *Fractals*, 11(2), 183, 2003.)

availability of multitemporal, high-resolution interferometrically generated DEMs at different timescales. Certain geodynamic problems such as the morphological evolutionary behavior of a sand dune can be better modeled by using the multi-date DEMs, derived from high-resolution remotely sensed data. From such a study, one can understand the distribution of avalanches of real-world sand dunes of various sizes undergoing dynamics.

References

Abraham, R. H. and C. D. Shaw, 1982, *Dynamics—The Geometry of Behaviours*, Vols. 1–4, Aerial Press, Santa Cruz, CA.

Allegre, C. J., L. Le Mouel, and A. Provist, 1982, Scaling rules in rock fracture and possible implications for earthquake prediction, *Nature*, 297, 47–49.

Alonso, J. J. and H. J. Hermann, 1996, Shape of the tail of a two dimensional sandpile, *Physical Review Letters*, 76(26), 4911–4914.

Bagnold, R. A., 1941, *Physics of Wind Blown Sand and Sand Dunes*, Methuen, London, U.K.

Bak, P., C. Tang, and K. Wiesenfeld, 1987, Self-organized criticality: An explanation of 1/f noise, *Physical Review Letters*, 59, 381.

Beauvais, A. and J. Dubois, 1995, Attractor properties of a river discharge dynamical system, *Eos Transactions AGU* 73, 46, F234.

Chapple, M., 1968, A mathematical theory of finite amplitude folding, *Geological Society of America Bulletin*, 79, 47–68.

Chockalingam, L. and B. S. D. Sagar, 2003, Automatic generation of subwatershed map from Digital Elevation Model: A morphological approach, *International Journal of Pattern Recognition and Artificial Intelligence*, 17(2), 269–274.

Davy, P., A. Sornette, and D. Sornette, 1990, Some consequences of a proposed fractal nature of continental faulting, *Nature*, 348, 56–58.

Devaney, R. L., 1986, *An Introduction to chaotic Dynamical Systems*, Benjamin/Cumings, New York.

Dhar, D., 1990, Self-organized critical state of sandpile automaton models, *Physical Review Letters*, 64, 1613.

Dieterich, J. H., 1970, Computer experiments on mechanics of finite amplitude folds, *Canadian Journal of Earth Science*, 7, 467–476.

Dieterich, J. H. and N. L. Carter, 1969, Stress history of folding, *American Journal of Science*, 267, 129–154.

Feder, J., 1988, *Fractals*, Plenum, New York.

Feigenbaum, M. J., 1980, *Universal Behavior in Nonlinear Systems*, Vol. 1, Los Alomas Science, Los Alomas, CA, pp. 4–27.

deGennes, P. G., 1998, Reflections on the mechanics of granular matter, *Physica A*, 261, 267.

Harris, A. R., 1994, Time series remote sensing of a climatically sensitive lake, *Remote Sensing of Environment*, 50, 83–94.

Herrmann, H. J., 1999, Shapes of granular surfaces, *Physica A*, 270, 82.

Hobbs, B. E., W. D. Means, and P. F. Williams, 1976, *An Outline of Structural Geology*, John Wiley & Sons, New York.

Jayawardena, A. W. and F. Lai, 1994, Analysis and prediction of chaos in rainfall and streamflow time series, *Journal of Hydrology*, 153, 23–52.

Jenson, R. V., 1987, Classical chaos, *American Scientist*, 16, 168–181.

Kadanoff, L. P., S. R. Nagel, L. Wu, and S. M. Zhou, 1989, Scaling and universality in avalanches, *Physical Review A*, 39, 6524–6537.

Maden, T. R., 1983, Microcrack connectivity in rocks: A renormalization group approach to the critical phenomena of conduction and failure in crystalline rocks, *Journal of Geophysical Research*, 88, 585–592.

Mandelbrot, B. B., 1982, *Fractal Geometry of Nature*, Freeman & Co., San Francisco, CA.

Manna, S. S., 1991, Critical exponents of sandpile models in two dimensions, *Physica A*, 179, 249–268.

Matheron, G., 1975, *Random Sets and Integrated Geometry*, Wiley, New York.

May, R. M., 1976, Simplified mathematical models with very complicated dynamics, *Nature*, 261, 459–467.

Means, W. D., 1976, *Stress and Strain: Basic Concepts of Continuous Mechanics for Geologists*, Springer Verlag, New York, p. 339.

Means, W. D., 1990, Kinematics, stress, deformation and material behavior, *Journal of Structural Geology*, 12, 953–971.

Murthy, T. V. R., M. V. Muley, M. Chakraborty, V. Tamilarasan, E. Amminedu, G. Meher Baba, A. Krishna, and S. R. Rao, 1988, Water quality studies in the Chilka lake using Landsat data (unpublished), Presented in *Workshop on Remote Sensing Applications ill Water Resources Management, India*, all 28–30, December, Orissa Remote Sensing Application Centre, Bubaneswar, India.

Nedermann, R., 1992, *Statics and Kinematics of Granular Materials*, Cambridge University Press, London, U.K.

Newman, W. I. and L. Knopof, 1982, Crack fusion dynamics: A model for large earthquakes, *Geophysical Research Letters*, 9, 735–738.

Parrish, D. K., (1973), A nonlinear finite element fold model, *American Journal of Science*, 273, 318–334.

Pasternack, G. B., 1999, Does the river run wild? Assessing chaos in hydrological systems, *Advances in Water Resources*, 23, 253–260.

Price, N. J. and J. W. Cosgrove, 1990, *Analysis of Geological Structures*, Cambridge University Press, Cambridge, U.K., p. 502.

Priezzhenev, V. B., A. Dhar, S. Krishnamurthy, and D. Dhar, 1996, Eulerian walkers as a model of self-organized criticality, *Physical Review Letters*, 77, 5079.

Ramsay, J. G. and M. I. Huber, 1987, The techniques of modern structural geology. In: *Folds and Fractures*, Vol. 2, Academic Press, London, U.K., p. 391.

Ruelle, D., 1987, Diagnosis of dynamical system with fluctuating parameters, *Proceedings of the Royal Society of London A*, 413, 5–8.

Sagar, B. S. D., 1996, Fractal relations of a morphological skeleton, *Chaos, Solitons & Fractals*, 7(5), 1871–1879.

Sagar, B. S. D., 1998, Numerical simulations through first order nonlinear difference equation to study highly ductile symmetrical fold (HDSF) dynamics: A conceptual study, *Discrete Dynamics in Nature and Society*, 2, 181–198.

Sagar, B. S. D., 1999a, Estimation of number-area-frequency dimensions of surface water bodies, *International Journal of Remote Sensing*, 20(13), 2491–2496.

Sagar, B. S. D., 1999b, Morphological evolution of a pyramidal sandpile through bifurcation theory: A qualitative model, *Chaos, Solitons & Fractals*, 10(9), 1559–1566.

Sagar, B. S. D., 2000, Fractal relation of medial axis length to the water body area, *Discrete Dynamics in Nature and Society*, 4(1), 97.

Sagar, B. S. D., 2001a, Generation of self organized critical connectivity network map (SOCCNM) of randomly situated surface water bodies, Letters to Editor, *Discrete Dynamics in Nature and Society*, 6(1), 225–228.

Sagar, B. S. D., 2001b, Hypothetical laws while dealing with effect by cause in discrete space, Letter to the Editor, *Discrete Dynamics in Nature and Society*, 6(1), 67–68.

Sagar, B. S. D., 2005, Discrete simulations of spatio-temporal dynamics of small water bodies under varied streamflow discharges (invited paper), *Nonlinear Processes in Geophysics (American Geophysical Union)*, 12, 31–40.

Sagar, B. S. D. and L. Chockalingam, 2004, Fractal dimension of non network space of a catchment basin, *Geophysical Research Letters*, 31(6), L12502.

Sagar, B. S. D., G. Gandhi, and B. S. P. Rao, 1995a, Applications of mathematical morphology on water body studies, *International Journal of Remote Sensing*, 16(8), 1495–1502.

Sagar, B. S. D. and K. S. R. Murthy, 2000, Generation of fractal landscape using nonlinear mathematical morphological transformations, *Fractals*, 8(1), 267–272.

Sagar, B. S. D., M. B. R. Murthy, and P. Radhakrishnan, 2003a, Avalanches in numerically simulated sand dune dynamics, *Fractals*, 11(2), 183–193.

Sagar, B. S. D., M. B. R. Murthy, C. B. Rao, and B. Raj, 2003b, Morphological approach to extract ridge-valley connectivity networks from Digital Elevation Models (DEMs), *International Journal of Remote Sensing*, 24(1), 573–581.

Sagar, B. S. D., C. Omoregie, and B. S. P. Rao, 1998a, Morphometric relations of fractal-skeletal based channel network model, *Discrete Dynamics in Nature and Society*, 2(2), 77–92.

Sagar, B. S. D. and B. S. P. Rao, 1995a, Ranking of lakes: Logistic models, *International Journal of Remote Sensing*, 16, 365–368.

Sagar, B. S. D. and B. S. P. Rao, 1995b, Possibility on usage of return maps to study the dynamical behavior of lakes: A hypothetical study, *Current Sciences*, 68, 950–954.

Sagar, B. S. D., D. Srinivas, and B. S. P. Rao, 2001, Fractal skeletal based channel networks in a triangular initiator basin, *Fractals*, 9(4), 429–437.

Sagar, B. S. D. and T. L. Tien, 2004, Allometric power-law relationships in a Hortonian Fractal DEM, *Geophysical Research Letters*, 31(2), L06501.

Sagar, B. S. D. and M. Venu, 2001, Phase space maps of a simulated sand dune: A scope, *Discrete Dynamics in Nature and Society*, 6(1), 64.

Sagar, B. S. D., M. Venu, G. Gandhi, and D. Srinivas, 1998b, Morphological description and interrelationship between force and structure: A scope to geomorphic evolution process modeling, *International Journal of Remote Sensing*, 19(7), 1341–1358.

Sagar, B. S. D., M. Venu, and K. S. R. Murthy, 1999, Do skeletal network derived from water bodies follow Horton's laws? *Mathematical Geosciences*, 31(2), 143–154.

Sagar, B. S. D., M. Venu, and B. S. P. Rao, 1995b, Distributions of surface water bodies, *International Journal of Remote Sensing*, 16(16), 3059–3067.

Savard, C. S., 1990, Correlation integral analysis of South Twin River streamflow, central Nevada: Preliminary application of chaos theory, *Eos Transactions AGU*, 71(43), 1341.

Savard, C. S., 1992, Looking for chaos in streamflow discharge derivative data, *Eos Transactions AGU*, 73(14), 50.

Serra, J., 1982, *Image Analysis and Mathematical Morphology*, Academic Press, London, U.K., p. 610.

Sivakumar, B., 2004, Chaos theory in geophysics: Past, present and future, *Chaos, Solitons & Fractals*, 19(2), 441–462.

Smalley, Jr. R. F., D. L. Turcotte, and S. A. Solla, 1985, A renormalization group approach to study stick-slip behavior, *Journal of Geophysical Research*, 90(B2), 1894–1900.

Tay, L. T., B. S. D. Sagar, and H. T. Chuah, 2007, Granulometric analysis of basin-wise DEMs: A comparative study, *International Journal of Remote Sensing*, 28(15), 3363–3378.

Takayasu, H., 1989, Steady-state distribution of generalized aggregation system with injection, *Physical Review Letters*, 63, 2563.

Tsonis, A. A., G. N. Triantafyllou, and J. B. Elsner, 1994, Searching for determinism in observed data: A review of the issue involved, *Nonlinear Processes in Geophysics*, 1, 12–25.

Vandewalle, N., 1999, Phase segregation and avalanches in multispecies sandpiles, *Physica A*, 272, 450–458.

Vandewalle, N. and M. Ausloos, 1996, Static and dynamic epidemics on looped chains and looped trees, *Computers & Graphics*, 20, 921.

Zhang, Y. C., 1989, Scaling theory of self-organized criticality, *Physical Review Letters*, 63, 470.

10

Quantitative Spatial Relationships and Spatial Reasoning

In quantitative spatial reasoning, spatial relationships such as adjacency, betweenness, directional, distances, shape–size, and centrality are important aspects. Many studies available have dealt with such spatial relationships through qualitative reasoning. However, mathematical morphological transformations offer several insights to provide quantitative approaches in handling the spatial reasoning tasks. Chapters 10 through 13 provide details on how mathematical morphology could be employed in addressing the following aspects of relevance to spatial reasoning studies: (1) directional spatial relationship, (2) between space, (3) adjacency and touch relationship, (4) distance-based relationships, (5) relationships based on shape–size complexity measures, and (6) centrality relationship.

The ability to recognize strategically important set(s) within a cluster has interesting applications in geographic information science (GISci). Using techniques and principles borrowed from mathematical morphology, we introduce geometrically based criteria that serve as indicators of the strategic importance of sets within a cluster. We have applied a morphology-based approach developed on data derived from a spatial map of India and on a theme depicting water bodies traced from geo-coded remotely sensed satellite data.

Spatial Reasoning and Mathematical Morphology

Spatial information theory provides theoretical basis in general to GISci (Rhind 1973, Tobler 1976, Tomlin 1983, Goodchild 1992, Wilson 2008). An advanced concept theory with more geometrical rigor that revolutionizes the subject of spatial data analysis includes mathematical morphology (Serra 1982). The representative works with indirect relevance to spatial information science appeared, during recent past, on applications of these concepts either individually or combinedly on retrieval (e.g., Beucher and Meyer 1992), analysis, and characterization (e.g., Rosenfeld and Pal 1988, Maragos 1989) of certain features. Many operations, involved in GISci-related analysis—that fall under the name "Map Algebra" (Tomlin 1983)—can be performed via mathematical morphology and fuzzy set theory (Pullar 2001, Stell 2007). Making logical inference about spatial aspects of the environment through humans and computers is a topic of spatial reasoning, which is a branch of spatial information science. One of the important aspects of spatial reasoning is computation besides cognition and formalism (Egenhofer and Mark 1995).

Spatial relationships based on topology, direction, and distances between the sets (objects) are some of the important ingredients of spatial reasoning. The identification of strategically important zones within a set of spatial objects is directly related to the field of spatial reasoning. Quantitative spatial reasoning requires quantitative spatial relations among the sets (e.g., water bodies, states of a country) under investigation. In addition, there are several context-dependent parameters that can be derived between sets taking one as the source (origin set) and the other (destination sets) as the target (e.g., Wilson 2009). Here, the sets and the zones interchangeably used are areal type of objects (e.g., states, districts, water bodies that are spatially spread). The geometric criteria are based on parameters such as distance, boundary being shared, and the geometric similarity (in terms of both shape and size) between each zone and other zones.

It is common to use from the geometric point of view the combination of qualitative and quantitative information to derive an object that is strategically important within a set of spatial objects. But it is straightforward if one uses quantitative information. The importance of spatial relations, based on topology, direction, and distance among sets, has been recognized in various domains that include spatial reasoning (Jiang and Yao 2006, Yao and Thill 2006, Kwan and Ding 2008, Liu et al. 2008, 2010, Cidell 2010, Gao et al. 2010, Ocalir et al. 2010). A number of related studies deal with spatial reasoning and analysis, including approximating shortest routes by global navigation and local search using multiagent models that operate in cellular space (Batty and Jiang 2000), spatial agents for geographic information processing (Rodrigues et al. 1996), and computational solutions to zone designing problems using a simple geographical framework built from regular grid squares (Openshaw 1977, Martin 2000) and automation of the design of zones using various computational approaches (Aarts and Korst 1989, Beucher and Meyer 1992, Horn 1995, Mehrotra et al. 1998).

Mathematical morphology (Serra 1982) is a science of shapes, forms, and structures, and has been employed in the context of spatial relationships, reasoning, and spatiotemporal modeling. Some representative studies include Serra (1982), Bouzy (2003), Bloch et al. (2007), Stell (2007), Sagar (2010), Sagar and Serra (2010), and Rajashekara et al. (2012). Since the early twentieth century, the flavor of morphological transformations is obvious from the seminal studies (Cayley 1889, Hausdorff 1914, Coxeter 1950, 1961, Alexandroff 1961, Serra 1982). Thompson (1992) has looked at biological features to establish relationships based on allometric measures that clearly possess morphological components. In the late twentieth century, mathematical concepts have been employed to deal with geo(spatial)graphic data and information by pioneering researchers (Rhind 1973, Batty 1976, Tobler 1976, Tomlin 1983, Goodchild 1992, Worboys, 1994, Egenhofer and Mark 1995, Spielman and Thill 2008, Wilson 2008). Fuzzy set-theoretic concepts have also proved especially robust in this area of study (Rosenfeld and Pal 1988, Chaudhuri 1990, Nafarieh and Keller 1991, Krishnapuram and Keller 1993). A detailed classification of satellite images into land use units of rural and urban regions is important in assisting government agencies in a number of ways including urban planning, transportation management, and rescue operations (Unsalan and Boyer 2005). Mathematical morphology that offers various map algebraic concepts (e.g., Tomlin 1983, Pullar, 2001) also provides insights for GISci, in general, and automatic map generalization, in particular. Many researchers proposed elegant computational approaches for automatic map generalization in discrete environment (Rhind 1973, 1988, Monmonier 1983, McMaster 1987, Shea and McMaster 1989, Muller 1991, McMaster and Shea 1992, Muller and Wang 1992, Schylberg 1993, Muller et al. 1995, Su and Li 1995, Weibel 1995, Su et al. 1997, Yao and Thill, 2006, Yan and Thill 2009). Mathematical morphological transformations, for the first time, were employed in the context of map generalization (e.g., map aggregation, select and eliminate, and coarsening) by Su and Li (1995) and Su et al. (1997). However, those transformations were employed in distributions, which are similar to select and eliminate process in map generalization, of surface water bodies of various sizes and shapes (Sagar et al. 1995b), simulation of areal spread (similar to feature aggregation) (Sagar et al. 1995b), and the interplay between numerics and graphics to visualize the spatiotemporal behavior of water bodies under perturbations caused due to drought and monsoon (Sagar et al. 1998, Sagar 2005). In the context of GISci, mathematical morphological operations have been earlier employed essentially to compute basic measures of planar features (e.g., surface water bodies [Sagar et al. 1995]), perform size and shape distributions of water bodies (Sagar et al. 1995), map-based coupled dynamical systems (Sagar 1994, 2005, Sagar et al. 1998), and map generalization (Su et al. 1997).

Background on Strategic Set Identification

The motivation of this chapter comes from spatial reasoning studies. We consider a system consists of zones (sets, states, watersheds) in planar form over a geographical space. We are interested in the geometric organization of several zones within a spatial system (Figure 10.1) and the quantitative spatial relationships among those zones. Recognizing strategically significant sets within such a system (cluster) composed of various sets is a part of the spatial reasoning process and can be accomplished both quantitatively and qualitatively. Identifying a strategic importance of a zone from the geometric point of view based on qualitative spatial relationship–based reasoning is nontrivial, and such identification process varies from person to person according to their own individual spatial perceptions. Alternately, detecting

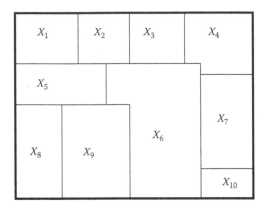

FIGURE 10.1
A spatial system with 10 zones. If X_i (e.g., X_1) is the origin zone, then all other zones (X_j) (e.g., $X_2 - X_{10}$) are treated as destination zones. Then computing the degree of strategic importance of X_i is subjected in this chapter. (From Sagar, B.S.D. et al., *IEEE Geosci. Remote Sens. Lett.*, 10(3), 2013.)

such strategically significant sets can be achieved through the use of geometric operations, as strategic importance in this sense can be defined as "an object (e.g., state of a country) from which it is easy to reach all of its neighboring objects, has similar shapes and sizes with other objects, and possesses a longer shared boundary with its neighboring objects".

Strategically significant sets within a cluster may possess one or a combination of the following characteristics: greater size, greater number of neighbor sets, greater length of the boundary being shared with adjacent sets, greater proximity to other sets and greater contextuality with other sets, significance of location, the spatial complexity involved in reaching all other sets, the degrees of contextuality between the set in question and sets that surround it, the degrees of similarity, and shape–size relationships between sets." Thus, topology, distance, and direction form the geometric dimensions used to designate a zone as strategically important. A zone X_i is chosen as strategically most significant to establish a facility as it is (1) bigger than X_js (other zones in the system), (2) possessing longer boundary being shared with adjacent zones, (3) located in a place closer to all X_js, and also from all X_js reaching X_i requires shorter distance (minimum energy expenditure involved), (4) possessing less contorted boundary and shape while relating X_js, and (5) possessing high contextual relationship with X_js. No other zone from a pool of X_js match with X_i with respect to these characteristic(spatial) relationships, and hence X_i is chosen as the best zone and is termed strategically the most important zone.

Strategic importance of each zone is usually designated according to the economic activity of a zone or by its qualitative spatial relationship with other zones in terms of topology (e.g., adjacency, neighborhood), direction (e.g., north, south), and distance (e.g., close, far). The spatial relationships employed in designating X_i as strategically significant are qualitative and hence the way X_i is chosen as strategically significant via qualitative reasoning (description). There is important literature on this subject when the spatial significance is modeled by qualitative spatial relationships. However, qualitative spatial relationships in identifying spatially significant set may lead to results that are subjective.

Description via quantitative spatial relationships is of use for two reasons: (1) to have homogeneous result that is highly objective and (2) to handle data sets providing a large number of zones in georeferenced mode. Map algebraic concepts were employed in modeling and in spatial reasoning domains (e.g., Tomlin 1983). Despite the success of these concepts, their applicability in quantitative spatial relationship studies has not been fully feasible.

Within an urban and regional planning context, streets, cities, districts, states, etc., all possess varying degrees of strategic importance that can be computed according to these quantitative geometric criteria that provide insights into facility allocation and spatial planning studies. For example, states within a country possess a certain degree of strategic importance, and determining the importance of all states within a country is important to decide a suitable location to establish a facility (e.g., defense facility). Based on a raster setting, the mathematical models presented have a sound mathematical basis, offer the flexibility of being applicable over raster maps, and are conceptually, fairly intuitive.

Modeling Concepts

Cluster of sets: Let a cluster of sets (I) be composed of a number of nonempty, compact sets denoted by $X_1, X_2, X_3, ..., X_N$, such that $I = \bigcup_{i=1}^{N} X_i$. If we select any pair of sets X_i and X_j from this cluster such that $i \neq j$, the following spatial relations hold true:

1. $X_i \cap X_j = \varnothing$

2. $X_i \cap \left(\bigcup_{\substack{j=1 \\ j\neq i}}^{N} X_j \right) = \varnothing, \forall i, j = 1 - N$

3. $(X_i \oplus B) \cap \left(\bigcup_{\substack{j=1 \\ j\neq i}}^{N} X_j \right) = \left(\left(\left(\bigcup_{\substack{j=1 \\ j\neq i}}^{N} X_j \right) \oplus B \right) \cap X_i \right) \neq \varnothing$

We define an ideal system as one that is composed of spatial objects that are identical to each other in terms of shapes and sizes. The definition also requires that such spatial objects be arranged in a systematic way in such a way that the distances between any spatial object to any other spatial object should be the same. A nonideal system is the one in which the adjacent spatial objects are of varied sizes and shapes and are also arranged heterogeneously. An example is shown in Figure 10.1. The main variations of these measures between an ideal case and a nonideal case are shown in this chapter.

Recognition and Visualization of Strategically Significant Spatial Sets via Morphological Analysis

The ability to recognize strategically important set(s) within a cluster has interesting applications in GISci. Using techniques and principles borrowed from mathematical morphology, we introduce geometrically based criteria that serve as indicators of the strategic importance of sets within a cluster. We have applied a morphology-based approach developed on data derived from a spatial map of India and on a theme depicting water bodies traced from geocoded remotely sensed satellite data. A host of transformations available with mathematical basis offered by mathematical morphology—that have already shown potential in map generalization and map aggregation studies—may address several aspects of spatial reasoning via quantitative means. Hence, the main purpose of this chapter is to determine the degree of strategic importance of zones that are part of a collection of zones through development of a framework, based on mathematical morphology, the science of shapes, forms, and structures (Serra 1982). The goal of this chapter is to provide the following:

- Measures, for quantitative spatial relationships between origin (X_i) and destination sets (X_j), which include dilation distances, length of boundary being shared, shape–size similarities, and spatial complexities
- The applications of these measures between every set and other sets of a cluster to choose measure-specific strategic importance
- Finally, a framework (based on geometric criteria) to show how mathematical morphology has a role to select or identify "strategic regions within a collection of regions" by employing the measures computed

Morphology-based distances such as dilation distances and Hausdorff dilation distances have been used to address the focused topic of identifying spatially significant sets(s) from a cluster consisting of several connected components (objects) of varied shapes and sizes. In a vector-based network setting, determining the influence of a node utilizing these metrics is more straightforward as spatial relationships between origins and destinations can be depicted. However, when dealing with areal features, determining spatial relationships based on Euclidean metrics is a greater challenge and is the focus of this chapter.

The organization of this chapter is as follows: the "Strategically Significant Set" section provides an overview on the existing spatial reasoning and link with mathematical morphology literature, followed by modeling concepts, rationale, and methodology in the "Experimental Results on Clusters of Sets" section. The "Discussion and Open Problems" section describes the application of the techniques described in previous sections and provides concluding observations and remarks.

Distance between the Sets

The iterative dilation is a better choice to provide a mathematical description of distances between sets. To do so, we consider a set X to have subsets X_i and X_j. Then,

$$d(x, X_i) = \inf\{d(x, y), y \in X_j\}, \quad x \in E;\ X_j \in K' \tag{10.1}$$

and introduce the mapping $K' \times K' \to R_+$

$$\rho(X_i, X_j) = \max\left\{\underset{x \in X_i}{\mathrm{Sup}}\, d(x, X_j);\ \underset{y \in X_j}{\mathrm{Sup}}\, d(y, X_i)\right\} \tag{10.2}$$

Mapping ρ turns out to be a distance, called the *Hausdorff distance*, which holds on K' and no longer on E. Equation 10.2 can equivalently be written as the dilation by the balls of the space E. According to Serra (1994, 1998), the Hausdorff dilation distance $\rho(X_{ij})$, between X_i and X_j (Figure 10.2a)—a pair of disjoint compact sets is given by $X_i \cap X_j = \varnothing$—is given in (10.2) (Figure 10.2c).

$$\rho(X_{ij}) = \inf\left(n : \left(X_i \subseteq (X_j \oplus nB)\right), \left(X_j \subseteq (X_i \oplus nB)\right)\right) \tag{10.3}$$

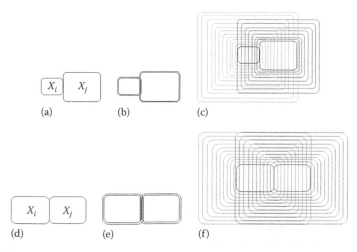

FIGURE 10.2
Illustration of how the shape and size characteristics of the sets influence the three measures explained. (a) Two homothetic adjacent sets X_i and X_j of different sizes, (b) length of the boundary being shared $P(X_{ij}) = P(X_{ji})$, (c) the dilation distances $d(X_{ij}) = 11$, and $d(X_{ji}) = 7$, and in turn the $\rho(X_{ij}) = 7$ further attributing that the $C(X_{ij}) = 7/11 = 0.64$, (d) two geometrically similar homothetic adjacent sets X_i and X_j of similar sizes, (e) length of the boundary being shared $P(X_{ij}) = P(X_{ji})$, (f) the dilation distances $d(X_{ij}) = 10$, and $d(X_{ji}) = 10$, and in turn the $\rho(X_{ij}) = 10$ further attributing that the $C(X_{ij}) = 10/10 = 1$.

Strategic Sets

The task at hand is to determine sets, X_i within a cluster, the presence of which keeps the cluster integrated, while their absence results in the disintegration of the same. In order to do so, one needs to define appropriate measures of the strategic importance of a set. Our concept of strategic importance of a set stresses on the property of a set being central with respect to spatial relationships with other entities in the cluster. We maintain that in a cluster of adjacent, nonempty compact, and nonoverlapping sets, it is possible to compare strategic importance on the basis of an assortment of parameters. These include lengths of the boundary being shared between pairs of sets, degree of reachability to other sets with minimum expenditure of energy, degree of contextuality between pairs of sets, and spatial complexities with respect to contextuality, distance, and perimeter. We shall now attempt to explain one of these ideas, i.e., selecting a set that is strategically important with respect to minimum expenditure of energy, in other words distance-based spatial relationship with the following analogy: consider X_i and X_j to be nonoverlapping sets in a cluster occupied by regiments of soldiers. A confrontation between the two regiments will be influenced to a large extent by the boundary shared by their parent sets with their adjacent sets, if any. For the sake of simplicity, we assume a uniform distribution of soldiers across the whole set by ignoring spatial aspects such as type of terrain and transportation facilities, as well as nonspatial aspects such as the relative strength of the regiments. We assume these factors to be constant and consider a set that shares the maximum length of boundaries with its adjacent regiments to possess a greater degree of strategic significance.

If a regiment of soldiers from X_i walks toward the regiment X_j, considered to be stationary, the minimum time required from regiment X_i to occupy regiment X_j is related to the distance $d(X_{ij})$. If we interchange the roles of the regiments, the distance is given by $d(X_{ji})$. Now, if X_i wants to successively conquer other regiments X_j that are stationary, under the condition that having conquered one regiment, it will move back to its original position before proceeding to conquer the next and a significant amount of expenditure of energy is involved. The higher the energy expenditure involved for X_i to conquer a stationary regiment X_j, the greater is the distance between them. At this point, we may want to determine that regiment for which such an expenditure of energy is minimum. Such a regiment that conquers all other regiments with minimum expenditure of energy is a strategically significant regiment. Please see the "Spatial Significance Index of a Zone" section in Chapter 11 for details on how this analysis has been put in perspective.

A regiment that is larger than any other regiments within the cluster of regiments may also be strategically important. This fact, however, depends to a large extent on its degree of contextuality with other regiments. In turn, the spatial position of such a regiment plays an important role, whether or not it can be designated as strategically important. The degree of contextuality between the regiments X_i and X_j may be computed by finding the ratio between $\min(d(X_{ij}),d(X_{ji}))$ and $\max(d(X_{ij}),d(X_{ji}))$. The ratio ranges between 0 and 1. A value of 1 indicates that $d(X_{ij})$ and $d(X_{ji})$ are identical.

Length of Boundary Being Shared between Origin Set and Destination Sets

The length of the boundary shared between an origin set X_i and a destination set X_j and *vice versa* may be expressed as follows:

$$P(X_{ij}) = P((X_i \oplus B) \cap (X_j)) \tag{10.4}$$

$$P(X_{ji}) = P((X_j \oplus B) \cap (X_i)) \tag{10.5}$$

where $i \neq j$, $P(X_{ii}) = P(X_i \setminus X_i \ominus B)$ and $P(X_{ij}) = P(X_{ji})$.

The sum of the lengths $TP(X_{ij})$ of the boundary shared between the source set and every adjacent set as a target set may be expressed as follows:

$$TP(X_{ij}) = \sum_j P(X_{ij}) = \sum_{\substack{j=1 \\ j \neq i}}^{N} P(X_{ij}) = P\left((X_i \oplus B) \cap \left(\bigcup_{\substack{j=1 \\ j \neq i}}^{N} X_j\right)\right) \tag{10.6}$$

$$TP(X_{ji}) = \sum_i P(X_{ji}) = \sum_{\substack{j=1 \\ j \neq i}}^{N} P(X_{ji}) = P\left(\left(\left(\bigcup_{\substack{j=1 \\ j \neq i}}^{N} X_j\right) \oplus B\right) \cap X_i\right) \tag{10.7}$$

where $P[\cdot]$ is cardinality of P, and $TP(X_{ij}) = TP(X_{ji})$. The maximum length of the boundary, P_{max}, that X_i shares with adjacent sets X_j is

$$P_{max} = \max_{\forall i}\left(P : P((X_i \oplus B) \cap (X_j))\right), \quad i,j = 1 - N; \; i \neq j \tag{10.8}$$

The normalized length of the boundary shared between X_i and X_j may be computed as

$$NP(X_{ij}) = \frac{P(X_{ij})}{P_{max}} \tag{10.9}$$

The mean sum of normalized perimeter is computed as

$$\overline{\sum_j NP(X_{ij})} = \frac{\sum_j NP(X_{ji}) = \sum_i NP(X_{ij})}{\text{Number of adjacent sets}} \tag{10.10}$$

Dilation Distance between Origin Set and Destination Set(s)

The distance from X_i to X_j is represented by

$$d(X_{ij}) = \min_{i \neq j} \left(n : X_j \subseteq (X_i \oplus nB) \right) \tag{10.11}$$

Similarly, the distance from X_j to X_i is represented by

$$d(X_{ji}) = \min_{i \neq j} \left(n : X_i \subseteq (X_j \oplus nB) \right) \tag{10.12}$$

Considering X_i and X_j to be the origin and the destination sets, respectively, we may state the following:

- $d(X_{ii}) = 0$
- $i \neq j$, then $d(X_{ij}) \neq d(X_{ji})$ and $d(X_{ij}) = d(X_{ji})$ if and only if both X_i and X_j possess identical size, shape, and orientation.

The minimum of $d(X_{ij})$ and $d(X_{ji})$ yields Hausdorff (dilation) distances. This may be represented as

$$\rho(X_{ij}) = \min \left(d : d(X_{ij}), d(X_{ji}) \right) \tag{10.13}$$

This approach for estimating the dilation distance between the origin and destination sets is justified as such a dilation distance is essential to compute further the degree of contextuality between the sets. Such a measure cannot be computed if one takes Euclidean distance between the centroids of the two sets under investigation into account. This is due to fact that the Euclidean distance between the centroids of the two sets does not explain the morphological properties of the sets under consideration. However, the limitation of this distance is that it is essentially affected by points of the object's boundary points that are farthest out with respect to other spatial objects.

The sum of the distances between the origin set X_i and every other destination set is $\sum_j d(X_{ij}) = Td(X_{ij})$, while the sum of the distances between each of the destination sets X_j and the origin set is $\sum_i d(X_{ji}) = Td(X_{ji})$. The morphological approach followed in the computation of these sums may be described as follows:

$$Td(X_{ij}) = \sum_j d(X_{ij}) = \sum_{\substack{j=1 \\ j \neq i}}^{N} d(X_{ij}) = \sum_{n \forall j} \left(\min_{i \neq j} \left(n : X_j \subseteq (X_i \oplus nB) \right) \right) \tag{10.14}$$

$$Td(X_{ji}) = \sum_i d(X_{ji}) = \sum_{\substack{j=1 \\ j \neq i}}^{N} d(X_{ji}) = \sum_{n \forall i}\left(\min_{i \neq j}\left(n : X_i \subseteq (X_j \oplus nB)\right)\right) \qquad (10.15)$$

$$Td(X_{ij}) \neq Td(X_{ji}) \qquad (10.16)$$

The maximum distance, $d_{\max}$, between any two sets of a cluster is represented as

$$d_{\max} = \max_{\forall i}\left(\min\left(n : \left(\left(\bigcup_{\substack{j=1 \\ j \neq i}}^{N} X_j\right) \subseteq (X_i \oplus nB)\right)\right)\right); \quad i \neq j \qquad (10.17)$$

Normalized distance between the source and target states is expressed as

$$Nd(X_{ij}) = \frac{d(X_{ij})}{d_{\max}} \qquad (10.18)$$

The normalized Hausdorff distance is represented as

$$N\rho(X_{ij}) = \min\left(Nd : Nd(X_{ij}), Nd(X_{ji})\right), \quad i \neq j \qquad (10.19)$$

The mean sum of normalized distances is defined as

$$\overline{\sum_j Nd(X_{ij})} = \left(\frac{\sum_{\substack{j=1 \\ j \neq i}}^{N} d(X_{ij})/d_{\max}}{N-1}\right) = \left(\frac{\sum_{\substack{i,j-1 \\ i \neq j}}^{N} Nd(X_{ij})}{N-1}\right) \qquad (10.20)$$

$$\overline{\sum_i Nd(X_{ji})} = \left(\frac{\sum_{\substack{j=1 \\ j \neq i}}^{N} d(X_{ji}/d_{\max})}{N-1}\right) = \left(\frac{\sum_{\substack{i,j-1 \\ i \neq j}}^{N} Nd(X_{ji})}{N-1}\right) \qquad (10.21)$$

$$\overline{\sum_j Nd(X_{ij})} \neq \overline{\sum_i Nd(X_{ji})} \qquad (10.22)$$

The minimum number of dilations required from the set X_i to cover $\bigcup_{\substack{j=1 \\ j \neq i}}^{N} X_j$ is

$$\rho = \min\left(n : \left(\bigcup_{\substack{j=1 \\ j \neq i}}^{N} X_j \right) \subseteq (X_i \oplus nB) \right)$$

Shape–Size Similarities between Sets

Consider X_i and X_j to be nonempty disjoint compact sets, where X_i is larger than X_j. From Equations 10.11 and 10.12, it is evident that a smaller object, to completely occupy a relatively larger one, requires a greater number of dilation cycles than that in the converse scenario. If there exists a shape–size dissimilarity between the two sets under investigation, one can observe that $d(X_{ij}) \neq d(X_{ji})$. This observation is true with all the sets, between which the distances are computed accordingly, further facilitating a way to compute an index that describes shape–size relationship between sets. $d(X_{ij}) = d(X_{ji})$ if $X_i = X_j$. The utility of these distances further extends to compute the degree of contextuality.

Contextuality between Origin and Destination Sets

The normalized degree of contextuality between X_i and X_j, denoted by $C(X_{ij})$, is computed as

$$C(X_{ij}) = C(X_{ji})$$

$$= \left(\frac{\rho(X_{ij})}{\max\left(d(X_{ij}), d(X_{ji})\right)} \right)$$

$$= \left(\frac{\rho\left(Nd(X_{ij})\right)}{\max\left(Nd(X_{ij}), Nd(X_{ji})\right)} \right) \tag{10.23}$$

The mean sum of contextuality of a set X_i denoted by $\overline{TC(X_{ij})}$ or $\overline{\sum_j C(X_{ij})}$ is computed by summing up the individual degrees of contextuality between X_i and every other set X_j by the total number of sets.

$$\overline{\sum_i C(X_{ij})} = \left(\frac{\sum_{\substack{j=1 \\ j \neq i}}^{N} C(X_{ij})}{N} \right), \tag{10.24}$$

where $C(X_{ij})$ ranges between 0 and 1.

Also,

$$\overline{TC(X_{ij})} = \sum_j C(X_{ij}) = \overline{TC(X_{ji})} = \sum_i C(X_{ji})$$

We have $C_{ii} = 0$, which satisfies the condition that the distance between the set and itself is zero, and hence the degree of contextuality is also zero. It is worth mentioning here that the adjacency, contextuality, direction, and distance between a set and itself are always zero.

Spatial Complexity between Origin Set and Destination Sets

Spatial complexities between pairs of sets: The idea of proposing spatial complexity in terms of various measures stems from the following heuristic argument: let X be with a straight path of length L and Y be with a curved path of the same length L. The complexity involved in traversing the latter is more than that in the case of the former. This further extends to spatial objects with symmetric properties in case of differentiable boundaries and asymmetric properties in case of non-differential boundaries.

The spatial complexity, $H/P(X_{ij})$ with respect to $P(X_{ij})$ or $P(X_{ji})$, is expressed as

$$H/P(X_{ij}) = -\sum_{\substack{\forall j \\ i \neq j}} \Pr\left(P(X_{ij})\log\Pr(P(X_{ij}))\right) \tag{10.25}$$

where $\Pr[P(X_{ij})] = P(X_{ij})/TP(X_{ij})$ and $\Pr\left(P(X_{ji})\right) = \Pr\left(P(X_{ij})\right)$ that range between 0 and 1, respectively, and $\sum_j \Pr(P(X_{ij})) = 1$ and $\sum_i \Pr(P(X_{ji})) = 1$. Finally, $H/P(X_{ij}) = H/P(X_{ji})$.

The spatial complexity, $H/C(X_{ij})$ with respect to $C(X_{ij})$ or $C(X_{ji})$, is expressed as

$$H/C(X_{ij}) = -\sum_{\substack{\forall j \\ i \neq j}} \Pr\left(C(X_{ij})\log\Pr\left(C(X_{ij})\right)\right), \tag{10.26}$$

where $\Pr\left(C(X_{ij})\right) = \dfrac{C(X_{ij})}{TC(X_{ij})}$ and $\Pr\left(C(X_{ji})\right) = \Pr\left(C(X_{ij})\right)$ that range between 0 and 1, respectively, and $\sum_j \Pr\left(C(X_{ij})\right) = 1$ and $\sum_i \Pr\left(C(X_{ji})\right) = 1$. Finally, $H/C(X_{ij}) = H/C(X_{ji})$.

The spatial complexities, $H/d(X_{ij})$ with respect to $d(X_{ij})$ and $H/d(X_{ji})$ with respect to $d(X_{ji})$, are

$$H/d(X_{ij}) = -\sum_{\substack{\forall j \\ i \neq j}} \Pr\big(d(X_{ij})\log\Pr\big(d(X_{ij})\big)\big) \tag{10.27}$$

$$H/d(X_{ji}) = -\sum_{\substack{\forall i \\ i \neq j}} \Pr\big(d(X_{ji})\log\Pr\big(d(X_{ji})\big)\big) \tag{10.28}$$

where $\Pr(d(X_{ij})) = d(X_{ij})/Td(X_{ij})$ and $\Pr(d(X_{ji})) = d(X_{ji})/Td(X_{ji})$ that range between 0 and 1, respectively, and $\sum_{j} \Pr\big(d(X_{ij})\big) = 1$ and $\sum_{i} \Pr\big(d(X_{ji})\big) = 1$. Finally, $H/d(X_{ij}) \neq H/d(X_{ji})$, and $H/d(X_{ij}) = H/d(X_{ji})$ if X_i and X_j are compact, identical, and nonoverlapping.

In the case of homogeneously distributed sets with similar sizes and shapes within a collection of sets (a cluster set) from which the lengths of the boundary being shared, distances to other sets and the degree of contextuality with other sets yield lower complexity measures. The higher the uniformity in the distribution pattern, the lower is the complexity with respect to the length of the boundary being shared, distances, and contextuality degree. A low-complexity measure points to geometrical similarity, with respect to both size and shape, of subsets in a cluster.

Strategically Significant Set

The strategically significant sets with respect to perimeter (P), distance (d), and contextuality degree (C) between every origin set and other destination sets are denoted by SX_i^P, SX_i^d, and SX_i^C, respectively. SA_i^P is the set that shares the maximum length of boundary with other adjacent sets. This property can be expressed mathematically as

$$(SX_i^P) = \max_{\forall i}\left(\overline{\sum_i NP(X_{ij})}\right) \tag{10.29}$$

SX_i^d is the set that is closest to all destination sets and it satisfies the following property:

$$(SX_i^d) = \min_{\forall i,j}\left(\min\left(\overline{\sum_i Nd(X_{ij})}, \overline{\sum_j Nd(X_{ji})}\right)\right) \tag{10.30}$$

The physical meaning of Equation 10.30 is that it provides a set, within a cluster of sets, which is closest to all the other sets. SX_i^c is the set that possesses the following property:

$$(SX_i^c) = \max_{\forall i}\left(\overline{\sum_i C(X_{ij})}\right) \tag{10.31}$$

The physical meaning of Equation 10.31 is that it provides a set that possesses maximum contextuality with other sets in the cluster of sets. It is worth mentioning that in a spatially ideal system (e.g., Figure 10.3), all the sets in the cluster possess similar strategic importance with respect to contextuality.

Strategically significant set(s) with respect to spatial complexity in terms of perimeter (P), distance (d), and contextuality (C) between every origin and all destination sets are respectively denoted as SH_i^P, SH_i^d, and SH_i^C.

The set(s) with a relatively lower spatial complexity measure, with respect to each of the three aforementioned parameters, exhibits properties, which may be mathematically described as follows:

$$SH_i^P = \min_{\forall i}\left(H/P(X_{ij})\right) \tag{10.32}$$

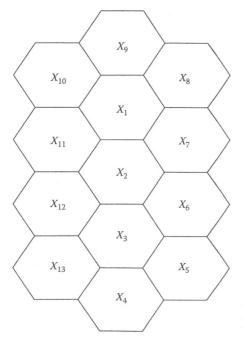

FIGURE 10.3
An ideal case, as a cluster set with multiple adjacent homothetic sets of similar sizes, which explains the measures proposed.

$$SH_i^d = \min_{\forall i}\left(\min\left(H/d(X_{ij}), H/d(X_{ji})\right)\right) \qquad (10.33)$$

$$SH_i^C = \min_{\forall i}\left(H/C(X_{ij})\right) \qquad (10.34)$$

Experimental Results on Clusters of Sets

Ideal Spatial System

Figure 10.3 represents an ideal spatial system composed of hexagonal cells of the same size, arranged in a regular fashion. These may be categorized into exterior and interior cells. Exterior cells are the ones having less than six adjacent cells. The rest are treated as interior cells. In case of such a system, the distance, contextuality, and length of boundary being shared between a cell and every other cell are the same, as long as interior cells considered. Thus, all interior cells are strategically significant with respect to distance, contextuality, and length of the boundary being shared. Having stated that, we realized the influence of shape and size in the process of deriving strategically significant sets. Hence, we incorporated techniques and transformations that capture the characteristics of shape and size. For an ideal case, it is evident that

1. $d(X_{ij}) = d(X_{ji})$
2. $C(X_{ij}) = C(X_{ji}) = 1$
3. $\sum_j P(X_{ij}) = \sum_j P(X_{ji})$, if X_i is an interior cell

Nonideal Spatial System: Planar Forms of States of India

The techniques explained previously are demonstrated in this section on a geographical data set. A total of 28 sets that denote states of the Indian peninsula are represented in a subset of the two-dimensional discrete space, Z^2, of dimensions 480×480 pixels.

We use the terms "set" and "state" interchangeably in this study. Sets representing various states are denoted by X_i, the indices i being assigned in alphabetical order with respect to the name of the state (see the caption of Figure 10.4 for a complete list). These sets, along with the sets adjacent to them, the lengths of the boundary that they share with other sets, and the minimum and maximum dilation distances between themselves and every other sets, are represented in the form of arrays in Figure 10.5a, b, and d.

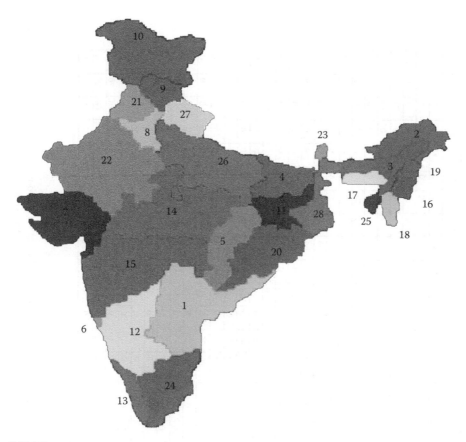

FIGURE 10.4

Map of India (spatial system) with its constituent 28 states (subsets), indexed according to alphabetical order, are shown: Andhra Pradesh (X_1), Arunachal Pradesh (X_2), Assam (X_3), Bihar (X_4), Chhattisgarh (X_5), Goa (X_6), Gujarat (X_7), Haryana (X_8), Himachal Pradesh (X_9), Jammu & Kashmir (X_{10}), Jharkhand (X_{11}), Karnataka (X_{12}), Kerala (X_{13}), Madhya Pradesh (X_{14}), Maharashtra (X_{15}), Manipur (X_{16}), Meghalaya (X_{17}), Mizoram (X_{18}), Nagaland (X_{19}), Orissa (X_{20}), Punjab (X_{21}), Rajasthan (X_{22}), Sikkim (X_{23}), Tamil Nadu (X_{24}), Tripura (X_{25}), Uttar Pradesh (X_{26}), Uttarakhand (X_{27}), and West Bengal (X_{28}). Union territories and Delhi (capital state) that are parts Indian peninsular are not included in the figure.

Two states, say X_i and X_j, are defined as adjacent if they satisfy the following conditions:

1. $(X_i \cap X_j) = \emptyset$
2. $(X_i \oplus B) \cap X_j \neq \emptyset$, $\forall i, j \in I$, $i \neq j$

We determined the states adjacent to each state and recorded the results in a square matrix of dimensions 28 × 28. Every nonzero entry in the matrix denotes an adjacency relationship between a pair of states—the values themselves represent the length of the boundary being shared by the pair. A value

	X_1	X_2	X_3	X_4	X_5	X_6	X_7	X_8	X_9	X_{10}	X_{11}	X_{12}	X_{13}	X_{14}	X_{15}	X_{16}	X_{17}	X_{18}	X_{19}	X_{20}	X_{21}	X_{22}	X_{23}	X_{24}	X_{25}	X_{26}	X_{27}	X_{28}
X_1	513	0	0	0	26	0	0	0	0	0	0	130	0	0	75	0	0	0	0	91	0	0	0	41	0	0	0	0
X_2	0	336	116	0	0	0	0	0	0	0	0	0	0	0	0	0	0	0	10	0	0	0	0	0	0	0	0	0
X_3	0	112	343	0	0	0	0	0	0	0	0	0	0	0	0	17	82	16	49	0	0	0	0	0	2	0	0	19
X_4	0	0	0	265	0	0	0	0	0	0	88	0	0	0	0	0	0	0	0	126	0	0	0	0	0	78	0	20
X_5	26	0	0	0	317	0	0	0	0	0	35	0	0	79	52	0	0	0	0	0	0	0	0	0	0	10	0	0
X_6	0	0	0	0	0	34	0	0	0	0	0	16	0	0	7	0	0	0	0	0	0	0	0	0	0	0	0	0
X_7	0	0	0	0	0	0	396	0	0	0	0	0	0	21	55	0	0	0	0	0	0	74	0	0	0	0	0	0
X_8	0	0	0	0	0	0	0	206	10	0	0	0	0	0	0	0	0	0	0	0	54	77	0	0	0	55	15	0
X_9	0	0	0	0	0	0	0	11	174	74	0	0	0	0	0	0	0	0	0	0	40	0	0	0	0	0	39	0
X_{10}	0	0	0	0	0	0	0	0	76	417	0	0	0	0	0	0	0	0	0	0	14	0	0	0	0	0	0	84
X_{11}	0	0	0	89	35	17	0	0	0	0	280	391	0	0	100	0	0	0	0	64	0	0	0	0	0	14	0	0
X_{12}	128	0	0	0	0	0	0	0	0	0	0	49	51	0	0	0	0	0	0	0	0	0	0	0	0	0	0	0
X_{13}	0	0	0	0	0	0	0	0	0	0	0	0	214	0	0	0	0	0	0	0	0	0	0	0	0	0	0	0
X_{14}	0	0	0	0	78	0	20	0	0	0	0	0	0	596	140	0	0	0	0	0	0	0	0	0	0	0	0	0
X_{15}	75	0	0	0	53	0	56	0	0	0	100	101	0	139	493	0	0	0	0	0	0	0	0	0	0	0	0	0
X_{16}	0	0	17	0	0	0	0	0	0	0	0	0	0	0	0	99	0	0	28	0	0	163	0	0	0	202	0	0
X_{17}	0	0	79	0	0	0	0	0	0	0	0	0	0	0	0	0	120	0	0	0	0	0	0	0	15	0	0	0
X_{18}	0	11	16	0	0	0	0	0	0	0	0	0	0	0	0	0	0	110	0	0	0	0	0	0	0	0	0	0
X_{19}	0	0	46	0	0	0	0	0	0	0	66	0	0	0	0	0	0	0	96	0	0	0	0	0	0	0	0	0
X_{20}	90	0	0	0	126	0	0	0	0	0	0	0	0	0	0	0	0	0	0	368	0	0	0	0	0	0	0	0
X_{21}	0	0	0	0	0	0	0	54	41	12	0	0	0	166	0	0	0	0	0	0	162	12	0	0	0	26	0	21
X_{22}	0	0	0	0	0	0	75	80	0	0	0	0	0	0	0	0	0	0	0	0	12	526	0	0	0	0	0	0
X_{23}	0	0	0	0	0	0	0	0	0	0	0	0	70	0	0	0	0	0	0	0	0	0	57	0	0	0	0	0
X_{24}	40	0	0	78	0	0	0	0	0	0	0	67	0	204	0	0	0	0	0	0	0	0	0	308	73	0	0	0
X_{25}	0	0	2	0	0	0	0	54	0	0	0	0	0	0	0	0	0	16	0	0	0	0	0	0	0	0	0	0
X_{26}	0	0	0	78	10	0	0	0	0	0	14	0	0	0	0	0	0	0	0	0	0	26	0	0	0	518	70	0
X_{27}	0	0	0	0	0	0	0	15	39	0	0	0	0	0	0	0	0	0	0	0	0	0	0	0	0	68	189	0
X_{28}	0	0	19	21	0	0	0	0	0	84	86	0	0	0	0	0	0	0	0	21	0	0	0	12	0	0	0	326

(a)

FIGURE 10.5

Matrices denoting (a) adjacent states existing between each origin state and all other destination states with length of the boundaries being shared between each origin state and its corresponding adjacent states. If there exists a value 0 between X_i and X_j, then such a state is termed a nonadjacent state.

	X_1	X_2	X_3	X_4	X_5	X_6	X_7	X_8	X_9	X_{10}	X_{11}	X_{12}	X_{13}	X_{14}	X_{15}	X_{16}	X_{17}	X_{18}	X_{19}	X_{20}	X_{21}	X_{22}	X_{23}	X_{24}	X_{25}	X_{26}	X_{27}	X_{28}
X_1	0	225	200	161	69	172	163	210	246	274	129	104	126	116	63	247	202	240	255	72	238	146	204	87	225	145	225	146
X_2	178	0	20	135	189	351	342	289	269	255	137	283	302	213	242	54	62	73	39	139	301	282	125	253	74	175	240	105
X_3	158	36	0	115	168	329	320	269	248	229	115	261	280	191	220	50	41	53	54	119	280	260	104	232	54	165	219	88
X_4	110	120	95	0	55	215	207	154	134	125	39	159	209	78	106	147	57	135	150	69	164	147	70	197	120	51	107	45
X_5	65	165	140	94	0	159	152	140	214	204	63	103	160	48	53	187	147	180	195	28	168	85	134	149	165	85	155	87
X_6	50	270	250	155	110	0	72	179	217	244	155	8	43	87	12	297	257	290	305	122	207	116	224	57	275	130	194	197
X_7	132	355	330	234	188	130	0	107	143	169	233	105	173	90	65	381	331	367	383	201	135	45	302	162	352	140	148	276
X_8	158	245	225	130	98	213	82	0	37	64	129	175	255	54	123	277	227	265	280	125	28	64	200	243	249	35	45	172
X_9	195	228	208	110	133	250	119	36	0	29	124	213	293	93	160	259	205	247	261	153	29	180	230	280	180	42	30	154
X_{10}	248	270	249	158	185	303	173	89	53	0	175	266	346	145	215	301	251	288	304	206	67	99	220	336	273	95	80	196
X_{11}	89	120	95	31	54	215	199	154	129	146	0	156	188	75	104	147	57	135	150	39	164	145	74	168	119	50	105	42
X_{12}	45	265	245	180	105	68	121	228	265	294	149	0	79	134	61	177	247	285	300	117	261	164	222	69	269	174	244	192
X_{13}	69	265	245	230	138	99	166	279	315	342	199	49	0	185	111	279	246	273	289	138	306	214	272	19	258	223	293	193
X_{14}	102	260	239	142	95	157	129	93	129	157	142	119	200	0	70	290	240	277	294	109	119	69	210	188	262	65	108	184
X_{15}	68	290	265	168	124	108	99	168	204	231	168	51	130	75	0	315	265	298	318	135	190	104	237	119	289	119	183	210
X_{16}	145	38	15	103	157	317	309	258	236	216	104	199	269	179	209	0	29	20	19	108	273	254	91	219	34	155	209	77
X_{17}	115	30	9	72	127	287	278	228	106	186	74	218	238	150	179	50	0	39	53	78	233	219	62	188	25	124	179	47
X_{18}	129	64	30	84	140	299	289	239	219	200	85	232	249	160	190	27	42	0	44	89	249	230	74	204	20	135	190	59
X_{19}	149	19	12	104	160	322	313	262	240	205	61	254	273	183	213	14	34	33	0	112	272	253	95	223	39	158	208	79
X_{20}	45	154	129	93	50	209	199	150	178	34	140	166	213	69	100	174	130	168	183	0	164	140	134	128	157	86	154	78
X_{21}	180	255	235	140	120	235	105	25	30	127	208	200	279	79	148	285	235	275	289	140	0	32	209	268	264	4	55	180
X_{22}	149	330	305	209	164	204	75	79	94	125	39	167	248	88	115	354	305	343	359	174	91	0	308	234	338	114	123	249
X_{23}	129	52	26	24	69	225	217	165	145	125	199	169	219	184	120	78	29	65	80	79	174	157	0	208	52	60	117	13
X_{24}	65	265	236	231	139	100	172	280	316	343	68	51	50	184	113	259	244	248	263	139	307	211	273	0	233	226	294	194
X_{25}	110	52	17	65	120	282	273	222	200	180	93	215	233	144	173	26	29	15	34	73	232	213	55	184	0	225	168	40
X_{26}	152	215	190	94	89	207	155	104	91	118	98	170	249	49	119	240	185	229	244	109	113	93	163	238	214	0	69	135
X_{27}	168	205	180	85	104	222	101	49	32	52	27	184	264	64	134	226	182	220	235	124	58	46	149	253	204	20	0	125
X_{28}	116	80	59	39	79	240	228	179	159	152	42	183	213	103	132	105	55	93	109	67	190	173	78	194	78	78	132	0

(b)

FIGURE 10.5 (continued)

Matrices denoting (b) dilation distances and Hausdorff dilation distances in pixels between the origin and destination states.

(continued)

(c)

FIGURE 10.5 (continued)
Matrices denoting (c) geodesic distance contours at five-pixel interval between set X_1 and rest of India as a set with thick lines indicating set boundaries.

	X_1	X_2	X_3	X_4	X_5	X_6	X_7	X_8	X_9	X_{10}	X_{11}	X_{12}	X_{13}	X_{14}	X_{15}	X_{16}	X_{17}	X_{18}	X_{19}	X_{20}	X_{21}	X_{22}	X_{23}	X_{24}	X_{25}	X_{26}	X_{27}	X_{28}
X_1	0	0.79	0.79	060	0.94	0.29	0.81	0.75	0.79	0.91	0.69	0.43	0.55	0.88	0.93	0.59	0.57	0.54	0.58	0.63	0.76	0.98	0.63	0.75	0.49	0.95	0.75	0.8
X_2	0.79	0	0.56	0.89	0.87	0.77	0.96	0.85	0.85	0.94	0.88	0.94	0.88	0.82	0.83	0.7	0.48	0.88	0.49	0.9	0.85	0.85	0.42	0.95	0.7	0.81	0.85	0.76
X_3	0.79	0.56	0	0.83	0.83	0.76	0.97	0.84	0.84	0.92	0.83	0.94	0.88	0.8	0.83	0.3	0.22	0.57	0.22	0.92	0.84	0.85	0.25	0.98	0.31	0.87	0.82	0.67
X_4	0.68	0.89	0.83	0	0.59	0.72	0.88	0.84	0.82	0.79	0.79	0.88	0.91	0.8	0.63	0.7	0.79	0.62	0.69	0.74	0.85	0.7	0.34	0.85	0.54	0.54	0.79	0.87
X_5	0.4	0.87	0.83	0.59	0	0.69	0.81	0.7	0.62	0.91	0.86	0.98	0.86	0.51	0.43	0.84	0.86	0.78	0.82	0.56	0.71	0.7	0.51	0.93	0.73	0.96	0.67	0.91
X_6	0.29	0.77	0.76	0.72	0.69	0	0.55	0.84	0.87	0.81	0.72	0.12	0.43	0.55	0.11	0.94	0.9	0.78	0.95	0.58	0.88	0.57	1	0.28	0.98	0.63	0.67	0.91
X_7	0.81	0.96	0.97	0.88	0.81	0.55	0	0.77	0.83	0.98	0.85	0.87	0.96	0.7	0.66	0.81	0.84	0.79	0.82	0.99	0.78	0.6	0.72	0.94	0.78	0.9	0.68	0.82
X_8	0.75	0.85	0.84	0.84	0.7	0.84	0.77	0	0.97	0.72	0.84	0.77	0.91	0.58	0.73	0.93	1	0.9	0.94	0.83	0.89	0.34	0.83	0.87	0.89	0.9	0.68	0.96
X_9	0.79	0.85	0.84	0.82	0.62	0.87	0.83	0.97	0	0.72	0.84	0.77	0.91	0.58	0.66	0.81	0.9	0.79	0.82	0.99	0.78	0.34	0.6	0.87	0.78	0.9	0.68	0.83
X_{10}	0.91	0.94	0.92	0.79	0.91	0.81	0.98	0.72	0.97	0	0.96	0.8	0.93	0.72	0.55	0.91	0.52	0.89	0.92	0.86	0.97	0.53	0.81	0.98	0.66	0.81	0.65	0.78
X_{11}	0.69	0.88	0.83	0.79	0.86	0.72	0.85	0.84	0.84	0.96	0	0.96	0.94	0.53	0.62	0.72	0.74	0.63	0.72	0.64	0.51	0.78	0.57	0.84	0.57	0.54	0.93	0.64
X_{12}	0.43	0.94	0.94	0.88	0.98	0.12	0.87	0.77	0.8	0.9	0.96	0	0.62	0.89	0.84	0.71	0.77	0.81	0.85	0.82	0.77	0.98	0.76	0.74	0.8	0.98	0.75	0.95
X_{13}	0.55	0.88	0.88	0.91	0.86	0.43	0.96	0.91	0.91	0.99	0.91	0.62	0	0.93	0.85	0.96	0.97	0.91	0.94	0.83	0.91	0.86	0.81	0.77	0.9	0.9	0.9	0.91
X_{14}	0.88	0.82	0.8	0.8	0.51	0.55	0.7	0.58	0.58	0.72	0.53	0.89	0.93	0	0.93	0.62	0.63	0.58	0.62	0.63	0.66	0.97	0.42	0.98	0.55	0.75	0.59	0.56
X_{15}	0.93	0.83	0.83	0.63	0.43	0.11	0.66	0.73	0.66	0.55	0.62	0.84	0.85	0.93	0	0.66	0.68	0.64	0.67	0.74	0.78	0.9	0.86	0.85	0.6	1	0.73	0.63
X_{16}	0.59	0.7	0.3	0.7	0.84	0.91	0.84	0.68	0.68	0.68	0.71	0.89	0.96	0.62	0.68	0	0.58	0.64	0.64	0.6	0.78	0.72	0.86	0.85	0.6	1	0.92	0.5
X_{17}	0.72	0.48	0.22	0.79	0.86	0.9	0.84	1	0.9	0.52	0.74	0.77	0.97	0.63	0.68	0.58	0	0.93	0.64	0.6	0.99	0.72	0.47	0.77	0.86	0.67	0.98	0.85
X_{18}	0.54	0.88	0.57	0.62	0.78	0.78	0.79	0.9	0.79	0.89	0.63	0.81	0.91	0.58	0.64	0.74	0.93	0	0.75	0.53	0.91	0.67	0.88	0.82	0.75	0.59	0.86	0.63
X_{19}	0.58	0.49	0.22	0.69	0.82	0.95	0.82	0.94	0.82	0.92	0.72	0.85	0.94	0.62	0.67	0.64	0.64	0.75	0	0.61	0.94	0.7	0.84	0.85	0.87	0.65	0.89	0.72
X_{20}	0.63	0.9	0.92	0.74	0.56	0.58	0.99	0.83	0.86	0.72	0.64	0.82	0.83	0.63	0.74	0.74	0.6	0.53	0.61	0	0.85	0.8	0.59	0.92	0.46	0.79	0.81	0.86
X_{21}	0.76	0.85	0.84	0.85	0.71	0.88	0.78	0.89	0.97	0.51	0.85	0.77	0.91	0.66	0.78	0.96	0.99	0.91	0.94	0.85	0	0.35	0.83	0.87	0.88	0.4	0.95	0.95
X_{22}	0.98	0.85	0.85	0.7	0.7	0.57	0.6	0.34	0.34	0.78	0.7	0.98	0.86	0.97	0.9	0.72	0.72	0.67	0.7	0.8	0.35	0	0.51	0.9	0.63	0.82	0.37	0.69
X_{23}	0.63	0.42	0.25	0.34	0.51	1	0.72	0.85	0.6	0.81	0.42	0.85	0.81	0.57	0.53	0.76	0.47	0.88	0.84	0.59	0.83	0.8	0	0.76	0.95	0.37	0.79	0.17
X_{24}	0.75	0.95	0.98	0.85	0.93	0.28	0.94	0.87	0.87	0.98	0.84	0.74	0.77	0.98	0.85	0.77	0.82	0.85	0.92	0.87	0.9	0.76	0.77	0	0.79	0.95	0.86	1
X_{25}	0.49	0.7	0.31	0.54	0.73	0.98	0.78	0.89	0.8	0.66	0.57	0.8	0.9	0.55	0.6	0.86	0.86	0.75	0.87	0.46	0.88	0.63	0.95	0.79	0	0.95	0.82	0.51
X_{26}	0.95	0.81	0.87	0.54	0.96	0.63	0.9	0.9	0.94	0.81	0.54	0.98	0.9	0.75	1	0.65	0.67	0.59	0.65	0.79	0.4	0.82	0.37	0.95	0.95	0	0.29	0.58
X_{27}	0.75	0.85	0.82	0.79	0.67	0.87	0.68	0.68	0.94	0.65	0.93	0.75	0.	0.59	0.73	0.92	0.98	0.86	0.89	0.81	0.95	0.37	0.79	0.86	0.82	0.29	0	0.95
X_{28}	0.79	0.76	0.67	0.87	0.91	0.91	0.82	0.83	0.97	0.78	0.64	0.95	0.91	0.56	0.63	0.5	0.85	0.63	0.72	0.86	0.95	0.69	0.17	1	0.51	0.58	0.95	0

(d)

FIGURE 10.5 (continued)

Matrices denoting (d) degree of contextuality between X_i and every other destination state(s) X_r.

of zero indicates that the states are nonadjacent. For instance, the states adjacent to X_1 include X_5, X_{12}, X_{15}, X_{20}, and X_{24}.

We also present the minimum dilation distances, $d(X_{ij})$ and $d(X_{ji})$, between pairs of sets. These are represented in Figure 10.5b. To compute these distances, we opted for a square structuring element that is symmetric about the origin, with a primitive size of 3×3. We considered a pair of sets and recorded the number of dilation cycles required, with respect to the structuring element for the dilated version of one of the states to completely contain the other and *vice versa*. The smaller and larger values between these denote (in pixels) the minimum and maximum dilation distances respectively between the two sets. The values represented in Figure 10.5b are these distances between each set X_i and every other set X_j, where i and j range from 1 to 28, and $i \neq j$. The higher the value, the larger is the dilation distance between the sets. From Figure 10.5b, it can easily be inferred that the two closest, nonadjacent, disjoint sets are X_6 and X_{12}, while the farthest ones are X_7 and X_{19}. It is also possible to infer from Figure 10.5b the closest and farthest sets to any specific set. The following inferences can be made from the same:

- The largest distance computed from the innermost to the outermost extremities for any two states is for the states X_{19} (Nagaland) and X_7 (Gujarat), in the order mentioned. The distance is 383 pixels.

- The smallest distance computed from the innermost to the outermost extremities for any two states is for the states X_{12} (Karnataka) and X_6 (Goa), in the order mentioned. The distance is 8 pixels.

- States X_6, X_{16}, X_{23}, and X_{25} are significantly smaller than most other states. Values of ρ for states X_1, X_2, and X_{28} are 248, 355, and 276 pixel units, respectively. We may, therefore, state that X_1 is strategically more important than X_2 and X_{28}, since 248 is the lowest out of 248, 355, and 276. Thus, a minimum of 248 dilations of X_1 are required to cover the other 27 states. Geodesic distances between each set and $I = \bigcup_{i=1}^{28} X_i$ can be visualized using the following equation:

$$\bigcup_{\substack{i=1 \\ n=1}}^{\substack{N \\ 28}} \left((X_i \oplus nB) \cap \left(\bigcup_{j=1}^{28} X_j \right) \right) \qquad (10.35)$$

As an example, we show the geodesic contours (Figure 10.5c) within India (considered a set), depicting the equal distance lines from a set X_1. The contextuality between each state to every other state is computed according to Equation 10.23 and is represented in Figure 10.5d. C_{ij} is proportional to shape–size similarities of the two sets under investigation.

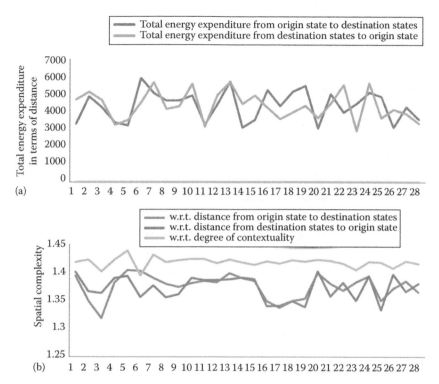

FIGURE 10.6

Graphs depicting (a) total energy expenditure in terms of total distance between origin state to all other destination states, and total energy expenditure between each destination state and an origin state, and (b) spatial complexity with respect to (i) distance from origin state to destination states, (ii) distance from destination states to an origin state, and (iii) degree of contextuality between each state and every other state.

The total expenditure of energy computed in terms of distance between source and destination states and *vice versa* is plotted as functions of the state (Figure 10.6a). It is conspicuous that the energy required to visit all other destination states from one of the extreme exterior states as an origin state is more from extreme exterior states than that of interior states. States at the extreme exterior include X_2, X_6, X_{10}, X_{13}, and X_{24}. The expenditure of energy is also observed to be more in case of source states of smaller size. By considering the data tabulated in the arrays represented in Figure 10.5a, b, and d, the spatial complexities in terms of the parameters, computed both between source and destination states, and *vice versa*, according to Maragos (1989, 2005), Huttenlocher et al. (1993), and Serra (1994, 1998) are represented in Figure 10.6b.

Twenty-eight rankings of states in decreasing order in terms of 10 parameters are represented in 10 rows (Figure 10.7). In terms of boundary being shared, the ranked states in order are represented in the first row. Rankings in terms of minimum energy required either from source to destination states and *vice versa* are represented in the second and third rows respectively.

SX_i^P	X_{14}	X_{26}	X_{15}	X_{12}	X_1	X_{22}	X_5	X_{20}	X_3	X_{11}	X_8	X_4	X_9	X_{28}	X_7	X_{27}	X_2	X_{13}	X_{21}	X_{19}	X_{10}	X_{17}	X_{16}	X_{18}	X_{24}	X_6	X_{25} X_{23}
$SX_i^{d(X_{ij})}$	X_{20}	X_{14}	X_{11}	X_5	X_{26}	X_1	X_4	X_{15}	X_{22}	X_{28}	X_3	X_{27}	X_{17}	X_{12}	X_{23}	X_8	X_9	X_{25}	X_2	X_{10}	X_{21}	X_7	X_{24}	X_{18}	X_{16}	X_{19}	X_{13} X_6
$SX_i^{d(X_{ij})}$	X_{23}	X_{11}	X_4	X_{28}	X_5	X_{17}	X_{20}	X_{25}	X_{27}	X_{18}	X_{26}	X_8	X_{16}	X_9	X_{19}	X_{14}	X_{21}	X_6	X_3	X_1	X_{15}	X_{12}	X_2	X_{22}	X_{10}	X_{24}	X_7 X_{13}
SX_i^C	X_{13}	X_{24}	X_7	X_{12}	X_{21}	X_9	X_8	X_{10}	X_2	X_{27}	X_{28}	X_5	X_{11}	X_{18}	X_{16}	X_{19}	X_{20}	X_{17}	X_4	X_{25}	X_{15}	X_3	X_1	X_{26}	X_{22}	X_{14}	X_6 X_{23}
$SX_i^{\min(d(X_{ij}),\,d(X_{ij}))}$	X_{23}	X_{20}	X_{14}	X_{26}	X_{11}	X_5	X_4	X_1	X_{28}	X_{15}	X_{17}	X_{25}	X_{27}	X_{18}	X_{22}	X_8	X_{16}	X_3	X_9	X_{19}	X_{12}	X_{21}	X_6	X_2	X_{10}	X_7	X_{24} X_{13}
SH_i^P	X_{23}	X_{17}	X_2	X_{25}	X_{10}	X_6	X_{13}	X_{19}	X_{27}	X_4	X_7	X_{24}	X_{16}	X_{18}	X_{21}	X_9	X_{20}	X_{22}	X_{28}	X_8	X_{14}	X_{12}	X_{11}	X_1	X_5	X_3	X_{26} X_{15}
$SH_i^{C(X_{ij})}$	X_{23}	X_6	X_{14}	X_{22}	X_{26}	X_3	X_1	X_{15}	X_{25}	X_4	X_{17}	X_{20}	X_{19}	X_{16}	X_{28}	X_{18}	X_5	X_{11}	X_{27}	X_2	X_8	X_{10}	X_{12}	A_{21}	X_9	X_{10}	X_{24} X_{13}
$SH_i^{d(X_{ij})}$	X_3	X_{17}	X_{16}	X_{18}	X_2	X_{25}	X_{19}	X_{28}	X_{22}	X_{26}	X_9	X_8	X_{21}	X_{10}	X_4	X_{12}	X_{23}	X_{27}	X_{15}	X_{11}	X_{14}	X_7	X_{24}	X_1	X_{13}	X_{20}	X_6 X_5
$SH_i^{d(X_{ij})}$	X_{25}	X_{19}	X_{16}	X_{17}	X_{18}	X_{23}	X_8	X_6	X_{21}	X_9	X_3	X_{27}	X_2	X_7	X_{28}	X_{22}	X_{11}	X_{12}	X_{13}	X_{15}	X_4	X_{10}	X_{14}	X_{24}	X_5	X_{26}	X_1 X_{20}
$SH_i^{\min(d(X_{ij}),\,d(X_{ij}))}$	X_3	X_{25}	X_{17}	X_{19}	X_{16}	X_{18}	X_{23}	X_2	X_6	X_8	X_{21}	X_9	X_{28}	X_{27}	X_{22}	X_{26}	X_7	X_{10}	X_4	X_{12}	X_{15}	X_{11}	X_{13}	X_{14}	X_{24}	X_5	X_1 X_{20}

FIGURE 10.7
Twenty-eight rankings in the decreasing order denoting the strategic importance of each state in terms of 10 parameters of spatial importance.

Rankings in terms of degree of contextuality and Hausdorff distances are represented in the fourth and fifth rows. Rows from 6th to 10th denote rankings in decreasing order according to minimum spatial complexities involved in the distribution patterns computed based on the length of boundary being shared, contextuality, and distance between source and target states.

With respect to the maximum length of the boundary being shared, minimum distance (energy) required to reach from source to destination states (and *vice versa*), maximum contextuality, and Hausdorff distance between source and destination states, X_{14}, X_{20}, X_{23}, X_{13}, and X_{23} are designated as strategically important (Figure 10.8a through e). Nine other strategically important states, followed by strategically significant states in terms of five mentioned parameters, are also shown (Figures 10.7 and 10.8a through e).

The degree of homogeneity in the distribution pattern of the lengths of boundary being shared, distance, and contextuality between origin and destination states and *vice versa* determines the degree of spatial complexity. It is conjectured that the smaller the spatial complexity of a state, the better are its spatial relationships with neighboring states and hence the better is its rank. A rank of 1 indicates that a state is strategically most important. The first 10 strategically important states, determined on the basis of spatial complexities with respect to boundaries being shared between origin and destination states, are represented in Figure 10.8f. State X_{23} is strategically most important with respect to spatial complexity in terms of perimeter distribution. In terms of distance (energy required) from origin to destination states, ranks have been assigned as outlined in Figure 10.8g. The state that is strategically most important in terms of spatial complexity in the distribution pattern of distance from origin to destination states is X_3 (Assam). X_{25} is designated as strategically most important in terms of spatial complexity involved in the distribution pattern of distance from destination states to origin state. However, if we consider the minimum spatial complexity involved in the distribution pattern, Hausdorff distances between origin and destination states and *vice versa*, X_3 is designated as strategically most significant.

In terms of spatial complexity involved in the distribution pattern of contextuality from origin to destination states and *vice versa*, states X_6 and X_{23}—smaller, exterior states—are designated as strategically important. The nine strategically important states that follow the most significant states mentioned earlier are shown in maps (Figure 10.8f through j).

The first 10 categories of strategically significant states with respect to the length of the boundary being shared are predominantly occupied by larger and/or interior states (Figure 10.8a).

This inference is also true with respect to the parameters such as distance between origin to destination states (Figure 10.8b), Hausdorff distances between source and target states (Figure 10.8e), and minimal spatial complexity with respect to contextuality involved between destination states and origin state (Figure 10.8). In terms of the least amount of energy required to

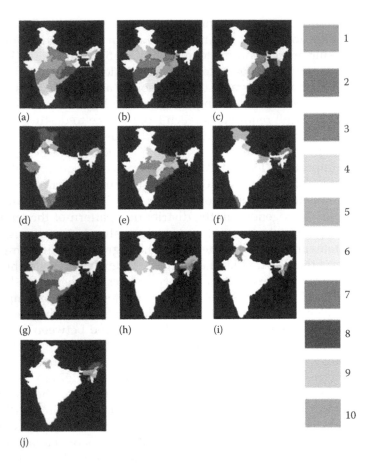

FIGURE 10.8

Spatial representation of strategically important states in the order from 1 to 10 is carried out in terms of 10 different parameters shown in Figure 10.7. In each panel of this figure, first 10 strategically significant states (please refer to the legend on each panel) are shown in different gray shades. These strategically significant sets are with respect to (a) boundary being shared, (b) shortest distance from origin to destination states, (c) shortest total distance from destination states to origin state, (d) contextuality, (e) Hausdorff dilation distance, (f) spatial complexity involved in the length of the boundary being shared, (g) spatial complexity in terms of contextuality, (h) spatial complexity in terms of distance from origin to destination states, (i) spatial complexity in terms of distance from destination states to origin state, and (j) spatial complexity in terms of Hausdorff dilation distance from origin state to destination states. States with gray shades denote first 10 strategically significant states, and the region with white space represents the states that are strategically nonsignificant with ranks starting from 11 to 28.

travel from destination states to origin state, the first 10 categories of strategically important states are found to be situated mostly in the northeastern region (Figure 10.8c). This is also true with respect to minimal spatial complexity involved in the perimeter distribution (Figure 10.8f). Exterior states occupy the first 10 categories of strategic importance with respect to maximum contextuality observed between source and target states (Figure 10.8d). This further supports the fact that interior states possess relatively less contextuality with other states. Certain smaller states occupy the first 10 categories of strategic importance with respect to minimal spatial complexity involved in the distribution pattern of distance between destination states and an origin state (Figure 10.8i). To a certain degree, this inference is also valid if we consider minimum spatial complexity involved in the distribution pattern of Hausdorff distances between origin and destination states (Figure 10.8j). Latitudinally central states occupy first 10 categories of strategic importance with respect to minimal spatial complexity involved in the distribution pattern of contextuality between origin and destination states (Figure 10.8g) and distance between origin and destination states (Figure 10.8h).

Maps depicting the degree of strategic importance of various states can be obtained by an intelligent assignment of weights to each state, combined with map algebra and GIS overlays. The applications of this case study can be foreseen in planning and logistics, and in facility planning and allocation.

A nonideal spatial system: small water bodies and their zones of influence. This study deals with finding strategically significant water bodies and their zones of influence, and is of relevance to environmental planning. The data have been sourced from a map depicting 66 water bodies situated in the region between the geographical coordinates 18°00′–18°07′N latitudes and 83°22′–83°30′E longitudes. The water bodies were manually traced from paper products of geo-coded remotely sensed satellite data with topographic map reference. The traced water bodies were discretized (Figure 10.9a) and their corresponding influence zones (Figure 10.9b) were computed using the technique of skeletonization by zones of influence (SKIZ).

The proposed framework was utilized to analyze the water bodies and their zones of influence, in order to identify those that are strategically important. The first 10 ranks categorized on the basis of 8 parameters for both

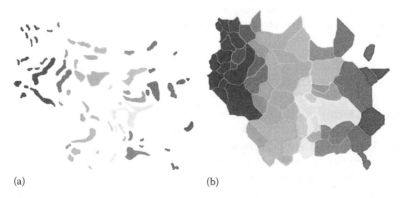

(a)   (b)

FIGURE 10.9
(a) Discretized small water bodies traced from geo-coded remotely sensed data with the help of topographic map, and (b) zones of influence of corresponding water bodies computed via SKIZ transformation.

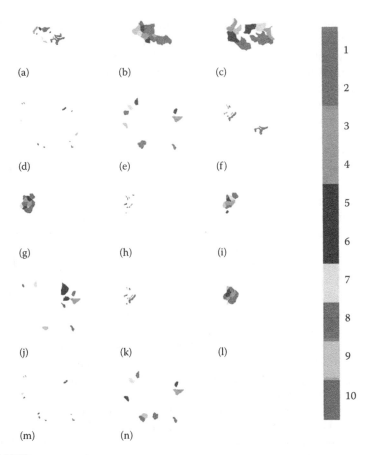

FIGURE 10.10
(a–n) First 10 ranks of strategically important water bodies and zones of their influence from
the points of (i) minimum energy required to reach out all other water bodies (zones), (ii) maximum boundary being shared with adjacent zones, (iii) degree of contextuality between water
bodies (zones), and spatial complexity involved with respect to the three mentioned points.
Detailed explanation on these panels can be seen in text.

water bodies and zones of influence are represented in Figure 10.10a through
n. Larger water bodies that are located in the middle of the region, with a
greater number of adjacent water bodies, are categorized as strategically
important in terms of minimum energy required to reach out all other water
bodies (Figure 10.10a). More or less, the zones of influence of corresponding
10 water bodies that are designated as strategically significant are also considered strategically significant (Figure 10.10b). With respect to maximum
boundary being shared with adjacent zones, the first 10 strategically significant water bodies are shown in Figure 10.10c. The three panels, Figures 10.10a
through c, have fallen in the central region of the space. In terms of the maximum sum of contextuality computed for both water bodies and zones, the
strategically significant water bodies are found to be at outer periphery of the

region (Figure 10.10d and e). It is also found that the water bodies that possess maximum contextuality with other water bodies possess the interesting characteristic that they are least complex with respect to contextuality. This observation is obvious while making comparison between Figure 10.10d and e, and between Figure 10.10h and i. According to spatial complexity in terms of distance from source to target, the strategically significant water bodies are found to be at the extreme exterior (Figure 10.10f and g). A more or less similar observation has been made in strategically significant water bodies in terms of spatial complexity involved with respect to distances between all target water bodies and the source (Figure 10.10h and i). Certain exterior zones of water bodies' influence are categorized in the first 10 ranks in terms of spatial complexity involved with respect to perimeter (Figure 10.10j), which further attributes the fact that extreme exterior zones share their boundaries with adjacent zones in more or less uniform way. It is found that strategically significant water bodies in terms of spatial complexity involved with respect to the minimum of the distances between source and targets fall in the northwestern sides (Figure 10.10k and l). The 10 strategically significant water bodies in terms of spatial complexity involved with respect to contextuality of water bodies are found at extreme exterior positions (Figure 10.10m and n).

It is found out that this process of recognition of strategically significant sets is sensitive to shape, size, location, distance, and adjacency between the sets under investigation. In turn, this recognition process yields different sets of strategically significant water bodies and their zones of influence, as the geometries of water bodies (and their zones of influence) evolve with time due to the fact that they are climatically sensitive.

Discussion and Open Problems

The framework outlined here can be performed with the help of computer-assisted techniques and map algebraic concepts adapted from mathematical morphology. The approaches to compute the measures, explained in the "Recognition and Visualization of Strategically Significant Spatial Sets via Morphological Analysis" section, on raster images do not show any impact with changes in spatial scales. This approach can be further extended by incorporating various other parameters employed in conventional location theory, where the main concern is with the geographic location of economic activity. One of the parameters, besides the economic activity, is the transportation cost that relies on the energy expenditure involved to reach destination locations. In the approach outlined in this study, transportation costs could be related to the distance between source and target states. The extreme-value-based measure, namely, the Hausdorff distance that is adapted in the analysis of this chapter, is a proxy for the cost of travel between regions (or for the expenditure of energy). The method is sensitive to variations in rotations and translations and to geometric distortions, but insensitive to variations in scale of the considered sets.

When we have several nonempty compact sets available, arranging them to form a cluster with a set X_i that is strategically important is a challenging open problem. This problem requires choosing appropriate spatial position for X_i, in addition to other parameters that explain the spatial relationships that X_i needs to possess with other sets, X_j, in the final cluster. Then, among the sets in the cluster, the task of determining which set has the potential to act as strategic set can be based on the following conditions: all the sets need to be considered to form the cluster set, each set must have at least one adjacent set, no set overlaps even partially with any other set within the cluster, all the sets are nonempty compact sets, and there must exist connectivity between any two pairs of sets within the cluster. If all the sets being used to arrange the cluster are of equal size and shape, only then does the spatial position of a set plays role in defining and designating the strategic set. In practice, however, several other spatial relationships between sets also play a role in determining the strategic set. Solving this problem should lead to a solution that would be of use in facility allocation and facility planning studies. Finally, although their application, as proposed in this study, is restricted to raster representations, the theory underlying these techniques can be extended to a wide class of metric spaces and to other representations (such as objects bounded by 2-D vectors), without significant computational difficulty.

References

Aarts, E. H. L. and J. Korst, 1989, *Simulated Annealing and Boltzman Machines: A Stochastic Approach to Combinatorial Optimization and Neural Computing*, Wiley, Chichester, U.K.

Alexandroff, P., 1961, *Elementary Concepts of Topology*, Dover Publications, Inc., New York.

Batty, M., 1976, *Urban Modelling: Algorithms, Calibrations, Predictions*, Cambridge University Press, Cambridge, U.K.

Batty, M. and B. Jiang, 2000, Multi-agent simulation: Computational dynamics within GIS, In: *GIS and Geocomputation*, eds. P. Atkinson and D. Martin, Taylor & Francis, London, pp. 55–71.

Beucher, S. and F. Meyer, 1992, The morphological approach to segmentation: The watershed transformation. In: *Mathematical Morphology in Image Processing*, ed. E. R. Dougherty, Marcel Dekker, Inc., New York.

Bloch, I., H. A. M. Heijmans, and C. Ronse, 2007, Mathematical morphology. In: *Handbook of Spatial Logics*, eds. M. Aiello, I. Pratt-Hartmann, and J. van Benthem, Springer, Amsterdam, the Netherlands, pp. 857–944.

Bouzy, B., 2003, Mathematical morphology applied to computer go, *International Journal of Pattern Recognition and Artificial Intelligence*, 17(2), 257–268.

Cayley, A., 1889, A theorem on trees, *Quarterly Journal of Mathematics*, 23, 376–378.

Chaudhuri, B. B., 1990, Fuzzy set theoretic interpretation of object shape and relational properties for computer vision, *International Journal of Systems Science*, 21(7), 1169–1184.

Cidell, J., 2010, Content clouds as exploratory qualitative data analysis, *Area*, 42(4), 514–523.

Coxeter, H. S. M., 1950, Self-dual configurations and regular graphs, *Bulletin of the American Mathematical Society*, 56, 413–455.

Coxeter, H. S. M., 1961, *Introduction to Geometry*, Wiley, New York.

Egenhofer, M. J. and D. M. Mark, 1995, Modeling conceptual neighborhoods of topologic line-region relation, *International Journal of Geographical Information Science*, 9(5), 555–565.

Gao, Y. et al., 2010, A semantic geographical knowledge wiki system mashed up with Google Maps, *Science China-Technological Sciences*, 53(Suppl. 1), 52–60.

Goodchild, M. F., 1992, Geographic data modeling, *Computers & Geosciences*, 18(4), 401–408.

Hausdorff, F., 1914, *Grundzuge der Mengenlehre*, Viet and Co., (Gekurzte) Auft, Chi Minh City, Vietnam.

Horn, M. E. T., 1995, Solution for large regional partitioning problems, *Geographical Analysis*, 27, 230–248.

Huttenlocher, D. P., G. A. Klunderman, and W. J. Rucklidge, 1993, Comparing images using the Hausdorff distance, *IEEE Pattern Analysis and Machine Intelligence*, 15(9), 850–863.

Jiang, B. and X. B. Yao, 2006, Location-based services and GIS in perspective, *Computers Environment and Urban Systems*, 30(6), 712–725.

Krishnapuram, R. and J. Keller, 1993, A probabilistic approach to clustering, *IEEE Transactions on Fuzzy Systems*, 1(2), 98–110.

Kwan, M. P. and G. X. Ding, 2008, Geo-narrative: Extending geographic information systems for narrative analysis in qualitative and mixed-method research, *Professional Geographer*, 60(4), 443–465.

Liu, Y. et al., 2010, A point-set-based approximation for areal objects: A case study of representing localities, *Computers Environment and Urban Systems*, 34(1), 28–39.

Liu, Y., Q. H. Guo, and M. Kelly, 2008, A framework of region-based spatial relations for non-overlapping features and its application in object based image analysis, *ISPRS Journal of Photogrammetry and Remote Sensing*, 63(4), 461–475.

Maragos, P. A., 1989, Pattern spectrum and multiscale shape representation, *IEEE Transactions on Pattern Analysis and Machine Intelligence*, 11(7), 701–716.

Maragos, P., 2005, Morphological filtering for image enhancement and feature detection. In: *The Image & Video Processing Handbook*, ed. A. C. Bovik, Elsevier Academic Press, Amsterdam, the Netherlands, pp. 135–156.

Martin, D., 2000, Automated zone designing in GIS. In: *GIS and Geocomputation*, eds. P. Atkinson and D. Martin, Taylor & Francis, London, pp. 103–113.

McMaster, R. B., 1987, Automated line generalisation, *Cartographica*, 24, 74–111.

McMaster, R. B. and K. S. Shea, 1992, *Generalisation in Digital Cartography*, Association of American Geographers, Washington, D.C., p. 134.

Mehrotra, A., E. L. Johnson, and G. L. Nemhauser, 1998, An optimization based heuristic for political districting, *Management Science*, 44, 1100–1114.

Monmonier, M., 1983, Raster-mode area generalisation for land use and land cover maps, *Cartographica*, 20, 65–91.

Muller, J.-C., 1991, The cartographic agenda of the 1990s: Updates and prospects, *ITC Journal* 1992, 2, 55–62.

Muller, J.-C. and Z.-S. Wang, 1992, Area-patch generalization: A competitive approach, *Cartographic Journal*, 29, 137–144.

Muller, J.-C., R. Weibel, J. P. Lagrange, and F. Salge, 1995, Generalization: State of the art and issues. In: *GIS and Generalization*, eds. J.-C. MuÈ ller, J. P. Lagrange, and R. Weibel, Taylor & Francis Group, London, U.K., pp. 3–17.

Nafarieh, A. and J. Keller, 1991, A new approach to inference in approximate reasoning, *Fuzzy Sets and Systems*, 41(1), 17–37.

Ocalir, E. V., O. Y. Ercoskun, and R. Tur, 2010, An integrated model of GIS and fuzzy logic (FMOTS) for location decisions of taxicab stands, *Expert Systems with Applications*, 37(7), 4892–4901.

Openshaw, S., 1977, A geographical solution to scale and aggregation problems in region-building, partitioning and spatial modeling, *Transactions of the Institute of British Geographers NS*, 2, 459–472.

Pullar, D., 2001, MapScript: A map algebra programming language incorporating neighborhood analysis, *Geoinformatica*, 5, 145–163.

Rajashekara, H. M., P. Vardhan, and B. S. D. Sagar, 2011, Generation of zonal map from point data via weighted skeletonization by influence zone, *IEEE Geoscience and Remote Sensing Letters*, 9(3), 403–407.

Rhind, D., 1973, Generalization and realism within automated cartographic system, *Canadian Cartographer*, 10, 51–62.

Rhind, D., 1988, A GIS research agenda. *International Journal of Geographic Information Systems*, 2, 22–28.

Rodrigues, A., C. Grueau, J. Raper, and N. Neves, 1996, Environmental planning using spatial agents, In: *Innovations in GIS 5*, ed. S. Carver, Taylor & Francis, London, pp. 108–118.

Rosenfeld, A. and S. K. Pal, 1988, Image enhancement and thresholding by optimization of fuzzy compactness, *Pattern Recognition Letters*, 7, 77–86.

Sagar, B. S. D., 1994, Applications of mathematical morphology and fractal geometry to study small water bodies, PhD thesis, Andhra University, Visakapatnam, India.

Sagar, B. S. D., 2005, Discrete simulations of spatio-temporal dynamics of small water bodies under varied streamflow discharges, *Nonlinear Processes in Geophysics, (American Geophysical Union)*, 12, 31–40.

Sagar, B. S. D., 2010, Visualization of spatiotemporal behavior of discrete maps via generation of recursive median elements, *IEEE Transactions on Pattern Analysis and Machine Intelligence*, 32(2), 378–384.

Sagar, B. S. D., G. Gandhi, and B. S. P. Rao, 1995a, Applications of mathematical morphology on water body studies, *International Journal of Remote Sensing*, 16(8), 1495–1502.

Sagar, B. S. D., N. Rajesh, S. A. Vardhan, and P. Vardhan, 2013, Metric based on morphological dilation for the detection of spatially significant zones, *IEEE Geoscience and Remote Sensing Letters*, 10(3), 500–504, DOI: 10.1109/LGRS.2012.2211565.

Sagar, B. S. D. and J. Serra, 2010, Spatial information retrieval, analysis, reasoning and modeling, *International Journal of Remote Sensing*, 31(22), 5747–5750.

Sagar, B. S. D., M. Venu, G. Gandhi, and D. Srinivas, 1998, Morphological description and interrelationship between force and structure: A scope to geomorphic evolution process modelling, *International Journal of Remote Sensing*, 19(7), 1341–1358.

Sagar, B. S. D., M. Venu, and B. S. P. Rao, 1995b, Distributions of surface water bodies, *International Journal of Remote Sensing*, 16(16), 3059–3067.

Schylberg, L., 1993, Computational methods for generalization of cartographic data in a raster environment, Doctoral thesis, Royal Institute of Technology, Stockholm, Sweden, 137pp.

Serra, J., 1982, *Image Processing and Mathematical Morphology*, Academic Press, New York, 610pp.

Serra, J., 1994, Interpolations et distance de Hausdorff, Technical Report N-15/94/MM, Ecole des Mines de Paris, Paris, France.

Serra, J., 1998, Hausdorff distances and interpolations. In: *Mathematical Morphology and Its Applications to Images and Signal Processing*, eds. Henk. J. A. M. Heijmans and Jos B. T. M. Roerdink, Kluwer Academics Publishers, Dordrecht, the Netherlands.

Shea, K. and R. McMaster, 1989, Cartographic generalization in a digital environment: When and how to generalize, *Proceedings of Auto-Carto 9*, Baltimore, MD, March 1989, ACSM-APSRS, Bethesda, MD, pp. 57–67.

Spielman, S. E. and J. C. Thill, 2008, Social area analysis, data mining, and GIS, *Computers, Environment, and Urban Systems*, 32(2), 110–122.

Stell, J. G., 2007, Relations in mathematical morphology with applications to graphs and rough sets. *Proceedings of Conference on Spatial Information Theory, COSIT07*, Melbourne, Australia, eds. S. Winter et al., *Springer Lecture Notes in Computer Science*, Vol. 4736, pp. 438–454.

Su, B. and Z. Li, 1995, An algebraic basis for digital generalization of area-patches based on morphological techniques, *Cartographic Journal*, 32, 148–153.

Su, B, Z. Li, G. Lodwick, and J.-C. Muller, 1997, Algebraic models for the aggregation of area features based upon morphological operators, *International Journal of Geographical Information Science*, 11(3), 233–246, Systems, 2, 23–28.

Thompson, D. W., 1992, *On Growth and Form*, Dover reprint of 1942, 2nd edn. (1st edn. 1917), Cambridge University Press, Cambridge, p. 346.

Tobler, W., 1976, Spatial interaction patterns, *Journal of Environmental Systems*, VI(4), 1976/77, 271–301.

Tomlin, C. D., 1983, A map algebra. In: *Proceedings of Harvard Computer Graphics Conference*, Cambridge, MA, pp. 127–150.

Unsalan, C. and K. L. Boyer, 2005, A theoretical and experimental investigation of graph theoretical measures for land development in satellite imagery, *IEEE Transactions on Pattern Analysis and Machine Intelligence*, 27(4), 575–589.

Weibel, R., 1995, Summary report: Workshop on progress in automated map generalization, Technical Report, ICA Working Group on Automated Map Generalization, Barcelona, Spain, September 1995.

Wilson, A. W., 2008, Boltzmann, lotka and volterra and spatial structural evolution: An integrated methodology for some dynamical systems, *Journal of the Royal Society, Interface*, 5, 865–871.

Wilson, A., 2009, Remote sensing as the 'X-Ray Crystallography' for urban 'DNA.' *Proceedings of Annual Seminar on Spatial information Retrieval, Analysis, Reasoning and Modeling*, ed. B. S. D. Sagar, March 18–20, 2009, Bangalore, India, pp. 1–12.

Worboys, M. F., 1994, A unified model of spatial and temporal information, *Computer Journal*, 37, 26–34.

Yan, J. and J. C. Thill, 2009, Visual data mining in spatial interaction analysis with self-organizing maps, *Environment and Planning B—Planning and Design*, 36(3), 466–486.

Yao, X. B. and J. C. Thill, 2006, Spatial queries with qualitative locations in spatial information systems, *Computers Environment and Urban Systems*, 30(4), 485–502.

11

Derivation of Spatially Significant Zones from a Cluster

The ability to derive spatially significant zones (e.g., water bodies, zones of influence) within a cluster of zones has interesting applications in understanding commonly sharing physical mechanisms. Using morphological dilation distance technique, we introduce geometrically based criteria that serve as indicator of the spatial significance of zones within a cluster of zones. This chapter focuses on the problem of identifying zones that are "strategic" in the sense that they are the most central or important based on their proximity to other zones. We have applied this technique to a task aiming at detecting spatially significant water body from a cluster of water bodies retrieved from IRS LISS-III multispectral satellite data.

Background on Derivation of Spatially Significant Zones from a Cluster

High-resolution remotely sensed satellite data and digital elevation models are of immense use to map spatial entities such as water bodies (Sagar et al. 1995), zones of influence (Sagar 2007, Rajashekara et al. 2012), watersheds (Tay et al. 2005, 2007), and urban features (Pesaresi and Benediktsson 2001, Benediktsson et al. 2003, Barata and Pina 2006, Chanussot et al. 2006, Taubenbock et al. 2006, Mering et al. 2010, Trianni et al. 2010, Wilson 2010, Dalla Mura et al. 2011) that could be represented as areal objects on specific thematic maps. Understanding the spatial organization of such spatial entities (zones) by involving distances between all the zones of a cluster of zones is important from the point of spatial reasoning. Derivation of spatial significance of each zone within a cluster of zones is important to decide a suitable facility (e.g., reservoir). Spatial significance of a zone is defined as "a zone from which it is easy to reach all of its neighboring zones" (Sagar et al. 2013). A watershed (cluster of zones) consists of sub-watersheds (zones), and sub-watersheds consist of still minor watersheds, and so on. A main watershed that consists of sub-watersheds is treated as a spatial system (Figure 10.1) with sub-watersheds being subsystems. Spatially significant zone within a cluster of zones possesses a geometric characteristic that is in greater proximity to other zones highlighting the significance of location. Identifying a spatial significance of a zone from geometric point of view based on qualitative spatial reasoning is nontrivial when a spatial system includes a large number of zones, and such identification process varies from person to person according to their own individual spatial perceptions. Recognizing spatially significant zones within such a spatial system composed of various zones could be accomplished quantitatively. This section attempts to provide geometric criteria to identify spatially significant zones within a cluster of zones.

Spatial System and Its Subsystems

Let a cluster of zones (X) be composed of a number of nonempty compact sets (zones) denoted by X_1, X_2, X_3, ..., X_N, such that $X = \bigcup_{i=1}^{N} X_i$. These sets are like possible partitions of an image. A better analogy is that a DEM is an image and possible partitions of a DEM are subbasins (zones). For any pair of zones X_i and X_j, from this cluster, such that $i \neq j$,

the following spatial relations hold true: (1) $X_i \cap \left(\bigcup_{\substack{j=1 \\ j \neq i}}^{N} X_j \right) = \varnothing$ and

(2) $(X_i \oplus B) \cap \left(\bigcup_{\substack{j=1 \\ j \neq i}}^{N} X_j \right) = \left(\left(\left(\bigcup_{\substack{j=1 \\ j \neq i}}^{N} X_j \right) \oplus B \right) \cap X_i \right) \neq \varnothing.$

For instance, for the cases of water bodies, nodes, and point-specific data (noncontiguous form), the relation (1) would be satisfied. In many cases, where the zones are in noncontiguous form, relation (2) may not be satisfied. This relation (2) would be satisfied if all the zones of a cluster are in contiguous form (e.g., zones of influence of water bodies). In this chapter, we consider both the cases that respectively satisfy relations (1) and (2).

Dilation Distances between Origin and Destination Zones

Determining distances between spatial objects (zones) based on Euclidean metric is a challenge. If all the zones in a cluster considered are identical such that the shapes and sizes of zones are similar, then the simple Euclidean distances between all the possible pairs of centroids of such zones would suffice to detect the spatially significant centroid corresponding to a zone. Euclidean distance of centroids of zones possessing dissimilar shapes and sizes would lead to a problem detecting precise spatially significant zone due to following reasons: (1) computation of centroids of zones requires an additional step perhaps based on "minimal skeletal point" that is computationally expensive, and (2) Euclidean distance between the centroids of the two zones does not explain the morphological (geometric) properties of the zones under consideration. However, the iterative dilation is a better choice to compute distances between zones. Dilation distance is employed to address the topic of identifying spatially significant zone(s) from a cluster of zones of varied shapes and sizes either in a contiguous or in a noncontiguous way.

Let nonempty disjoint compact zones X_i and X_j be the origin and the destination zones. X_i is smaller than X_j (Figure 10.2a). According to Equations 10.11 and 10.12, the distance from X_i to X_j (Figure 10.2c) is represented by $d(X_{ij}) = \min_{i \neq j}(n : X_j \subseteq (X_i \oplus nB))$ and the distance from X_j to X_i is represented by $d(X_{ji}) = \min_{i \neq j}(n : X_i \subseteq (X_j \oplus nB))$. We may state the following: $d(X_{ii}) = 0$, $d(X_{ij}) \neq d(X_{ji})$, and $d(X_{ij}) = d(X_{ji})$ if both X_i and X_j possess identical size, shape, and orientation (Figure 10.2d and f). From Figure 10.2d and f, it is evident that a smaller object, to completely occupy a relatively larger one, requires a greater number of dilation cycles than that in the converse scenario (Figure 10.2a and c). If there exists a shape–size dissimilarity between the two sets, one can observe that $d(X_{ij}) \neq d(X_{ji})$, and the minimum of $d(X_{ij})$ and $d(X_{ji})$ is Hausdorff dilation distance (Equations 10.3 and 10.13), which is mentioned for clarity $\rho(X_{ij}) = \min(d : d(X_{ij}), d(X_{ji}))$.

Estimation of the dilation distance between the origin and destination zones is justified as such a dilation distance is essential to compute distances between the zones. The limitation of this distance is that it is essentially affected by the object's boundary points that are farthest out with respect to other spatial objects. The maximum distance d_{max} between an origin zone (X_i) and destination zones (X_j) of a cluster is computed as per Equation 10.17. Similarly, d_{max} between the destination zones and an origin zone is computed as $d_{max}(X_{ji}) = \max_{\forall j}(\min(n : (X_i \subseteq (X_j \oplus nB))))$.

Spatial Significance Index of a Zone

A zone X_i is designated as spatially most significant to establish a facility if (1) it is located in a place closer to all X_js, and (2) reaching X_i from all X_js required shorter distance (minimum energy expenditure involved). No other zone from a cluster of X_js matches with X_i with respect to these two characteristic (spatial) relationships, and hence X_i is chosen as the best zone and is termed spatially the most important zone. Keeping these characteristics in view, we propose (Equation 11.1) involving dilation distances between origin (X_i) and destination zones (X_j).

$$SSI = \min_{\forall i}(d_{max}(X_{ij})) \tag{11.1}$$

Minimum of all the maximum values of the corresponding origin zones would explain about the zone from which it is easier to reach out all other zones with minimum energy expenditure (dilation distance). The spatial significance index (SSI) of a zone (X_i) is a dimensionless unit. The lower the SSI of a zone (X_i) in a cluster of zones, the higher is its significance. Equation 11.2 to compute Normalized Spatial Significance Index (NSSI) that ranges between 0 and 1 takes the form of

$$\text{NSSI} = \left(\frac{\min_{\forall i}(d_{\max}(X_{ij}))}{\max_{\forall i}(d_{\max}(X_{ij}))} \right) \tag{11.2}$$

Low value of SSI or NSSI enables the location significance/importance of a zone X_i from which every other zone could be reached or the zone X_i could be reached from every other zone with minimum expenditure of energy. If the zones of a cluster are not similar in shape and/or size wise, then $\min_{\forall i}(d_{\max}(X_{ij}))$ and $\min_{\forall j}(d_{\max}(X_{ji}))$ are not equal. They are equal if the shapes and sizes of zones of a cluster are identical to each other. When all zones in a cluster are similar in terms of both size and shape, the following relationship holds good:

$$\left(\frac{\min_{\forall i}(d_{\max}(X_{ij}))}{\max_{\forall i}(d_{\max}(X_{ij}))} \right) = \left(\frac{\min_{\forall j}(d_{\max}(X_{ji}))}{\max_{\forall j}(d_{\max}(X_{ji}))} \right) \tag{11.3}$$

This relationship holds good also for cases where centroids of zones are considered. See the synthetic example that follows for more details.

Synthetic Example

For clarity, a toy example is given to explain Equations 11.1 through 11.3. Let X_1, X_2, and X_3 be three spatial objects in a cluster (Figure 11.1a). The assumed distances between all possible pairs of these three spatial objects are shown in Figure 11.1b. Its corresponding matrix is shown in Figure 11.1b from which $d_{\max}(X_{ij})$, $d_{\max}(X_{ji})$, SSI, NSSI, and homogeneity degree of spatial objects explained in Equations 10.11 and 10.12 could be easily understood.

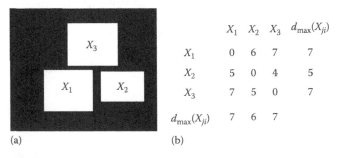

	X_1	X_2	X_3	$d_{\max}(X_{ji})$
X_1	0	6	7	7
X_2	5	0	4	5
X_3	7	5	0	7
$d_{\max}(X_{ij})$	7	6	7	

(a)   (b)

FIGURE 11.1
(a) Synthetic example consisting of three spatial objects and (b) dilation distances between every possible pair are shown in a matrix form besides the values obtained according to Equations 10.11 and 10.12. (From Sagar, B.S.D. et al., *IEEE Geosci. Remote Sens. Lett.*, 10(3), 2013.)

As per the SSI and NSSI (i.e., 6 and 0.857) computed according to Equations 11.1 and 11.2—where the considered data include assumed dilation distances (Figure 11.1b)—X_2 is designated as spatially significant zone.

Experimental Results

Cluster of Zones of Water Body Influence

Small water bodies and their zones of influence of varied sizes and shapes arranged heterogeneously (Figure 11.2a through d) are good examples of spatial systems. The data—sourced from an IRS LISS-III multispectral data of 23.5 m spatial resolution (Figure 11.2a) and a topographic map of

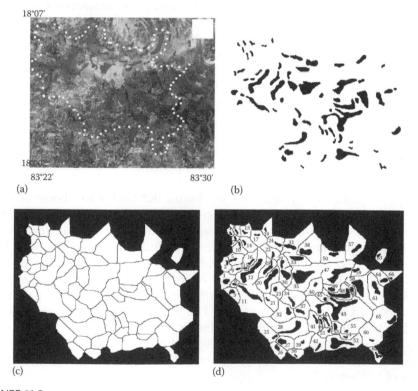

(a)   (b)

(c)   (d)

FIGURE 11.2
(a) Indian Remote Sensing satellite (IRS LISS-III) multispectral image of the study area and the black objects are water bodies traced from IRS LISS-III image with topographic map reference superposed on IRS LISS-III image, and white dots indicate the boundary of the considered cluster, (b) small water bodies, (c) zones of influence of corresponding water bodies, and (d) water bodies and zones of influence with labeling. (From Sagar, B.S.D. et al., *IEEE Geosci. Remote Sens. Lett.*, 10(3), 2013.)

a region situated in between the geographical coordinates 18°00′–18°07′N and 83°22′–83°30′E—have been employed. Sixty-six water bodies were traced from IRS LISS-III multispectral data with topographic map reference (Figure 11.2b). The corresponding 66 influence zones, defined as the catchment basins of the corresponding water bodies (markers), computed by using the technique of *skeletonization by zones of influence* are shown in Figure 11.2c. Since the region considered is in the slope category of <2° slope, the elevation differences across the region considered are minimal, and hence the region is treated as flat. In view of this fact, DEM has not been used. Water bodies and zones respectively representing markers and catchment basins are denoted by X_i with proper labeling (Figure 11.2d). Dilation distances, which are essential parameters of Equations 11.1 and 11.2, between each water body (zone) and all other destination water bodies (zones) in a cluster of water bodies (zones) are computed respectively according to Equations 10.11 and 10.12.

Maximum dilation distances observed from the distances computed between every water body and every other water body belonging to a cluster of 66 water bodies are plotted as functions of water bodies (Figure 11.3a). Similarly maximum dilation distances observed from the estimated distances between every zone of influence to every other zones of influence are also plotted as functions of zones of influence (Figure 11.3b). The observed minimum distances among 66 maximum distances for both water bodies (Figure 11.3a) and zones (Figure 11.3b) include 53 and 52, respectively. This table also provides details of other four spatially significant water bodies and zones of influence. The maximum distances among 66 maximum distances for both water bodies and zones observed include 109 and 110, respectively.

As per Equations 11.1 and 11.2, we found out that water body labeled with 35 and zone labeled with 35 are spatially significant. The corresponding water body and zone of influence are shown in Figure 11.4a and b. SSIs of 66 water bodies and 66 zones of water body influences could be seen. The lower the index, the higher is the spatial significance. The zones labeled with 35, 41, 43, 37, and 46 are the five best zones that have SSIs respectively 52, 55, 57, 60, and 62. The corresponding NSSIs for these zones respectively include 0.47, 0.50, 0.51, 0.54, and 0.56. The SSIs and NSSIs for water bodies are also shown in the "Experimental Results" section. Interestingly, these spatially significant water body (Figure 11.4a) and zone of influence (Figure 11.4b) possess longer boundary being shared with neighboring water bodies/zones of influence.

States of India

The approach demonstrated on a cluster of water body has been extended to recognize spatially significant state from a cluster of 28 states of India (Figures 10.4 and 11.5a). The dilation distances between every state to every

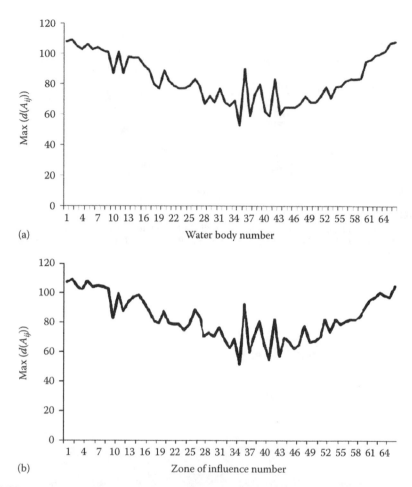

(a)

(b)

FIGURE 11.3

Maximum dilation distances observed from the estimated distances between every water body (zone of influence) and every other water body (zones of influence) of a cluster of 66 water bodies (zones of influence). (a) $Max(d(XZ_{ij}))$, and (b) $Max(d(X_{ij}))$ for all i's. SSI of 66 zones of water body influences, and 66 water bodies could be seen. The lower the index, the higher is the spatial significance. See Figure 11.2a through d for spatial organization of water bodies and their zones of influence. The zones labeled with 35, 41, 43, 37, and 46 are the five best zones that have SSIs respectively 52, 55, 57, 60, and 62. The corresponding normalized SSIs for these zones respectively include 0.47, 0.50, 0.51, 0.54, and 0.56. Such SSIs for water bodies are shown in Table 11.2. (From Sagar, B.S.D. et al., *IEEE Geosci. Remote Sens. Lett.*, 10(3), 2013.)

other state are estimated, and origin-state-specific maximum distances are computed (Table 11.1). Maximum dilation distances observed from the estimated distances between every state and every other state of a cluster of 28 states of India are considered, and minimum of these maximum distances are considered to detect spatially significant state. Minimum of all these maximum distances is 189, followed by 206, 213, 226, and 233.

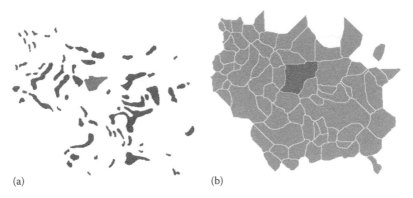

(a)   (b)

FIGURE 11.4
Spatially significant (a) water body with label 35 (lighter shade), and (b) zone of water body influence labeled with 35 (darker shade). (From Sagar, B.S.D. et al., *IEEE Geosci. Remote Sens. Lett.*, 10(3), 2013.)

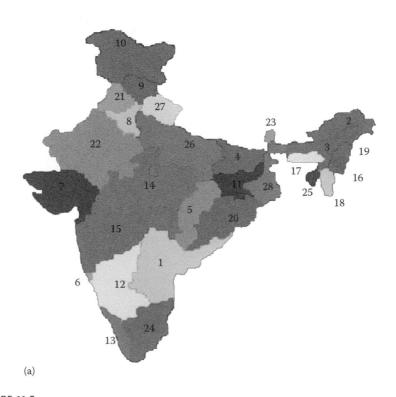

(a)

FIGURE 11.5
(a) The 28 states of India and their indexes given as per the alphabetical order can also be seen in Figure 10.4. (From Sagar, B.S.D. et al., *IEEE Geosci. Remote Sens. Lett.*, 10(3), 2013.)

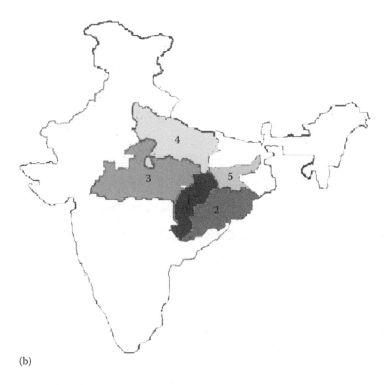

(b)

FIGURE 11.5 (continued)
(b) first five spatially significant states with their ranks. (From Sagar, B.S.D. et al., *IEEE Geosci. Remote Sens. Lett.*, 10(3), 2013.)

The maximum of maximum distances estimated between each origin-state and all destination-states is 383. Normalized SSI is computed according to Equation 11.2. First five states that possess minimum of maximum distances are shown in Figure 11.5b, and their corresponding SSIs and NSSIs, which were already shown for 66 water bodies and their zones of influence (Table 11.2), are shown in Table 11.3.

On Intel Core 2 Duo T5850 @ 2.17 GHz with 3GB RAM on 32-bit Operating System, it took 30, 31, and 17 min respectively to compute the dilation distances for the cases of (1) 66 water bodies (400 × 400 pixels), (2) 66 zones of influence (400 × 400 pixels), and (3) 28 states of India (400 × 400 pixels). The number of dilation distances required to be computed increases with the number of spatial objects and the sizes of the individual spatial objects. Hence, the computational complexity increases with increasing (1) number of spatial objects and (2) spatial resolution. This time could be significantly reduced by rescaling the data such that the zones of the cluster do not lose their shape characteristics.

TABLE 11.1

Dilation Distances between Every State to Every Other State of India

Dilation Distances between Every State and Every Other State of India (Cluster) Containing 28 States (Size of the Image is 400 × 400 Pixels)

	X_1	X_2	X_3	X_4	X_5	X_6	X_7	X_8	X_9	X_{10}	X_{11}	X_{12}	X_{13}	X_{14}	X_{15}	X_{16}	X_{17}	X_{18}	X_{19}	X_{20}	X_{21}	X_{22}	X_{23}	X_{24}	X_{25}	X_{26}	X_{27}	X_{28}
X_1	0	225	200	161	69	172	163	210	246	274	129	104	126	116	63	247	202	240	255	72	238	146	204	87	225	145	225	146
X_2	178	0	20	135	189	351	342	289	269	255	137	283	302	213	242	54	62	73	39	139	301	282	125	253	74	175	240	105
X_3	158	36	0	115	168	329	320	269	248	229	115	261	280	191	220	50	41	53	54	119	280	260	104	232	54	165	219	88
X_4	110	120	95	0	55	215	207	154	134	125	39	159	209	78	106	147	57	135	150	69	164	147	70	197	120	51	107	45
X_5	65	165	140	94	0	159	152	140	214	204	63	103	160	48	53	187	147	180	195	122	168	85	134	149	165	85	155	87
X_6	50	270	250	155	110	0	72	179	217	244	155	8	43	87	12	297	257	290	305	201	207	116	224	57	275	130	194	197
X_7	132	355	330	234	188	130	0	107	143	169	233	105	173	90	65	381	331	367	383	122	135	45	302	162	352	140	148	276
X_8	158	245	225	130	98	213	82	0	37	64	129	175	255	54	123	277	227	265	280	125	28	27	200	243	249	35	45	172
X_9	195	228	208	110	133	250	119	36	0	29	124	213	293	93	160	259	205	247	261	153	29	50	180	280	230	42	30	154
X_{10}	248	270	249	158	185	303	173	89	53	0	175	266	346	145	215	301	251	288	304	206	67	99	220	336	273	95	80	196
X_{11}	89	120	95	31	54	215	199	154	129	146	0	156	188	75	104	147	57	135	150	39	164	145	222	168	119	50	105	42
X_{12}	45	265	245	180	105	68	121	228	265	294	149	0	79	134	61	177	247	285	300	117	261	164	222	69	269	174	244	192
X_{13}	69	265	245	230	138	99	166	279	315	342	199	49	0	185	111	279	246	273	289	138	306	214	272	19	258	223	293	193
X_{14}	102	260	239	142	95	157	129	93	129	157	142	119	200	0	70	290	240	277	294	109	119	69	210	188	262	65	108	184
X_{15}	68	290	265	168	124	108	99	168	204	231	168	51	130	75	0	315	265	298	318	135	190	104	237	119	289	119	183	210
X_{16}	145	38	15	103	157	317	309	258	236	216	104	199	269	179	209	0	29	20	19	108	273	254	91	219	34	155	209	77
X_{17}	115	30	9	72	127	287	278	228	106	186	74	218	238	150	179	50	0	39	53	78	233	219	62	188	25	124	179	47
X_{18}	129	64	30	84	160	299	289	228	219	200	85	232	249	160	190	27	42	0	44	89	249	230	74	204	20	135	190	59
X_{19}	149	19	12	104	160	322	313	262	240	220	108	254	273	183	213	14	34	33	0	112	272	253	95	223	39	158	208	79
X_{20}	45	154	129	93	50	209	199	150	178	205	61	142	166	69	100	174	130	168	183	0	164	140	134	128	157	86	154	78
X_{21}	180	255	235	140	120	235	105	25	30	34	140	200	279	79	148	285	235	275	289	140	0	32	209	268	264	45	55	180
X_{22}	149	330	305	209	164	204	75	79	94	127	208	167	248	67	115	354	305	343	359	174	91	0	308	234	338	114	123	249
X_{23}	129	52	26	24	69	225	217	165	145	125	39	169	219	88	120	78	29	65	80	79	174	157	0	208	52	60	117	13
X_{24}	65	265	236	231	139	100	172	280	316	343	199	51	50	184	113	259	244	248	263	139	307	211	273	0	233	226	294	194
X_{25}	110	52	17	65	120	282	273	222	200	180	51	215	233	144	173	26	29	15	34	73	232	213	55	184	0	225	168	40
X_{26}	152	215	190	120	89	207	155	104	91	118	93	170	249	49	119	240	185	229	244	109	113	93	163	238	214	0	69	135
X_{27}	168	205	180	94	104	222	101	49	32	52	98	184	264	64	134	226	182	220	235	124	58	46	149	253	204	20	0	125
X_{28}	116	80	59	39	79	240	228	179	159	152	27	183	213	103	132	105	55	93	109	67	190	173	78	194	78	78	132	0
Max(dX_{ij})	248	355	330	234	189	351	342	289	316	343	233	283	346	213	242	381	331	367	383	206	307	282	308	336	352	226	294	276

Min(Max(X_{ij})) = 189

Max(Max(X_{ij})) = 383

Source: Sagar, B.S.D. et al., *IEEE Geosci. Remote Sens. Lett.*, 10(3), 2013.

TABLE 11.2

SSI of Top Five Water Bodies and Zones

Rank	Water Body (W) Label	D-Dist	Zone (Z) Label	D-Dist	NSSI (W)	NSSI (Z)
1	35	53	35	52	0.48	0.47
2	41	59	41	55	0.54	0.50
3	43	59	43	57	0.54	0.51
4	49	60	37	60	0.55	0.54
5	46	62	46	62	0.56	0.56

Source: Sagar, B.S.D. et al., *IEEE Geosci. Remote Sens. Lett.*, 10(3), 2013.

TABLE 11.3

Spatial Significance Indexes of Top Five States

Rank	State Label	D-Dist	NSSI
1	5	189	0.49
2	20	206	0.53
3	14	213	0.55
4	26	226	0.59
5	11	233	0.60

Source: Sagar, B.S.D. et al., *IEEE Geosci. Remote Sens. Lett.*, 10(3), 2013.

Conclusions

Technique proposed here provides SSI for zones of a cluster of zones. Identifying spatially significant zone via Equation 11.1 is a scale-invariant process. This equation is sensitive to variations in rotations and translations and to geometric distortions, but insensitive to variations in scale of the considered zones. Mostly, a larger interior zone that could be reached by other destination zones of a cluster would stand as a spatially significant zone. Although the application of this technique is shown for data represented in raster format, without significant computational difficulty, this technique can be extended (1) to a wide class of metric spaces and to other representations (such as objects bounded by 2-D vectors), and (2) to 3-D case by replacing dilation distance with grayscale geodesic distances. This approach provides useful insights in (1) clustering-classification frameworks, (2) detecting the spatially significant segmented zones (spatial objects in 2-D case) obtained via various segmentation approaches, (3) automatically deriving a central node from a large number of nodes, (4) determining the influence of a node in a vector-based network setting, and (5) deciding on nodal center(s) to establish an administrative facility, from a cluster of cadastral zones mapped from remotely sensed satellite data.

References

Barata, T. and P. Pina, 2006, A morphological approach for feature space partitioning, *IEEE Geoscience and Remote Sensing Letters*, 3(1), 173–177.

Benediktsson, J. A., M. Pesaresi, and K. Arnason, 2003, Classification and feature extraction for remote sensing images from urban areas based on morphological transformations, *IEEE Transaction on Geoscience Remote Sensing*, 41, 1940–1949.

Chanussot, J., J. A. Benediktsson, and M. Fauvel, 2006, Classification of remote sensing images from urban areas using a fuzzy possibilistic model, *IEEE Geoscience and Remote Sensing Letters*, 3(1), 40–44.

Dalla Mura, M. et al., 2011, Classification of hyperspectral images by using extended morphological attribute profiles and independent component analysis, *IEEE Geoscience and Remote Sensing Letters*, 8(3), 542–546.

Mering, C., J. Baro, and E. Upegui, 2010, Retrieving urban areas on Google Earth images: Application to towns of West Africa, *International Journal of Remote Sensing*, 31(22), 5867–5878.

Pesaresi, M. and J. A. Benediktsson, 2001, A new approach for the morphological segmentation of high-resolution satellite imagery, *IEEE Geoscience and Remote Sensing Letters*, 39, 309–320.

Rajashekara, H. M., P. Vardhan, and B. S. D. Sagar, 2012, Generation of zonal map from point data via weighted skeletonization by influence zone, *IEEE Geoscience and Remote Sensing Letters*, 9(3), 403–407.

Sagar, B. S. D., 2007, Universal scaling laws in surface water bodies and their zones of influence, *Water Resources Research*, 43(2), W02416.

Sagar, B. S. D., G. Gandhi, and B. S. P. Rao, 1995, Applications of mathematical morphology on water body studies, *International Journal of Remote Sensing*, 16(8), 1495–1502.

Sagar, B. S. D., N. Rajesh, S. A. Vardhan, and P. Vardhan, 2013, Metric based on morphological dilation for the detection of spatially significant zones, *IEEE Geoscience and Remote Sensing Letters*, 10(3), 500–504.

Taubenbock, H. et al. 2006, Automated allocation of highly structured urban areas in homogeneous zones from remote sensing data by Savitzky-Golay filtering and curve sketching, *IEEE Geoscience and Remote Sensing Letters*, 3(4), 532–536.

Tay, L. T., B. S. D. Sagar, and H. T. Chuah, 2005, Analysis of geophysical networks derived from multiscale digital elevation models: A morphological approach, *IEEE Geoscience and Remote Sensing Letters*, 2(4), 399–403.

Tay, L. T., B. S. D. Sagar, and H. T. Chuah, 2007, Granulometric analysis of basin-wise DEMs: A comparative study, *International Journal of Remote Sensing*, 28(15), 3363–3378.

Trianni, G., F. Dell'Acqua, and P. Gamba, 2010, Geographic information system (GIS)-aided per-segment scene analysis of multi-temporal spaceborne synthetic aperture radar (SAR) series with application to urban area, *International Journal of Remote Sensing*, 31(22), 6005–6014.

Wilson, A., 2010, Remote sensing as the 'X-Ray Crystallography' for urban 'DNA', *International Journal of Remote Sensing*, 31(22), 5993–6004.

12

Directional Spatial Relationship

We provide an approach to compute origin-specific morphological dilation distances between planar sets to further determine the directional spatial relationship between sets. The origin chosen for a structuring element (B) that yields shorter dilation distance than that of the other possible origins of B determines the directional spatial relationship between X_i (origin set) and X_j (destination set). We demonstrate this approach on a cluster of spatial sets (states) decomposed from a spatial map depicting the country of India. This approach has potential to extend to any number (type) of sets on Euclidean space.

Background on Directional Spatial Relationship

In remote sensing images, the spatial objects are embedded in a complex environment, and their spatial arrangement impacts information for reasoning and interpolation tasks (Bloch 1999, Bloch and Ralescu 2003, Bloch et al. 2006). Drawing spatial relationship between such spatial objects that possess several attributes is a challenging task. In the thematic information retrieved from remotely sensed satellite data of various spatial and spectral resolutions, understanding the organization of spatial objects is a basic requisite to computer vision. Spatial relationships between such objects are partly denoted by qualitative spatial relationships. In the modeling and analysis of such spatial objects (e.g., Figure 12.1), spatial position and the direction between origin sets and destination sets play important roles. Qualitative spatial reasoning (QSR) is an important branch of geographic information science (GISci). Spatial relationships—topological (adjacency), directional (N, W, E, S), and metric (near, far)—are important aspects of spatial reasoning (Keller and Sztandera 1990, Gapp 1994). Various models to determine directional spatial relationships automatically between spatial objects (origin and destination sets (Wilson 2009) were addressed (Matsakis and Wendling 1999, Cinbis and Aksoy 2007, Aksoy and Cinbis 2010). Mathematical morphology (Serra 1982, Sagar and Serra 2010) has been earlier employed in understanding the relative position–based relationship (Cinbis and Aksoy 2007, Aksoy and Cinbis 2010). To deal with directional spatial relationships between the origin and destination spatial objects, methods based on angle measurements between points of objects of interest or the angle between objects' centroids, or the histogram of angles

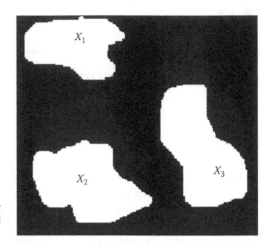

FIGURE 12.1
Three disjoint objects (X_1, X_2, X_3) possessing different directional spatial relationship.

between all pairs of points (Matsakis and Wendling 1999) have been used. By assimilating spatial objects to elementary entities (e.g., barycenter or a bounding rectangle), several researchers have proposed spatial relationships between spatial objects (Keller and Sztandera 1990, Gapp 1994). However, these methods have several limitations when the objects under consideration—possess strong concavities and complex spatial arrangements of the objects—do not have compact shapes, and in turn are computationally expensive. It was reported in Cinbis and Aksoy (2007) and Aksoy and Cinbis (2010) that mathematical morphology provides a basis to handle the complex shapes also. Bloch (1999) proposed an approach where fuzzy structuring elements are employed to encode the semantics of the directional relations. Mathematical morphology (Serra 1982) with more geometrical rigor offers insights for spatial data analysis (Sagar and Serra 2010, Rajashekara et al. 2012).

Directional Spatial Relationship via Origin-Specific Dilation Distances

Thematic maps generated from remotely sensed satellite data consist of spatial objects (planar sets) of varied degrees of spatial complexity. We provide an approach to compute origin-specific morphological dilation distances between planar sets to further determine the directional spatial relationship between sets. The origin chosen for a structuring element (B) that yields shorter dilation distance than that of the other possible origins of B determines the directional spatial relationship between X_i (origin set) and X_j (destination set). We demonstrate this approach on (1) a cluster of spatial sets (states) decomposed from a spatial map depicting the country of India, and (2) water bodies traced from remotely sensed satellite data. This chapter provides an intuitive idea

demonstrated for applications in spatial reasoning based on geographic data. In this chapter, we deal with the determination of directional spatial relationship between sets (e.g., Figure 12.1) of varied degrees of spatial complexities entirely by using mathematical morphology. The basis for this study is dilation distance (Serra 1982, Sagar 2010) between two sets (e.g., origin set and destination set). Such a morphological distance was earlier used in spatial interpolations (Sagar 2010). For this purpose, recursive dilations with respect to a primitive structuring element by varying origins are considered in this study to determine the directional spatial relation between sets automatically. This chapter is organized as follows: Directional dilation and dilation distance are explained in the "Methods to Derive the Directional Spatial Relationship" section; Experimental results are provided in "Experimental Results and Discussion" section; and conclusions drawn are given in the "Conclusion" section.

Methods to Derive the Directional Spatial Relationship

Directional Dilation and Dilation Distance

The circle (Figure 2.4a) is hereafter referred to as structuring element (B). Eight connectivity circle (Figure 2.4a) in discrete space possess eight neighbors to the center. This B (Figure 2.4a) is symmetric about the origin if the origin is positioned at the center (Figure 2.4a). If the chosen origin is at the center of the structuring element (B^O), then B^O is treated as symmetric B such that B is equivalent to its transpose (i.e., $\hat{B}$). Other eight possible origins (O) involved with a 3×3 square structuring element are shown in Figure 12.2a. The superscript O denotes the position of origin of B. Origin chosen for B plays vital role in the morphological transformation. Different origins of B (Figure 12.2) are

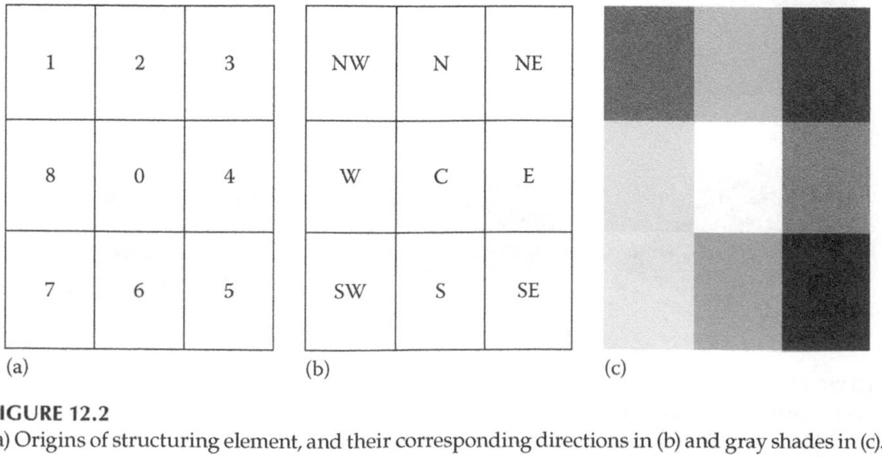

(a) | (b) | (c)

FIGURE 12.2
(a) Origins of structuring element, and their corresponding directions in (b) and gray shades in (c).

employed to determine the direction between X_i and X_j. It should be noted that $B^O \neq \widehat{B^O}$ and hence it is an asymmetric structuring element. In effect, morphological dilation is not equivalent to Minkowski addition as was the case with symmetric structuring element, where $B = \hat{B}$. These eight neighbors are assigned with indexes (Figure 12.2a) to denote eight origins ranging from 1 to 8 excluding the center. Further, these origins of eight connectivity circles are respectively related to eight directions (Figure 12.2b), namely, northwest (NW), north (N), northeast (NE), east (E), southeast (SE), south (S), southwest (SW), west (W). However, there are eight other possible coordinate positions of origins of B shown in Figure 12.2 that include (1, 1), (1, 2), (1, 3), (2, 3), (3, 3), (3, 2), (3, 1), and (2, 1). The gray shade representation shown for these eight directions are shown in Figure 12.2c. The structuring element with any of these eight possible origins is not a symmetric B, and it does mean that $B \neq \hat{B}$. Asymmetric B with these eight origins dilates set X in an origin-specific direction. Such directional dilations provide crucial information required in automatically determining the directional spatial relationship between the objects. The dilation impacts by structuring elements with different origins shown in Figure 12.2a are illustratively shown in Figure 12.3a through i. It should be noted that dilation of a set by B^0 (Figure 12.2a) is uniform because $B^0 = \widehat{B^0}$. We need to perform morphological dilation transformation on set X_i with respect to $\widehat{B^O}$.

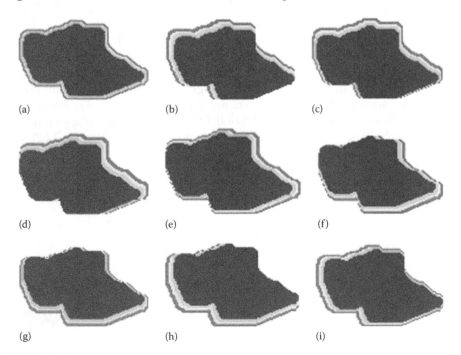

(a) (b) (c)

(d) (e) (f)

(g) (h) (i)

FIGURE 12.3
Directional dilations on objects X by all nine origins: (a) $X \oplus B^0$, (b) $X \oplus B^1$, (c) $X \oplus B^2$, (d) $X \oplus B^3$, (e) $X \oplus B^4$, (f) $X \oplus B^5$, (g) $X \oplus B^6$, (h) $X \oplus B^7$, and (i) $X \oplus B^8$.

In effect, Equation 12.1 takes the form of $X_i \oplus \widehat{B^O}$. For instance, if we choose origin $O = 1$, B^1, which is in northwest corner point in 3×3 size structuring element, the portion that is dilated is in the direction of northwestern side (Figure 12.3b). When the structuring element (B) is not equivalent to its transpose with respect to origin (i.e., $B \neq \hat{B}$), we treat such a B as asymmetric B. In this chapter, all B^Os considered are asymmetric as $B^O \neq \widehat{B^O}$. By definition, morphological dilation will be performed by transposing B first (i.e., B^O rotated 180° about origin O) and then dilating set X. If $B = \hat{B}$—which is the only possibility that exists with symmetric B, which is equivalent to its transpose with respect to a rotation for 180° about the origin—then Minkowski sum and morphological dilation are the same. In this chapter, symmetric B's have no role. By saying this, all the results and illustrations mentioned are correct. For instance, the dilation of set X by B^1 (i.e., B with NW origin $O = 1$; Figure 12.2b) will be performed by first transposing B^1 (i.e., $\widehat{B^1}$), and then we translate all the elements of set X (as per the definition of morphological dilation) set X with respect to $\widehat{B^1}$. In turn, the dilated portion will be seen in the NW direction only. With origin $O = 3$, if we perform the dilation, the dilated portion would be seen in the northeastern direction of the set (Figure 12.3d).

Dilation Distance

To compute origin-dependent dilation distance between the sets, an asymmetric structuring element needs (Figure 12.2) to be employed. To find out the directional relationship between objects, we compute shortest dilation distance, between the origin-object and the destination-object with respect to B with a unique origin, which is one of the eight origins. The unique origin is the origin chosen for B with which the shortest dilation yields while satisfying a property according to Equation 12.1. The procedure is as follows: A structuring element (B) with which origin (O) can be used to perform minimum number of dilations such that $\left(X_j \subseteq \left(X_i \oplus n\widehat{B^O} \right) \right)$ will be determined is given as

$$\Delta(X_i, X_j) = O \left| \min_{\forall O} \left\{ n : X_j \subseteq \left(X_i \oplus n\widehat{B^O} \right) \right\} \right.$$ (12.1)

In Equation 12.1, the letter "O" in B^O, X_i, and $O \left| \min_{\forall O} \left\{ n : X_j \subseteq \left(X_i \oplus n\widehat{B^O} \right) \right\} \right.$, respectively, denote origin for structuring element B, source set X_i, and the directional relationship between the source set (X_i) and target set (X_j). It is worth mentioning that the dilation distance $d(X_i, X_j)$, computed by satisfying the property $\left(X_j \subseteq \left(X_i \oplus n\widehat{B^O} \right) \right)$, between X_i and X_j is not equivalent to $d(X_j, X_i)$. In the section that follows, we provide a morphological framework to determine the directional spatial relation between the sets X_i and X_j separated by finite distance such that the intersection of which is an empty set, $\left(X_i \cap X_j \right) = \varnothing$.

Directional Spatial Relationship

The origin chosen for B to perform the dilation operation according to Equation 12.1 decides the directional spatial relationship between X_i and X_j. Let X_1, X_2, and X_3 (Figure 12.1) be three compact disjoint spatial objects, the intersections of which yield empty sets. From Figure 12.1, it is obvious that X_2 is in southern direction of X_1, and X_3 is in southeastern direction of X_1. We demonstrate origin-specific dilation-distance-based approach to compute directional spatial relationship on these three objects X_1, X_2, and X_3. The shortest dilation distance between X_i and X_j by B with unique origin essentially yields shortest distance than that of other seven origins. Such a unique origin decides the directional relationship between X_i and X_j. We show computations of shortest dilation distances obtained by B with unique origins between the pairs of three objects (Figure 12.1). Figure 12.4 shows these distances, unique origins, and the directional spatial relationships between X_i and X_j. The directions in terms of origin (O) between a set and every other set are depicted in Figure 12.4.

Animation shows how the dilation distances and directional relationships are automatically determined. This animation is also available at http://www.isibang.ac.in/~bsdsagar/AnimationOfDirectionalSpatial Relationship.wmv. Figure 12.4 shows minimum dilation distances computed by means of structuring element with corresponding unique origins, and the corresponding directional spatial relationships with gray-shaded codes between the pairs of three areal objects shown in Figure 12.1. Distance between a state and to itself is zero, and the direction between a state and to itself is central, and gray shade denoting such a central directional relationship is white. The minimum dilation distance between X_1 and X_2 is 53 cycles of dilations with respect to B with origin 2, and hence X_1 is in the northern (N) direction to X_2. Such a directional spatial relationship is shown with gray shade for better visualization. This approach is extended to automatically determine the directional spatial relationships between the pairs of 29 states (areal objects) of India (Figure 12.5a).

Minimum dilation distances			Unique origins				Directional relations				Visualization of directional relations		
	X_1	X_2	X_3		X_1	X_2	X_3		X_1	X_2	X_3		X_1	X_2	X_3
X_1	0	53	50	X_1	0	2	1	X_1	C	N	NW	X_1			
X_2	46	0	36	X_2	6	0	7	X_2	S	C	SW	X_2			
X_3	52	49	0	X_3	5	3	0	X_3	SE	NE	C	X_3			

FIGURE 12.4
Directional spatial relationships via origin-specific dilations. Distances, unique origins, and directional relationships of three objects shown in Figure 12.1.

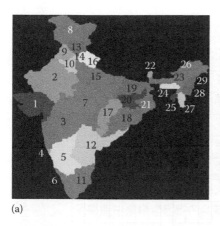 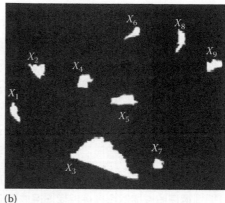

(a)   (b)

FIGURE 12.5

(a) Twenty-nine sets (states of India) indexed are shown—Gujarat (X_1), Rajasthan (X_2), Maharashtra (X_3), Goa (X_4), Karnataka (X_5), Kerala (X_6), Madhya Pradesh (X_7), Jammu & Kashmir (X_8), Punjab (X_9), Haryana (X_{10}), Tamil Nadu (X_{11}), Andhra Pradesh (X_{12}), Himachal Pradesh (X_{13}), Delhi (X_{14}), Uttar Pradesh (X_{15}), Uttaranchal (X_{16}), Chhattisgarh (X_{17}), Orissa (X_{18}), Bihar (X_{19}), Jharkhand (X_{20}), West Bengal (X_{21}), Sikkim (X_{22}), Assam (X_{23}), Meghalaya (X_{24}), Tripura (X_{25}), Arunachal Pradesh (X_{26}), Mizoram (X_{27}), Manipur (X_{28}), and Nagaland (X_{29}). Union territories are not considered. (b) Surface water bodies (Sagar et al. 1995) traced from remotely sensed satellite data.

Experimental Results and Discussion

In this section, we demonstrate the application of dilation-distance-based directional spatial relations on (a) 29 states of India (planar sets) (Figure 12.5a), treating each state as a set (areal object), and (b) surface water bodies (Figure 12.5b) traced from remotely sensed satellite data (Figure 3.19).

States of India as Planar Sets

A total of 29 sets X_i (planar objects) that denote states of Indian peninsula are represented in two-dimensional Euclidean discrete space Z^2 of size 480 × 480 pixels (Figure 12.5a). States and sets are interchangeably used in this study. These sets are considered to explain the proposed approach here. Each state (set) is assigned with index. For instance, the state Gujarat is assigned with index 1, and hence referred to as X_1. Set assigned with index 2 is Rajasthan, X_2. The unique origins, with which the shortest dilation distances (Figure 12.6) between all pairs of sets are computed according to Equation 12.1, are shown in an array of 29 rows and 29 columns (Figure 12.6a). By employing the corresponding data given in Figure 12.6a—also denote the origin of structuring element B with certain characteristic information be chosen between

	X1	X2	X3	X4	X5	X6	X7	X8	X9	X10	X11	X12	X13	X14	X15	X16	X17	X18	X19	X20	X21	X22	X23	X24	X25	X26	X27	X28	X29
X1	0	49	58	54	89	122	80	119	70	55	124	94	80	43	95	68	94	121	127	125	147	135	199	160	167	214	177	189	193
X2	30	0	74	81	116	149	44	70	21	20	151	104	32	9	60	31	58	86	91	90	111	99	164	133	132	178	141	153	158
X3	40	78	0	7	42	75	45	148	90	83	77	41	109	67	80	89	46	63	69	67	89	78	141	111	109	156	119	131	135
X4	92	141	135	0	70	69	107	211	162	146	70	104	172	130	143	152	103	131	136	135	156	144	209	178	177	223	186	198	203
X5	68	117	61	6	0	34	83	187	138	122	35	64	148	106	119	128	68	91	101	99	121	109	169	138	137	183	146	158	163
X6	122	171	94	30	56	0	137	241	192	175	38	82	202	159	173	181	109	110	142	123	144	147	183	152	151	197	160	172	177
X7	48	42	54	60	95	129	0	104	55	39	130	84	65	23	36	45	37	42	47	46	67	55	120	89	88	134	97	109	114
X8	114	87	161	168	203	236	108	0	25	41	238	191	16	36	82	31	144	142	81	102	106	75	136	105	109	150	119	125	130
X9	92	82	139	146	181	214	94	62	0	19	216	169	15	15	63	32	122	120	94	93	114	102	167	136	135	181	144	156	161
X10	73	46	120	127	162	195	87	66	17	0	197	150	27	4	59	28	103	101	91	89	111	99	163	133	131	178	141	153	157
X11	115	164	90	36	47	16	130	235	185	168	0	98	195	152	156	174	102	86	135	111	134	140	151	120	119	165	128	140	145
X12	75	103	41	28	27	45	69	173	124	107	47	0	134	91	105	113	44	36	74	61	84	86	105	75	73	120	83	95	99
X13	98	71	145	152	187	220	92	46	18	25	222	175	0	20	66	20	128	126	80	86	100	88	152	122	120	167	130	142	146
X14	81	133	126	132	167	201	94	82	33	42	202	156	43	0	93	32	108	107	94	93	114	102	167	136	135	181	144	156	161
X15	81	67	83	89	121	154	81	69	27	22	156	109	29	4	0	12	62	60	32	30	52	40	104	74	72	119	82	94	98
X16	90	75	130	137	172	205	83	60	34	30	207	160	21	15	62	0	113	111	63	71	83	71	135	105	103	150	113	125	129
X17	103	91	71	67	62	93	55	132	83	65	94	50	93	52	75	72	0	28	34	32	53	44	106	75	74	120	83	95	100
X18	114	100	82	78	73	94	66	148	99	82	96	80	109	66	80	88	32	0	49	26	53	54	78	48	46	93	56	68	72
X19	138	123	106	101	101	135	90	99	82	78	157	111	69	57	62	52	63	62	0	21	47	15	73	42	41	87	50	62	67
X20	158	123	106	101	101	130	110	123	82	78	136	90	89	57	63	70	43	41	32	0	46	29	74	44	42	89	52	64	68
X21	158	144	126	122	117	184	132	125	103	98	132	94	95	77	77	76	55	50	32	21	0	6	53	22	28	67	38	42	47
X22	180	165	148	143	151	184	114	75	91	82	186	139	106	99	99	90	92	82	55	50	54	0	65	34	42	79	49	54	59
X23	180	178	160	156	154	160	144	134	137	120	159	128	119	116	111	103	89	83	55	55	43	13	0	180	16	16	22	11	8
X24	192	182	165	160	167	162	149	142	141	137	164	133	123	116	116	107	94	83	59	59	64	37	64	0	20	45	27	20	25
X25	208	194	176	172	169	161	160	158	153	148	183	146	135	127	127	119	105	94	71	71	62	37	64	20	0	55	16	22	27
X26	210	196	178	174	175	197	134	150	181	178	165	120	167	143	119	150	120	93	87	89	67	79	49	45	55	0	74	50	41
X27	217	202	185	180	183	160	97	119	144	141	128	83	130	136	82	113	83	56	50	52	38	49	64	28	9	74	0	18	32
X28	224	210	192	188	183	172	109	125	156	153	140	95	142	156	94	125	95	68	62	64	42	45	11	20	22	29	13	0	15
X29	229	215	197	193	188	177	114	130	161	157	145	99	146	161	98	129	100	72	67	68	47	50	74	33	26	41	32	15	0

(a)

FIGURE 12.6

Directional spatial relationships visualized shown in terms of unique origins, and directional relationships for 29 states shown in Figure 12.5a. (a) Dilation distances.

	X_1	X_2	X_3	X_4	X_5	X_6	X_7	X_8	X_9	X_{10}	X_{11}	X_{12}	X_{13}	X_{14}	X_{15}	X_{16}	X_{17}	X_{18}	X_{19}	X_{20}	X_{21}	X_{22}	X_{23}	X_{24}	X_{25}	X_{26}	X_{27}	X_{28}	X_{29}
X_1	NA	7	1	1	1	1	8	7	7	7	1	1	7	7	7	7	1	1	8	8	8	8	8	8	8	8	8	8	8
X_2	3	NA	2	1	1	1	1	7	7	7	1	1	7	7	8	7	1	1	7	1	8	8	8	7	1	8	1	1	1
X_3	5	6	NA	1	1	1	7	7	7	7	1	1	7	7	7	7	8	8	7	7	7	7	7	7	8	7	8	8	7
X_4	6	6	0	NA	8	1	7	7	6	7	1	8	7	7	7	7	8	8	7	7	7	7	7	7	7	7	8	8	7
X_5	5	6	6	3	NA	1	7	6	6	6	1	7	7	7	7	7	7	7	7	7	7	7	7	7	7	7	7	7	7
X_6	6	6	6	2	6	NA	7	7	6	7	8	7	6	6	6	7	1	1	8	8	8	7	8	8	8	7	7	8	8
X_7	4	5	3	2	2	2	NA	7	6	5	1	1	6	6	6	1	1	1	1	1	1	1	1	1	1	1	1	1	1
X_8	3	3	2	2	2	2	2	NA	2	3	2	2	3	3	1	1	1	1	1	1	1	1	1	1	1	1	1	1	1
X_9	3	3	2	2	2	2	2	6	NA	1	1	1	6	1	1	1	1	1	1	1	1	1	1	1	1	1	1	1	1
X_{10}	3	3	2	2	2	2	2	6	5	NA	2	1	6	6	8	8	1	1	1	1	1	1	1	1	1	1	1	1	1
X_{11}	5	6	2	2	2	5	6	6	6	6	NA	1	6	6	7	7	7	7	7	7	7	7	7	7	7	7	7	7	7
X_{12}	5	5	6	2	2	2	6	6	6	6	2	NA	6	6	7	1	7	7	7	7	7	7	7	7	7	7	7	7	7
X_{13}	3	3	2	2	2	2	2	5	3	3	2	2	NA	2	1	1	1	1	1	1	1	1	1	1	1	1	1	1	1
X_{14}	3	0	2	2	2	2	2	6	6	0	2	2	6	NA	8	1	1	1	1	1	1	1	1	1	1	1	1	1	1
X_{15}	3	3	2	3	3	2	3	6	5	5	2	2	8	3	NA	7	2	1	1	1	1	8	1	1	1	4	1	1	8
X_{16}	3	3	3	3	2	2	2	5	4	3	2	2	5	3	2	NA	3	1	7	1	1	1	1	1	1	8	1	1	1
X_{17}	3	4	3	3	2	3	5	5	5	5	3	3	5	5	5	5	NA	1	7	7	7	7	8	8	8	8	8	8	8
X_{18}	5	5	4	3	3	3	5	5	5	5	3	3	5	5	5	5	0	NA	7	7	7	7	1	7	7	8	8	7	7
X_{19}	3	4	3	3	3	3	3	5	5	4	3	3	5	5	5	5	2	7	NA	2	1	1	8	1	8	1	1	1	8
X_{20}	4	4	3	3	3	3	4	5	5	4	3	3	5	5	5	5	3	3	7	NA	8	7	8	8	8	8	8	8	8
X_{21}	4	4	3	3	3	3	4	5	5	4	3	3	5	5	5	5	3	5	5	8	NA	5	8	1	8	1	1	8	8
X_{22}	3	4	3	3	3	3	3	5	5	4	3	3	5	4	4	4	3	3	3	3	2	NA	8	8	8	8	8	8	8
X_{23}	3	4	3	3	3	3	4	5	5	4	3	3	5	4	4	5	3	3	4	3	3	5	NA	6	1	7	1	1	1
X_{24}	3	4	3	3	3	3	4	5	4	4	3	3	5	4	4	4	3	3	4	3	4	5	6	NA	3	7	1	1	1
X_{25}	4	4	3	3	3	3	4	5	5	4	3	3	5	4	4	4	3	3	4	3	4	5	6	3	NA	2	1	1	1
X_{26}	4	4	3	3	3	3	3	5	4	4	3	3	4	4	5	4	3	3	4	4	4	4	6	6	2	NA	1	7	7
X_{27}	4	4	3	3	3	3	4	5	4	4	3	3	5	4	4	4	3	4	4	3	4	5	0	3	4	7	NA	7	7
X_{28}	4	4	3	3	3	3	3	5	4	4	3	3	4	4	4	4	3	3	4	3	4	5	0	6	6	2	2	NA	2
X_{29}	3	4	3	3	3	3	3	5	4	4	3	3	4	4	4	4	3	4	4	3	4	4	0	3	3	3	2	2	NA

(b)

FIGURE 12.6 (continued)

Directional spatial relationships visualized shown in terms of unique origins, and directional relationships for 29 states shown in Figure 12.5a.
(b) Unique origins.

(continued)

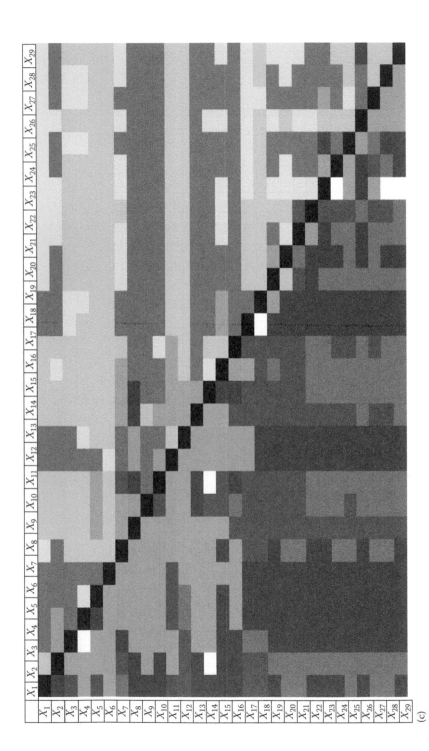

FIGURE 12.6 (continued)

Directional spatial relationships visualized shown in terms of unique origins, and directional relationships for 29 states shown in Figure 12.5a.

(c) Gray shades representing directional relationships between the states.

any two sets and minimum distance between such sets—one can derive the direction between sets of the given order according to Equation 12.1.

It is worth mentioning here that the shortest dilation distance and the direction between set and itself would always be zero. The minimum dilation distance, $d(X_{i,j})$, between each set and the other set is computed in terms of B with unique origin are only shown (Figure 12.6b). To compute $d(X_{i,j})$, we opted a primitive size of 3×3 structuring element with specific origin shown in Figure 12.2a that depicts the directional relationship between the sets in question. The quantitative direction between sets of the given order between each set and every other set can be derived for states of X_i and X_j. The direction between sets (X_1) and (X_2) is obtained by dilating (X_1) with respect to B^7 for 49 cycles such that (X_2) is contained in the version (X_1) dilated for 49 cycles. In this case, X_1 is in the southwestern direction of X_2. Directional relationship can be derived in terms of minimum dilation distances obtained by the unique origin of B. For instance, the minimum number of dilations required, which is only with respect to B with origin $i = 7$, for set X_1 such that set X_2 is completely contained in the dilated version of X_1 is 49. It should be noted that the minimum dilation distance between (X_1) and (X_2) is achieved by B with origin $O = 7$, and hence $d(X_{1,2})$ according to Equation 12.1 is 49, and is in the southwestern direction. The distances between (X_1) and (X_2) by other origins such as $O = 1, 2, 3, 4, 5, 6, 8$ are much higher than 49 obtained by $O = 7$. The required minimum number of dilations, which is only with respect to B with origin $O = 8$, for set X_1 such that set X_7 is completely contained in the dilated version of X_1 is 80.

The higher the values, the larger are the distances between the sets that are subjected to dilation distance computations. For instance, compare to $B \neq \hat{B}$ with other origins, the required minimum number of dilations of set X_1 by B with origin $O = 7$ such that sets $X_2, X_8, X_9, X_{10}, X_{13}, X_{14}, X_{15}, X_{16}$ are contained within the dilated versions with minimum dilations, with B^7, that include 49, 119, 70, 55, 80, 43, and 95 respectively. This process is explained in Equation 12.1. It is obvious from Figure 12.6a that the two closest nonadjacent disjoint sets include (X_3) and (X_4), whereas the two farthest nonadjacent sets include (X_8) and (X_{11}). For instance, the state X_1 is in the southwestern direction of the states $X_2, X_8, X_9, X_{10}, X_{13}, X_{14}, X_{15}$, and X_{16} and is in the western direction of the states $X_7, X_{19}, X_{20}, X_{21}, X_{22}, X_{23}, X_{24}, X_{25}, X_{26}, X_{27}, X_{28}$, and X_{29}. These directional spatial relationships between the states are given in Figure 12.6a through c.

Surface Water Bodies

Nine surface water bodies (Sagar et al. 1995) are traced from remotely sensed satellite data (Figure 3.19) of a small area (5×5 km). By following the approach demonstrated on the model and on 29 states of India (Figures 12.4 and 12.6), the directional spatial relationships are visualized (Figure 12.7).

Unique origins										Directional relations										Visualization of directional relations
	X_1	X_2	X_3	X_4	X_5	X_6	X_7	X_8	X_9		X_1	X_2	X_3	X_4	X_5	X_6	X_7	X_8	X_9	
X_1	0	7	1	7	8	7	1	7	7	X_1	C	SW	NW	SW	W	SW	NW	SW	SW	
X_2	3	0	1	8	1	7	1	7	8	X_2	NE	C	NW	W	NW	SW	NW	SW	W	
X_3	5	5	0	6	6	6	8	7	7	X_3	SE	SE	C	S	S	S	W	SW	SW	
X_4	3	4	2	0	1	7	1	7	8	X_4	NE	E	N	C	NW	SW	NW	SW	W	
X_5	4	5	2	5	0	6	2	7	7	X_5	E	SE	N	SE	C	S	N	SW	SW	
X_6	3	3	2	3	2	0	2	8	1	X_6	NE	NE	N	NE	N	C	N	W	NW	
X_7	5	5	4	5	6	6	0	6	7	X_7	SE	SE	E	SE	S	S	C	S	SW	
X_8	3	3	3	3	3	4	2	0	1	X_8	NE	NE	NE	NE	NE	E	N	C	NW	
X_9	3	4	3	4	3	5	3	5	0	X_9	NE	E	NE	E	NE	SE	NE	SE	C	

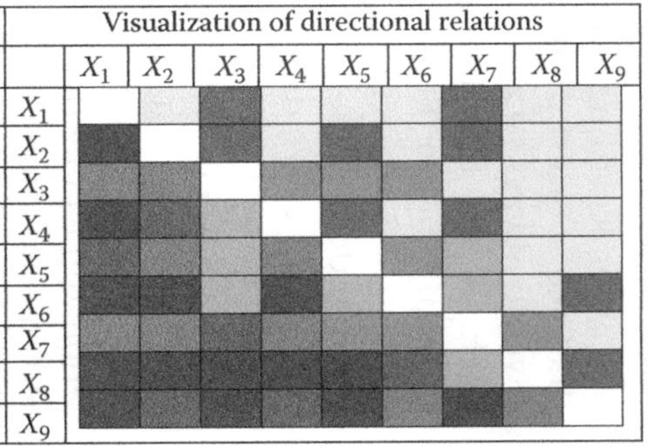

FIGURE 12.7
Directional spatial relationships visualized shown in terms of unique origins, and directional relationships for nine surface water bodies traced from remotely sensed satellite data.

Conclusion

A new approach to compute origin-specific morphological dilation distances between planar sets has been developed to further determine the directional spatial relationship between sets. This approach has been demonstrated on a cluster of states decomposed from a spatial map depicting the country of India and on nine water bodies traced from remotely sensed satellite data. However complex the spatial objects are, this proposed method will work fine.

References

Aksoy, S. and R. G. Cinbis, 2010, Image mining using directional spatial constraints, *IEEE Geoscience and Remote Sensing Letters*, 7(1), 33–37.

Bloch, I., 1999, Fuzzy relative position between objects in image processing: A morphological approach, *IEEE Transactions on Pattern Analysis and Machine Intelligence*, 21(7), 657–664.

Bloch, I., O. Colliot, and R. M. Cesar, 2006, On the ternary spatial relation between, *IEEE Transactions on Systems, Man, and Cybernetics*, 36(2), 312–327.

Bloch, I. and A. Ralescu, 2003, Directional relative position between objects in image processing: A comparison between fuzzy approaches, *Pattern Recognition*, 36(7), 1563–1582.

Cinbis, R. G. and S. Aksoy, 2007, Relative position-based spatial relationships using mathematical morphology, *Proceedings of IEEE International Conference on Image Processing*, San Antonio, TX, September 16–19, Vol. II, pp. 97–100.

Gapp, K. P., 1994, Basic meaning of spatial relations: Computation and evaluation in 3D space, *Proceedings of 12th National Conference on Artificial Intelligence (AAAI-94)*, Seattle, WA, pp. 1393–1398.

Keller, J. M. and L. Sztandera, 1990, Spatial relations among fuzzy subsets of an image, *Proceedings of First International Symposium on Uncertainty Modeling and Analysis*, University of Maryland, College Park, MD, pp. 207–211.

Matsakis, P. and L. Wendling, 1999, A new way to represent the relative position between areal objects, *IEEE Transactions on Pattern Analysis and Machine Intelligence*, 21(7), 634–643.

Rajashekara, H. M., P. Vardhan, and B. S. D. Sagar, 2012, Generation of zonal map from point data via weighted skeletonization by influence zones, *IEEE Geoscience and Remote Sensing Letter*, 9(3), 403–407.

Sagar, B. S. D., 2010, Visualization of spatiotemporal behavior of discrete maps via generation of recursive median elements, *IEEE Transactions on Pattern Analysis and Machine Intelligence*, 32(2), 378–384.

Sagar, B. S. D., G. Gandhi, and B. S. P. Rao, 1995, Applications of mathematical morphology in surface water body studies, *International Journal of Remote Sensing*, 16(8), 1495–1502.

Sagar, B. S. D. and J. Serra, 2010, Spatial information retrieval, analysis, reasoning and
    modelling, *International Journal of Remote Sensing*, 31(22), 5747–5750.
Serra, J., 1982, *Image Analysis and Mathematical Morphology*, Academic Press, London, U.K.
Wilson, A. G., 2009, Remote sensing as the 'X-Ray Crystallography' for urban
    'DNA', *Proceedings of Annual Seminar on Spatial information Retrieval, Analysis,
    Reasoning and Modeling*, ed. B. S. D. Sagar, Bangalore, India, March 18–20,
    pp. 1–12.

13

"Between" Space

To analyze and reason the spatial relationships between planar sets—such as spatially represented countries, states, cities, lakes, and/or binary thresholded images decomposed from function-like grayscale images—across either spatial or temporal scales, mathematically viable explanations are proposed. Sets that represent such varied geographic phenomena possess varied spatiotemporal complexities. Hausdorff dilation distance between such sets is considered to compute spatially significant parameters such as (1) the "between" space, (2) minimum (Hausdorff) and maximum distances, and (3) degree of adjacency, which together provide insights to derive (1) visibility regions, (2) shortest paths between any "two points of two disconnected sets" (e.g., cities of states, states of countries) via the "between" space, (3) shape–size similarity measure between sets, and (4) degree of contextuality between disjoint sets. Superficially simple mathematical morphological operators and certain logical operations are employed to derive these parameters. The model study is shown for trivial and nontrivial cases by incorporating the concavities on visible and/or non-visible sides of the sets. Results drawn—by applying the proposed framework on two case studies that involve (1) spatial sets (states) decomposed from a spatial map depicting country India and (2) two elevation structures of grayscale form— are discussed and demonstrated. The framework thus developed and demonstrated has potential to extend to any number (type) of sets on Euclidean space.

Background on "Between" Space

Drawing spatial relationship between entities that possess several attributes is a challenging task. Entities and their attributes in this section respectively denote sets represented on Z^2 and their shape properties. Let X_1 and X_2 be two compact disjoint spatial objects of which the intersection yields an empty set, and X be the third spatial input of which the degree of containedness in the "between" space determined between X_1 and X_2 is addressed as one of the important parts of this study. What degree of object Y is contained in the "between" space? To answer this seemingly trivial question of GIS relevance, one should have (1) an approach to automatically derive the "between" space of X_1 and X_2 and (2) an approach to quantify the degree of containedness. In Chaudhuri (1990), authors investigated the latter aspect. In this study (Chaudhuri 1990), using fuzzy set theoretic concepts, (1) computation of geometrical and topological properties of planar sets and (2) spatial relationships between sets are addressed. Besides several geometric and topological properties, one of the important spatial relationships considered in this study (Chaudhuri 1990) includes "degree of betweenness." A basic prerequisite to study the set X's degree of betweenness in between the sets X_1 and X_2 (e.g., Figure 13.1a) is to automatically determine the "between" space $\beta(X_1, X_2)$ of sets X_1 and X_2. Here, "between" denotes spatial relation. Studying the spatial relations "between," such sets come under the topic of spatial reasoning. To some extent, initial impetus to such problem was given in Aiello and van Bentha (2002) and Larvor (2004), where coliniariti between points

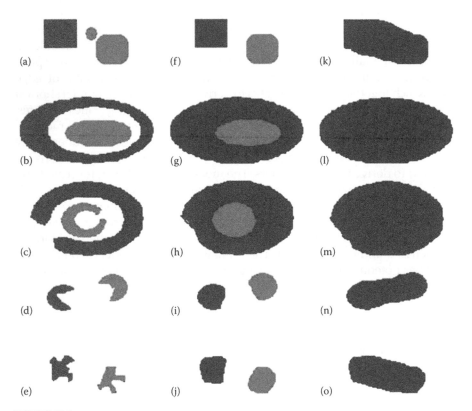

FIGURE 13.1
(a–e) Sets X_1 (in black) and X_2 (in gray) of categories (a–c) 1 and (d and e) 2 and (f–j) their corresponding convex hulls. Note that the pink object (set X in between sets A_1 and A_2 lies on "between" space) and the convex hulls of sets X_1 and X_2 (Figure 13.1f) are respectively similar to their corresponding sets in (a). The sets with concavities that are parts of $\beta(X_1, X_2)$ are subsets of their corresponding convex hulls that are shown in (g–j). Convex hulls of sets A_1 and A_2—that depict these sets contain hidden concavities beside the concavities that are parts of $\beta(X_1, X_2)$—are shown in (j). Convex hulls of individual sets are shown for the figures shown in Figure 13.1f through j. This figure facilitates understanding different categories briefed in the text and (k–o) convex hulls of category-wise set union $(X_1 \cup X_2)$. Note that the unions of sets X_1 and X_2 are contained within their corresponding convex hulls.

or center of spheres was considered, without taking the shape of the objects into account. Definitions of the degree of adjacency of two regions in the plane, and the degree of surroundedness of one region by another, are proposed by Rosenfeld and Klette (1985). Of late, this study is further extended in a series of papers (Bloch 1999, Serra 1982, Beucher and Meyer 1992, Bloch and Ralescu 2003, Bloch et al. 2006). Recursive dilations with certain logical operations are considered in these studies in determining the "between" space of sets. However, sets with concavities on both visible and non-visible sides pose limitations to deal with the approach proposed in these studies. Besides, authors dealt with by means of mixed approaches such as fuzzy morphology (Bloch et al. 2006).

In this chapter, we deal with the determination of "between" space of two sets of varied degrees of spatial complexities entirely by using mathematical morphology. The basis for this study is Hausdorff dilation distance (Beucher 1994, Meyer 1994, Serra 1994, 1998) between two sets. Such distance was earlier used to determine the degree of resemblance between two objects that are superimposed on one another (Huttenlocher et al. 1993) and in spatial interpolations (Sagar 2010). How Hausdorff dilation distance between the sets of varied degrees of spatial complexities forms the basis to analyze and reason the spatial relationships is thoroughly investigated by potential applications and open problems.

This chapter is organized as follows. In the "Methods to Derive the 'Between' Space" section, spatial relationships between sets with different categories of spatial extensions are explained along with morphology-based frameworks to determine the "between" space between sets of varied categories; extension of this framework briefed for binary sets to grayscale fields for the sake of generalization is explained in the "Extension to Grayscale Features" section; experimental results are provided in the "Experimental Results and Discussion" section; in the "Potential Applications" section, once "between" space is properly determined, derivations of the zone of prominent visibility, shortest path between two points via "between" space, shape–size similarity measures, and degree of contextuality between the sets are explained as potential applications; and conclusions drawn are given in the "Conclusion" section.

Spatial Analysis and Reasoning via Hausdorff Distance–Based Morphological Closing

Hausdorff Dilation Distance between Disjointed Compact Sets

According to Serra (1994, 1998), Hausdorff dilation distance, $\rho(X_1, X_2)$, between X_1 and X_2—compact disjointed sets such that $X_1 \cap X_2 = \varnothing$—is given in Equation 10.13, which is $\rho(X_1, X_2) = \inf \left\{ n : [X_1 \subseteq (X_2 \oplus nB)], [X_2 \subseteq (X_1 \oplus nB)] \right\}$.

Methods to Derive the "Between" Space

This section explores a morphological framework that can be applied to determine the "between" space between the two sets X_1 and X_2 separated by finite distance such that the intersection of which is an empty set, $(X_1 \cap X_2) = \varnothing$. Determining the "between" space between such compact disjoint sets relies on the shape–size complexities of sets. Based on spatial extensions between the sets, there are three categories of spatial relationships between the sets that one can intuitively visualize (Figure 13.1a through e). They include the following: (1) sets are simple in the sense that there exists (a) an empty set while we subtract the original set from its convex hull (Figure 13.1a), and (b) no hidden concavities (Figure 13.1b and c); (2) sets are spatially complex in the sense that there exist concavities, on visible and/or hidden sides, associated with both the sets involved, and sets are also dissimilar in sizes (Figure 13.1d and e); and (3) one of the sets is unbounded and is of infinite size with reference to the other set. The convex hulls of individual sets (connected components), for example, Figure 13.1a, and also of the union of the connected components are shown in Figure 13.1a through e, in Figure 13.1f through j, and in Figure 13.1k through o. To verify automatically whether the sets come under which of the three categories mentioned, the following validity tests need to be performed:

Category 1:

$$[CH(X_1)] \setminus (X_1) = \varnothing\dagger \quad \text{and} \quad [CH(X_2)] \setminus (X_2) = \varnothing \qquad (13.1)$$

Category 2:

$$[CH(X_1)] \setminus (X_1) \neq \varnothing, \quad [CH(X_2)] \setminus (X_2) \neq \varnothing,$$

$$\text{and} \quad X_1 \approx X_2 \text{ or } X_2 \approx X_1; \quad X_1 > X_2 \text{ or } X_2 > X_1 \qquad (13.2)$$

Category 3:

$$X_1 \gg X_2 \quad \text{or} \quad X_2 \gg X_1. \quad \text{Either } X_1 = \infty \text{ or } X_2 = \infty \qquad (13.3)$$

In the sections that follow, morphology-based approaches to compute "between" spaces between the companion compact but disjoint sets that fall under these three categories are given. The three categories considered to compute "between" spaces are respectively named as (1) sets without concavities, (2) sets with concavities on visible sides and also sets with concavities on both visible and invisible sides, and (3) sets of contextual type.

Sets without Concavities

If $[CH(X)] \setminus (X) = \varnothing$ (e.g., Figure 13.1a), it is said that there exist no concavities. Under such a circumstance, the "between" space between X_1 and X_2 can be determined by two steps:

Step 1: By simulating Equation 13.4,

$$(X_1 \oplus N_1 B) \quad \text{and} \quad (X_2 \oplus N_2 B) \tag{13.4}$$

By satisfying the following conditions (1)–(10):

1. $(X_1 \cup X_2) \subseteq (X_1 \oplus N_1 B)$

2. $(X_1 \cup X_2) \cap (X_1 \oplus N_1 B) \neq \varnothing$

3. $A[(X_1 \cup X_2) \cap (X_1 \oplus N_1 B)] = A[(X_1 \cup X_2) \cap (X_1 \oplus (N_1 + 1)B)]$

4. $(X_1 \cup X_2) \not\subset (X_1 \oplus (N_1 - 1)B)$

5. $(X_1 \oplus (N_1 - 1)B) \not\subset (X_1 \cup X_2)$

6. $(X_1 \cup X_2) \subseteq (X_2 \oplus N_2 B)$

7. $(X_1 \cup X_2) \cap (X_2 \oplus N_2 B) \neq \varnothing$

8. $A[(X_1 \cup X_2) \cap (X_2 \oplus N_2 B)] = A[(X_1 \cup X_2) \cap (X_2 \oplus (N_2 + 1)B)]$

9. $(X_1 \cup X_2) \not\subset (X_2 \oplus (N_2 - 1)B)$

10. $(X_2 \oplus (N_2 - 1)B) \not\subset (X_1 \cup X_2)$

Step 2: $\min(N_1, N_2)B$ is nothing but the Hausdorff dilation distance $\rho(X_1, X_2)$ between the sets X_1 and X_2. Equation 13.5 yields the "between" space of sets X_1 and X_2 by suppressing the sets X_1 and X_2 through the subtraction process (Figure 13.2a):

$$\beta(X_1, X_2) = \left[(X_1 \cup X_2) \bullet \frac{\rho}{2} B \right] \setminus [(X_1 \cup X_2)] \tag{13.5}$$

Sets with Concavities within "Between" Space

Step 1: If $[CH(X)] \setminus (X) \neq \varnothing$ (e.g., Figure 13.1b through d and g through i), the between space can be identified automatically by a set of morphological equations. Let $CH(X_1 \cup X_2)$ be the convex hull of $(X_1 \cup X_2)$ (e.g., Figure 13.1k through o), and medial axes [5, 7] of sets X_1 and X_2, respectively, be MX_1 and MX_2. Hausdorff distance between MX_1 and MX_2 is computed as follows.

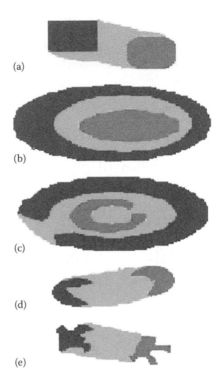

FIGURE 13.2
(a–c) Obtained after suppressing sets X_1 and X_2 from corresponding convex hulls of $(X_1 \cup X_2)$ to directly determine "between" space (in lighter shade). Hence these categories are treated as category 1. This direct determination is possible as no concavities exist in the hidden side(s) of set(s) X_1 and/or X_2 and (d) concavities are present in the visible sides to each other, but not on the hidden sides. Hence medial axes of X_1 and X_2 are closed—by means of $\rho/2(B)$—from which X_1 and X_2 are suppressed to obtain "between" space $\beta(X_1, X_2)$ shown in lighter shade, and (e) concavities exist on both visible and hidden (invisible) sides of X_1 and X_2, and hence directional (geodesic) closing is performed with certain conditions to obtain "between" space (in lighter shade).

Step 2: Hausdorff distance between the medial axes of sets: Dilate set (MX_1) by (B) for (N_1) cycles as shown in Equation 13.6, such that its intersection with (MA_2) yields a nonempty set as shown in the following by satisfying conditions depicted in (1)–(4):

$$(MX_1 \oplus N_1 B)\dagger \ \text{ and } \ (MX_2 \oplus N_2 B) \tag{13.6}$$

1. $[(MX_1) \oplus N_1 B] \cap (MX_2) \neq \varnothing$
2. $\mathrm{A}\left\{[(MX_1) \oplus (N_1 + 1)B] \cap (MX_2)\right\} = \mathrm{A}\left\{[(MX_1) \oplus N_1 B] \cap (MX_2)\right\}$
3. $[(MX_2) \oplus N_2 B] \cap (MX_1) \neq \varnothing$
4. $\mathrm{A}\left\{[(MX_2) \oplus (N_2 + 1)B] \cap (MX_1)\right\} = \mathrm{A}\left\{[(MX_2) \oplus N_2 B] \cap (MX_1)\right\}$

$\min(N_1, N_2)B$ is nothing but the Hausdorff distance $\rho(MX_1, MX_2)$ between the medial axes of sets and satisfies the following properties:

1. $\left\{ \left[(MX_1) \cup (MX_2) \right] \cdot \dfrac{\rho(MX_1, MX_2)}{2} B \right\} \subseteq CH(X_1 \cup X_2)$

2. $\left\{ \left[(MX_1) \cup (MX_2) \right] \cdot \dfrac{\rho(MX_1, MX_2)}{2} B \right\} \subseteq CH(MX_1 \cup MX_2)$

Step 3: Equation 13.7 yields the "between" space of sets X_1 and X_2 of category 2 sets that possess concavities within the "between" space:

$$\beta(X_1, X_2) = \left\{ \left[(MX_1) \cup (MX_2) \right] \cdot \frac{\rho(MX_1, MX_2)}{2} B \right\} \cap CH(MX_1 \cup MX_2) \quad (13.7)$$

One can obtain the between space by suppressing the common information, if any exists, between the outcome yielded from Equation 13.7 and $(X_1 \cup X_2)$. This process takes care of finding out the "between" space of the sets with concavities (Figure 13.2b through d) and also of dissimilar sizes. However, this approach fails if the $\rho(MX_1, MX_2)$ is greater than or equal to the distance between any two outer extremities of any of these two medial axes of the sets as $\left[(MX_1) \cup (MX_2) \right] \cdot \dfrac{\rho(MX_1, MX_2)}{2} B$ transformation closes the concavities.

Sets with Concavities in All the Sides

In the situation (e.g., Figure 13.1e) that earlier three-step approach fails to compute the "between" space due to the reason mentioned, an asymmetric structuring element needs to be derived and considered. The procedure is as follows.

Step 1: A structuring element (B_1) with which origin (i) can be used to perform minimum number (N_1) of dilations such that $\left[X_2 \subseteq \left(X_1 \oplus N_1 \widehat{B_1^i} \right) \right]$ will be determined. Similarly, a structuring element (B_2) with which i can be used to perform minimum number (N_2) of dilations such that $\left[X_1 \subseteq \left(X_2 \oplus N_2 \widehat{B_2^i} \right) \right]$ will also be determined as shown in Equation 13.8:

$$\left[X_2 \subseteq \left(X_1 \oplus N_1 \widehat{B_1^i} \right) \right] \quad \text{and} \quad \left[X_1 \subseteq \left(X_2 \oplus N_2 \widehat{B_2^i} \right) \right] \quad (13.8)$$

Note that $\left(B_1^i \right)$ and $\left(B_2^i \right)$ associated with (N_1) and (N_2) are not necessarily with the similar origin and size.

Step 2: $\min \left(N_1, N_2 \right) B_n^i$ is nothing but the Hausdorff distance $\rho(X_1, X_2)$ between the sets X_1 and X_2 that possess concavities in all the sides.

Step 3: $\left\{ \left[(X_1) \cup (X_2) \right] \bullet \rho(X_1, X_2) \widehat{B_n^i} \right\}$ yields the convex hull $CH(X_1 \cup X_2)$ of $(X_1 \cup X_2)$.

Step 4: Derivation of partial convex hulls $(CH^p(X_1))$ and $(CH^p(X_2))$ is done to avoid closing the concavities, if any exist, that are parts of the "between" space in the following manner:

$$CH^p(X_1) = \left(X_1 \bullet K_1 \widehat{B_2^i} \right) \cap [CH(X_1)] \neq \varnothing$$

such that

$$A\left\{ \left(X_1 \bullet (K_1+1)\widehat{B_2^i} \right) \cap [CH(X_1)] \right\} = A\left\{ \left(X_1 \bullet K_1 \widehat{B_2^i} \right) \cap [CH(X_1)] \right\}.$$

Similarly, $CH^p(X_2) = \left(X_2 \bullet K_2 \widehat{B_1^i} \right) \cap [CH(X_2)] \neq \varnothing$

such that

$$A\left\{ \left(X_2 \bullet (K_2+1)\widehat{B_1^i} \right) \cap [CH(X_2)] \right\} = A\left\{ \left(X_2 \bullet K_2 \widehat{B_1^i} \right) \cap [CH(X_2)] \right\}$$

Step 5: Equation 13.9 yields the "between" space of sets X_1 and X_2 that possess the concavities both within "between" space and hidden spaces (e.g., Figure 13.2e):

$$\beta(X_1, X_2) = CH(X_1 \cup X_2) \setminus [CH^p(X_1) \cup CH^p(X_2)] \tag{13.9}$$

This process takes care of finding out the "between" space of the sets with concavities, by protecting the concavities that should be the part of "between" space of sets X_1 and X_2.

Derivation of Contextual Type of "Between" Space between Sets of Category 3

Step 1: Let $\left(B_i^i \right)$ be an asymmetric structuring element with origin i (i.e., at one of the eight neighborhood positions of B, [Figure 13.3]) of primitive size 3×3, and X_1 and X_2 be sets, respectively, of bounded and unbounded compact disjoint sets (e.g., Figure 13.4a). Then determine B with which origin (i) requires least number of dilations (Equation 13.10).

$$\left(X_1 \oplus K_1 \widehat{B_1^i} \right) \tag{13.10}$$

1	2	3
8	0	4
7	6	5

FIGURE 13.3

Indexes (i) of eight possible origins chosen to perform directional dilations appropriately at different contexts in this work.

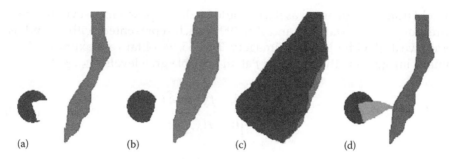

(a)   (b)   (c)   (d)

FIGURE 13.4

(a) Sets X_1 (in black) and X_2 (in dark gray) of which the set X_2 is significantly larger than the other—category 3, (b) convex hulls of X_1 and X_2, and (c) convex hulls of union of X_1 and X_2; the zones represented with contrasting shade are parts of convex hulls, and they depict the hidden concavities in Figure 13.4a and d contextual-type "between space" (in lighter gray) obtained between X_1 and X_2.

to satisfy the following three conditions (i–iii):

i. $\left(X_1 \oplus K_1 \widehat{B_1^i}\right) \cap (X_2) \neq \varnothing$

ii. $X_2 \not\subset \left(X_1 \oplus K_1 \widehat{B_1^i}\right)$

iii. $\left(X_1 \oplus (K_1 - 1)\widehat{B_1^i}\right) \cap X_2 = \varnothing$

Step 2: Then the portion(s) obtained through intersection of set (X_2) and the set (X_1) dilated by $\left(B_1^i\right)$ for (K_1) times—in other words $\left(X_1 \oplus K_1 \widehat{B_1^i}\right)$—is (are) considered. Here, $\left(B_1^i\right)$ with origin ($i = 1, 2, ..., 8$) is crucial to further perform the directional (conditional) dilation to derive the "between" space of contextual category.

Step 3: The "between" space of contextual category is computed by Equation 13.11:

$$\beta_C(X_1, X_2) = \left\{\left[\left(X_1 \oplus K_1 \widehat{B_1^i}\right) \cap (X_2)\right] \bullet \left(K_1 \widehat{B_1^{i\pm4}}\right)\right\} \backslash (X_1 \cup X_2) \qquad (13.11)$$

The direction of B to perform the closing operation will be determined by adding or subtracting 4 with ith origin determined at step 1 of B. We add 4, if $(i \leq 4)$; on the contrary (i.e., if $i \geq 5$), we subtract 4.

Extension to Grayscale Features

A field (function), denoted as $f(x, y)$, could be decomposed into several binary thresholded sets (Margaos and Ziff 1990), each represented with 1s and 0s. Since we deal with 8 bit/pixel imagery, $T = 255$, we obtain a maximum of 255 binary images—by thresholding f at all possible gray levels $0 \leq t \leq T$:

$$f^t(x, y) = \begin{cases} 1, & f(x, y) \geq t \\ 0, & f(x, y) < t \end{cases} \tag{13.12}$$

For notational simplicity, we denote threshold f as f^t, and f can be reconstructed from binary thresholded images:

$$f(x, y) = \sum_{t=1}^{T} f^t(x, y) \tag{13.13}$$

$$= \max\{t : f^t(x, y) = 1\} \tag{13.14}$$

For more details of threshold superposition, see Maragos and Schafer (1986). Let f_1 and f_2 (Figure 13.5a) be two functions (e.g., digital elevation maps, buildings, temperature, and rainfall fields), which could respectively be decomposed into maximum of $T = 255$ (for 8 bit/pixel images) thresholded sets. Let the *infimum* of these two functions be zero (empty set). Then to compute the "between" region, $\beta(f_1, f_2)$, the functions f_1 and f_2 need to be threshold decomposed. For simplicity, we denote binary threshold sets decomposed from f_1 and f_2, respectively, as f_1^t and f_2^t, where $0 \leq t \leq T$. Depending upon the spatial relationships between the corresponding threshold-decomposed sets f_1^t and f_2^t—of varied degrees of spatial complexities—of f_1 and f_2, one can figure out the appropriate category according to Equations 13.1 through 13.3. Depending upon the category, $\beta(f_1^t, f_2^t)$ could be derived according to respective equations mentioned in the "Sets without Concavities," "Sets with Concavities within 'Between' Space," "Sets with Concavities in All the Sides," and "Derivation of Contextual Type of 'Between' Space between Sets of Category 3" sections. Once "between" spaces are determined between all the corresponding

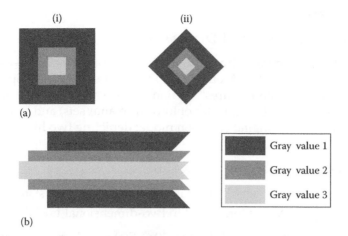

(i)   (ii)

(a)

(b)

Gray value 1

Gray value 2

Gray value 3

FIGURE 13.5
(a) Two grayscale functions, namely, f_1 (i) and f_2 (ii), in which three spatially distributed regions are shown with shades, and (b) the superposed "between" spaces, between the corresponding binary thresholded images obtained according to Equations 13.12 and 13.13, yield "between" region between the two functions shown in (a).

binary threshold sets respectively decomposed from f_1 and f_2, $\beta(f_1, f_2)$ can be computed according to

$$\beta(f_1, f_2) = \bigvee_{\forall t \geq t} \left[\beta\left(f_1^t, f_2^t \right) \right] \tag{13.15}$$

This process is demonstrated on two synthetic functions containing three gray levels (Figure 13.5a, i and ii). The gray levels for both f_1 and f_2 range from 1 to 3. f_1 and f_2 are convex functions, as all the three binary threshold images decomposed from each function are convex attributing to the category 1 type of sets. As these thresholded sets f_1^1, f_1^2, and f_1^3 of ordered form and their corresponding ordered sets from f_2 that include f_2^1, f_2^2, and f_2^3, respectively, are of similar size, Equation 13.5 is adapted to determine $\beta\left(f_i^t, f_2^t \right)$. The "between" region (Figure 13.5b) is obtained according to Equations 13.15. However, for better visualization, one can color code these binary threshold-region-wise "between" spaces systematically and superpose them. This process is explained for two synthetic functions (Figure 13.5). It is worth mentioning here that if gray-level ranges of f_1 and f_2 significantly differ, the "between" region would be of contextual type (category). "Between" space and "between" region, respectively, deal with sets and functions. Air corridor is similar to "between" region. "Between" region, $\beta(f_1, f_2)$, is well within the convex hull of supremum of two grayscale functions, $(f_1 \vee f_2)$, $CH(f_1 \vee f_2)$. One can compare the "between" region with the region that is obtained from suppressing $(f_1 \vee f_2)$ from $CH(f_1 \vee f_2)$.

Experimental Results and Discussion

In this section, we demonstrate the application of formalism—explained on synthetic cases in the "Methods to Derive the 'Between' Space" and "Extension to Grayscale Features" sections—on realistic cases that respectively include planar set-case (states of India as planar sets) and function-like grayscale image-case (digital elevation model depicting two hills).

States of India as Planar Sets

A total of 28 number of sets X_i (planar objects) that denotes states of Indian peninsula, $I = \bigcup_{i=1}^{28} X_i$, is represented in two-dimensional Euclidean discrete space Z^2 of size 480 by 480 pixels (Figure 10.4). States and sets are interchangeably used in this study. These sets are considered to explain the proposed approach here. Each state (set) index is assigned according to alphabetical order (perhaps there is a better approach via scanning mechanism from top to bottom). In turn, the state Andhra Pradesh is assigned with index 1, and hence referred to X_1. Set assigned with index 2 is Arunachal Pradesh, X_2. The sets with their corresponding indexes are illustrated in Figure 10.4. It is observed that most of the set relationships fall under the type-2 of category 2, explained in the "Sets with Concavities in All the Sides" section, due to the fact that the concavities exist in all the sides. Hence, we followed the five-step approach to compute the "between" space (Figure 13.6). These indexed sets and each set's adjacent set(s), minimum and maximum Hausdorff distances by dilation between the pairs of sets—with an exception that pair of sets are not adjacent to each other—"between" spaces between the all possible pairs of sets, and the degrees of containedness of the adjacent sets within the computed "between" space(s) are shown in array forms (Figure 10.5). It is worth mentioning here that distance, direction, adjacency, and contextuality between set and itself would always be zero.

At first instance, the adjacent states are defined in such a way that the following two conditions are satisfied: (1) $\left(X_i \cap X_{i\pm n}\right) = \varnothing$ and (2) $\left[\left(X_i \oplus B\right) \cap \left(X_{i\pm 1}\right)\right] \neq \varnothing,\ \forall i \in I$. The states that possess these two conditions are considered as adjacent states. For 28 states of India, we computed first the adjacent states to each state and to every other state by taking the two conditions into account and recorded in an array of 28 rows and 28 columns (Figure 10.5a). If there exists a numerical value 1, it denotes that there exists adjacency relationship between the states. Otherwise, they are treated as nonadjacent (the sets are situated far apart). Each state is indexed with set notation $X_1, X_2, \ldots, X_{28}$. For instance, for state X_1, the adjacent states include $X_5, X_{12}, X_{15}, X_{20}, X_{24}$ (Figure 10.5a).

Eight possible origins (i) involved with a 3×3 square structuring element are shown in Figure 13.3. Different origins of B (Figure 13.3) employed to

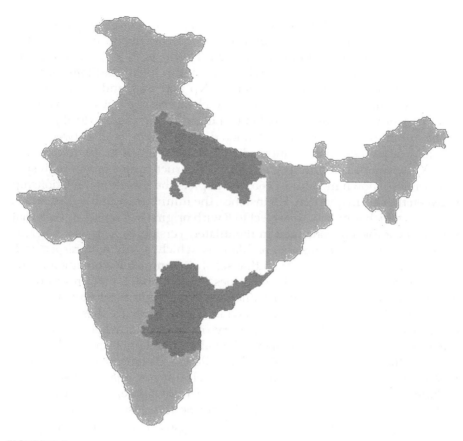

FIGURE 13.6
"Between" space (in white shade) computed between sets X_1 and X_{26} (in dark gray shades). For X_1 and X_{26}, refer to Figure 10.4.

generate $\beta(X_i, X_{i+n})$ range from $i = 1, 2, ..., 8$. It should be noted that if $B^i \neq \widehat{B^i}$, then it is asymmetric structuring element. In effect, morphological dilation is not equivalent to Minkowski addition as was the case with symmetric structuring element. We need to perform morphological dilation transformation on set X_i with respect to $\widehat{B^i}$. In effect, the equation takes the form of $X_i \oplus \widehat{B^i}$. For instance, if we choose origin $i = 3$, B^3, which is in the northeast corner point in 3×3 size structuring element, the portion that is dilated is in the direction of northeastern side. With origin $i = 1$, if we perform the dilation, the dilated portion would be seen in the northwestern direction of the set. The directions in terms of origin (i) between each set and every other set are depicted in Figure 10.5b. For instance, compared to B with other origins, the required minimum number of dilations of set X_1 by B with origin $i = 1$—such that set X_{10} is contained within the dilated version—includes 248 (N_1), whereas the required number of dilations of set X_{10} by B

with origin $i = 5$ is 274 (N_2). In turn, the direction relationships between sets X_1 and X_{10}, and X_{10} and X_1 in terms of i, respectively, are $i = 2$ and 6. This process is explained in Equation 12.1. Hence, $\beta(X_1, X_{10})$ is obtained by following steps (3, 4) of the "Sets with Concavities in All the Sides" section. The origins associated with B to derive (N_1) and (N_2), and the minimum number of dilations required to use it to close the gap between every set and every other set, are represented in a matrix form (Figure 10.5b).

The minimum and maximum dilation distances, $\rho(X_i, X_{i \pm n})$, between each set and every other set are computed and shown (Figure 10.5b). To compute $\rho(X_i, X_{i \pm n})$, we opted a primitive size of 3×3 structuring element with specific origin—shown in Figure 12.6b that depicts the directional relationship between the sets in question. For instance, the minimum number of dilations required, which is only with respect to B with origin $i = 2$, for set A_1 such that set X_{10} is completely contained in the dilated version of X_1 is 248, whereas the required minimum number of dilations, which is only with respect to B with origin $i = 6$, for set X_{10} such that set X_1 is completely contained in the dilated version of X_{10} is 274. These values 248 (N_1) and 274 (N_2) respectively denote minimum and maximum dilation distances in pixel units between sets X_1 and X_{10}. The values represented in Figure 10.5b are these Hausdorff distances between each set X_i and every other set $X_{i \pm n}$, where "i" and "n" range between 1 and 28. The higher the values, the larger are the distances between the sets that are subjected to dilation distance computations. It is obvious from Figure 10.5b that the two closest nonadjacent disjoint sets include X_{12} and X_6, whereas the two farthest nonadjacent sets include X_{19} and X_7. From Figure 10.5b, one can determine the closest (farthest) sets to any specific set. Geodesic distances between each set and $I = \bigcup_{i=1}^{28} X_i$ can be visualized according to Equation 13.16:

$$\bigcup_{\substack{i=1 \\ n=1}}^{\substack{N \\ 28}} [(X_i \oplus nB) \cap (I)] \tag{13.16}$$

As an example, we show the geodesic contours (Figure 10.5e) within India (considered as a set) depicting the equal-distance lines from a set X_1.

Hausdorff distance (ρ) is the minimum of N_1 and N_2 and hence (ρ) is considered as 248. This (ρ) is taken as the basis to perform morphological closing of image $M = (X_1 \cup X_{10})$. Closing of $\left[(X_1 \cup X_{10}) \bullet \rho \widehat{B^3} \right] = \left\{ \left[(X_1 \cup X_{10}) \oplus \rho \widehat{B^3} \right] \ominus \left(\rho \widehat{B^3} \right) \right\}$ determines the close-hull of the union of sets X_1 and X_{10}. By doing so, only the region in between the sets X_1 and X_{10} would be closed by ignoring the invisible concavities that exist in the sets X_1 and X_{10}. By suppressing sets X_1 and X_{10} from $\left[(X_1 \cup X_{10}) \bullet \rho \widehat{B^3} \right]$, $\beta(X_1, X_{10})$ is determined. The Hausdorff dilation distances between the sets can be seen in an array form (Figure 10.5b). Then, the regions between all possible sets are closed by taking the structuring elements of diameter equivalent to Hausdorff dilation distance with corresponding origins (Figure 12.6b). Precisely, this process is explained in steps 3 and 4 of category 3 (the "Sets with Concavities in All the Sides" section). Corresponding sets are further suppressed from such closed sets to obtain the "between" space. As there are 28 sets (states of India), there are 784 possible set combinations, and hence similar number of possible "between" spaces.

By employing the corresponding data given in Figures 12.6b and 10.5b—respectively denote the origin of structuring element B with certain characteristic information be chosen between any two nonadjacent sets and Hausdorff distance between such nonadjacent sets—one can derive the "between" space according to Equation 13.5. The "between" spaces between each set and every other set, with an exception to adjacent set(s), can be extracted as illustrated for one case of nonadjacent states of X_1 and X_{26} in Figure 13.6. In the process of determining the "between" space between the nonadjacent disjoint states, we avoided the states that are adjacent to certain states. However, the "between" space, for instance, between X_1 and X_{26}, traverses some of the adjacent sets that include X_5, X_{11}, X_{14}, X_{15}, X_{20}, and X_{22}. The "between" space between sets X_1 and X_{26} is obtained by choosing $\widehat{B^2}$ of diameters 152 pixels in Equation 13.5. The Hausdorff dilation distance computed by B with origin $i = 2$, between sets X_1 and X_{26}, is 152 pixels; this distance by B with origin $i = 6$ between sets X_{26} and X_1 is 145 pixels. The distance 145 being the minimum of 152 and 145, we have chosen $\widehat{B^6}$ of diameter 145 pixels to obtain the "between" space (Figure 13.6f). In a similar way, by making use of the relationships between sets in terms of direction (determined by origin of B) and Hausdorff distance, one can obtain the "between" space between any two nonadjacent disjoint sets.

The largest distance computed between the innermost extremity to outermost extremity of any two sets is obvious (Figure 10.5b) for the sets (states) in the order mentioned A_{19} (Nagaland) and X_7 (Gujarat). This distance is 383 pixels. This distance between the states of Jammu & Kashmir (X_{10}) and Tamil Nadu (X_{24}) is 343 pixels, whereas the smallest distance computed between the innermost extremity to outermost extremity of any two sets is for the states in the order mentioned X_{12} (Karnataka) and X_6 (Goa). This distance is 8 pixels. For instance, sets X_6, X_{16}, X_{23}, and X_{25} are significantly smaller than the many other sets. The "between" space that could be derived between one of these smaller sets and any other significantly larger set yields contextual type "between" space explained in the "Derivation of Contextual Type of 'Between' Space between Sets of Category 3" section and Figure 13.4.

Elevation Structures as Grayscale Functions

To demonstrate the application of procedure to obtain the "between" region, explained on synthetic functions (Figure 13.5) in the "Extension to Grayscale Features" section, in realistic context, we chose a part of raster

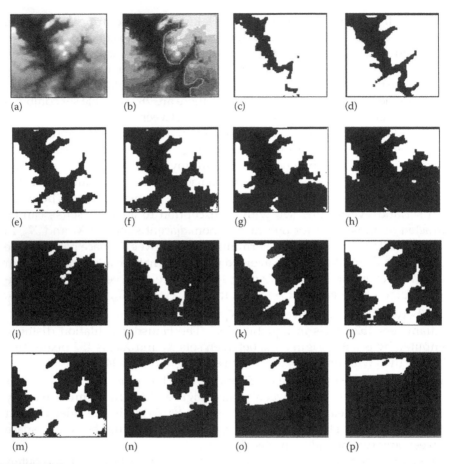

FIGURE 13.7
(a) DEM of size 127 × 113 pixels, where the elevations are depicted in terms of equalized gray levels ranging from 0 to 255; (b) density-sliced image that depicts only seven spatially distributed elevation regions—this is done to demonstrate the "between region" computation with minimum-threshold-decomposed sets—the regions embedded within dark-line boundary and brighter line boundary, respectively, depict functions 1 and 2, in other words f_1 and f_2, (c–i) threshold elevation regions decomposed from both the functions with threshold values ranging from 1 to 7—(c) f_1^1, f_2^1; (d) f_1^2, f_2^2; (e) f_1^3, f_2^3; (f) f_1^4, f_2^4; (g) f_1^5, f_2^5; (h) f_1^6, f_2^6; (i) f_1^7, f_2^7; and (j–p) the "between" space computed respectively for the threshold regions between the corresponding two threshold elevation regions of the two functions shown in Figure 13.7c through i.

DEM of size 127×113 pixels (Figure 13.7a)—taken from Shuttle Radar Topographic Mission (SRTM) Digital Elevation Model (Jarvis et al. 2008) situated between the geographical coordinates of 20°–25°N latitudes and 80°–85°E longitudes—depicting two functions that respectively represent two isolated hilly regions. The region between these two elevation structures, which are parts of two distinct hills separated by a valley, contains no-elevation regions. The region between such elevation structures, here referred to as functions, is a kind of corridor.

The "between" space between the corresponding binary thresholded images decomposed (Equations 13.12 through 13.14) respectively from the two functions are computed. There are 7 and 7 spatially distributed binary thresholded images that could be respectively decomposed from the two functions. These thresholded sets are ordered sets as always f_1^{i+1} are contained within f_1^i. It is also true with f_2. The spatial relations, between the seven companion threshold elevation sets (TESs) decomposed respectively from the two functions shown in Figure 13.7b with clear demarcation, enable that these TESs belong to category 2 (Figure 13.7c through i). The concavities exist only in the visible sides of corresponding TESs— f_1^1, f_2^1; f_1^2, f_2^2; f_1^3, f_2^3; f_1^4, f_2^4; f_1^5, f_2^5; f_1^6, f_2^6; and f_1^7, f_2^7. The Hausdorff dilation distances computed in pixel units respectively for the companion TESs include 20, 32, 36, 36, 40, 76, and 68. Hence, the "between" spaces between companion TESs of these two functions are computed by following Equation 13.5. The "between" spaces are shown in Figure 13.7j through p and also in Figure 13.8.

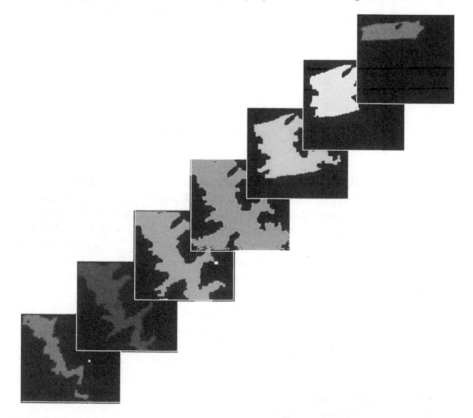

FIGURE 13.8
"Between" spaces, computed between the corresponding threshold elevation regions decomposed (Figure 13.7c through i) respectively, denote for f_1^1, f_2^1; f_1^2, f_2^2; f_1^3, f_2^3; f_1^4, f_2^4; f_1^5, f_2^5; f_1^6, f_2^6; and f_1^7, f_2^7 shown in different gray shades for better visualization.

These "between" spaces are superposed according to Equation 13.15 to finally obtain the "between" region (Figure 13.8) of two functions. The "between" region between the two functions is always a part contained within the grayscale convex hull of maximum (supremum) of two functions.

Potential Applications

Visibility Region (Line of Sight) within "Between" Space

$[CH(X_1)] \setminus (X_1)$ and $[CH(X_2)] \setminus (X_2)$ provide the concavity zones where one should not stand to miss the sight of a person searching for other person from either of the sets. These steps take care of all the concavities of both visible and non-visible sides of each set; hence one can apply this algorithm that relies entirely on mathematical morphological transformations.

$$V(X_1, X_2) = \left[\bigcup_{n=0}^{N} (CH(X_1) \oplus nB) \cap (\beta(X_1, X_2)) \right] \qquad (13.17)$$

$$V(X_2, X_1) = \left[\bigcup_{n=0}^{N} (CH(X_2) \oplus nB) \cap (\beta(X_1, X_2)) \right] \qquad (13.18)$$

Iterative dilations that are needed to perform will be terminated the moment the intersection of $(CH(X_1) \oplus nB)$ and $\beta(X_1, X_2)$ yields empty set (Equations 13.17 and 13.18). The maximum number of iterative dilations involved is denoted as N. The smaller the N involved in obtaining $V(X_1, X_2)$ and $V(X_2, X_1)$, the smaller the visible region. If N involved in obtaining $V(X_1, X_2)$ is smaller than that of $V(X_2, X_1)$, then it is easier to locate a person—standing somewhere on $\beta(X_1, X_2)$—from set X_1 to X_2 than the contrary. However, the zones where visibility is rather obscure, perhaps due to concavities and shadow zones, can be derived by subtracting $(X_1 \cup X_2)$ from $V(X_1, X_2)$ and $V(X_2, X_1)$. The visibility depends on the (1) Hausdorff distance between the two sets and (2) set from which one is viewing. The higher the Hausdorff distance between the sets, the larger the zone of visibility, and *vice versa*.

Path between Two Points (Cities) of Nonadjacent
Sets (States) via "Between" Space

Once "between" space between X_i and $X_{i \pm n}$ is determined—by assuming that X_i and $X_{i \pm n}$ are the two states of a country, from any point (city) (pX_i) of set

X_i to any point $(pX_{i\pm n})$ of set $X_{i\pm n}$—the shortest path (R) is always a geodesic distance that can be derived according to Equation 13.19:

$$R_{CH^P(X_i, X_{i\pm n})}(pX_i) = \min_{\lambda \geq 0}\left\{\lambda : \begin{bmatrix} (pX_{i\pm n}) \subseteq [(pX_i \oplus \lambda B) \cap [CH^P(X_i, X_{i\pm n})]]; \\ [(pX_i) \subseteq [(pX_{i+n} \oplus \lambda B) \cap [CH^P(X_i, X_{i\pm n})]]] \end{bmatrix}\right\}$$

(13.19)

where (pX_i) and (pX_{i+n}) denote points from sets X_i and $X_{i\pm n}$, respectively, $CH^P(X_i, X_{i\pm n}) = \left\{[\beta(X_i, X_{i+n})] \cup [X_i \cup X_{i+n}]\right\}$. From the application context, they respectively denote cities of states X_i and X_{i+n}. The aforementioned equation yields a geodesic path between the two cities, which belong to two states, via $\beta(X_i, X_{i\pm n})$. This phase of work offers insights to deal with route planning strategies.

Further, some foreseen applications include the following. $\beta(X_1, X_2)$ between any two countries X_1 and X_2 on a map represented in Cartesian coordinate system helps planners to (1) decide which are the countries or parts of the countries that are parts of the "between" space and (2) find out the shortest and/or cost-effective route (e.g., air, sea, land) between X_1 and X_2. Shortest route—between the two objects (X_1, X_2) of which the intersection yields an empty set—should traverse $\beta(X_1, X_2)$. Between the two closely spaced channel segments, say X_1 and X_2, there exists a ridge path that is the part of $\beta(X_1, X_2)$. If a robot is instructed to place an object between two other objects in natural language, such quantification may be useful (assuming that the robot has a language-understanding module). Some of the other potential applications of this study in brief include transportation studies to determine corridors, seaport and airport location, pipeline corridor studies, floodplain and floodway determination, tsunami inundation zones, bluff erosion studies, line of sight for communication tower location, water and sewer system design in cities, aircraft navigation and safety, and geological studies in mineralized areas.

Shape–Size Similarities between Sets

Let mutually exclusive X_i and $X_{i\pm n}$ denote smaller and bigger nonempty compact sets, respectively. One can compute N_1 and N_2 according to Equations 13.20 and 13.21:

$$N_1 = \inf_{\forall n \geq 0}\{n : (X_i \oplus nB) \supseteq X_{i\pm n}\}$$

(13.20)

$$N_2 = \inf_{\forall n \geq 0}\{n : (X_{i\pm n} \oplus nB) \supseteq A_i\}$$

(13.21)

From the aforementioned equations, one can understand that for a smaller object to completely occupy a relatively bigger object, it requires more number of dilation cycles than that of bigger object to completely occupy the smaller object. If there exists shape–size dissimilarities between the two sets under investigation, one can observe that $N_1 > N_2$. This observation is true with all the states, between which the distances are computed according to Equations 13.20 and 13.21, further facilitating a way to compute an index that describes shape–size relationships between sets. $N_1 = N_2$ if and only if $X_i = X_{i\pm n}$. This further extends to compute the degree of contextuality.

Degree of Contextuality between Nonadjacent Disjoint Sets

The degree of contextuality, $(DC_{i,j})$, for all $i = 1$–28 and $j = 1$–28, between the two sets, say X_i and X_j (i and j being columns and rows), can be computed according to

$$DC_{i,j} = \frac{d_{\min} = N_1 = \rho}{d_{\max}} \tag{13.22}$$

where $d_{\min} = \min\{d : d(X_i, X_j), d(X_j, X_i)\}$ and $d_{\max} = \max\{d : d(X_i, X_j), d(X_j, X_i)\}$; $d_{\min}$ is nothing but the Hausdorff dilation distance, $\rho(X_i, X_j)$. $(DC_{i,j})$ ranges between 0 and 1. For those states with $i = j$, $(DC_{i,j})$ is zero, which satisfies the condition that the distance between the set and to itself is zero, and hence the degree of contextuality is also zero. For instance, from the distances given in pixel units for the 28 states (Figure 10.5b), $(DC_{i,j})$ computed for X_2 and X_1 according to Equation 13.22 is $178/225 = 0.79$. The higher the $(DC_{i,j})$, the larger is the "between" space, and $(DC_{i,j})$ is proportional to shape–size similarities of the two sets under investigation.

Conclusion

This section deals with derivation of "between" space or region between the two nonadjacent disjoint sets (objects). Sets, planar objects, are with different spatial complexities. In order to understand the spatial relationships between disjoint compact sets of varied categories, we provide a morphological treatment to derive the "between" space. The derivation of such "between" space is entirely based on deriving the following two aspects: (1) the directional dilation involved between the sets and (2) the Hausdorff dilation distance between the sets. Once the direction and Hausdorff dilation distance are computed, one can perform the closing transformation by means of a directional (conditional) structuring element of diameter equivalent to Hausdorff dilation distance between the sets under question.

Spatial reasoning between the two binary sets is shown by application of mathematical morphological transformations and certain logical operations. In particular, derivation of "between" space between sets of both simple and complex types via closing by means of structuring element of diameter equivalent to Hausdorff dilation distances is shown. This framework is demonstrated on (1) synthetic sets and function to explain the procedures involved at understandable level, (2) 28 states of India (as planar compact sets) to show the real-world application, and (3) two function-like images depicting spatially distributed elevation regions, between which the "between" region is mapped to show the real-world application (e.g., air, sea, land corridor derivation) in a generalized way. This study would supplement with various topics already involved in the subjects of geographic information science (GISci), spatial information theory. With the known morphological transformations such as morphological closing and dilation, we propose a framework to derive "between" space between disjoint sets (states, continents, countries, etc.) via Hausdorff distance–based closing transformation (or) closing transformation w.r.t. structuring element of diameter equivalent to Hausdorff distance between the two sets. A few potential applications include derivation of (1) line of sight (prominent visibility region) between the two sets and (2) shortest (geodesic) path between any two points (cities) of any two disconnected sets (states), via "between" space of corresponding sets, (3) shape–size similarities between sets, and (4) degree of contextuality between sets. An open problem lies in the form of making use of quantitative parameters such as Hausdorff dilation distances, degrees of contextuality, "betweenness," and adjacency of the sets (states) to determine the strategically important sets with proper basis. An intuitive reader can make use of this morphological study to apply in application-specific studies.

References

Aiello, M. and J. van Bentha, 2002, A modal walk through space, *Journal of Applied Non-Classical Logics*, 12(3–4), 319–364.

Beucher, S., 1994, Interpolation d'ensembles, de partitions et de fonctions, Tech. Rep. N-18/94/MM, Centre de Morphologie Mathematique, Ecole des Mines de Paris.

Beucher, S. and F. Meyer, 1992, The morphological approach to segmentation: The watershed transformation. In: *Mathematical Morphology in Image Processing*, eds. Edward R. Dougherty, Marcel Dekker, Inc., New York.

Bloch, I., July 1999, Fuzzy relative position between objects in image processing: A morphological approach, *IEEE Transactions on Pattern Analysis and Machine Intelligence*, 21(7), 657–664.

Bloch, I., O. Colliot, and R. M. Cesar, 2006, On the ternary spatial relation "between," *IEEE Transactions on System, Man, Cybernet Part B: Cybernetics*, 36(2), 312–327.

Bloch, I. and A. Ralescu, 2003, Directional relative position between objects in image processing: A comparison between fuzzy approaches, *Pattern Recognition*, 36(7), 1563–1582.

Chaudhuri, B. B., 1990, Fuzzy set theoretic interpretation of object shape and relational properties for computer vision, *International Journal of Systems Science*, 21(7), 1169–1184, http://srtm.csi.cgiar.org (accessed on October 30, 2008).

Huttenlocher, D. P., R. M. Klunderman, and W. J. Rucklidge, 1993, Comparing images using the Hausdorff distance, *IEEE Pattern Analysis and Machine Intelligence*, 15(9), 850–863.

Jarvis, A., W. J. Reuter, A. Nelson, and E. Guevara, 2008, Hole-filled seamless SRTM data V4, International Centre for Tropical Agriculture (CIAT), available from http://srtm.csi.cgiar.org (last accessed January 17, 2013).

Larvor, Y., 2004, Notion de mereogeometries: Description qualitative de propertietes geometriques, du mouvement et de la forme d'objets tridimensionnels, PhD thesis, Universite Paul Sabatier, Toulouse, France.

Maragos, P. A. and R. W. Schafer, 1986, Morphological skeleton representation and coding of binary images, *IEEE Transactions on Acoustics, Speech and Signal Processing*, ASSP-34(5), 1228–1244.

Maragos, P. A. and R. D. Ziff, 1990, Threshold superposition in morphological image analysis systems, *IEEE Pattern Analysis and machine Intelligence*, 12(5), 498–504.

Meyer, F., 1994, Interpolations, Tech. Rep. N-16/94/MM Centre de Morphologie Mathematique, Ecole des Mines de Paris.

Rosenfeld, A. and R. Klette, 1985, Degree of adjacency or surroundedness, *Pattern Recognition*, 18(2), 169–177.

Sagar, B. S. D., 2010, Visualization of spatio-temporal behavior of discrete thematic data via Hausdorff distances and interpolations, *IEEE Pattern Analysis and Machine Intelligence*, 32(2), 378–384.

Serra, J., 1982, *Image Analysis and Mathematical Morphology*, Academic Press, London, U.K.

Serra, J., 1994, Interpolations et distance de Hausdorff, Tech. Rep N-15/94/MM, Ecole des Mines de Paris, France.

Serra, J., 1998, Hausdorff distances and interpolations. In: *Mathematical Morphology and Its Applications to Images and Signal Processing*, eds. Henk J. A. M. Heijmans and Jos B. T. M. Roerdink, Kluwer Academics Publishers, Dordrecht, the Netherlands, pp. 107–115.

14

Spatial Interpolations

Two techniques have been explained in this chapter. These techniques address spatial interpolation problems. First technique is to convert point-specific variable data into contiguous zonal map forms. Second technique provides interpolated maps between the source and target maps, which are also termed as maps at two periods. The latter approach provides a set of equations to generate interpolated maps between the time-dependent maps of varied complexities.

Introduction

To prepare domain-specific thematic maps by applying digital image processing techniques (e.g., filtering, segmentation, classification), remotely sensed satellite data act as source. However, several important variable data that are available over a geographic space are in point (location-specific) form. Evidently, a procedure is required to convert such point-specific data into zonal form for better visualization. Such conversion approach by using computer-assisted techniques and spatial statistical tools is important to (1) integrate thematic information retrieved from multiscale multitemporal remotely sensed satellite data with other variables for which only location-specific (point) data are available, (2) develop spatiotemporal models for various phenomena and processes, and (3) visualize relationships between the geographic variables in terms of spatial form.

Thiessen polygon construction, where the space is divided into polygons with the point data in the middle of each polygon assumed to be representative for the rainfall on the area of land included in its polygon, is a conventional approach to convert such point data into polygonal forms. These polygons are made by drawing lines between gauges and then making perpendicular bisectors of those lines that form the polygons. This method was adapted to analyze space use in geographic information science (GISci) (Casaer et al. 1999, Cao and Glover 2010). The zone (area) of influence map that could be generated via Thiessen polygon method resembles a convex polygon. Traditional geostatistical interpolation method such as simple kriging (Cressie 1991) is available.

In the development of an information system that is both scale and time independent, spatial interpolation techniques are important. The concepts from GISci provide new insights to develop information systems in spatial form further facilitating to visualize the relations between "layered information" (Worboys and Duckam 2004). From various sources of data acquired by remote sensing, field surveys, demographic surveys, historical records, etc., thematic layers depicting variable specific information will be prepared—by computer-assisted mapping or by digitizing manually mapped information. Integration of spatiotemporal information available as snapshots of the ever-changing phenomena at discrete intervals is an important problem posed to the geographic information system (GIS) community (Snodgras 1992, Frank 1998). By using such snapshots as input layers depicting thematic information, the way of mapping algebraic concepts (Tomlin 1990) extended with category theory (Lane and Birkhoff 1967) in order to generate an output layer is explained in Frank (2005).

Spatially represented thematic maps are essential in the development of a theme-specific information system. Such maps, derived from data acquired either physically or remotely, are usually stored in layered forms. Each layer represents a theme (foreground) and a no-theme (background) in noise-free binary form. The layered information is available at different spatiotemporal scales. A usual limitation is that this information is available in a discrete form, i.e., at discrete spatiotemporal resolutions. A procedure is required to derive layers in continuous form from a limited set of layers available at discrete intervals.

A spatial interpolation procedure is required to predict a spatial structure between two other spatial structures—which may be represented at two different spatial and/or temporal resolutions. "Spatial structure" and "spreading of a phenomenon" are interchangeably used here, though the phenomenon may evolve with time. One needs to generate (interpolate) the intermediary sequence of phenomena between the known time periods in order to predict and visualize the spatiotemporal dynamics.

The available popular spatial interpolation techniques include kriging (Cressie 1993), shape-based interpolation (Raya and Udupa 1990, Herman et al. 1992), and Hausdorff distance–based interpolation (Serra 1982, 1994, 1998, 2010, Beucher 1994, Meyer 1994a, Iwanowski 2000, Vidal et al. 2005). Other interpolation methods for binary objects include elastic dynamic interpolation (Burr 1981, Chen et al. 1990) and directional interpolation (Werahera et al. 1995). When dealing with non-convex objects, these algorithms (Burr 1981, Chen et al. 1990, Werahera et al. 1995) are computationally and algorithmically expensive and have limitations. While kriging yields promising results (Cressie 1993), this interpolation technique, in the context of geoscience and/or GISci, has been used only for spatial sets (layers). Moreover, kriging techniques that ignore the connectivity of components involved in the two input sets are meant for global transforms. To make use of a spatial interpolation technique in the context of spatiotemporal visualization, the companion-connected components that belong to two input sets of different spatial and/or temporal scales need to be categorized based on the spatial relationships. The material that follows "Visualization of Spatiotemporal Behavior of Discrete Maps *via* Generation of Recursive Median Elements section" deals with (1) the categorization of the connected components by means of Hausdorff erosion and dilation distances and (2) the computation of category-specific median set recursively.

Generation of Zonal Map from Point Data via Weighted Skeletonization by Influence Zone

This section presents an algorithm using mathematical morphology (Serra 1982), in particular weighted skeletonization by influence zone (WSKIZ) transformation, to construct contiguous zonal maps from variable-specific point data. Mathematical morphology has been employed in the contexts of GISci (Su et al. 1997, Pullar 2001, Tay et al. 2005, Aksoy and Cinbis 2010, Sagar 2010), geosciences (Sagar and Chockalingam 2004, Sagar and Tien 2004, Tay et al. 2007), and remote sensing (Sagar et al. 1995a,b, Pesaresi and Benediktsson 2001, Benediktsson et al. 2003, Sagar et al. 2003, Barata and Pina 2006, Chanussot et al. 2006, Taubenbock et al. 2006, Dalla Mura et al. 2008, Huang et al. 2009, Pan et al. 2010, Sagar and Serra 2010, and Dalla Mura et al. 2011). This WSKIZ approach—an alternative to Voronoi diagrams that are used in geophysics and meteorology to analyze spatially distributed data (such as rainfall measurements)—can be used to describe the area of influence of a point in a set of points possessing varied values (rainfall values, etc.).

Data about many variables are available as numerical values at specific geographic locations. To convert point-specific data into zonal map, a methodology based on mathematical morphology has been explained. WSKIZ—that determines the points of contact of multiple frontlines propagating, from various points spread over the space, at the traveling rates depending upon the variable's strength—is the principle involved in the methodology. Rainfall data available at specific rain gauge locations (points) have been considered to demonstrate this approach to generate spatially distributed zonal map. Such a contiguous zonal map generated suggests zones of equal rainfall.

The organization of this section is as follows: in "Conversion of Point-Specific Values into Zonal Map via WSKIZ" section, model and algorithm concepts, motivation, and methodology to convert point-specific value data into zonal map form are explained; the results drawn in terms of zonal map for a variable (e.g., rainfall) available as numerical values at specific points over the geographic space out of demonstrations and the respective discussion of the significance of the obtained results have been provided in the "Experimental Results" section; and the "Conclusion on Conversion of Point Data into Zonal Map" section presents general inferences.

Conversion of Point-Specific Values into Zonal Map via WSKIZ

Location-Specific Data over Geographic Space

Let S be the underlying space, endowed with a distance d, and X (mask, Figure 14.1a) be a subset of S that consists of several locations as points (e.g., Figure 14.1b) where the time-varying data such as rainfall and temperature values are available. Figure 14.1b depicts four points (gauge stations), and each point (X_i) possesses a value that denotes the strength of a variable. Such a map with points satisfies the following morphological relationship: $(X_i \cap X_j) = \varnothing$ for $i \neq j$.

Model to Generate Zonal Map from Point Data

To generate zonal map from point data, the following algorithm based on mathematical morphology has been proposed. This algorithm consists of the following steps:

1. Consider the location (point)-specific values (e.g., rainfall, temperatures) as points that act as markers from which the dilation propagations compete to fill the geodesic space (i.e., global mask, X).

2. Sort and rank the points according to the corresponding strengths of a geographic variable (e.g., rainfall values). Such ranking allows assigning to each marker a specific rate at which it competes to fill the available space.

3. Multiple markers (locations) over a geographic space assigned with different strengths (weights) of a variable (the rates at which those markers) need to be dilated with propagation speeds that are proportional to the assigned weights to obtain the final sizes and shapes of the influence zones in zonal map constructed.

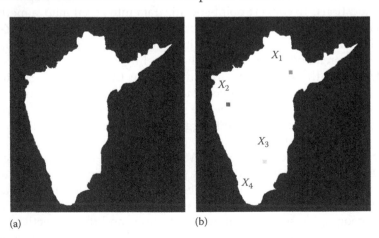

(a)   (b)

FIGURE 14.1
(a) Region considered is South India and (b) gauge station locations (X_1, X_2, X_3, X_4). (From Rajashekara, H.M. et al., *IEEE Geosci. Remote Sens. Lett.*, 9(3), 403, 2012.)

Simulating flood propagation process, by treating each point as a lake (marker), the water frontlines generated from corresponding lakes that are spatially distributed over a Cartesian space would extinguish (meet) each other at various places. By preserving all such extinguishing points, while suppressing all other details, we obtain *skeletonization by influence zones* (SKIZ). When the propagation speed of floodwaters originating from spatially distributed lakes is uniform, it is easy to visualize the SKIZ; this process can be simulated with ease as there are no constraints imposed on flood propagation speed. However, for the purpose of construction of zonal map from point-specific data by synchronizing variable's strength at each point, the propagation speed of dilation (flood) frontlines needs to be made point dependent. Treating the original map (X) as the mask (Figure 14.1a) and the points in that map as multiple markers (X_i) (e.g., Figure 14.1b), recursive geodesic dilations (Equation 14.1) with marker-dependent propagation speeds simultaneously from multiple markers would provide a WSKIZ. Such a WSKIZ is the zonal map, where the specified variable strength determines the dilation propagation speeds. The zonal map generation from point-specific data requires following steps:

Step 1: Let points X_i denote locations (e.g., gauge stations) at which the values of a variable (e.g., rainfall) are available.

Step 2: By means of primitive structuring element B of size λ_i that is uniquely dependent on the variable value at location "i," compute recursive geodesic dilations of each point X_i.

Step 3: By systematically performing recursive geodesic dilations with location-dependent propagation speeds simultaneously from multiple points to compute all possible extinguishing points (Equations 14.2 and 14.3), WSKIZ for X with X_i locations would be obtained.

Step 4: Represent each zone of zonal map obtained at step 3 with a specific gray shade such that no two neighboring zones have similar gray shades. If there are N number of points, then there would be N-zones in the zonal map. See Figure 14.2 for more details about sequential steps involved in the implementation of WSKIZ. In the sections that follow, details of the aforementioned sequential steps involved in converting point data into zonal map have been provided.

Computation of Point-Dependent Recursive Geodesic Dilations

According to the variable strength at point X_i, dilation propagation speed is assigned to points. Higher the variable strength, the faster will be the dilation propagation speed of the point per unit time step. Each point X_i to be dilated is assigned with a structuring element of primitive size of λ_i, where λ denotes the primitive size of B and i denotes the index of the point for the

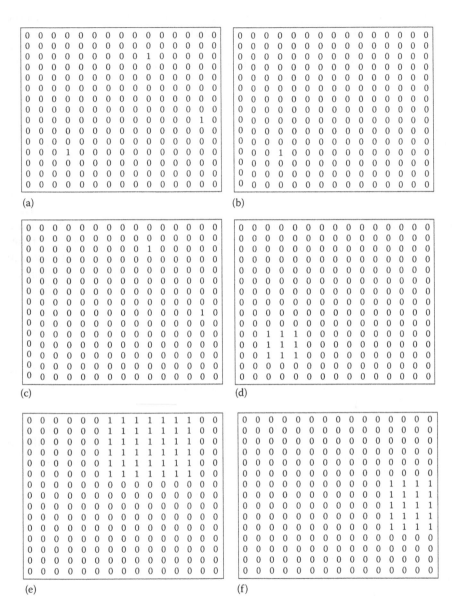

(a)

(b)

(c)

(d)

(e)

(f)

FIGURE 14.2

(a) Original map with three points (shown with 1s) for (X_1), (X_2), and (X_3), (b) ith point $(X_i) = (X_1)$, (c) union of jth points, $\bigcup_{\forall j\ j\neq i} X_j = (X_2)\cup(X_3)$, (d) first cycle of dilation of ith point by B (square in shape) with the propagation speed of $\lambda = 1$, denoted by $\delta^{n=1}_{\lambda=1}(X_1)$, (e) first cycle of dilation of jth point (X_2) by B with the propagation speed of $\lambda = 3$, $\delta^{n=1}_{\lambda=3}(X_2)$, (f) first cycle of dilation of ith point (X_3) by B with the propagation speed of $\lambda = 2$, $\delta^{n=1}_{\lambda=2}(X_3)$.

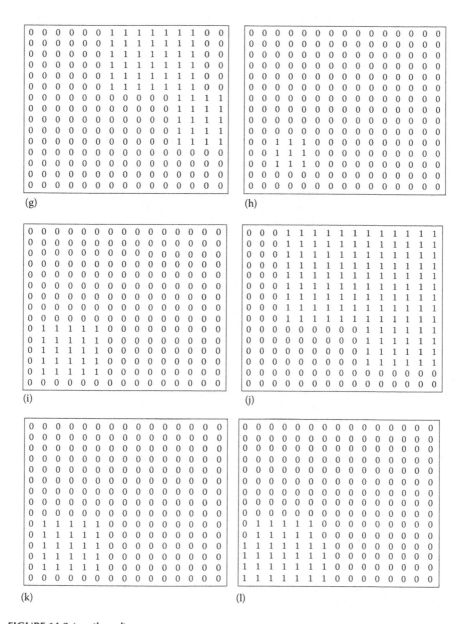

FIGURE 14.2 (continued)
(g) union of $\delta^{n=1}_{\lambda=3}(X_2)$ and $\delta^{n=1}_{\lambda=2}(X_3)$, (h) $\delta^{n=1}_{\lambda=1}(X_1) \setminus \delta^{n=1}_{\lambda=3}(X_2) \cup \delta^{n=1}_{\lambda=2}(X_3)$, (i) $\delta^{n=2}_{\lambda=1}(X_1)$, (j) similarly for next iteration: $\delta^{n=2}_{\lambda=3}(X_2) \cup \delta^{n=2}_{\lambda=2}(X_3)$, (k) $\delta^{n=2}_{\lambda=1}(X_1) \setminus \delta^{n=2}_{\lambda=3}(X_2) \cup \delta^{n=2}_{\lambda=2}(X_3)$, (l) $Z(X_1) = \bigcup_n \left[\delta^n_{\lambda=1}(X_1) \setminus \delta^n_{\lambda=3}(X_2) \cup \delta^n_{\lambda=2}(X_3) \right]$.

(continued)

```
(m)
1 1 1 1 1 1 1 1 1 1 1 1 1 1 1
1 1 1 1 1 1 1 1 1 1 1 1 1 1 1
1 1 1 1 1 1 1 1 1 1 1 1 1 1 1
1 1 1 1 1 1 1 1 1 1 1 1 1 1 1
1 1 1 1 1 1 1 1 1 1 1 1 1 1 1
1 1 1 1 1 1 1 1 1 1 1 1 1 1 1
1 1 1 1 1 1 1 1 1 1 0 0 0 0
1 1 1 1 1 1 1 1 1 1 0 0 0 0
1 1 1 1 1 1 1 1 1 0 0 0 0 0
1 0 0 0 0 0 1 1 1 0 0 0 0 0
1 0 0 0 0 0 1 1 1 0 0 0 0 0
0 0 0 0 0 0 1 1 1 0 0 0 0 0
0 0 0 0 0 0 0 0 0 0 0 0 0 0
0 0 0 0 0 0 0 0 0 0 0 0 0 0
0 0 0 0 0 0 0 0 0 0 0 0 0 0
```

```
(n)
0 0 0 0 0 0 0 0 0 0 0 0 0 0 0
0 0 0 0 0 0 0 0 0 0 0 0 0 0 0
0 0 0 0 0 0 0 0 0 0 0 0 0 0 0
0 0 0 0 0 0 0 0 0 0 0 0 0 0 0
0 0 0 0 0 0 0 0 0 0 0 0 0 0 0
0 0 0 0 0 0 0 0 0 0 1 1 1 1
0 0 0 0 0 0 0 0 0 0 1 1 1 1
0 0 0 0 0 0 0 0 0 1 1 1 1 1
0 0 0 0 0 0 0 0 1 1 1 1 1 1
0 0 0 0 0 0 0 1 1 1 1 1 1 1
0 0 0 0 0 0 0 1 1 1 1 1 1 1
0 0 0 0 0 0 1 1 1 1 1 1 1 1
0 0 0 0 0 0 1 1 1 1 1 1 1 1
0 0 0 0 0 0 1 1 1 1 1 1 1 1
0 0 0 0 0 0 1 1 1 1 1 1 1 1
```

```
(o)
2 2 2 2 2 2 2 2 2 2 2 2 2 2
2 2 2 2 2 2 2 2 2 2 2 2 2 2
2 2 2 2 2 2 2 2 2 2 2 2 2 2
2 2 2 2 2 2 2 2 2 2 2 2 2 2
2 2 2 2 2 2 2 2 2 2 2 2 2 2
2 2 2 2 2 2 2 2 2 2 2 2 2 2
2 2 2 2 2 2 2 2 2 2 3 3 3 3
2 2 2 2 2 2 2 2 2 2 3 3 3 3
2 2 2 2 2 2 2 2 2 3 3 3 3 3
2 1 1 1 1 1 2 2 2 3 3 3 3 3
2 1 1 1 1 1 2 2 2 3 3 3 3 3
1 1 1 1 1 1 2 2 2 3 3 3 3 3
1 1 1 1 1 1 1 3 3 3 3 3 3 3
1 1 1 1 1 1 1 3 3 3 3 3 3 3
1 1 1 1 1 1 1 3 3 3 3 3 3 3
```

FIGURE 14.2 (continued)
(m) similarly follow the steps from (b–l); by changing the *i*th point from (X_1) to (X_2), and by treating (X_1) and (X_3) as *i*th points, $Z(X_2)$ is obtained, (n) obtained $Z(X_3)$, and (o) three zones $Z(X_1)$, $Z(X_2)$, and $Z(X_3)$ are shown with 1s, 2s, and 3s.

purpose of performing point-dependent geodesic dilations. Then the geodesic dilation of X_i by B of primitive size λ_i takes the form

$$(X_i \oplus nB_{\lambda_i}) \cap X = \delta^n_{\lambda_i}(X_i) \tag{14.1}$$

where nB_{λ_i} denotes the structuring element B of primitive size λ_i that depends upon the point-dependent variable strength (propagation speed) for n-cycles (n ranging from 1, 2, …, N). The intersection between mask (X) and the version of X_i dilated by nB_{λ_i} for $n = 1$ time yields the first level of geodesic dilation of X_i, $\delta^n_{\lambda_i}(X_i)$.

Zonal map computation via WSKIZ: Let $Z(X_i)$ be a zone of (X_i). In the process of converting the point data into zonal map, the two steps involved include the following:

$$Z(X_i) = \bigcup_n \left(\delta^n_{\lambda_i}(X_i) \cap X \right) \setminus \bigcup_{\forall j} \left(\delta^n_{\lambda_{j \neq i}}(X_j) \cap X \right) \tag{14.2}$$

$$Z(X) = \left(\bigcup_i \left(Z(X_i) \right) \right)^c \qquad (14.3)$$

where
   $Z(X_i)$ is a zone constructed for the point (X_i)
   (X_j) are the other points
   X denotes mask (e.g., Figure 14.1a)

Each zone of map $Z(X_i)$ is computed by a three-step approach: (1) Subtracting the union of markers (X_j) zones other than (X_i), $(X_{j \neq i})$, geodesically dilated simultaneously by the structuring element of point-dependent primitive size from the geodesic dilated version of (X_i) by structuring element of primitive size dependent on (X_i); (2) Consider the union of such subtracted versions for all n values ranging from 0 to N to obtain the map-zone for the zone of the map (X_i); and (3) Taking the complement of the union of all obtained map-zones yields WSKIZ, in other words a zonal map converted from point data. If the regions occupied by mountains are subtracted from the mask (X), then by employing Equation 14.2, zonal map only within the non-mountainous regions could be generated. Such a mask that excludes mountainous regions (geographic hurdles) is essential while employing this approach to simulate flood propagation.

Gray-shade assignment to ZX_i: ZX_i denotes a zone obtained for a point X_i. Let us assume, for example, that there are four zones obtained for four different points (Figure 14.1b), and each zone possesses spatially distributed values (Figure 14.1b). According to the positions of these zones, gray shades will be assigned to different zones such that those zones that have been generated with similar dilation propagation speeds λ_i would be assigned with similar shades. Such a gray-shading scheme is used to easily identify zones with equal variable strengths. However, in realistic cases, similar weights may be needed to assign to multiple points. Under such circumstance, zones obtained need to be further merged based on the similar weights adopted. All these steps explained in the "Location-Specific Data over Geographic Space" and the "Model to Generate Zonal Map from Point Data" sections are demonstrated in the four points shown in Figure 14.1b by assigning arbitrarily different propagation speeds.

Model Demonstration

A total of N number of points X_i (locations) that denote, for instance, in two-dimensional discrete space, Z^2 (Figure 14.1a), are considered. To demonstrate the proposed approach, four gauge stations (X_1, X_2, X_3, X_4) shown (Figure 14.1b) are considered. Within the mask (Figure 14.1a), the four possible zones that one can visualize by imposing varied propagation speeds of dilation of (X_1, X_2, X_3, X_4) are shown in Figure 14.3a through d. If these four point data depict four different strengths of some time-varying parameters such as rainfall and temperature for a specified time, then generation of zonal maps requires point data–specific geodesic-dilation propagations. Four zonal maps, each zone assigned with a gray shade (Figure 14.3a through d), are generated for four different point-specific sequences with the following order: (1) $X_2 > X_4 > X_1 > X_3$, (2) $X_2 > X_1 > X_3 > X_4$, (3) $X_1 > X_3 > X_2 > X_4$, and (4) $X_1 > X_4 > X_2 > X_3$. In the model, four zones—which are obtained after

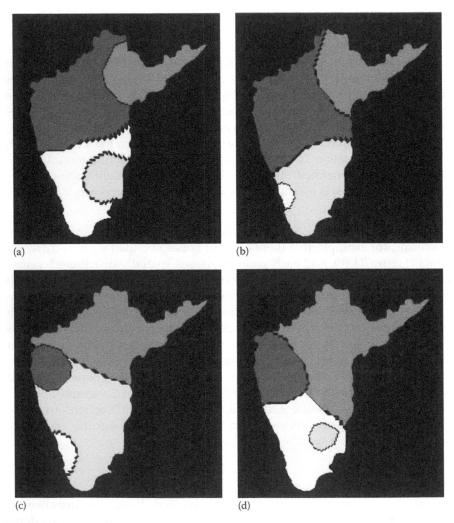

(a)   (b)

(c)   (d)

FIGURE 14.3
Variable strengths (in terms of propagation speeds) are given as (a) $X_2 > X_4 > X_1 > X_3$, (b) $X_2 > X_1 > X_3 > X_4$, (c) $X_1 > X_3 > X_2 > X_4$, and (d) $X_1 > X_4 > X_2 > X_3$. (From Rajashekara, H.M. et al., *IEEE Geosci. Remote Sens. Lett.*, 9(3), 403, 2012.)

assigning gray shades to zones in each panel (Figure 14.3a through d) based on the four aforementioned weighting schemes—have been shown.

Zonal maps generated by WSKIZ approach can be compared with those of polygons constructed based on Thiessen polygon and Voronoi diagram construction approaches. WSKIZ approach has advantages over the other approaches for three reasons: (1) straightforward implementation of algorithms, (2) weights could be assigned to generate weight-based zonal maps, and (3) it could be fully automated.

Experimental Results

A map depicting 34 locations (gauge stations; Figure 14.4a), spread across India, with rainfall values for these locations recorded for the period of March–April 2011, has been considered to demonstrate the applicability of the algorithm explained in the "Conversion of Point-Specific Values into Zonal Map via WSKIZ" section. Weights are assigned, according to the rainfall values for 34 locations (Table 14.1), for dilation propagation speed for each gauge station. The higher the rainfall recorded, the larger is the assigned weight. The faster the dilation propagation speed, the larger the weight.

Primitive size of structuring element (λ_i) assigned for respective points of (X_i) by assigning propagation speeds, and by allowing those points (X_i) to dilate for n times satisfying the involved processes according to Equations 14.1 through 14.3, the point data have been converted into zonal map (Figure 14.4b). This zonal map (Figure 14.4b) suggests that rainfall at all locations within each zone belongs to a particular station within the zone and, hence, has the same values. In the non print material, available at http://www.isibang.ac.in/~bsdsagar/AnimationOfPointPolygonConversion.wmv, animation of rainfall zone map generated by using the proposed approach has been shown. It is obvious that the boundaries separated by different distances in the zonal map are due to the fact that the dilation propagation speeds are rainfall (weight) dependent. In the zonal maps (Figure 14.4b), the gray shades are assigned to each zone to have better demarcation. But the gray shades in Figure 14.4b have no significance. The weights are highest for locations (X_1, X_{15}), and hence their corresponding zones shown in zonal map (Figure 14.4b) are the zones of high rainfall. The lowest ranked locations in terms of rainfall include X_{13}, X_{26}, and those zones could be seen in Figure 14.4b. The zones of zonal map generated with similar propagation speed (weights) are merged (Figure 14.4c) as six broad zones, and those merged zones are assigned with similar gray shades to visualize as a broader zonal map. In this broader zonal map (Figure 14.4c), if there are n number of weights employed to generate WSKIZ map, then there will be n number of zones with n-gray shades depicting broader zones (e.g., Figure 14.4c). The broad zones from the merged zonal map (Figure 14.4c) depicted in six gray shades suggest that (1) there are multiple zones (Figure 14.4b) obtained with similar weights (propagation speeds), and (2) they belong to six different zones depicting six different ranges of rainfall patterns. Kriged map

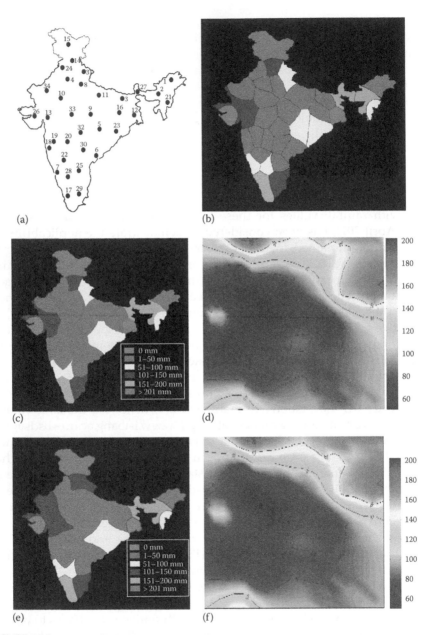

FIGURE 14.4
(a) Thirty-four points (locations) of rain gauge stations spread over India indexed ($X_1 - X_{34}$), (b) rainfall zonal map generated by having various possible propagation speeds and the variable strengths in terms of propagation speeds are given in Table 14.1, (c) broader zones obtained after merging the zones (Figure 14.4b) obtained with similar propagation speeds, (d) kriged map for 34 gauge station data, and (e and f) WSKIZ and kriged map for 29 gauge station data, where last 5 stations are dropped from Figure 14.4a. (From Rajashekara, H.M. et al., *IEEE Geosci. Remote Sens. Lett.*, 9(3), 403, 2012.)

TABLE 14.1

Ranks according to Variable Strengths for the Points I, II, and III, Respectively, Denote Point Index, Rainfall Values, and Weights

I	II	III	I	II	III
X_1	214.5	6	X_{18}	4.4	2
X_2	181.3	5	X_{19}	9.2	2
X_3	31.8	2	X_{20}	11.6	2
X_4	9.2	2	X_{21}	96.6	3
X_5	56.1	3	X_{22}	45.9	2
X_6	35.3	2	X_{23}	58.6	3
X_7	69.2	3	X_{24}	18.9	2
X_8	8.9	2	X_{25}	30.1	2
X_9	6.8	2	X_{26}	0.0	1
X_{10}	1.4	2	X_{27}	168.1	5
X_{11}	8.6	2	X_{28}	92.5	3
X_{12}	129.2	4	X_{29}	100.3	4
X_{13}	0.0	1	X_{30}	34.7	2
X_{14}	119.0	4	X_{31}	52.8	3
X_{15}	252.6	6	X_{32}	33.9	2
X_{16}	36.9	2	X_{33}	1.6	2
X_{17}	179.1	5	X_{34}	3.6	2

Source: Rajashekara, H.M. et al., *IEEE Geosci. Remote Sens. Lett.,* 9(3), 403, 2012.

Note: More details about the boundaries of dilation propagations from the points to the state of reaching the convergence (Fig. S2, S3) may be found at http://www.isibang. ac.in/~bsdsagar/GRSL-00335-2011-FIG-S1-S3-Supporting-Material.pdf

(Figure 14.4d) and WSKIZ map data, generated for the 34 gauge stations, are visually in good agreement. WSKIZ map (Figure 14.4e) and kriged map (Figure 14.4f) have also been generated by considering first 29 gauge station data from Figure 14.4a. It is obvious from merged WSKIZ maps (Figure 14.4c and e) generated respectively for 34 and 29 gauge station data that there is significant visual agreement in the spatial distribution of rainfall.

Intensity of the rainfall at the gauge stations enables a zone of influence, the size of which does not necessarily be proportional to the intensity. Gauge stations with high rainfall intensity recorded may possess smaller zone, and *vice versa*. Size and shape of a zone ZX_i are collectively governed by the following factors: (1) intensity (weight) of rainfall at location (X_i), (2) number of gauge stations (X_j) in the proximity of a station (X_i) under question, and (3) intensities of rainfall at those adjacent gauge stations (X_j).

If the rainfall values at the gauge stations are relatively similar in pattern (phase synchronous), then the zonal maps generated by dividing (multiplying) the weights by two are also similar to that of the zonal maps shown.

Such generated zonal maps are invariant under dividing (multiplying) weights. However, if the rainfall values recorded at rain gauge stations are with different patterns, then the shapes and sizes of zones in the zonal maps generated from such point-specific rainfall values will significantly differ.

For better results, other important approaches such as watershed transform (Vincent and Soille 1991, Meyer 1994b, Najman and Schmit 1994, Cousty et al. 2009) and Euclidean distance transform (Saito and Toriwaki 1994, Hirata 1996, Meijster et al. 2000), which are treated as generalization of SKIZ to arbitrary metrics, could be applied in the context of converting point-specific data, of geoscientific interest, into zonal map. The underlying physical principle of this proposed technique would work well to generate spatial maps of several phenomena of climatological, ecological, and geomorphological relevance. Such maps generated for a time-dependent phenomenon (e.g., rainfall, temperature) yield maps that possess varied forms of zones that are also time dependent. In spatiotemporal modeling, such maps generated for different time periods are of immense use. This WSKIZ could be performed in a geodesic manner within the mask that is without the hurdles such as mountains. However, this approach may not be an appropriate one for certain numerical variables such as population densities. This approach provides insights into studies on location analysis, flood modeling, epidemic spread, etc. By using this approach, location-specific variable data in numerical form of relevance to various terrestrial phenomena and processes available as point data could be mapped into continuous display of the sampling patterns. Such continuous maps of a time-varying variable generated via this approach for different time periods provide insights (1) to understand the spatiotemporal behavior of a phenomenon and (2) to establish spatial relationships between the phenomena. The main advantage of this approach over the Thiessen polygonal approach is that the structure of each influence zone does not look like a polygon.

Conclusion on Conversion of Point Data into Zonal Map

An approach based on WSKIZ to generate contiguous zonal maps from point-values available at fixed geographic locations has been proposed and demonstrated. In this approach, points (markers) were dilated geodesically with propagation speed that is proportional to the strength of the variable at that location; the outline of each zone would be much smoother in the zonal maps generated via WSKIZ approach as we choose primitive structuring element circle in shape. This way of converting point data into zonal map is stable and could be easily programmed. The variable specific zonal maps thus generated with the proposed approach could be spatially integrated with thematic information retrieved from remotely sensed data. This section deals with the generation of similar polygonal map that mimics point-dependent diffusion process through mathematical morphology-based WSKIZ transformation. Such maps generated via WSKIZ for time-dependent phenomena (e.g., rainfall, temperature) yield maps that possess varied forms of zones that are also time dependent. Maps across time periods can be used to generate continuous interpolating sequential maps. To generate interpolating sequences, we adopted a technique to compute recursive median maps that provide the changes that happen in continuous sequence.

Visualization of Spatiotemporal Behavior of Discrete Maps via Generation of Recursive Median Elements

Generation of contiguous zonal maps, from variable-specific point data, via WSKIZ has been explained in the previous section. However, to generate interpolated maps in a continuous manner by using two discrete spatial and/or temporal thematic maps evidently requires different types of spatial interpolation techniques that have high demand in GISci. Noise-free data (thematic layers) depicting a specific theme at varied spatial or temporal resolutions consist of connected components either in aggregated or in disaggregated forms. This section provides a simple framework (1) to categorize the connected components of layered sets of two different time instants through spatial relationships between the companion-connected components, quantified via Hausdorff distances between the companion-connected components; and (2) to generate sequential maps (interpolations) between the discrete thematic maps. How median maps (interpolated map) could be computed between the source and target map by employing dilation and erosion has been demonstrated on lake geometries mapped at two different times, and also on the bubonic plague epidemic spread data available for 11 consecutive years. Significantly fair quality of the median maps could be computed for epidemic data between alternative years that are visually in good agreement between the interpolated maps and actual maps. To visualize (animate) the spatiotemporal behavior of a specific theme in a continuous sequence of the interpolated maps computed via morphological interpolations is of immense use.

Subsequent information provides details on Hausdorff distance and median set computation between two sets as a global transformation in the "Hausdorff Erosion Distance and Hausdorff Dilation Distance" section. In the "Computation of the Median Set" section, spatially represented themes with different categories are identified with reference to logical relationships and Hausdorff distances. In the "Layered Information as Sets: Spatial Interpolation" section, the application of morphologic interpolation to generate sequential interpolated layered information is explained along with experimental results drawn for two cases: small water bodies at two different time periods and bubonic plague data. The "Description of Categories Using Hausdorff Distances" and "Morphologic Interpolation via Median

Element Computation" sections contain a brief discussion on the potential applications of the proposed framework in the context of GISci and some other conclusions.

Hausdorff Erosion Distance and Hausdorff Dilation Distance

(X^t) and (X^{t+1}) denote nonempty compact sets at two time instants t and $t + 1$. According to Serra (1982), the Hausdorff erosion distance $\sigma(X^t, X^{t+1})$ and the dilation distance $\rho(X^t, X^{t+1})$ between X^t and X^{t+1} are defined respectively as

$$\sigma(X^t, X^{t+1}) = \inf\left\{n : \left[(X^t \ominus nB) \subseteq X^{t+1}\right], \left[(X^{t+1} \ominus nB) \subseteq X^t\right]\right\} \quad (14.4)$$

$$\rho(X^t, X^{t+1}) = \inf\left\{n : \left[X^t \subseteq (X^{t+1} \oplus nB)\right], \left[X^{t+1} \subseteq (X^t \oplus nB)\right]\right\} \quad (14.5)$$

Algebraically, these two distances yield metrics, which are dual to each other with respect to the "complement" operation. The classic concept of "Hausdorff distance" (Hausdorff 1914) and the Hausdorff dilation distance (Serra 1998) are similar.

Computation of the Median Set

By employing multiscale erosions and dilations along with certain logical operations, the median set (Serra 1998), which is central to the theme of this section, can be computed. If there exists a bijection between the sets (X^t) and (X^{t+1})—such that (X^t) is completely contained in (X^{t+1}), $(X^t \subseteq X^{t+1})$—the equation for computing the median set $M(X^t, X^{t+1})$ between (X^t) and (X^{t+1}) takes the following form:

$$M(X^t, X^{t+1}) = \bigcup_{n \geq 0}\left((X^t \oplus nB) \cap (X^{t+1} \ominus nB)\right) \quad (14.6)$$

Equation 14.6 takes the following form if (X^t) is only partially contained in (X^{t+1}):

$$M(X^t, X^{t+1}) = \bigcup_{n \geq 0}\left(\left[(X^t \cap X^{t+1}) \oplus nB\right] \cap \left[(X^t \cup X^{t+1}) \ominus nB\right]\right) \quad (14.7)$$

$M(X^t, X^{t+1})$ satisfies a more symmetrical property (see Serra 1998, Iwanowski and Serra 2000):

$$\mu = \inf\left\{n : n \geq 0, (X^t \oplus nB) \supseteq (X^{t+1} \ominus nB)\right\} = \rho(X^t, M) = \sigma(M, X^{t+1}) \quad (14.8)$$

$M(X^t, X^{t+1})$ is at Hausdorff dilation distance μ from (X^t), while $M(X^t, X^{t+1})$ is at Hausdorff erosion distance μ from (X^{t+1}). This further implies, for the case of $(X^t \subseteq X^{t+1})$, that $X^t \subseteq M \subseteq X^{t+1}$, and one has strictly $\rho(X^t, M) = \inf_{n \geq 0} \{n : M \subseteq (X^t \oplus nB)\}$ and $\sigma(M, X^{t+1}) = \inf_{n \geq 0} \{n : (X^{t+1} \ominus nB) \subseteq M\}$.

Equations 14.4 through 14.8, originally meant for global transforms that ignore connectivity, will be extended to companion-connected components X_i^t and X_i^{t+1} with index i of sets (X^t) and (X^{t+1})—with four possible spatial relationships. The four possible spatial relationships between the time-dependent themes X^t and X^{t+1} (sets) are treated as four different categories that require four different ways to compute interpolated maps. We assume that there exists a *bijection* between the connected components X_i^t of set (X^t) and the connected components X_i^{t+1} of set (X^{t+1}), for the various indexes i. The categorization based on spatial relationships explained via both logical relationships and Hausdorff erosion dilation distances between the companion-connected components of X^t and X^{t+1} would be clear after the "Limited Layered Sets" section.

Layered Information as Sets: Spatial Interpolation

A procedure for generating continuous layers starting out from a discrete set of layers is proposed in this section. Discrete sets (maps with connected components) here refer to two input sets. To achieve the objective, the sequential steps involved include (1) extraction and description of layered information available at two different time periods, (2) establishing spatial relationships between the sets and also between the corresponding companion subsets of the two main sets, as well as categorization of the subsets based on the spatial relationships, (3) computation of the median set between the two input sets, and (4) generation of a sequence of interpolated sets based on the two input sets and the median sets thus generated.

Limited Layered Sets

Three types of the layered information depicting a specific phenomenon available for static systems or for a time-dependent (dynamic) system are ordered, semi-ordered, or disordered. $X_1^t, X_2^t, \ldots, X_n^t$ and $X_1^{t+1}, X_2^{t+1}, \ldots, X_n^{t+1}$ denote connected components (e.g., lakes) at time periods "t" and "$t + 1$" represented on Z^2 (Figure 14.5 a and b). (X^t) and (X^{t+1}) that are denoted as sets (layers) for notational simplicity represent the connected components $X_1^t, X_2^t, \ldots, X_n^t$ and $X_1^{t+1}, X_2^{t+1}, \ldots, X_n^{t+1}$ as the subsets of (X^t) and (X^{t+1}), respectively, X_i^t and X_i^{t+1}, $\forall i \in I$ are assumed always to be nonempty and compact. "Sets" and "layered information," as well as "subsets" and "connected components," are interchangeably used here.

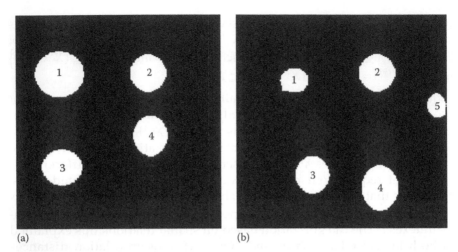

(a)   (b)

FIGURE 14.5

Two sets depicted in (a) and (b), respectively, represent corresponding subsets at two different time instants. These two sets are the inputs to generate a sequence of interpolated sets. The four categories explained in the text reflect the four and five subsets, respectively, from the two input sets (a and b). Corresponding subsets of each panel are indexed with numerals. It should be noted that blobs with the same index from the two panels (a and b) belong to two different periods. If these two panels are superimposed, blobs with indexes 1 and 2 of panel (b) will be contained in blobs 1 and 2 of panel (a) further indicating that the spatial relationships fall under category 1. Spatial relationships between the blobs with indexes 3 and 4 from the panels (a) and (b), respectively, fall under categories 2 and 3 whereas blob 5, in panel (b), that does not possess companion subset in panel (a) falls under category 4. (From Sagar, B.S.D., *IEEE Trans. Pattern Anal. Mach. Intell.*, 32(2), 378, 2010.)

Spatial Relationships between Sets and Their Categorization

Sets are treated as *ordered sets* if $X^t \subseteq X^{t+1}$ or $X^{t+1} \subseteq X^t$. The subsets embedded within each set at a respective time instant follow the relationship: $X_i^t \cap X_j^t = \varnothing; \forall t; i, j = 1, 2, \ldots, N; i \neq j$. (X^t) and (X^{t+1}) are in semi-ordered form, if subsets of X^t (resp. X^{t+1}) are only partially contained in the other set X^{t+1} (resp. X^t). (X^t) and (X^{t+1}) are considered as *disordered sets* if there arises an empty set while taking the intersection of (X^t) and (X^{t+1}), or of their corresponding subsets.

Description of Categories via Logical Relations

Let us assume that there exists a bijection between the connected components X_i^t and X_i^{t+1} of sets X^t and X^{t+1}, respectively. The evolution of subsets of X^t over a period of time depends upon some controlling factors, e.g., temperature is one such factor of lake evolution. The corresponding subsets in X^{t+1} explain the effect of controlling factor that causes changes in the subsets of X^t. One can study the spatial behavior of such subsets by investigating the spatial relationship between each subset of X^t and its companion subset of

X^{t+1} by assuming that these subsets are lakes at time t and $t + 1$ respectively embedded in sets X^t and X^{t+1}. Based on the relationships between the corresponding subsets of two layered data, different possibilities have been classified broadly into three groups.

Companion subsets fall under different categories if the following logical relationships are satisfied:

Group I

Category 1: $X_i^t \cap X_i^{t+1} \neq \varnothing$ for all $i \in I$ $X_i^t \subseteq X_i^{t+1}$ or $X_i^t \supseteq X_i^{t+1}$; $X_i^t = X_i^{t+1}$, $\left(X_i^t \cap X_i^{t+1} \right) = \left(X_i^t \cup X_i^{t+1} \right) \neq \varnothing$.

Category 2: If $X_i^t \not\subset X_i^{t+1}$ and/or $X_i^{t+1} \not\subset X_i^t$; $X_i^t \cap X_i^{t+1} \neq \varnothing$.

Due to the fact that the intersections between the connected components of X^t and X^{t+1} yield nonempty sets, the first two categories have been grouped as group I.

Group II

Category 3: The sets X^t and X^{t+1} have the same number of connected components, but the conditions mentioned in categories 1 and 2 (group I) are not always satisfied in a situation (category 3) where $X_i^t \cap X_i^{t+1} = \varnothing$.

As the intersection between the nonempty compact connected components yields an empty set, the logical relationship explained as category 3 has been grouped in group II.

Group III

Category 4: Assume that X^t has less connected components than X^{t+1} (category 4), but each X_i^t has a companion X_i^{t+1} such that one of the conditions given in the two categories of group I is satisfied. In some situations, for example, at the drought season, t, a lake may become empty. Then one can complete X^t by adding to it the ultimate erosions of those connected components of X_i^{t+1} that have no companion sets in X^t. Under the assumption that the intersection between the connected component(s), one of which is an empty set, yields the empty set, such a situation is explained via category 4 logical relationship grouped as group III.

Based on set theoretical and logical relations, the four categories under the three groups have been conceived. Clear distinction between categories 3 and 4 of groups II and III can be seen in terms of variations in the properties of connected components of the companion sets. There exist some methods, in GISci literature, strictly based on logical reasoning (Lane and Birkhoff 1967, Tomlin 1990, Snodgrass 1992, Worboys and Duckam 2004) that deal with the categorization of the corresponding connected components of layered data, whether or not they are connected components.

Description of Categories Using Hausdorff Distances

Companion-connected components of X^t and X^{t+1} can be quantitatively categorized into four categories in terms of Hausdorff distances.

Category 1: $\rho\left(X_i^t, X_i^{t+1}\right)$ and $\sigma\left(X_i^t, X_i^{t+1}\right)$ are zero for three companion-connected components of category 1 of group I, and such connected components follow one of the following three conditions: (1) $X_i^t \subseteq X_i^{t+1}$, $\forall\, i \in I$, (2) $X_i^t \supseteq X_i^{t+1}$, and (3) $\left(X_i^t \cap X_i^{t+1}\right) = \left(X_i^t \cup X_i^{t+1}\right) \neq \varnothing$. For all these three conditions, $\sigma\left(X_i^t, X_i^{t+1}\right)$ and $\rho\left(X_i^t, X_i^{t+1}\right)$ yield zero. By Equation 14.4, why $\sigma\left(X_i^t, X_i^{t+1}\right)$ yields zero for condition (1) is provided in three steps: (a) the erosion by zeroth size B of X_i^t (i.e., $(X_i^t \ominus 0B)$) would become a subset of X_i^{t+1}, (b) X_i^{t+1} would become a subset of X_i^t only after erosions of X_i^{t+1} by nB, where $n \geq 1$, and (c) the minimum of n obtained from steps (a, b), involved in the erosion process, is zero. According to Equation 14.5, for condition (1), the following three steps (p–r) explain why $\rho\left(X_i^t, X_i^{t+1}\right)$ yields zero: (p) X_i^t would become a subset of zeroth dilated version of X_i^{t+1} (i.e., $\left(X_i^{t+1} \oplus 0B\right)$), (q) X_i^{t+1} would become a subset of some finite dilated version of X_i^t, and (r) the minimum of n obtained from steps (p, q), involved in the dilation process, is zero. These three-step explanations are also true with conditions (2) and (3).

Category 2: If both $\sigma\left(X_i^t, X_i^{t+1}\right)$ and $\rho\left(X_i^t, X_i^{t+1}\right)$ are finite distances (i.e., $n \geq 1$), then the involved connected components, X_i^t and X_i^{t+1}, are considered to be of category 2 of group I. Such categorized connected components strictly follow the conditions: (1) $X_i^t \cap X_i^{t+1} \neq \varnothing$ and (2) $X_i^t \not\subset X_i^{t+1}$.

Category 3: If only $\rho\left(X_i^t, X_i^{t+1}\right)$ yields a finite distance (i.e., ≥ 1) between X_i^t and X_i^{t+1}, and there is no possibility to compute $\sigma\left(X_i^t, X_i^{t+1}\right)$—as no degree of erosion of either of the involved (corresponding) subsets makes the eroded set being contained in the other corresponding subset—then the spatial relationship between X_i^t and X_i^{t+1} is considered to be of category 3 of group II. Such category appears only when $X_i^t \cap X_i^{t+1} = \varnothing$.

Category 4: Those nonempty connected components of X^{t+1} that have no companion subsets in X^t and for which neither of the Hausdorff distances exists are categorized as category 4 of group III. To compute recursive median elements under such unique situation, one can complete X^t by adding to it the ultimate erosions $\left(UX_i^{t+1} \neq \varnothing\right)$ of those connected components of X_i^{t+1} that have no companion sets in X^t. The ultimate erosion of X_i^{t+1} $\left(UX_i^{t+1}\right)$ retains the final pixel value just before the last erosion that changed X_i^{t+1} to 0. Note that both $\sigma\left(X_i^{t+1}, UX_i^{t+1}\right)$ and $\rho\left(X_i^{t+1}, UX_i^{t+1}\right)$ yield zero distance according to Equations 14.4 and 14.5.

The relationships categorized earlier are dependent on the controlling factor. For instance, if the changes in the areal extent of lakes over a period

TABLE 14.2

Category-Wise Hausdorff Distances

Group	Category	$s\left(X_i^t, X_i^{t+1}\right)$	$r\left(X_i^t, X_i^{t+1}\right)$
I	1	0	0
I	2	≥ 1	≥ 1
II	3	Does not exist	≥ 1
III	4	Does not exist	Does not exist

Source: Sagar, B.S.D., *IEEE Trans. Pattern Anal. Mach. Intell.*, 32(2), 378, 2010.

of time have occurred due to a change in the temperature regimes, then the geometric evolution of lakes with evolving temperature fields can be computed by constructing interpolated layered maps. Table 14.2 depicts the possible Hausdorff distances both by erosion and by dilation for the different categories.

The categorization of connected components based on spatial relations between the connected components (X_i^t) and their companions (X_i^{t+1}) of X^t and X^{t+1} can be done directly by computing $\sigma\left(X_i^t, X_i^{t+1}\right)$ and $\rho\left(X_i^t, X_i^{t+1}\right)$ without checking for logical relationships (conditions). Checking the duality property of metrics $\sigma\left(X_i^t, X_i^{t+1}\right)$ and $\rho\left(X_i^t, X_i^{t+1}\right)$, the intersections of the sets— between which these distances need to be computed—must be within the four categories mentioned earlier. This validation further provides a basis (1) to properly categorize the sets and/or their corresponding subsets and (2) to test the quality of interpolations. In order to generate a sequence of interpolations between the category-wise connected components, Equations 14.6 and 14.7 form the basis to compute median elements between X_i^t and X_i^{t+1} of sets X^t and X^{t+1}.

Morphologic Interpolation via Median Element Computation

The median set $M(X^t, X^{t+1})$ between two sets X^t and X^{t+1}, according to Equations 14.6 and 14.7, is a global transform that ignores connectivity. But in some situations, it is required to introduce a bijection between the connected components X_i^t and X_i^{t+1} of X^t and X^{t+1}, respectively. For a better visualization of these spatially evolving subsets, interpolated sequences of subsets need to be generated at successive time instants in layered forms. The computation of the interpolated (median) layer $M\left(X_i^t, X_i^{t+1}\right)$ between the connected components X_i^t and X_i^{t+1} belonging to layers X^t and X^{t+1} is category dependent. To get an intermediary set between X^t and X^{t+1}, i.e., $X^{t+1/2}$, Equations 14.6 and 14.7 need to be used. Category-wise median element computations between the

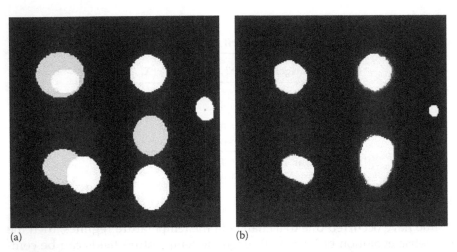

(a)   (b)

FIGURE 14.6
(a) Two layers (sets) consisting of subsets with various spatial relationships shown in the two panels of Figure 14.5 are superimposed, and (b) computed median sets between the corresponding input subsets shown in Figure 14.5a and b. (From Sagar, B.S.D., *IEEE Trans. Pattern Anal. Mach. Intell.*, 32(2), 378, 2010.)

category-wise companion-connected components are illustrated in Figure 14.6a and b and summarized in Equations 14.9 through 14.14:

Category 1 with $X_i^t \subseteq X_i^{t+1}$:

$$M\left(X_i^t, X_i^{t+1}\right) = \bigcup_{n=0}^{N}\left(X_i^t \oplus nB\right) \cap \left(X_i^{t+1} \ominus nB\right) \qquad (14.9a)$$

Category 1 with $X_i^t \supseteq X_i^{t+1}$:

$$M\left(X_i^{t+1}, X_i^t\right) = \bigcup_{n=0}^{N}\left(X_i^t \ominus nB\right) \cap \left(X_i^{t+1} \oplus nB\right) \qquad (14.9b)$$

Category 1 with $\left(X_i^t \cap X_i^{t+1}\right) = \left(X_i^t \cup X_i^{t+1}\right) \neq \varnothing$:

$$M\left(X_i^t, X_i^{t+1}\right) = X_i^t = X_i^{t+1} \qquad (14.10)$$

Category 2 with either $X_i^t \not\subset X_i^{t+1}$ or $X_i^{t+1} \not\subset X_i^t$:

$$M\left(X_i^t, X_i^{t+1}\right) = \bigcup_{n=0}^{N}\left(\left(X_i^t \cup X_i^{t+1}\right) \ominus nB\right) \cap \left(\left(X_i^t \cap X_i^{t+1}\right) \oplus nB\right) \qquad (14.11)$$

If X_i^t and X_i^{t+1} possess partial overlapping, such that $X_i^t \cap X_i^{t+1} \neq \emptyset$ (e.g., categories 1 and 2), then all possible median elements that could be generated through morphologic interpolation between the subsets X_i^t and X_i^{t+1}, and also between recursively generated median sets, are at least partially overlapping. A variant of Equation 14.11 was proposed and discussed in Iwanowski and Serra (2000). Since there is a *bijection* between the connected components X_i^t and X_i^{t+1} of (X^t), (X^{t+1}), which fall under categories 1 and 2 of group I, one can prove that

$$M(X^t, X^{t+1}) = \bigcup_{\forall i} M\left(X_i^t, X_i^{t+1}\right) \tag{14.12}$$

Category 3: Equation 14.12 does not hold for this category. One can modify the data by using the construction that results in Equation 14.13. $X_i^t, X_i^{t+1}, CH\left(X_i^t \cup X_i^{t+1}\right)$, respectively, denote the subsets of (X^t), (X^{t+1}), and the convex hull of the union of subsets X_i^t and X_i^{t+1}. Let these subsets and/or sets be nonempty compact and satisfy the property $\left(X_i^t, X_i^{t+1}\right) \subseteq CH\left(X_i^t \cup X_i^{t+1}\right)$. $M_1\left(X_i^t, CH\left(X_i^t \cup X_i^{t+1}\right)\right)$ and $M_2\left(X_i^{t+1}, CH\left(X_i^t \cup X_i^{t+1}\right)\right)$ are median sets, respectively, between X_i^t and $CH\left(X_i^t \cup X_i^{t+1}\right)$ and X_i^{t+1} and $CH\left(X_i^t \cup X_i^{t+1}\right)$. Such median sets M_1 and M_2 as well as the median set $M_s = M(M_1, M_2)$ satisfy the following conditions: (1) $M_1 \neq \emptyset$; $M_2 \neq \emptyset$, (2) $M_1 \cap M_2 \neq \emptyset$, and (3) $M(M_1, M_2) \neq \emptyset$, so that

$$M_s = M(M_1, M_2) = \bigcup_{n=0}^{N} [(M_1 \cap M_2) \oplus nB] \cap [(M_1 \cup M_2) \ominus nB] \tag{14.13}$$

Category 4: Median elements that under unique situation arise in category 4 of group III can be computed as

$$M\left(X_i^t, X_i^{t+1}\right) = \bigcup_{n=0}^{N} \left(UX_i^{t+1} \oplus nB\right) \cap \left(X_i^{t+1} \ominus nB\right) \tag{14.14}$$

where UX_i^{t+1} is the ultimate eroded version of X_i^{t+1}.

Sequence of Interpolated Sets

Using the median set, the interpolation sequence can be obtained recursively. Let X_i^t and X_i^{t+k} denote the input subsets, k the time gap between the two successive maps, and n the recursion level (always a power of 2), and let $X_i^{t+(k/2^n)}$ or $M\left(X_i^t, X_i^{t+k}\right)$ denote the median element between the two input subsets. The median element between the two input companion-connected components X_i^t and X_i^{t+k}, respectively, belonging to the two time instants t and $t + k$, is also denoted by $X_i^{t+(k/2^1)} = X_i^{t+(k/2)}$. Then, the sequence of interpolated sets, between the two inputs X_i^t and X_i^{t+k}, is defined as

$$X_i^{t+(k/2^1)} = M\left(X_i^t, X_i^{t+k}\right); \quad X_i^{t+(k/2^2)} = M\left(X_i^t, X_i^{t+(k/2^1)}\right); \quad \ldots \quad (14.15)$$

In $X_i^{t+(k/2)}$, the superscript $t + (k/2)$ denotes the intermediary time. For instance, the two successive maps X^t and X^{t+k} available are for years $t = 1896$ and $t + k = 1898$, where the time gap (k) is 2. Then, the superscript for the median element, generated by taking these two input maps, should be 1897. The median element at intermediary time is $X_i^{1896+(2/2^1)} = X_i^{(1897)}$.

The maximum possible number of layers (N_{max}) that can be generated (interpolated) between the input layers (sets) X^t and X^{t+1} including the two input layers is given by

$$(N_{max}) = \max\left\{\left[\min(n : X^{t+1} \subseteq (X^t \oplus nB))\right], \left[\min(n : X^t \subseteq (X^{t+1} \ominus nB))\right]\right\} \quad (14.16)$$

Experimental Results

The ideas in the previous section have been employed to practical examples to demonstrate their applicability. To demonstrate the applications of the Hausdorff distance for understanding the spatial relations between two discrete maps in general, and between the companion-connected components in particular, and to generate maximum possible sequential maps, we consider spatially represented (1) small water bodies, retrieved from remotely sensed data of peak drought (X^t) and peak monsoon (X^{t+1}) periods as basic inputs, and (2) bubonic plague epidemic data available at annual intervals during the period 1896–1906.

Case Study on Small Water Bodies

We chose two input synthetic sets, depicting small water bodies represented by 512 × 512 binary pixels, at peak drought period (Figure 14.7a) and peak flood period (Figure 14.7b), respectively. These two input slices fall under category 1 (see the "Spatial Relationships between Sets and Their Categorization" section), so we employ Equation 14.9a to generate an interpolated slice (median set) (Figure 14.7c).

By using the median set thus generated, a sequence of interpolated slices is recursively generated, as shown in Figure 14.8. Including two input slices, a total of five slices are generated as (N_{max}) computed according to Equation 14.16 is 5.

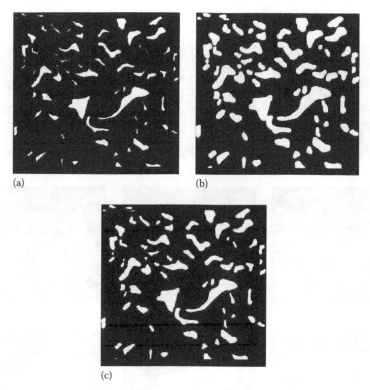

(a)

(b)

(c)

FIGURE 14.7
(a) Input slice 1, lakes (in peak drought time) in binary format, (b) lakes (in peak flood time) as input slice 2, and (c) median set computed between the two input slices shown in Figure 14.7a and b. (From Sagar, B.S.D., *IEEE Trans. Pattern Anal. Mach. Intell.*, 32(2), 378, 2010.)

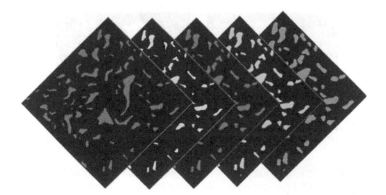

FIGURE 14.8
Sequence of interpolated sets (slices) in between the two input slices shown in Figure 14.7a and b. Equations 14.9a and 14.15 are used to recursively generate the interpolated slices. The third (middle) layer depicting water bodies is the median set shown in Figure 14.7c. (From Sagar, B.S.D., *IEEE Trans. Pattern Anal. Mach. Intell.*, 32(2), 378, 2010.)

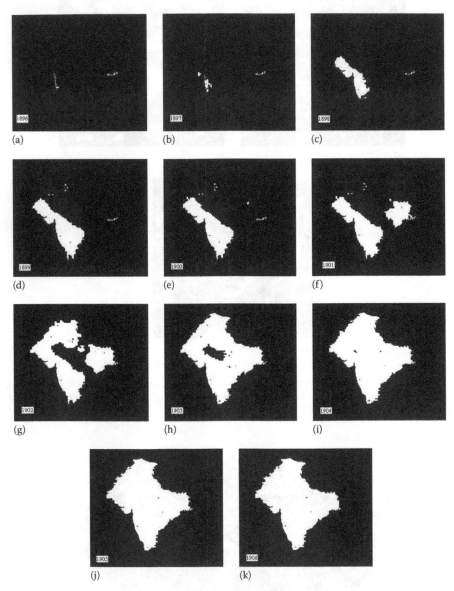

FIGURE 14.9
(a–k) Spatial–temporal maps that represent the geographic spread of bubonic plague in India between 1896 and 1906 at intervals of 1 year (Yu and Christakos 2006). The 11 spatial maps depicting the spread of plague were sequentially used to generate the maximum possible number of interpolated maps. (From Sagar, B.S.D., *IEEE Trans. Pattern Anal. Mach. Intell.*, 32(2), 378, 2010.)

Case Study on Spatial Maps of Epidemic

Spatial maps depicting bubonic plague data—available for a period of 11 years during 1896–1906 at annual intervals, generated by Yu and Christakos (2006)—were chosen. We applied the methodology explained in the previous case study by choosing the spatial maps of successive years (Figure 14.9a through k) to generate a sequence of interpolated maps between the sets of maps of successive years to visualize the spread of epidemics in a continuous manner.

According to the spatial relation between the successive spatial maps, which are ordered and/or semi-ordered sets, both $\rho(X^t, X^{t+1})$ and $\sigma(X^t, X^{t+1})$ yield ≥ 1. Hence, they are classified as category 2. According to Equation 14.8, μ for the first-level median sets for both successive maps (i.e., X^t and X^{t+1}), and also for X^t and X^{t+2}, is computed for all t values (Table 14.3). This μ provides an estimate of the maximum number of interpolated maps that could be generated.

Validation of the Middle Elements as Interpolators

For this epidemics case, involving 11 yearly maps for the geographic spread of bubonic plague in India, while testing for the quality of the middle element as an interpolator (Figure 14.10), we made comparisons between $M(X^t, X^{t+2})$ with X^{t+1}, for all possible t.

These data further provide the distribution of the rate of spread in terms of μ computed for the data of successive years (Table 14.3). The higher the μ, the rapid is the rate of spread. For instance, between the years 1897

TABLE 14.3

μ Values Computed for X^t and X^{t+1} and X^t and X^{t+2}

	μ	
$t, t+1, t+2$	$M(X^t, X^{t+1})$	$M(X^t, X^{t+2})$
1896, 1897, 1898	3	9
1897, 1898, 1899	9	15
1898, 1899, 1900	11	11
1899, 1900, 1901	2	12
1900, 1901, 1902	12	15
1901, 1902, 1903	13	16
1902, 1903, 1904	9	14
1903, 1904, 1905	7	7
1904, 1905, 1906	2	2
1905, 1906, …	1	—

Source: Sagar, B.S.D., *IEEE Trans. Pattern Anal. Mach. Intell.*, 32(2), 378, 2010.

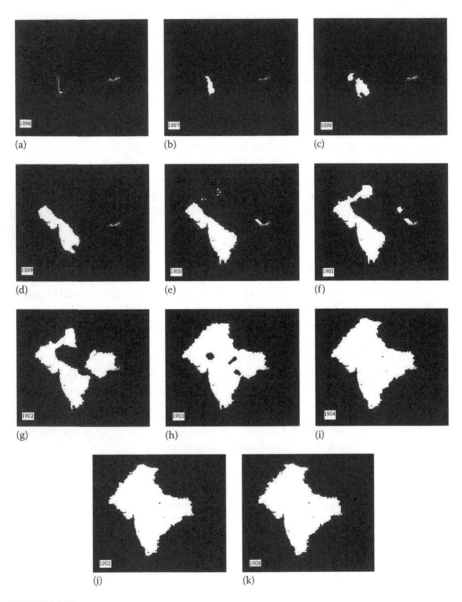

FIGURE 14.10

(a) Original spatial map of the bubonic plague during 1896, (b–j) the first level median sets computed for $M(X^t, X^{t+2})$ for all "t," ranging from 1896 to 1905, and (k) original spatial map during 1906. For validation, the maps of Figure 14.10b through j obtained as first-level median sets $M(X^t, X^{t+2})$ are respectively compared for all "t" with those t of Figure 14.9b through j. These first-level median sets show a reasonable matching with the actual sets (Figure 14.9b through j). (From Sagar, B.S.D., *IEEE Trans. Pattern Anal. Mach. Intell.*, 32(2), 378, 2010.)

and 1898, 1898 and 1899, 1900 and 1901, 1901 and 1902, as well as 1902 and 1903, the rates of spread of plague were significantly faster than that in the other periods. By considering the spatial maps of years 1896 and 1897, a maximum of six spatial maps could be generated by using Equations 14.9 through 14.15. The interpolated maps computed from 10 pairs of maps at annual intervals during the period 1896–1906 can be employed to generate an animation depicting the way the epidemic spread spatially in a continuous manner (animation is available as .avi file at http://www.isibang. ac.in/~bsdsagar/Epid-animate2.avi). This procedure paves the way to generate continuous slices between two input slices recorded at significantly different instants of time.

We found a significantly reasonable quality for the median sets, $M(X^t, X^{t+2})$, generated for the epidemic data between alternative years as interpolators. The median sets were tested by comparing $M(X^t, X^{t+2})$ with X^{t+1} for all possible t. There obviously exists a significant match between the interpolated median elements (Figure 14.10b through j)—computed between the real data X^t and X^{t+2} that belong to nine pairs of maps (Table 14.3)—and the real data (Figure 14.9b through j). For the sake of a better visual comparison, we also represent the original spatial maps (Figure 14.9a through k) and the spatial maps generated according to $M(X^t, X^{t+2})$ (Figure 14.10a through k) in a composite way by superimposing them on one another and assigning gray shade to each original spatial map as well as maps generated *via* median set computation (Figure 14.11a and b).

The Hausdorff dilation and erosion distances were computed between $M(X^t, X^{t+2})$ and X^{t+1} for all t values (Table 14.4) to test the quality of interpolation. These distances were compared with the respective Hausdorff dilation

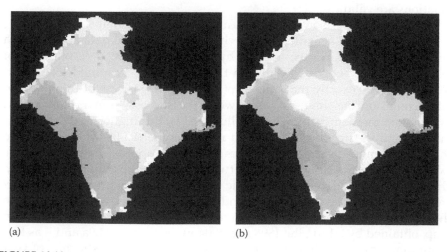

(a) (b)

FIGURE 14.11
Superimposed gray-coded (a) original spatial maps and (b) spatial maps generated *via* median set computations. (From Sagar, B.S.D., *IEEE Trans. Pattern Anal. Mach. Intell.*, 32(2), 378, 2010.)

TABLE 14.4

Hausdorff Distance Values

t	$\rho[M(X^t, X^{t+1}), X^{t+1}]$	$\sigma[M(X^t, X^{t+1}), X^{t+1}]$	$\rho(X^t, X^{t+1})$	$\sigma(X^t, X^{t+1})$
1896	8	2	7	1
1897	2	2	1	1
1898	1	1	1	1
1899	4	2	1	1
1900	12	9	1	1
1901	8	7	2	1
1902	8	8	1	1
1903	3	3	2	1
1904	2	2	1	1
1905	—	—	2	1

Source: Sagar, B.S.D., *IEEE Trans. Pattern Anal. Mach. Intell.*, 32(2), 378, 2010.

and erosion distances of X^t and X^{t+1} (Table 14.4). The lower the difference between the values of $\rho[M(X^t, X^{t+1}), X^{t+1}]$ or $\sigma[M(X^t, X^{t+1}), X^{t+1}]$ and $\rho(X^t, X^{t+1})$ or $\sigma(X^t, X^{t+1})$, the higher is the degree of matching. Mismatch between the interpolated and actual maps is observed in terms of these values for the interpolated maps for the t values of 1896 and 1901. This slight discrepancy in the values for the years 1896 and 1901 is due to the presence of a few spikes related to the connected component(s) of one of the two input sets. Other maps in the sequence have exhibited a significantly higher degree of matching. For all cases shown in the last four columns of Table 14.4, the Hausdorff erosion distances are found to be slightly less than the Hausdorff dilation distances. However, the fair overall matching further signifies that the interpolations are valid.

From this comparison of the real and interpolated maps, we conclude that (1) there exists a significant match between the real data and the data generated as median sets, and (2) by comparing $M(X^t, X^{t+2})$, one cannot expect to exactly obtain X^{t+1} due to the fact that μ computed for X^t and X^{t+2} may not be exactly the same as $\rho(X^t, X^{t+1})$. The success of this interpolation relies on the time gap (k) between the successive maps considered to generate median maps. The smaller the (k), the higher is the approximation, and in turn the interpolated maps geometrically well conform to the realistic maps as shown in the results. Better approximations could be expected when the time interval between the two input maps is smaller. On contrary, the degree of geometric similarity between the maps generated via median map computation and the realistic maps may be poor as in the case of many other interpolation techniques. For instance, the median map, obtained by taking the 1896 and 1904 maps (Figure 14.9a and i) as two input maps, may not show significant match with the realistic map of year 1900 (Figure 14.9e). This is due to the fact that the time gap (k) between the two input maps is 8 years.

Potential Applications: A Brief Discussion

Prediction is a challenging task to model/simulate/visualize the general spatiotemporal behavior of varied phenomena, such as spreads of rainfall, floodwater, lakes, epidemics, population, cities, elevation structures, temperatures, hurricanes, soil moisture, earth resistivity, clouds, the impact of sea level rise on the landmass, etc. The reader will certainly find several other phenomena where one can apply the framework shown here. Of geomorphological interest, two specific examples where one can apply this interpolation technique for discrete spatial and/or temporal themes are the following:

1. This approach could be applied to create spatially distributed elevation regions with dense elevation contours from sparse elevation contours. Successive elevation regions, which are usually in ordered form, would be threshold decomposed (Maragos and Ziff 1990). If spatial resolution of the data is coarse, then the spatial gap between the any two successive threshold decomposed elevation regions is significantly larger. One can follow the median set computation approach that is meant for category 1 to generate intermediary elevation regions between such coarser successive elevation regions. Kriging is an appropriate technique to deal with the maps of this category 1. In most of its applications, simple Euclidean distance, which is not always an appropriate metric to define the separation between the points, has been employed to define the separation between the sample points. But this framework employs non-Euclidean metrics such as erosion and dilation distances to develop median set(s).

2. Spatiotemporal behavior of lakes: One can easily observe the changing patterns in geometries of lakes of a group that could be mapped as layered information from remotely sensed data belonging to two different time instants. To analyze how the groups of lakes have geometrically evolved between the successive time periods, this morphological interpolation technique would be of use. Maps depicting lakes, retrieved from remotely sensed data at two significantly staggered time periods (e.g., peak monsoon time and peak drought time within a year), can be considered to generate several possible intermediary layers of lakes. Such a study would provide insight to have a better understanding of spatiotemporal organization of the lakes.

Conclusions

Between two input maps and/or median maps thus generated, the recursive generation of interpolated layers (sets) is a challenge in the context of GISci. One of the ways to address such a challenge is possible through morphologic interpolation *via* median set computation. A framework has been provided to describe the four possible spatial relations between the corresponding thematic units of two input thematic maps. This framework addresses the categorization of companion thematic units into four categories based on both logical relationships and Hausdorff (dilation and erosion) distances. To measure the Hausdorff erosion and dilation distances between the corresponding subsets in particular, and between the input sets in general, and to compute a sequence of interpolated maps, mathematical morphologic transformations are employed. Synthetic sets, small water bodies in two different seasons, and also spatial maps depicting the spread bubonic plague through 11 years have been considered as datasets to demonstrate the algorithm. Further, the quality of median elements as interpolators were evaluated. Besides several existing interpolation methods, this approach also provides potentially valuable insights into the context of GISci.

References

Aksoy, S. and R. G. Cinbis, 2010, Image mining using directional spatial constraints, *IEEE Geoscience and Remote Sensing Letters*, 7(1), 33–37.

Barata, T. and P. Pina, 2006, A morphological approach for feature space partitioning, *IEEE Geoscience and Remote Sensing Letters*, 3(1), 173–177.

Benediktsson, J. A., M. Pesaresi, and K. Arnason, 2003, Classification and feature extraction for remote sensing images from urban areas based on morphological transformations, *IEEE Transactions on Geoscience and Remote Sensing*, 41, 1940–1949.

Beucher, S., 1994, Interpolation d'Ensembles, de Partitions et de Fonctions. Technical Report N-18/94/MM, Ecole des Mines de Paris, Paris, France.

Burr, D. J., Nov 1981. Elastic matching of line drawings, *IEEE Transactions on Pattern Analysis and Machine Intelligence*, 3(6), 708–713.

Cao, B. and F. Glover, 2010, Creating balanced and connected clusters to improve service delivery routes in logistics planning, *Journal of System Sciences and System Engineering*, 19(4), 453–480.

Casaer, J., M. Hermy, P. Coppin, and R. Verhagen, 1999, Analyzing space use patterns by Thiessen polygon and triangulated irregular network interpolation: A non parametric method for processing telemetric animal fixes, *International Journal of Geographical Information Science*, 13(5), 499–511.

Chanussot, J., J. A. Benediktsson, and M. Fauvel, 2006, Classification of remote sensing images from urban areas using a fuzzy possibilistic model, *IEEE Geoscience and Remote Sensing Letters*, 3(1), 40–44.

Chen, S. Y., W. C. Lin, C. C. Liang, and C. T. Chen, 1990, Improvement on dynamic elastic interpolation technique for reconstructing 3-D objects from serial cross sections, *IEEE Transaction on Medical Imaging*, 9(1), 71–83.

Cousty, J., G. Bertrand, L. Najman, L., and M. Couprie, 2009, Watershed cuts: Minimum spanning forests and the drop of water principle, *IEEE Transactions on Pattern Analysis and Machine Intelligence*, 31(8), 1362–1374.

Cressie, N. A. C., 1991, *Statistics of Spatial Data*, John Wiley & Sons, New York.

Cressie, N. A. C., 1993, *Statistics for Spatial Data*, John Wiley and Sons, New York, p. 900.

Dalla Mura, M., J. A. Benediktsson, F. Bovolo, and L. Bruzzone, 2008, An unsupervised technique based on morphological filters for change detection in very high resolution images, *IEEE Geoscience and Remote Sensing Letters*, 5(3), 433–437.

Dalla Mura, M., A. Villa, J. A. Benediktsson, J. A. et al. 2011, Classification of hyperspectral images by using extended morphological attribute profiles and independent component analysis, *IEEE Geoscience and Remote Sensing Letters*, 8(3), 542–546.

Frank, A. U., 1998, GIS for politics, *Proceedings of Annual Conference GIS Planet '98*, Lisbon, Portugal.

Frank, A. U., 2005, Map algebra with functors for temporal data, *Proceedings of ER Workshop Conceptual Modeling for Geographic Information Systems*, Klagenfurt, Austria.

Hausdorff, F., 1914/2002, *Grundzuge der Mengenlchre*, Viet and Co. (Gekurzte) Auft Springer, Berlin, ISBN 3-540-42224-2.

Herman, G. T., J. Zheng, and C. A. Bucholtz, 1992, Shape-based interpolation, *IEEE Computer Graphics and Applications*, 12(3), 69–79.

Hirata, T., 1996, A unified linear-time algorithm for computing distance maps, *Information Processing Letters*, 58(3), 129–133.

Huang, X., L. P. Zhang, and L. Wang, 2009, Evaluation of morphological texture features for mangrove forest mapping and species discrimination using multispectral IKONOS imagery, *IEEE Geoscience and Remote Sensing Letters*, 6(3), 393–397.

Iwanowski, M., 2000, Application of mathematical morphology to image interpolation, PhD thesis, School of Mines Paris, Paris, France—Warsaw University of Technology, Warsaw, Poland.

Iwanowski, M. and J. Serra, 2000, The Morphological-affine object deformation. In: *Mathematical Morphology and Its Applications to Image and Signal Processing*, eds. J. Goutsias, L. Vincent, and D. S. Bloomberg, Kluwer Academic Publishers, Boston, MA, pp. 81–90.

Lane, S. M. and G. Birkhoff, 1967, *Algebra*, Macmillan, New York.

Maragos, P. and R. D. Ziff, 1990. Threshold superposition in morphological image analysis systems, *IEEE Transactions on Pattern Analysis and Machine Intelligence*, 12(5), 498–504.

Meijster, A., J. B. T. M. Roerdink, and W. H. Hesselink, 2000, A general algorithm for computing distance transform in linear time. In: *Mathematical Morphology and Its Applications to Image and Signal Processing*, Kluwer, Dordrecht, the Netherlands, pp. 331–340.

Meyer, F., 1994a, Interpolations, Technical Report N-16/94/MM, Ecole des Mines de Paris, Paris, France.

Meyer, F., 1994b, Topographic distance and watershed lines, *Signal Processing, Mathematical Morphology and Its Applications to Signal Processing*, 38(1), 113–125, ISSN 0165-1684.

Najman, L. and M. Schmitt, 1994, Watershed of a continuous function, *Signal Processing*, 8(1), 98–112.

Pan, J., M. Wang, D. R. Li, and J. L. Li, 2010, A network-based radiometric equalization approach or digital aerial orthoimages, *IEEE Geoscience and Remote Sensing Letters*, 7(2), 401–405.

Pesaresi, M. and J. A. Benediktsson, 2001, A new approach for the morphological segmentation of high-resolution satellite imagery, *IEEE Transactions on Geoscience and Remote Sensing*, 39, 309–320.

Pullar, D., 2001, MapScript: A map algebra programming language incorporating neighborhood analysis, *Geoinformatica*, 5, 145–163.

Rajashekara, H. M., P. Vardhan, and B. S. D. Sagar, 2012, Generation of zonal map from point data via weighted skeletonization by influence zone, *IEEE Geoscience and Remote Sensing Letters*, 9(3), 403–407.

Raya, S. P. and J. A. Udupa, 1990, Shape based interpolation of multidimensional objects, *IEEE Transactions on Medical Imaging*, 9(1), 32–42.

Sagar, B. S. D., 2010, Visualization of spatiotemporal behavior of discrete maps *via* generation of recursive median elements, *IEEE Transactions on Pattern Analysis and Machine Intelligence*, 32(2), 378–384.

Sagar, B. S. D. and L. Chockalingam, 2004, Fractal dimension of non-network space of a catchment basin, *Geophysical Research Letters*, 31(12), L12502.

Sagar, B. S. D., G. Gandhi, and B. S. P. Rao, 1995a, Applications of mathematical morphology on water body studies, *International Journal of Remote Sensing*, 16(8), 1495–1502.

Sagar, B. S. D., M. B. R. Murthy, C. B. Rao, and B. Raj, 2003, Morphological approach to extract ridge-valley connectivity networks from Digital Elevation Models (DEMs), *International Journal of Remote Sensing*, 24(3), 573–581.

Sagar, B. S. D. and J. Serra, 2010, Preface: Spatial information retrieval, analysis, reasoning and modelling, *International Journal of Remote Sensing*, 31(22), 5747–5750.

Sagar, B. S. D. and L. T. Tien, 2004, Allometric power-law relationships in a Hortonian Fractal DEM, *Geophysical Research Letters*, 31(6), L06501.

Sagar, B. S. D., M. Venu, and B. S. P. Rao, 1995b, Distributions of surface water bodies, *International Journal of Remote Sensing*, 16(16), 3059–3067.

Saito, T. and J. I. Toriwaki, 1994, New algorithms for Euclidean distance transformation of an n-dimensional digitized picture with applications, *Pattern Recognition*, 27, 1551–1565.

Serra, J., 1982, *Image Analysis and Mathematical Morphology*, Academic Press, London, U.K.

Serra, J., 1994, Interpolations et Distance de Hausdorff, Technical Report N-15/94/MM, Ecole des Mines de Paris, Paris, France.

Serra, J., 1998, Hausdorff distances and interpolations. In: *Mathematical Morphology and Its Applications to Images and Signal Processing*, eds. H. J. A. M. Heijmans and J. B. T. M. Roerdink, Kluwer Academic Publishers, Dordrecht, the Netherlands.

Snodgrass, R. T., 1992, Temporal databases. In: *Theories and Methods of Spatiotemporal Reasoning in Geographic Space*, eds. A. U. Frank, I. Campari, and U. Formentini, Springer-Verlag, New York, pp. 22–64.

Su, B., Z. Li, G. Lodwick, and J. C. Muller, 1997, Algebraic models for the aggregation of area features based upon morphological operators, *International Journal of Geographical Information Science*, 11(3), 233–246.

Taubenbock, H., M. Habermeyer, A. Roth, A. et al., 2006, Automated allocation of highly structured urban areas in homogeneous zones from remote sensing data by Savitzky-Golay filtering and curve sketching, *IEEE Geoscience and Remote Sensing Letters*, 3(4), 532–536.

Tay, L. T., B. S. D. Sagar, and H. T. Chuah, 2005, Analysis of geophysical networks derived from multiscale digital elevation models: A morphological approach, *IEEE Geoscience and Remote Sensing Letters*, 2(4), 399–403.

Tay, L. T., B. S. D. Sagar, and H. T. Chuah, 2007, Granulometric analysis of basin-wise DEMs: A comparative study, *International Journal of Remote Sensing*, 28(15), 3363–3378.

Tomlin, C. D., 1990, *Geographic Information Systems and Cartographic Modeling*, Prentice Hall, Englewood Cliffs, NJ.

Vidal, J., J. Crespo, and V. Maojo, 2005, Recursive interpolation technique for binary images based on morphological median sets, *Proceedings of International Symposium on Mathematical Morphology: 40 Years On*, eds. C. Ronse, L. Najman, and E. Decenciere, Springer, Dordrecht, The Netherlands, pp. 53–62.

Vincent, L. and P. Soille, 1991, Watersheds in digital spaces: An efficient algorithm based on immersion simulations, *IEEE Transactions on Pattern Analysis and Machine Intelligence*, 13(6), 583–598.

Werahera, P. N., G. J. Miller, G. D. Taylor, T. Brubaker, F. Danesghari, and E. D. Crawford, 1995, A 3-D reconstruction algorithm for interpolation and extrapolation of planar cross sectional data, *IEEE Transactions on Medical Imaging*, 14(4), 765–771.

Worboys, M. and M. Duckam, 2004, *GIS: A Computing Perspective*, CRC Press, Boca Raton, FL.

Yu, H. L. and G. Christakos, 2006, Spatio-temporal modeling and mapping of the bubonic plague epidemics in India, *International Journal of Health Geographics*, 5(12), doi: 10.1186/1476–072X-5-12.

Afterword

Most of our forefathers never traveled more than a few miles from where they were born and had a very limited knowledge of the Earth's surface beyond what they could see around them. Nowadays, facilities for travel exist like never before, and modern communications enable people to have a much greater knowledge of the geophysical environment in which we all live. Following the launch of *Sputnik* in 1957, remote sensing satellites have revolutionized the access that we have to remotely sensed data on various terrestrial, lunar, planetary surfaces, and atmospheric phenomena such as clouds on multiple spatiotemporal scales. The processing and analysis of such remotely sensed data acquired at various spatial and temporal scales has received wide attention during the last three decades primarily by geoinformation science (GISci) specialists, geophysicists, geomorphologists, geologists, environmentalists, ecologists, climatologists, and hydrologists. One of the important byproducts derivable from remotely sensed data is a digital elevation model (DEM) that provides rich clues about the physiographic constitution of the Earth and Earth-like planetary surfaces. The author of this book has guest edited (along with Jean Serra, the founder of Mathematical Morphology) a special issue on "Spatial Information Retrieval, Analysis, Reasoning and Modelling" for the *International Journal of Remote Sensing* (31(22), 5747–6032, 2010). It is interesting to see that this book considers various topics—like pattern retrieval, pattern analysis, spatial reasoning, and simulation and modeling—of geoscientific interest. To address these intertwined topics that are useful for understanding the spatiotemporal behavior of many terrestrial phenomena and processes, various original algorithms and modeling techniques that are mainly based on mathematical morphology, fractal geometry, and chaos theory have been presented in this book.

A quick study of the table of contents will show that, all in all, the journey through this book should provide geomorphologists and GISci specialists a new experience and exposition and a host of new ideas to explore further in the contexts of quantitative geomorphology and spatial reasoning. This specialized book should be of immense value to the postgraduates and to doctoral and postdoctoral students who would like to venture into the applications of mathematical morphology in geomorphology and GISci.

Arthur P. Cracknell
University of Dundee
United Kingdom

Index

A

Abstract network, 66, 139, 234
Abstract structures, 3, 61, 127
Accretion process, 362, 371
Active dunes, 375
Adjacency, 381, 384, 393, 397, 409, 427,
    441, 443, 452, 461
Adjacent zones, 384, 408, 409
Aerial photographs, 62
AIAs, 345–346, 349–355, 357
AIRSAR, 57
Algorithms, 1–3, 7, 19, 37, 44, 55, 63, 222,
    464, 473
Allometric, 5, 128, 140, 143, 145, 150–154,
    159, 160, 162, 164, 165, 167, 181,
    182, 189, 206, 234, 262, 273, 382
    power laws, 143, 152, 154, 160, 162
    power-law relationships, 5, 143, 151,
       160, 181, 182, 206
    relationships, 127, 145, 150, 151, 153,
       159, 160, 162, 165, 167
    scaling analyses, 153
    scaling relationships, 143, 145, 164
    universal scaling laws, 167
Allometry, 61, 78, 92, 127, 164, 167
Allometry of networks, 164
Allometry-based analysis, 92
Amplification, 6, 335, 337, 340, 346, 347,
    355, 356
Analysis, 1–4, 6, 7, 12, 18, 19, 44, 47, 52,
    57, 84, 92, 101, 103, 107, 123, 124,
    124, 141, 153, 164, 165, 189, 190,
    197, 205, 225, 235, 251, 261, 262,
    266, 270, 273, 274, 283, 286, 301,
    305, 316, 320, 340, 347, 352, 358,
    367, 381, 382, 386, 388, 409, 427,
    428, 443, 476
Angles of repose, 360, 361, 370, 371,
    373–375
Angular points, 28, 63, 66
Anti-extensive, 18
Anti-granulometric analysis, 270

Apollonian space, 215, 218, 229
Area-correlation sums, 211, 212
Area-correlational integrals, 196
Areal objects, 415, 432
Area-number relationship, 255
Area-perimeter relationship, 4, 141,
    142, 150
Area-radius, 251, 254
Asymmetric fractal basins, 38
Asymmetric generator, 38
Atmospheric phenomena, 61
Attracting inter-slipface angles, 362, 366
Attractor interlimb angles, 345
Attractor sand dune profiles, 366, 368
Attractors, 324, 325, 330, 331, 334, 345,
    357, 358, 370
Avalanches, 361, 365, 370–377
    distribution, 373, 376
    size distribution, 311, 359, 365,
       373, 374
Average roughness, 266, 268–273
Average size(s), 266, 268, 270, 272

B

Background, 13, 19, 20, 22, 23, 32, 85, 87,
    216, 261, 262, 265, 268–273, 464
Background pixels, 32, 87
Basic measures, 4, 5, 55, 123, 127, 130, 139,
    140, 144–147, 150, 151, 154, 167,
    168, 181, 190, 235, 240, 253, 263,
    264, 274, 275, 325, 383
Basic principles, 16
Basin function, 274–281, 283, 287,
    306, 307
Basin pixels, 87–90
Basins, 2, 4, 5, 37–40, 44, 45, 47–50, 62, 78,
    79, 88, 90, 91, 123, 125–128, 130,
    132, 137–141, 143–153, 155, 160,
    162, 163, 165, 189, 190, 215, 231,
    232, 235–257, 261–270, 272–281,
    283, 285–287, 294, 297, 299, 301,
    305–307, 324, 416, 420

Printed and bound by CPI Group (UK) Ltd, Croydon, CR0 4YY

21/10/2024

01777109-0004